TOWARDS 5G

TOWARDS 5G

APPLICATIONS, REQUIREMENTS AND CANDIDATE TECHNOLOGIES

Edited by

Rath Vannithamby and Shilpa Talwar
Intel Corporation, USA

WILEY

This edition first published 2017

Registered Office
John Wiley & Sons, Ltd, The Atrium, Southern Gate, Chichester, West Sussex, PO19 8SQ, United Kingdom

For details of our global editorial offices, for customer services and for information about how to apply for permission to reuse the copyright material in this book please see our website at www.wiley.com.

Library of Congress Cataloging-in-Publication Data

Names: Vannithamby, Rath, editor. | Talwar, Shilpa, editor.
Title: Towards 5G : applications, requirements & candidate technologies /
edited by Rath Vannithamby and Shilpa Talwar.
Description: Chichester, West Sussex, United Kingdom : John Wiley & Sons Inc., 2017. |
Includes bibliographical references and index.
Identifiers: LCCN 2016019944| ISBN 9781118979839 (cloth) | ISBN 9781118979914 (epub)
Subjects: LCSH: Mobile communication systems–Research.
Classification: LCC TK5103.2 .T6835 2017 | DDC 621.3845/6–dc23
LC record available at https://lccn.loc.gov/2016019944

A catalogue record for this book is available from the British Library.

Cover Image: Gettyimages/Prykhodov
Gettyimages/Robert Mandel
Gettyimages/BsWei
Gettyimages/cybrain
Gettyimages/d_arth

Set in 10/12pt Times by SPi Global, Pondicherry, India

Printed and bound in Malaysia by Vivar Printing Sdn Bhd

10 9 8 7 6 5 4 3 2 1

Contents

List of Contributors

Sergey Andreev
Tampere University of Technology, Finland

Ejder Baştuğ
CentraleSupélec, France

Anass Benjebbour
NTT DoCoMo, Inc., Japan

Mehdi Bennis
Centre for Wireless Communications, University of Oulu, Finland

Vincent Berg
CEA, LETI, France

Dinesh Bharadia
Stanford University, USA

Mateusz Buczkowski
IS-Wireless, Poland

Daoud Burghal
University of Southern California, USA

Nicolas Cassiau
CEA, LETI, France

Yejian Chen
Alcatel Lucent Bell Labs, Germany

Mérouane Debbah
CentraleSupélec, France

Jean-Baptiste Doré
CEA, LETI, France

Michael Faerber
Intel Corporation, USA

Gerhard Fettweis
Technische Universität Dresden, Germany

Olga Galinina
Tampere University of Technology, Finland

Ivan Gaspar
Technische Universität Dresden, Germany

Mikhail Gerasimenko
Tampere University of Technology, Finland

Amitava Ghosh
Nokia Networks, USA

Shuangfeng Han
Green Communication Research Center, China Mobile Research Institute, China

Monowar Hasan
University of Manitoba, Canada

Nageen Himayat
Intel Corporation, USA

Ekram Hossain
University of Manitoba, Canada

Chih-Lin I
Green Communication Research Center, China Mobile Research Institute, China

Yuki Inoue
NTT DoCoMo Inc., Japan

Mingyue Ji
University of Utah, USA

Kerstin Johnsson
Intel Corporation, USA

Peter Jung
Fraunhofer Heinrich Hertz Institute, Germany

Martin Kasparick
Fraunhofer Heinrich Hertz Institute, Germany

Sachin Katti
Stanford University, USA

Joongheon Kim
Chung-Ang University, Korea

Yoshihisa Kishiyama
NTT DoCoMo, Inc., Japan

Yevgeni Koucheryavy
Tampere University of Technology, Finland

Dimitri Kténas
CEA, LETI, France

Toni Levanen
Tampere University of Technology, Finland

Anxin Li
DoCoMo Beijing Communications Laboratories Co., Ltd, China

Geoffrey Ye Li
Georgia Institute of Technology, USA

Maximilian Matthé
Technische Universität Dresden, Germany

Luciano Mendes
Technische Universität Dresden, Germany

Nicola Michailow
Technische Universität Dresden, Germany

Andreas F. Molisch
University of Southern California, USA

Takehiro Nakamura
NTT DoCoMo, Inc., Japan

David Ott
Intel Corporation, USA

Slawomir Pietrzyk
IS-Wireless, Poland

Juho Pirskanen
Nokia Networks, Finland

Alexander Pyattaev
Tampere University of Technology, Finland

Ashok Sunder Rajan
Intel Corporation, USA

Kannan Babu Ramia
Intel Corporation, USA

Rapeepat Ratasuk
Nokia Bell Labs, USA

Frank Schaich
Alcatel Lucent Bell Labs, Germany

Arun Sridharan
Samsung Research America, USA

Kazuaki Takeda
NTT DoCoMo Inc., Japan

Shilpa Talwar
Intel Corporation, USA

Rakesh Taori
Samsung Research America, USA

Arash Saber Tehrani
University of Southern California, USA

Timothy A. Thomas
Nokia Networks, USA

Mikko Valkama
Tampere University of Technology, Finland

Rath Vannithamby
Intel Corporation, USA

Benny Vejlgaard
Nokia Networks, Denmark

Frederick W. Vook
Nokia Networks, USA

Thorsten Wild
Alcatel Lucent Bell Labs, Germany

Gerhard Wunder
Fraunhofer Heinrich Hertz Institute, Germany

Cong Xiong
Georgia Institute of Technology, USA

Shu-ping Yeh
Intel Corporation, USA

List of Acronyms

Chapter 1

1G	First Generation
2G	Second Generation
3G	Third Generation
4G	Fourth Generation
5G	Fifth Generation
CDMA	Code Division Multiple Access
TDMA	Time Division Multiple Access
OFDMA	Orthogonal Frequency Division Multiple Access
GSM	Global System for Mobile communications
IMT	International Mobile Telecommunications
ITU-R	International Telecommunication Union-Radio
WCDMA	Wideband CDMA
3GPP	Third Generation Partnership Project
HSPA	High Speed Packet Access
LTE	Long-Term Evolution
FDMA	Frequency Division Multiple Access
SC-FDMA	Single Career Frequency Division Multiple Access
M2M	Machine to Machine communications
IoT	Internet of Things
QoE	Quality of Experience
RAT	Radio Access Technology
MIMO	Multiple Input Multiple Output
SDN	Software Defined Network
NFV	Network Function Virtualization

Chapter 2

5GMF	5G Mobile Communications Promotion Forum
NGMN	Next Generation Mobile Networks
D2D	Device to Device
FHD	Full High Definition
UHD	Ultra High Definition
V2V	Vehicle-to-Vehicle
C2C	Car-to-Car
V2I	Vehicle-to-Road Infrastructure
C2P	Car-to-Pedestrian
V2D	Vehicle-to-Device
BYOD	Bring Your Own Device
SoLoMo	Social Local Mobile
HMI	Human-Machine Interface
CAGR	Compound Annual Growth Rate
WRC	World Radio Conference
AR	Augmented Reality
RTT	Round Trip Time
TTI	Transmission Time Interval
HARQ	Hybrid Automatic Repeat reQuest

Chapter 3

3GPP	3rd Generation Partnership Project
BS	Base Station
D2D	Device to Device
DL	Downlink
EE	Energy Efficiency
EEC	European Economic Union
EFTA	European Free Trade Association
EP	European Parliament
ETP	European Technology Platform
ETSI	European Telecommunications Standards Institute
EU	European Union
HetNet	Heterogeneous network
ICT	Information and Communication Technology
IST	Information Society Technology
LTE	Long-Term Evolution
LTE-A	Long-Term Evolution-Advanced
LSA	Licensed Shared Access
MIMO	Multiple Input Multiple Output
MTC	Machine Type Communication
PPP	Public Private Partnership
QoS	Quality of Service

RAT	Radio Access Technology
TDMA	Time-Division Multiple Access
UE	User Equipment
UL	Uplink
UMTS	Universal Mobile Telecommunications System

Chapter 4

ISRA	Intel Strategic Research Alliance
NTIA	National Telecommunications and Information Association
GHz	Gigahertz
THz	Terahertz
Gbps	Gigabits per second
MIMO	Multi Input Multi Output
MU-MIMO	Multi-User MIMO
VLM	Very Large MIMO
CP	Cyclic Prefix
OFDM	Orthogonal Frequency Division Multiplexing
RAN	Radio Access Network
RAT	Radio Access Technology
WAN	Wide Area Network
LAN	Local Area Network
PAN	Personal Area Network
IoT	Internet of Things
QoE	Quality of Experience
QoS	Quality of Service
RFP	Request For Proposals
OTT	Over-The-Top
ARQ	Automatic Repeat reQuest
PHY	Physical Layer
FFR	Fractional Frequency Reuse
LSA	Licensed Shared Access
REM	Radio Environment Map
PC	Personal Computer
GNU	GNUs Not Unix

Chapter 5

SE	Spectral Efficiency
EE	Energy Efficiency
LSAS	Large Scale Antenna System
NOMA	Non Orthogonal Multiple Access
C-RAN	Cloud Radio Access Network
ICT	Information and Communications Technologies

MTC	Machine Type Communications
QoS	Quality of Service
MAC	Medium Access Control
PA	Power Amplifier
CSI	Channel State Information
TDD	Time Division Duplex
FDD	Frequency Division Duplex
UDN	Ultra Dense Network
DAS	Distributed Antenna System
CoMP	Coordinated Multi-Point
IM	Instant Messaging
LAPI	Low Access Priority Indication
RRC	Radio Resource Control

Chapter 6

SCN	Small Cell Network
UT	User Terminal
ICIC	Inter-Cell Interference Coordination
TTT	Time to Trigger
SINR	Signal-to-Interference-plus-Noise Ratio
OPEX	Operational Expenditures
CF	Collaborative Filtering
SVD	Singular Value Decomposition
CDN	Content Delivery Network
ICN	Information Centric Networks
MAB	Multi-Armed Bandit
ADMM	Alternating Direction Method of Multipliers
DMT	Diversity-Multiplexing Tradeoff
SNR	Signal-to-Noise Ratio
PPP	Poisson Point Process

Chapter 7

D2D	Device-to-Device
QoS	Quality of Service
RAT	Radio Access Technology
UE	User Equipment
HetNets	Heterogeneous Networks
WLAN	Wireless Local Area Network
3GPP	Third Generation Partnership Project
UMTS	Universal Mobile Telecommunications System
LTE	Long-Term Evolution
RAN	Radio Access Network
ANDSF	Access Network Discovery and Selection Function

SINR	Signal-to-Interference-plus-Noise Ratio
DL	Downlink
UL	Uplink
MIMO	Multiple Input Multiple Output
PPP	Poisson Point Process
AP	Access Point
BS	Base Station
MP	Maximum Power
FU	Full Utilization
SNR	Signal-to-Noise Ratio
SLS	System Level Simulator

Chapter 8

LTE-A	Long-Term Evolution-Advanced
ABS	Almost Blank Subframe
RB	Resource Block
CSI	Channel State Information

Chapter 9

D2D	Device-Device
FCC	Federal Communications Commission
V2V	Vehicle to Vehicle
D2I	Device-to-Infrastructure
RMS	Root Mean Square
GSCM	Geometry-based Stochastic Channel Model
BS	Base Station
MAC	Medium Access Control
DVCS	Directional Virtual Carrier Sensing
DCF	Distributed Coordinated Function
CS	Compressed Sensing
ZC	Zhadoff–Chu
CSI	Channel State Information
TDMA	Time Division Multiple Access
CSMA/CS	Carrier Sense Multiple Access with Collision Sensing
LATS	Location Aware Training Scheme
NMSE	Normalized Mean Square Error
QoS	Quality of Service
SINR	Signal-to-Interference-plus-Noise Ratio
SNR	Signal-to-Noise Ratio
SIR	Signal-to-Interference ratio
INR	Interference-to-Noise Ratio
PPP	Poisson Point Processes
MINLP	Mixed-Integer Nonlinear Programming

NE	Nash Equilibrium
PSO	Particle Swarm Optimization
OFDMA	Orthogonal Frequency Division Multiple Access
FDMA	Frequency Division Multiple Access
ITIS	Information-Theoretic Independent Sets
CU	Cellular User
ZF	Zero-Forcing
MC	Mobile Cloud
PCH	Primary Cluster Head
SCH	Secondary Cluster Head
MR-D	Maximum Rate towards Destination
RTS	Request To Send
CTS	Clear To Send
SIB	System Information Block
QoE	Quality of Experience

Chapter 10

OFDM	Orthogonal Frequency Division Multiplexing
OFDMA	Orthogonal Frequency Division Multiple Access
EE	Energy Efficiency
QoS	Quality of Service
AWGN	Additive White Gaussian Noise
DOF	Degree(s) of Freedom
SE	Spectral Efficiency
CSI	Channel State Information
CNR	Channel gain to Noise Ratio
LDD	Lagrange Dual Decomposition
MDSA	Maximum Downlink Subcarrier Assignment
MUSA	Maximizing Uplink Subcarrier Assignment
BPA	Bisection Power search Algorithm
LT	Luby Transform
MIMO	Multiple Input Multiple Output
PA	Power Amplifier

Chapter 11

MIMO	Multiple Input Multiple Output
SV-MIMO	Smart Vertical MIMO
SIMO	Single Input Multiple Output
NOMA	Non-Orthogonal Multiple Access
FDMA	Frequency Division Multiple Access
TDMA	Time Division Multiple Access
CDMA	Code Division Multiple Access
OFDMA	Orthogonal Frequency Division Multiple Access

SDMA	Spatial Division Multiple Access
OMA	Orthogonal Multiple Access
LTE	Long-Term Evolution
SU-MIMO	Single User MIMO
MU-MIMO	Multi-User MIMO
RAT	Radio Access Technology
ICIC	Inter-Cell Interference Coordination
CoMP	Coordinated Multi-Point
IRC	Interference Rejection Combining
MMSE	Minimum Mean Squared Error
NAICS	Network-Assisted Interference Cancellation and Suppression
MLD	Maximum Likelihood Detection
SIC	Successive Interference Cancellation
AAS	Active Antenna System
FD-MIMO	Full Dimensional MIMO
LOS	Line-Of-Sight
NLOS	Non Line-Of-Sight
SINR	Signal to Interference plus Noise Ratio
BS	Base Station
UE	User Equipment
AWGN	Additive White Gaussian Noise
CSI	Channel State Information
CQI	Channel Quality Indicator
SLIC	Symbol-Level Interference Cancellation
CWIC	Codeword Level Interference Cancellation
LLR	Log-Likelihood Ratio
MRC	Maximal Ratio Combining
BLER	Block Error Rate
RS	Reference Signal
C-RS	Common Reference Signal
UE-RS	UE-specific Reference Signal
SCM	Spatial Channel Model
HARQ	Hybrid Automatic Repeat reQuest
MCS	Modulation and Coding Scheme
MCPS	Modulation, Coding, and Power Set
TPA	Transmit Power Allocation
FSPA	Full Search Power Allocation
SFBC	Space Frequency Block Coding
CDD	Cyclic Delay Diversity
CRS	Cell Specific Reference Signal
BF	Beamforming
BB	Base-Band
PSS	Primary Synchronization Signal
SSS	Secondary Synchronization Signal
PDCCH	Physical Downlink Control Channel
EPDCCH	Enhanced PDCCH

PBCH	Physical Broadcast Channel
PDSCH	Physical Downlink Shared Channel
DM-RS	Demodulation Reference Signal
MS	Mobile Station

Chapter 12

RFID	Radio Frequency Identification
EDGE	Enhanced Data rates for GSM Evolution
RAN	Radio Access Network
UE	User Equipment
BS	Base Station
MME	Mobility Management Entity
PLMN	Public Land Mobile Network
EAB	Extended Access Barring
ACB	Access Class Barring
eNB	Evolved Node B (base station)
RF	Radio Frequency
PMU	Power Management Unit
BOM	Bill of Material
FFT	Fast Fourier Transform
TBS	Transport Block Size
PRACH	Physical Random Access Channel
PUSCH	Physical Uplink Shared Channel
PUCCH	Physical Uplink Control Channel
PDSCH	Physical Downlink Shared Channel
PBCH	Physical Broadcast Channel
EPDCCH	Enhanced Physical Downlink Control Channel
PSS	Primary Synchronization Signal
SSS	Secondary Synchronization Signal
MIB	Master Information Block
SIB	System Information Blocks
MCL	Maximum Coupling Loss
PRB	Physical Resource Block
NB	Narrow-Band
NB-IoT	Narrow-Band Internet of Things
TDM	Time Division Multiplexing

Chapter 13

PHY	Physical layer
HARQ	Hybrid Automatic Repeat reQuest
AIC	Advanced Interference Cancellation
LOS	Line Of Sight
NLOS	Non Line Of Sight

CP	Cyclic Prefix
GP	Guard Period
TA	Timing Alignment
Tx	Transmission
Rx	Reception
WLAN	Wireless Local Area Network
FCC	Federal Communications Commission
BF	Beam-Forming
CRS	Common Reference Symbol
DLCRS	Downlink Common Reference Symbol
DLCCH	Downlink Control Channels
ACK	Acknowledgement
DLSCH	Downlink Shared Channel
DMRS	Demodulation Reference Symbols
ULCRS	Uplink Common Reference Symbols
ULSCH	Uplink Shared Channel
ULDCH	Uplink Data Channel
RACH	Random Access Channel
ULCCH	Uplink Control Channel
MCS	Modulation and Coding Scheme

Chapter 14

PHY	Physical layer
DFT	Discrete Fourier Transform
MTC	Machine-Type Communication
IoT	Internet of Things
RACH	Random Access Channel
CoMP	Coordinated Multi-Point
CP	Cyclic Prefix
CS	Cyclic Suffix
FBMC	Filter Bank Multi-Carrier
TTI	Transmission Time Interval
ICI	Inter-Carrier Interference
GI	Guard Interval
ISI	Inter-Symbol Interference
IDMA	Interleave-Division Multiple Access
PRACH	Physical Layer Random Access Channel
D-PRACH	Data PRACH
ATA	Autonomous Timing Advance
OFDM	Orthogonal Frequency Division Multiplexing
UFMC	Universal Filtered Multi-Carrier (also UF-OFDM)
FFT	Fast Fourier Transform
IFFT	Inverse Fast Fourier Transform
QAM	Quadrature Amplitude Modulation

CFO	Carrier Frequency Offset
MUD	Multi-User Detection
MPR	Multi Packet Reception
MMC	Massive Machine Communication
GFDM	Generalized Frequency Division Multiplexing
AWGN	Additive White Gaussian Noise
MF	Matched Filter
ZF	Zero-Forcing
MMSE	Minimum Mean Square Error
DZT	Discrete Zak Transform
STC	Space Time Coding
TR-STC	Time-Reversal Space Time Coding
GFDM	Generalized Frequency Division Multiple Access
BER	Bit Error Rate
OQAM	Offset Quadrature Amplitude Modulation
FS-FBMC	Frequency Spreading FBMC
PPN-FBMC	Poly-Phase Network FBMC
SINR	Signal to Interference plus Noise Ratio
MQAM	M-ary Quadrature Amplitude Modulation
QPSK	Quadrature Phase Shift Keying
BFDM	Bi-orthogonal Frequency Division Multiplexing
PUSCH	Physical Uplink Shared Channel
ACK/NACK	Acknowledgment/Negative Acknowledgment

Chapter 15

MIMO	Multiple Input Multiple Output
CoMP	Coordinated Multi-Point
FD-MIMO	Full Dimension MIMO
SU-MIMO	Single-User MIMO
MU-MIMO	Multi-User MIMO
CRS	Common Reference Signals
CSI-RS	Channel State Information Reference Signals
DMRS	Dedicated Modulation Reference Signals
UE	User Equipment
CS	Coordinated Scheduling
CB	Coordinated Beamforming
DPS	Dynamic Point Selection
JP	Joint Processing
JT	Joint Transmission
NIB	Non-Ideal Backhaul
FDD	Frequency Division Duplexing
TDD	Time Division Duplexing
LOS	Line-of-Sight
NLOS	Non-Line-of-Sight

SNR Signal-to-Noise Ratio
PMI Precoder Matrix Indicator
AP Access Point
RFIC RF Integrated Circuit
MMIC Monolithic Microwave Integrated Circuit
LTCC Low Temperature Co-fired Ceramic
LCP Liquid Crystal Polymer
QAM Quadrature Amplitude Modulation

Chapter 16

LTE Long-Term Evolution
SNR Signal-to-Noise Ratio
MIMO Multiple Input Multiple Output
PHY Physical Layer
OFDM Orthogonal Frequency Division Multiplexing
PCB Printed Circuit Board
WARP Wireless Open Access Research Platform
LO Local Oscillator
ADC Analog to Digital Converter
PAPR Peak to Average Power Ratio
QAM Quadrature Amplitude Modulation
AGC Automatic Gain Control
LNA Low Noise Amplifier
IQ Inphase/Quadrature
USRP Universal Software Radio Peripheral
RS Rohde–Schwarz
QPSK Quadrature Phase Shift Keying
FD Full Duplex
HD Half Duplex

Chapter 17

BS Base Station
MS Mobile Station
CoMP Coordinated Multi-Point
PMP Point-to-Multipoint
AGW Access Gateway
BL Backhaul Link
AL Access Link
UL Uplink
DL Downlink
ISD Inter-Site Distance
LOS Line-of-Sight
NLOS Non-Line-of-Sight

SDM	Spatial Division Multiplexing
TDM	Time Division Multiplexing
SDMA	Space Division Multiple Access
SIR	Signal to Interference Ratio
W-BS	Wired BS
U-BS	Unwired BS
TDD	Time Division Duplex

Chapter 18

SDN	Software Defined Networking
NFV	Network Function Virtualization
EPC	Evolved Packet Core
CSP	Communication Service Provider
KPI	Key Performance Indicator
BGR	Border Gateway Router
TOC	Total Cost of Ownership
SEGW	Service Edge Gateway
PCRF	Policy Rules Charging Function
PGW	Packet Gateway
UP	User Plane
NAS	Non-Access Stratum
HSS	Home Subscription Server
TEID	Tunnel End Point Identifier
VoIP	Voice over IP

About the Companion Website

This book is accompanied by a companion website:

www.wiley.com/go/vannithamby/towards5g

There you will find valuable material designed to enhance your learning, including:

- Abstract and Keywords
- List of Contributors

Scan this QR code to visit the companion website

Part I

Overview of 5G

1

Introduction

Shilpa Talwar and Rath Vannithamby
Intel Corporation, USA

1.1 Evolution of Cellular Systems through the Generations

The first large-scale commercial cellular communications systems were deployed in the 1980s and these became known as first-generation (1G) systems. 1G systems were built on narrowband analog technology, and provided a basic voice service. These were replaced by second-generation (2G) cellular telecom networks by the early 1990s. 2G networks marked the start of the digital voice communication era, and provided a secure and reliable communication channel. 2G systems use either time division multiple access (TDMA) or code division multiple access (CDMA) technologies, and provided higher rates. The European Global System for Mobile Communications system is based on TDMA technology while IS-95 (also known as CDMA One) is based on CDMA technology. These 2G digital technologies provide expanded capacity, improved sound quality, better security and unique services such as caller ID, call forwarding, and short messaging. A critical feature was seamless roaming, which let subscribers move across provider boundaries.

The third-generation (3G) – International Mobile Telecommunications-2000 (IMT-2000) – is a set of standards for mobile phones and mobile telecommunications services fulfilling the recommendations of the International Telecommunication Union-Radio (ITU-R). 3G mobile networks became popular due to ability of users to access the Internet over mobile devices and laptops. The speed of data transmission on a 3G network is up to 2 Mbps, and therefore the network enables voice and video calling, file transmission, internet surfing, online TV, playing of games and much more. 3G uses CDMA technology in various forms. Wideband CDMA and High Speed Packet Access technologies were developed as part of the Third Generation Partnership Project (3GPP) organization, and CDMA2000 was developed as part of the 3GPP2 organization.

Towards 5G: Applications, Requirements and Candidate Technologies, First Edition.
Edited by Rath Vannithamby and Shilpa Talwar.

Companion website: www.wiley.com/go/vannithamby/towards5g

Fourth-generation (4G) requirements – the International Mobile Telecommunications Advanced (IMT-Advanced) specification – were specified by ITU-R in March 2008. The key requirements specified 4G peak service speeds of 100 Mbps for high-mobility communication (such as from trains and cars) and 1 Gbps for low-mobility communication (such as pedestrians and stationary users). A 4G system not only provides voice and other 3G services but also provides ultra-broadband network access to mobile devices. Applications vary from IP telephony, HD mobile television, video conferencing to gaming services and cloud computing. There are two 4G technologies: Long-Term Evolution (LTE) and Worldwide Interoperability for Microwave Access (WiMAX). LTE was developed as part of 3GPP and WiMAX was developed as part of IEEE. LTE uses orthogonal frequency division multiple access (OFDMA) in the downlink and single carrier frequency division multiple access in the uplink whereas WiMAX uses OFDMA in both uplink and downlink.

1.2 Moving Towards 5G

4G standards were completed in 2011 and networks are currently being deployed. The attention of the mobile research community is now shifting towards what will be the next set of innovations in wireless communication technologies, which we will refer to collectively as 5G (fifth-generation technologies). Given a historical 10-year cycle for every generation of cellular advancement, it is expected that networks with 5G technologies will be deployed around 2020. Similar to 3G/4G, where ITU-R issued a recommendation for IMT-2000/IMT-Advanced [1], ITU-R has recently released a recommendation for the framework and overall objectives of the future development of systems for 2020 and beyond [2]. This highlights the emerging consensus on the use cases and requirements that systems deployed in 2020 must address. These include requirements for new services such as smart grids, e-health, autonomous transport, augmented reality, wireless industry automation, remote tactile control and so on, which cannot be met by IMT-2000 systems.

The usage scenarios envisioned for IMT for 2020 and beyond can be broadly classified as follows:

Enhanced Mobile Broadband The dramatic growth in the number of smartphones, tablets, wearables, and other data-consuming devices, coupled with the advent of enhanced multimedia applications, has resulted in a tremendous increase in the volume of mobile data traffic. According to industry estimates, this increase in data traffic is expected to continue in the coming years and around 2020 cellular networks might need to deliver as much as 100–1000 times the capacity of current commercial cellular systems [3, 4]. While the roll-out of 4G technologies with their expected enhancements will address some of capacity demands of future mobile broadband users, a mobile broadband user in 2020 will expect to be seamlessly connected all the time, at any location, to any device. This poses stringent requirements on the 5G network, which must provide users with a uniform and seamless connectivity experience regardless of where they are and what device/network they connect to.

Massive Machine-type Communications This use case refers to the growing interest in the area of machine-to-machine (M2M) communications and the Internet-of-Things (IoT). Together, these represent a future in which billions of everyday objects are connected and managed through wireless networks and management servers [5]. One can envisage creating an immensely rich set of applications by connecting the thousands of objects surrounding us. Examples include:

- smart homes, in which intelligent appliances autonomously minimize energy use and cost
- remote monitoring of expensive industrial or medical equipment
- remote sensing of environmental metrics such as water pressure, air pollution and so on.

These applications and services demand communication architectures and protocols that are different from traditional human-based networks. The integration of human and machine-type traffic in a single 5G network is therefore a challenge. In addition, IoT traffic can be quite diverse, from low to high bandwidth, from delay-sensitive to delay-tolerant, from error-tolerant to high reliability, which poses additional complexity. This use case focuses on applications where a very large number of connected devices transmit relatively low volumes of non-delay-sensitive data. The devices are typically low-cost and low-complexity, and require a very long battery life.

Ultra-reliable and Low-latency Communications. This use case addresses IoT applications that have stringent requirements for reliability, latency, and network availability. Examples include:

- connected cars, which react in real time to prevent accidents
- body area networks, which track vital signs and trigger an emergency response when life is at risk
- wireless control of industrial manufacturing or production processes.

As evidenced by diverse set of usages anticipated by 2020, the 5G system will require enhancements to performance metrics beyond the "hard" metrics of 3G/4G, which included peak rate, coverage, spectral efficiency, and latency. The 5G system will see expanded performance metrics centered on the user's quality of experience (QoE), including factors such as ease of connectivity with nearby devices, connection density, area traffic capacity, and improved energy efficiency. The eight parameters in Table 1.1 are considered to be key capabilities of IMT-2020 systems. Their target values are also summarized. These are currently recommendations, and subject to further research and technological development [2].

1.3 5G Networks and Devices

As it can be seen from the description above, 5G networks will have to accommodate diverse types of traffic, spectrum, and devices. The network itself is anticipated to consist of hierarchical nodes of various characteristics and capacities. The 5G network will support multiple radio access technologies (RATs), such as 3G/4G/5G, WiFi, and WiGig, and also multiple modes ranging from ultradense small cells, device-to-device (D2D) communications, and new sub-networks oriented toward wearable devices. Inevitably, the user experience and quality will need to be maintained as users move along various networks and get connected to the various types of node. 5G networks will likely use a multi-layer network

Table 1.1 Key parameters of IMT-2020 systems.

Parameter	Details	Target
Peak data rate	Maximum achievable data rate under ideal conditions per user/device	10–20 Gbps
User-experienced data rate	Achievable data rate that is available ubiquitously across the coverage area to a mobile user/device	100 Mbps–1 Gbps, depending on wide-area or hotspot coverage
Latency	Time contribution by the radio network from the time from when the source sends a packet to when the destination receives it	1 ms over-the-air latency
Mobility	Maximum speed at which a defined QoS and seamless transfer between radio nodes which may belong to different layers and/or radio access technologies (multi-layer/-RAT) can be achieved	To provide high mobility up to 500 km/h with acceptable QoS
Connection density	Total number of connected and/or accessible devices per unit area	To support a connection density of up to 10^6/km^2, for example in massive machine-type communication scenarios
Energy efficiency		
(a) Network side	Quantity of information bits transmitted to/received from users, per unit of energy consumption of the radio access network (RAN) (in bit/Joule)	Target is at least 10x on network energy efficiency The 5G network must not consume more energy, while providing enhanced features
(b) Device side	Quantity of information bits per unit of energy consumption of the communication module (in bit/Joule)	
Spectrum efficiency	Average data throughput per unit of spectrum resource and per cell (bit/s/Hz)	3–5× increase in spectrum efficiency
Area traffic capacity	Total traffic throughput served per geographic area	10 Mbit/s/m^2 in hotspot scenarios

architecture, where the macro layer provides coverage to users moving at high speeds or for secure control channels, while a lower layer comprising network nodes with smaller capabilities provides high data rates and connectivity to other RATS (say, WiFi or new mmWave RATs). Moreover, a 5G device may have simultaneous active connections to more than one network node, with the same or different RATs, each connection serving a specific purpose, for example one connection to a given node for data and a second connection to another node for control. In addition, the use of remote radio heads connected to central processing nodes with the aid of ultra-high-speed backhaul is expected to be extended to more areas. Fast and high-capacity backhaul will enable tighter coordination between network nodes in a larger area. All of these changes will require a high level of integration of different nodes in the network and of technologies located even within the same node. In short, the 5G system will need to provide a flexible technological framework in which networks, devices, and applications can be co-optimized to meet the great diversity of requirements anticipated by 2020.

As the 5G usage models and networks evolve, 5G device architectures will also be more complex than in 4G. Devices will be capable of operating in multiple spectrum bands, ranging from RF to mmWave, while being compatible with existing technologies such as 3G and 4G. The need to support several RATs with multiple RF-chains will impose tremendous challenges for 5G device chipset and front-end module suppliers, as well as system and platform integrators. Another key feature of 5G devices will be their advanced interference suppression capabilities. The dense deployment of network nodes and increasing sources of interference will require that the devices deployed autonomously detect, characterize, and suppress interference from any source: intra-cell, inter-cell, or D2D. The task of interference cancellation will be exacerbated by the existence of strong self-interference in the case of simultaneous transmission and reception. In addition, devices will be required to actively manage all the available network connections, including D2D links, as well as to share contextual information with network layers so that network resources can be efficiently utilized. All of these enhanced features will need to be implemented in such a way that energy consumption is optimized for a small wireless device platform.

1.4 Outline of the Book

In this book we bring together a group of visionaries and technical experts from academia and industry to discuss the applications and technologies that will comprise the 5G system. It is expected that some of the new technologies comprising 5G will be evolutionary, covering gaps and enhancements from 4G systems, while some of the technologies will be disruptive, covering fundamentally new waveforms, duplexing methods, and new spectrum. These technologies will encompass the end-to-end wireless system: from wireless network infrastructure to spectrum availability to device innovations.

The book is organized into three parts. Part I has four chapters. In Part I, we provide an overview of 5G, address trends in applications and services, and summarize 5G requirements that will be need to be addressed in next-generation technologies and system architectures. We also provide an overview of some 5G research programs around the world: Horizon 2020 in Europe and Intel's 5G University Research Program in USA.

Part II has nine chapters. In Part II, we address evolutionary technologies that will be needed to meet 5G requirements, including:

- co-operative radio access architectures to enable greater energy efficiency and network performance
- small-cell networks with in-built caching
- multiple RAT integration, which is inevitable to provide a seamless user experience
- distributed resource allocation
- advances in device-to-device communications
- energy-efficient network design
- multi-antenna processing and interference co-ordination techniques
- design for M2M communications
- design for ultra-low latency.

These technologies are already being developed in 3GPP Release 11 and beyond as part of the evolution of 4G systems.

Part III has five chapters. In Part III, we discuss "revolutionary" candidate technologies: those that are essentially disruptive and different from 4G. These include:

- new physical layer waveforms that offer enhanced flexibility and performance
- massive MIMO technologies that enable large numbers of simultaneous users
- mmWave technologies to harness new spectrum for access and backhaul
- simultaneous transmit and receive on the same time/frequency resource
- software defined networking and network function virtualization to enable software-based flexible infrastructures.

References

[1] ITU-R, Recommendation M.1645: "Framework and overall objectives of the future development of IMT-2000 and systems beyond IMT-2000", June 2003.
[2] ITU-R, Document 5D/TEMP/625-E: "IMT Vision – Framework and overall objectives of the future development of IMT for 2020 and beyond", 17 June 2015.
[3] Cisco, "Cisco Visual Network Index: Global mobile traffic forecast update", 2013.
[4] Ericsson, "Traffic and market data report," 2011.
[5] Ericsson, White paper "More Than 50 billion connected devices", 2011. URL: http://www.ericsson.com/res/docs/whitepapers/wp-50-billions.pdf.

2

5G Requirements

Anass Benjebbour, Yoshihisa Kishiyama, and Takehiro Nakamura
NTT DoCoMo, Inc., Tokyo, Japan

2.1 Introduction

Over the last few decades, mobile communications have significantly contributed to the economic and social development of both developed and developing countries. Today, mobile communications form an indispensable part of the daily lives of billions of people in the world, a situation that is expected to continue and become even more widespread in the future. Currently, the 4G radio access system using Long-Term Evolution (LTE) is being deployed by many operators worldwide in order to offer faster access with lower latency and more efficiency than 3G/3.5G. In the future, however, it is foreseen that demand for higher volumes of traffic, many more connected devices with diverse service requirements, and better and uniform quality of user experience will bring a need for evolved systems with extended capabilities.

In order to meet these evolving needs for mobile communications, discussions on visions, requirements, and technologies for the 5G mobile communications system have been initiated by many organizations. 5G-related discussions are ongoing in the ITU-R Study Group 5 Working Party 5D (WP5D), which issued a new recommendation, "IMT Vision – Framework and overall objectives of the future development of IMT for 2020 and beyond". Also, technical studies on 5G have gained attention worldwide as evidenced by the acceleration of efforts by governmental entities and research bodies from both academia and industry. Many special sessions are also being held on the topic of 5G in international conferences. Several governments and groups of commercial companies and academic institutions have established projects and fora to study and promote 5G mobile technology. Examples of projects and initiatives with focus on 5G include the METIS project in Europe, the ARIB 2020 and Beyond Ad-hoc (20B AH) group, and the 5G Mobile Communications Promotion Forum (5GMF) in Japan, the operators' alliance Next Generation Mobile Networks (NGMN), IMT-2020 in China, and the 5G Forum in Korea.

Towards 5G: Applications, Requirements and Candidate Technologies, First Edition.
Edited by Rath Vannithamby and Shilpa Talwar.

Companion website: www.wiley.com/go/vannithamby/towards5g

2.2 Emerging Trends in Mobile Applications and Services

More and more customers are expecting to have the same quality of experience from Internet applications anytime, anywhere, and through any means of connectivity. This expectation is now being better fulfilled as the gap of user experience between mobile and fixed environments becomes narrower and higher data rates are offered by mobile networks. In the future we can therefore expect a further shift of services from the fixed to the mobile network, with users making use of the added value of mobility and location/context awareness. Furthermore, the emergence of new applications and needs are constantly changing user behavior. The younger generation now uses the Internet for gaming, social networking, and online education, among other things. At the same time, the introduction of IMT-Advanced networks, which substantially reduce network latency, will in the future provide better user experiences and make possible more advanced real-time services. Technological developments, such as faster radio interfaces, advanced graphical processing, and multiprocessing units at the device, will also contribute to the increase in user demand for mobile data. Growth will also be accelerated by new types of communications and devices, such as device-to-device communications between mobile users in proximity (user-to-user), and machine-type communications such as user-controlled mobile devices (user-to-machine) and inter-machine communications (machine-to-machine). The future trends in services and applications will generally be shaped by the evolution of the needs of the new generation of users and progress in technology and services.

In the following sections, we explain the main market trends and new services that have been observed in recent years and have the potential to drive and change the landscape of the future mobile market. Note that future services include, but are not limited to, the mere interpolation of current trends.

2.2.1 New Types of Mobile Device

The transition to the Internet era has significantly contributed to the rapid rise of data services as a significant revenue source for businesses. This trend has been accelerated by the introduction of always-on smartphones and new types of conversation via social networks. In recent years, a wide range of new smart devices – smartphones, dongles, and tablets – have emerged and have been key drivers of increased mobile broadband traffic. With rapid advances in display technologies, these devices offer larger screen sizes and high resolution, and hence increase data consumption and encourage the use of traffic-intensive applications such as video streaming. This type of Internet access via mobile terminals is spreading very rapidly. As a result, the volume of smartphone data carried by cellular networks is growing rapidly, driven predominantly by increases in device penetration, but also by increases in average usage. In developed markets, a typical smartphone generates about 50 times more data per month than a typical feature phone [1]. In the future, one notable development will be full high definition (FHD) and ultra-high definition (UHD) displays, which are anticipated to become well established on smartphones; it is estimated that these future smartphones could generate many times more traffic than established user applications. In addition, open operating systems (OSs), such as Android, iOS, and HTML5, have been another key force in the mobile internet ecosystem. With open OSs, the development and commercialization of new applications has become much easier than before. Users are able to access a wide variety of new applications on diverse smart devices, resulting in increased opportunities, as well as challenges, for all players in the mobile Internet ecosystem. Operators are making great efforts to embrace these changes and challenges, although they represent a double-edged sword. On the one hand, the majority of mobile applications on smart devices are planned with the assumption that users are online and connected, consequently increasing both control signaling and user mobile broadband traffic: video, music, games, and so on. On the other hand, memory as well as processor technologies are expected to improve according to Moore's law, and with reduced energy consumption. This will bring huge potential for information storage and processing on mobile devices and increased user-generated content. Furthermore, new types of user-to-device interaction can be expected to be triggered by novel user interfaces such as 3D cameras, and movement and gesture recognition. These will increase the generation and flow of information and beyond that of traditional human audio and visual capabilities.

2.2.2 *Video Streaming and Download Services*

Video streaming and download are among the most dominant traffic generators in mobile networks. Currently, the majority of streaming services are based on progressive downloading technologies utilizing the HTTP protocol. Video streaming services can be classified into server-client unidirectional applications and bidirectional streaming services.

Bidirectional streaming services with high quality of service demands are expected to become a dominant source of traffic in the near future. One example is the virtual classroom, with video streamed between a remote teacher and students in a classroom. Moreover, video consumption for many users is no longer limited to streaming but also involves sharing it with the community. Uploading of videos on social networking sites is becoming a way to share them. This contributes to increasing video consumption, as community networks are also becoming video viewing sites. In the future, video streaming or downloading will be responsible for most mobile data traffic growth, with a cumulative average growth rate (CAGR) of 69% expected between 2013 and 2018. Furthermore, it is predicted that video will account for more than 69% of mobile data traffic by 2018 [1]. In the future, the introduction of advanced graphical processing units will enhance the performance of video applications and thus promote mobile video consumption. In addition, mobile services that require 3D video and higher-definition video will proliferate and thus create significantly increased traffic over mobile networks.

2.2.3 *Machine-to-machine Services*

One big wave that will to contribute to the increase in mobile data demand is machine-to-machine (M2M) applications and devices. M2M is rapidly growing and is expected to continue to be one of the fastest growing segments in the future [1]. The growth of the M2M market has been driven by sectors such as fleet management, industrial asset management, point of sales, security, and healthcare. The number of M2M connections could be several orders of magnitude larger than the world population. The market for M2M systems is expected to grow by 30–40% per year. Cisco IBSG predicts there will be 25 billion devices connected to the Internet by 2015 and 50 billion by 2020 [2]. In terms of traffic, M2M's share will depend on the related applications. For instance, smart utility meters in homes consume some hundreds of kilobytes per second while surveillance video monitoring consume tens of megabytes per second. In the future, agricultural science will also benefit from the ability to communicate information remotely. Another potential service is smart energy-distribution grid systems. For example, the European commission mandated that 80% of consumers in its member countries should be equipped with smart meters by the year 2020 [3].

Another set of applications for M2M is for communications in the transport sector:

- car-to-car (C2C)/vehicle-to-vehicle (V2V)
- car-to-road/vehicle-to-road infrastructure (V2I)
- car-to-pedestrian (C2P)/vehicle-to-device (V2D).

These are collectively referred to as C2X or V2X communications. They will improve traffic safety, both for drivers and pedestrians, provide in-car infotainment services, and bring new business opportunities, such as highly automated driving and augmented-reality head-up displays.

M2M services will be a big trend in 2020 and beyond. One issue, however, is the very wide range of requirements this trend will bring with it. For example, sensor-type applications will require the support of massive machine communications, while other safety and remote-control-related M2M applications will require ultra-low latency and/or ultra-reliable machine communication. In order to facilitate the study of such a wide variety of requirements, the principal market segments and categories of M2M services will need to be identified and defined.

2.2.4 Cloud Services

The demand for mobile cloud services is also expected to grow exponentially as users adopt services that must be ubiquitous. In particular, the rapid development of ICT technologies and mobile network capabilities will enable a wide range of cloud services to be available on mobile devices, for example cloud speech services, such as speech recognition and synthesis. Mobile cloud traffic will grow 12-fold from 2013 to 2018, a compound annual growth rate of 64%. Cloud applications will account for 90% of total mobile data traffic by 2018, compared to 82% at the end of 2013 [1]. It is expected that in the future health, education, and other government services will be accessible by mobile devices, which will contribute to improvements in social welfare. These services will require guaranteed reliability and security of data communications between the clients and the cloud data centers.

However, harnessing and extracting value from the "big data" stored in the cloud is seen by many operators as a route to enhance the customer experience and to generate new revenues from them. Via user data collection and mining, operators can enhance the user experience. They can also compile this data, selling it on in anonymized or aggregated form as business and marketing reports. For instance, data on customer footfall patterns could be sold to retailers, helping them target promotions according to store location and the buying patterns of consumers in that area. It will also help them decide where to open new shops, and in what format. Another recent trend for cloud services is termed "bring your own device", which enables employees to bring personally owned mobile devices (laptops, tablets, and smart phones) to their workplace, and use them to access company information and applications stored in the cloud.

2.2.5 Context-based and Location-based Services

Context/location awareness will be an important enabler for providing user-centered services in the future. With such capabilities, mobile devices will not only act as personal communication devices but also as gateways to services in diverse environments that support personalized interactions and proactive assistance tailored to the user preferences and behaviors. Context/location-aware applications and devices capture context information from multiple sources and learn the associations between context cues and personal preferences and behaviors in order to adapt the configuration of devices and the behavior of interfaces, or to offer personalized access to services. Learning the user's important locations, known as their semantic locations, will be one of the most important tasks involved. Examples of semantic locations are "Main campus, Kyoto University" or "City center of Tokyo".

Several location-aware applications for mobile devices have been developed recently. These applications make use of colloquial places and paths rather than just geographical coordinates, for example by accessing personal applications such as geo-reminders and location diaries. The combination of the cloud and location information will also create what is called the personal cloud, which will gradually replace the PC as the location where individuals keep their personal content and personal preferences, access services, and center their digital lives [4]. The personal cloud will shift the focus from the services delivered on client devices to cloud-based services delivered *across* devices. Examples of context-based and location-based services (LBS) include:

Augmented Reality. Augmented reality is a live – direct or indirect – view of a physical, real-world environment whose elements are augmented by computer-generated sensory input such as sound, video, graphics, or GPS data [5]. With the help of technologies such as computer vision and object recognition, information about the real world surrounding the user becomes interactive and digitally manipulable. Artificial information about the environment and its objects can be overlaid on the real world. Services based on these technologies are expected to expand in the future.

Proximity-based Services. As the number of mobile devices continues to increase, it becomes important to take advantage of the physical proximity of communicating devices and provide proximity services, such as social networking and proximity-based multiplayer games. To this end, peer-to-peer discovery and communication becomes an important enabler of such services. Such features will also enable new services, for example allowing direct communication between devices when the network is damaged in the aftermath of a natural disaster.

SoLoMo. Social local mobile (SoLoMo) is a new marketing concept that refers to the convergence of social, local, and mobile technologies. SoLoMo aims to "hyper-target", that is, to reach the right consumer, at the right time, in the right place. For example, retailers can utilize the mobile experience to their advantage, using location targeting, in-store mobile marketing, gamification, and so on. With SoLoMo, a specific retailer can broadcast offers – retail deals, coupons, consumer events, and shopping and dining opportunities – to a mobile user based on their geographic proximity, brand/retailer allegiance, and shopping/check-in history. In addition, the integration of location-based functions with social networks can lead to new applications on mobile networks that are expected to generate more mobile data traffic.

Implicit Communications. Owing to the rapid evolution of human-machine interfaces (HMI) and graphical display capabilities, it can be expected that in the future the way of communicating – in particular within social communities and individual (small) groups – will change dramatically. This new communication culture will be characterized by subtle and continuous information generation and dissemination in an autonomous way. The mobile device, comprising a 3D camera and intelligent image recognition technologies as well as different types of sensors, will capture the surrounding of individuals and enrich this content with context-aware and location-based information. Already today, instead of sending an SMS it is possible to send a picture with geographical information to inform family members about the successful at the end of a journey. Until now, this type of communication has been actively triggered by the user. In the future, the communication will happen implicitly, without cumbersome manual typing or pressing keys, based on pre-defined profiles and supported via voice-control for a user-friendly and continuous context sharing in the cloud. For example, assisted or even autonomous driving supported by inter-vehicle communication will bring the driver new degrees of freedom to consume and/or generate data, even when they are at the wheel. Their devices and smartphone will become their main communication gateway to the cyber world. This information dissemination in turn will open up opportunities for new services based on profiling and cloud-intelligence-based information processing. This type of data generation and storage will contribute to the "big data" trend and cloud computing trends, and will generate new traffic profiles and requirements for reliability and latency.

2.2.6 Broadcast Services

Internet TV over cable/fiber is an important emerging service because it can provide users not only with real-time TV content, but also personalized video on demand. For an operator that owns both mobile and broadcast networks, it will be feasible to provide users with a new type of mobile/portable Internet TV service. Such mobile TV services will allow the mobile network operator to gain a better knowledge of users' behavior that can be used to target personalized content, as well as to generate other revenue streams. Such broadcast services could be beneficial in several ways, but new business models would need to be developed in order to monetize them. One example is to use broadcast services to distribute or update software over the air for large numbers of devices, such as sensors or cars. This, however, would require guaranteed service reliability, which would need new features to be added to broadcasts. For instance, caching in the terminal combined with peer-to-peer transmission could augment broadcast services and help to provide higher quality of service.

On the other hand, broadcasting by individual users and small communities will become more popular in the future, which will contribute to a further increase in uplink traffic.

2.2.7 Summary

The main trends of services and applications explained above are categorized as:

- video streaming, including ultra-high definition video over new types of smart devices, and so on.
- cloud services, including context/location/proximity-based services, such as personal cloud, augmented reality, SoLoMo, and implicit communications

- M2M communications, including V2X, smart grid meters, e-health, and so on
- broadcast services and group communications.

These trends can be categorized and summarized to the following:

1. Everything will be connected by wireless to enable monitoring and collection of information and control of devices. Technologies based around remote monitoring and real-time control of a wide variety of devices will support M2M communication and the Internet of Things (IoT), enabling services such as connected cars, connected homes, moving robots, and sensors.
2. Wireless services will become more extensive and enriched through content being delivered in real-time, and with safety and lifeline communications being ensured. Examples of such emerging services, which may use new types of mobile device, include high-resolution video streaming, tactile Internet, media-rich social network services, augmented reality, and road safety.

2.3 General Requirements

The trends in mobile applications and services that were discussed in the last section are expected to impose new requirements on service levels. These will be explained and discussed next.

2.3.1 Capacity Requirements

There is a general consensus in the industry that recent data-traffic growth trends will continue into the future. In recent years, many forecasters have projected mobile data traffic will grow 24-fold between 2010 and 2015, which corresponds to a compound annual growth rate of almost 1.9 [6]. On NTT's DOCOMO network in Japan, mobile data traffic almost doubled during 2010 and a 12-fold traffic increase is expected between 2011 and 2015 [7]. This tremendous increase in the volume of mobile data traffic was not foreseen before the World Radio Conference 2007 (WRC-07). For instance, actual data traffic in 2010 was more than five times greater than some of the estimates in the ITU-R M.2072 report [6]. In 2011 alone, the volume of mobile data traffic grew 2.3-fold with a nearly 3-fold increase in the average smartphone usage rate [8]. In general, a stable trend of data traffic growth is still being observed even today, but with some seasonal variations [9].

The "Cisco VNI report" is one of the most-cited forecasts of mobile traffic [8]. Table 2.1 summarizes the facts and forecasts on global mobile traffic growth published by Cisco between 2009 and 2014. Data traffic growth between 2008 and 2013 was 45-fold. Combining this with the forecasts from 2014 to 2018, the growth between 2008 and 2018 will be almost 500 times. Assuming a similar growth rate is maintained in the future, mobile traffic in 2025 will easily be 1000 times the 2010 level.

According to a recent report by the Japanese Ministry of Internal Affairs and Communications [10], mobile traffic growth rate in Japan in the past three years was around 1.7% per year as a whole and almost 90% of the total traffic was downlink traffic. The growth rate of downlink traffic is around 1.7% per year while the uplink growth rate is slightly higher at around 1.8% per year. The growth rate in busy periods at around 23:00 also shows a trend similar to that of average traffic growth.

Table 2.1 Global mobile data traffic 2008–2018.

	2008	2009	2010	2011	2012	2013	2014	2015	2016	2017	2018	CAGR
2009 report	0.033	0.085	0.2	0.48	1.07	2.18	—	—	—	—	—	131%
2010 report	0.033	0.09	0.22	0.53	1.15	2.16	3.56	—	—	—	—	108%
2011 report	0.033	0.09	0.24	0.54	1.16	2.2	3.8	6.25	—	—	—	92%
2012 report	0.033	0.09	0.24	0.6	1.25	2.4	4.2	6.9	10.8	—	—	78%
2013 report	0.033	0.09	0.24	0.6	0.88	1.57	2.8	4.7	7.4	11.2	—	66%
2014 report	0.033	0.09	0.24	0.6	0.88	1.49	2.6	4.35	7.0	10.8	15.8	61%

Source: Cisco VNI reports (2009–2014). Exabytes(EB)/month, or 1,000 petabytes/month. Shaded figures are forecasts rather than data.

On the other hand, services will be more diversified in the future. As explained earlier, a wide range of services will be provided over the mobile network, ranging from small packet services, such as low data-rate M2M services and real-time remote control, to richer content services such as high-definition video streaming, augmented reality, and tactile Internet. In addition, data traffic over today's networks is not evenly distributed; it is becoming extremely high in superdense or hot spot areas, such as stations, shopping malls, and stadiums, where large numbers of users generate huge volumes of traffic. Therefore, besides the huge growth in total traffic, there will be more variations in traffic volume, depending on the times, locations, applications, and types of device involved. These trends will be greater in the 5G era, with more diversified services, ranging from small packet services to richer content services.

Regarding the uplink/downlink traffic ratio, downlink traffic will continue to be dominant in many locations, but the uplink growth rate is expected to be higher than that of downlink, and such trends are already being observed. For example, in some crowded events such as sporting events or concert venues, with many photo and video uploads taking place simultaneously, the uplink traffic volume is already extremely high and may exceed that of the downlink in some cases.

Given these trends and forecasts, 5G has to be able to manage traffic volumes that will be many orders of magnitude larger than those seen on today's networks. This is considered to be the most important and challenging requirement for future networks. Our target is to achieve a 1,000-fold system capacity per square kilometer compared to 2010 (that is, LTE Release 8). For some specific scenarios and applications, uplink traffic will become much higher than downlink traffic. Flexible uplink/downlink resource balancing is also required.

2.3.2 User Data-rate Requirements

Considering the rapidly emerging trends towards richer content and cloud services, 5G should aim to provide higher data-rate services along with a more uniform quality of user experience than LTE. This can be achieved through improvements in both the achievable data rates and fairness in user throughput. In LTE Release 12, with 4×4 MIMO 64QAM, for 20-MHz bandwidth, up to 300 Mbps, and with 100-MHz bandwidth up to 1.5 Gbps is achievable. Peak data rate is relevant for some scenarios, but consistent user experience over the mobile network will be much more important in the future. Mobile traffic will be expected even in high-speed vehicular environments, such as commuter trains and self-driving cars.

5G has to practically provide higher user data rates than today's networks. The target is set here to a 10 fold improvement in peak data rate, targeting more than 10 Gbps and a 100-fold increase in user-experienced throughput, delivering throughput rates of 1 Gbps to users everywhere. Higher peak data rates will also become important for new scenarios such as mobile backhauling for moving nodes: 5G will need to deliver higher data rates than 4G, even in high speed scenarios of up to 500 km/h.

2.3.3 Latency Requirements

Some future real-time applications, such as augmented reality or the tactile Internet, and time-critical M2M communications, such as remote control and monitoring and V2X, will impose very stringent requirements on end-to-end latency. From a service-level perspective, end-to-end latency includes radio, core, and backhaul latencies. However, the system-level requirements can only be derived up to a certain layer, say the MAC layer or the application layer. There are two types of latencies: user-plane latency and control-plane latency, as discussed in the following.

2.3.3.1 User-plane Latency

The round-trip time is defined as the time from when a data packet is sent from the transmitting entity until acknowledgements are received from a receiving entity, such as an Internet server. This includes the user equipment (UE) and enhanced Node B (eNB) processing delay, HARQ retransmission, and the data transmission time: the transmission time interval (TTI).

In 3GPP, the HARQ round-trip time is specified as having a maximum of 8 ms. The LTE one-way user-plane latency for a scheduled UE consists of the fixed-node processing delays (which includes radio frame alignment) and the TTI of 1 ms duration. The equipment latency figures largely consist of processing delays such as channel encoding/decoding, scheduling, and channel estimation, and are thus subject to various implementation choices. The processing delay at the UE and eNB are typically assumed to be 1.5 ms each. Considering that the number of HARQ processes is fixed to eight, the one-way latency can be calculated as follows for the frequency division duplex (FDD) case [11]:

$$D_{\mathrm{UP}}\left[\mathrm{ms}\right]=1.5+1+1.5+n*8=4+n*8 \tag{2.1}$$

where n is the number of HARQ retransmissions. Considering a typical case where there would be 0 or 1 retransmission, the average user-plane latency is

$$D_{\mathrm{UP,typical}}\left[\mathrm{ms}\right]=4+p*8 \tag{2.2}$$

where p is the error probability of the first HARQ retransmission. For a 10% HARQ block error rate (BLER), the user-plane one-way latency becomes 4.8 ms.

Thus in LTE, the one-way radio-access network (RAN) latency is about 5 ms and thus the two-way user-plane latency is around 10 ms. However, besides RAN latency, the end-to-end latency experienced by the user also includes core-network delay components. For example, the typical end-to-end latency becomes several tens of milliseconds for LTE [12]. The end-to-end latency can be generally reduced by reducing the latency related to:

- processing delays of the equipment (UE/eNB processing)
- TTI duration
- HARQ delays
- transport and core-network latency.

2.3.3.2 Control-plane Latency

The LTE-advanced requirement for control-plane latency is 100 ms. Control-plane latency can be classified into idle-to-connected or dormant-to-active latency.

Idle-to-connected. For LTE Release 10 and beyond (TR 36.912), the idle-to-connected state transition can take less than 50 ms. The improvements for the idle-to-connected case stated for LTE Release 10 and beyond come from reduced UE processing time and simultaneous RRC and NAS request setup handling, instead of a serial approach, allowing parallel RRC and NAS processing.

Dormant-to-active. This transition is between states when the UE is already synchronized and is thus significantly faster; it takes as little as 9.5 ms.

Control-plane requirements are enablers of user-plane requirements; thus they should be set up and improved in such a way that they enable the target end-to-end latency.

Improvements in delivery times are important for emergency warning notifications, for example operators distribute warning notifications to users simultaneously by utilizing the Earthquake and Tsunami Warning System. In LTE the primary notification is delivered within 4 s to users in the notification area, even when there is a congestion situation. In the future, further reductions in the delivery time of warning notifications will make it possible to reach more people with timely alerts and warning information and therefore save more lives [13].

5G has to provide not only higher data rates, but also end-to-end latency of less than 10 ms – say 5 ms – in order to enable future cloud services that require almost zero latency and new services such as tactile Internet, augmented reality, and real-time and dynamic control for M2M systems. To achieve such levels of end-to-end latency, there will be new requirements for TTI duration, HARQ signaling, transport and core latency, and network architecture. From a user-plane latency perspective, the one-way latency over the RAN should be less than 1 ms, a large leap from LTE's 5 ms. Improvements in terms of delivery time of emergency warning notifications are also important.

2.3.4 *Massive Device Connectivity*

In 2020 and beyond, mobile operators wanting to expand their business will need to become a total service provider by offering a greater range of services and providing a mobile "smart life" to every user. To this end, cloud services provided by operators need to be more diversified and customized to each user. Collaboration between network and mobile terminals will create, besides conventional voice and data services, a variety of new added-value cloud services: real-time interactive services, such as Google glasses, data storage and processing, and others. To support these future cloud services, it is important to provide connectivity to a larger number of devices. This becomes a challenge in particular in areas with high user density.

The daytime population density of Chiyoda ward, Tokyo is about 80,000/km^2, while the nighttime population density is about 4,000/km^2. This presents an example of an area where the traffic volume varies greatly between day and night. In addition, the population density of the 23 wards of Central Tokyo is around 15,000/km^2 and the average population density of Tokyo is around 6000/km^2. In office areas such as Chiyoda ward, the population density can go up to 80,000/km^2 in daytime. Assuming 25% of the users are active, the number of active users can go up to 20,000/km^2 [14].

5G will also need to be able to also support services in highly dense areas such as concerts and stadiums. For example, the user density in a stadium can go up to 2 million/km^2 (2 users/m^2). Even with just 10% of the users being active, we will need to support 200,000 active users/km^2.

Table 2.2 Examples of active user/device densities.

	User density	Activity factor	Active user density
Chiyoda ward, Tokyo	80,000/km^2	25%	20,000/km^2
Stadium	2 million/km^2	10%	200,000/km^2
Sensors	2 million/km^2	Application dependent	Application dependent

Accommodating the massive number of connected devices with a wide range of requirements that is expected to be introduced by M2M communications will be a key challenge for 5G radio access. The support of massive device connectivity is a fundamental requirement for the future IoT. Besides massive device connectivity, there will be other requirements specific to particular M2M use cases, such as super-long battery life and ultra-high reliability. Some M2M or device-to-device use cases such as transportation and safety/lifeline system will require very high reliability while supporting the required latency and mobility. Table 2.2 summarizes examples of active user/device densities.

In 3GPP LTE, up to 30 active users/sector can be scheduled (where the sector size is 0.07 km^2), thus up to 420 active users/km^2 is supported. Assuming 3–4 operators, this means that the density of active users supported is about 2,000/km^2. The number of active users depicted in Table 2.2 is thus about 10–100 times what LTE can support today.

5G has to allow massive numbers of devices to be connected simultaneously to the network in order to support always-on connected cloud services and more machine-type devices for the IoT. Our target is to achieve a 100-fold increase in the number of simultaneously connected users compared to LTE. Massive connectivity will impose new requirements on the design of control channels and will require new protocols for connecting devices. In addition, significant reductions of signaling impact on core networks will also be needed.

2.3.5 Energy Saving and Robustness against Emergencies

In order to make 5G a sustainable system, its total energy consumption should not be much larger than that of current systems. It therefore needs to consume less, or at most the same, energy in terms of energy/bit. In particular, the power consumption of the network should not increase in proportion with the traffic increase, which is expected to be 1,000-fold. As for mobile terminals, 5G should enable less power consumption to realize longer battery life. Battery life is important for some specific M2M devices such as sensors and smart meters, where lifetimes of the order of 10 years may be required.

5G should be able to provide lifeline communications in case of natural disasters such as earthquakes, tsunamis, floods, and hurricanes. Several basic types of communication, such as voice and text messages, are needed instantaneously and simultaneously by the survivors. Network robustness is important in order to avoid suspension of services because of network damage. In addition, low network and user-terminal energy consumption is critical in emergency cases.

5G has to provide increased capacity per unit network cost while being energy efficient and resilient to natural disasters. This is particularly important as the future network will need to support diverse environments and services simultaneously. While these requirements are not easy to quantify, they should be factored in as much as possible throughout the design of the 5G system.

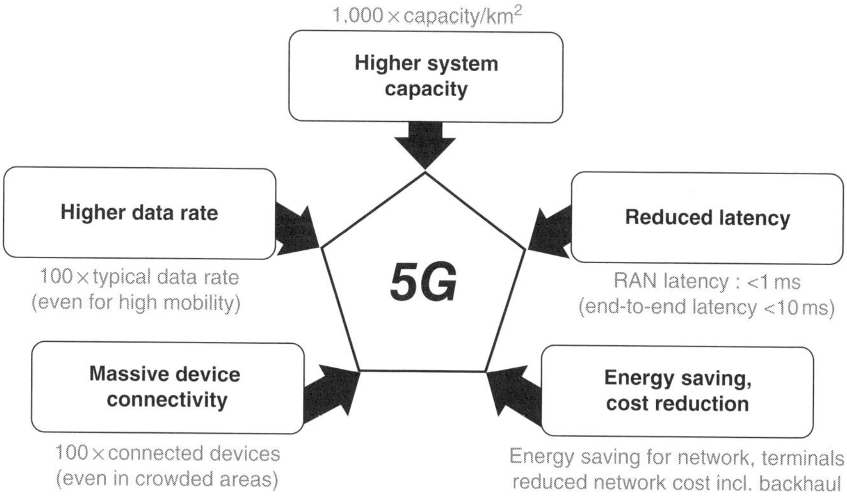

Figure 2.1 5G performance targets.

2.3.6 *Summary*

The high-level performance targets most relevant to 5G are summarized in Figure 2.1.

Luckily these performance targets do not need to be satisfied simultaneously. The 5G network should therefore be able to adapt in order to enable end-to-end flexibility and satisfy different sets of requirements in a cost-effective manner. Besides, the increasing need for new services, including user-centric services, that are easy to upgrade will also need to be taken into account. The flexibility and scalability of the network will be quite important design principles. Moreover, 5G should provide a high-level of security for all connected things.

References

[1] Cisco, White paper: "Cisco visual networking index: Global mobile data traffic forecast update", 2013–2018.

[2] Cisco Internet Business Solutions Group (IBSG), "The Internet of Things; How the next evolution of the internet is changing everything", 2011.

[3] EU Commission, Energy Community. "Study on smart meters rollout in the energy community", URL: https://www.energy-community.org/portal/page/portal/ENC_HOME/DOCS/2506178/0633975AB8B77B9CE053C92FA8C06338.PDF.

[4] Gartner press release, "The personal cloud will replace the personal computer as the center of users' digital lives by 2014", March 12, 2012. URL: www.gartner.com/newsroom/id/1947315 (accessed August 8, 2016).

[5] Wikipedia, "Augmented reality". URL: http://en.wikipedia.org/wiki/Augmented_reality.

[6] Recommendation ITU-R M.2083-0: "IMT Vision – Framework and overall objectives of the future development of IMT for 2020 and beyond," September 2015.

[7] DOCOMO Annual Report 2012, "Medium-Term Vision 2015". URL: www.nttdocomo.co.jp/english/corporate/ir/library/annual/fy2011/html/feature02/index.html (accessed August 8, 2016).

[8] Cisco, White paper: "Cisco visual networking index: Global mobile data traffic forecast update, 2012–2017", February 6, 2013.

[9] 2013 Ericsson Mobility Report: "https://www.ericsson.com/res/docs/2013/ericsson-mobility-report-november-2013.pdf", November 2013.

[10] MIC of Japan, "Data base of ICT statistics: Mobile communication traffic in Japan", December 2013. URL: www.soumu.go.jp/johotsusintokei/field/data/gt010601.xls (accessed August 8, 2016).

[11] METIS Deliverable D1.1. URL: https://www.metis2020.com/documents/deliverables/ (accessed August 8, 2016).
[12] 3GPP, TR 36.912 V11.0.0 "Feasibility study for further advancements for E-UTRA (LTE-Advanced) (Release 11)", Sept. 2012.
[13] 3GPP, TS 22.168 V9.0.0, "Earthquake and tsunami warning system (ETWS) requirements; Stage 1 (Release 9)", June 2008.
[14] Tokyo Metropolitan Government, "On wards system," March 16, 2012. URL: www.soumu.go.jp/main_content/000151653.pdf (in Japanese) (accessed August 8, 2016).

3

Collaborative 5G Research within the EU Framework of Funded Research

Michael Faerber
Intel Corporation, Santa Clara, CA, USA

3.1 Rationale for 5G Research and the EU's Motivation

Today's rollout of 4G networks makes us witnesses to the convergence of cloud computing, computing power, and connectivity at high speed. The growth of the digital society and economy is driving our imagination beyond the existing models and leads to the design imperatives for 5G: the next generation of ubiquitous ultra-high-speed broadband infrastructure that will support the future Internet.

The need to create a 5G mobile concept may not be obvious for the ordinary technology observer and from an end-user perspective, but we have to consider the new challenges of a digital society ahead of us. End users have enjoyed the fact that each mobile system generation has provided enhanced terminal capabilities for mass-market products. The scientific community considers each of these past generations of mobile technology as disruptive in nature, but whether they are disruptive or backward compatible is not of interest to most end users.

Technical experts have evolved the mobile technology from 2G – a simple voice service and cross-country roaming – into a mobile Internet service for wireless devices. Each generation evolved within the constraints of its fundamental design concepts.

The 5G network will be more than a new air interface; it is a holistic concept encompassing a multiplicity of existing and new air interfaces, new concepts for communication protocol structures, function splits, backhaul concepts, and RAN architectures that aims to achieve

Towards 5G: Applications, Requirements and Candidate Technologies, First Edition.
Edited by Rath Vannithamby and Shilpa Talwar.

Companion website: www.wiley.com/go/vannithamby/towards5g

maximum flexibility to meet service needs. 5G is therefore not a single technology; it is a vision of a highly flexible and adaptive service-oriented network. To achieve this aim, the integration of existing and new air interfaces is needed, alongside telecom function virtualization on generic computer platforms, new architecture concepts, a shift in content from central network nodes to edge nodes, and context-aware communication and service provision. In order to deal with the new requirements and key performance enhancements, it seems that a disruptive technology leap will be needed roughly every 10 years.

As 4G is still a commercially valuable growth field, we have to recognize that the time is right to lay the foundations for the network now. Past EU research programs have been instrumental in the search for new system concepts. For the Horizon 2020 work programme, the EU commission has made significant budget allocations, and seeks a strong and coordinated research on 5G driven by industry.

Why is collaborative research within the EU research ecosystem so attractive for organizations? Or in other words, why is it is attractive for the industry to be part of it? The main reason is that such collaborative research allows discussions, even among competitors, in the very early phases of the concept design. This phase can be seen as a pre-competitive phase, meaning that organizations benefit from sharing ideas rather than being harmed. This exchange of information helps in early consensus building and gives the jointly developed concepts a good start in the standards environment.

The EU commission has a vision of 5G information and communication technologies generating a new platform for services and creating benefits for all European citizens. This vision assumes that commercial off-the-shelf computing platforms will replace hardware specifically tailored for telecoms services. The new buzzword is "everything as a service" (XaaS), based around the idea that software solutions will take over what for decades has been the realm of customised hardware solutions.

Non-human communication devices have the potential to be the key driver for the development of a future 5G infrastructure that will cover increasing traffic demand. Individuals are breaking the barriers of data consumption, becoming the producers of large amounts of data via smart devices such as networked glasses, sensors, and so forth.

New terminals are going beyond the current smartphones, encompassing drones, robotic devices, and wearables. Together these represent local and personal network structures, using a multitude of devices and network access modes.

Vertical business segments are the new drivers of technology, envisaging the use of a unified wireless and wired telecommunication network for a high variety of services. This will require a telecom platform with flexible and adaptive performance as a basic enabler for new service provision.

Looking to the requirements of the new networks, we see that a 5G network is more than a new air interface. What is needed is a holistic approach dealing with new *and* existing air interfaces. To achieve the 5G design objectives, research is needed about topics such as integration of new and existing air interfaces, telecom function virtualization on generic computer platforms, new architecture concepts, new hardware/software function splits, shifting content from central network nodes to edge nodes, context-aware communication, and service provision features.

The EU has understood this fundamental challenge and, in its calls, set an attitude towards such new end-to-end network architectures. The 5G PPP Infrastructure Association will be the coordinating body that will create a research ecosystem that replaces the idea of isolated research projects with integrated research activity, with the aim of intense exchanges between projects, and a faster track towards the market. In the following, we will focus on the collaborative research programs funded by the EU commission.

3.2 EU Research

3.2.1 History

The EU is the successor to the European Economic Community (EEC), originally founded in 1957 to create a common market, remove customs barriers and to allow a free flow of money, services and people within the EEC. Through the Maastricht Treaty in 1992, the EEC became the EU. Already in the early phases of the EEC, the community started to create funded collaborative research activities. The first one was Euratom, a platform to do joint research on the civil use of nuclear power. Later, other independent research activities started, in the fields of steel and agriculture. This nicely represented an area of common interest in the early EEC. In the 1970s, the EU Commission identified that there was a significant lack of collaborative research within the union, especially in the fields of aerospace, nuclear technology, and ICT. However, it was not until 1982 that a first programme was created in response. The ESPRIT programme was later renamed the Information Society Technology programme (IST), and became part of the EU frame programmes structure. The ESPRIT programme was an ICT programme and already contained many important elements, such as industry and academic involvement, and the fact that the industry partners provided their own funding for the research, only partly reimbursed by the EU [1].

The first frame programme (FP1) was started in 1984. Early frame programmes lasted 4–5 years, but this timeframe was later enlarged to 6–7 years. Table 3.1 shows the history of previous programmes.

The funds for these programmes have been increased constantly, as shown in Figure 3.1. The longer duration of the later projects creates the impression of a strong increase in later programmes. Figure 3.2 shows a normalization of the budget on a yearly basis. From FP7 onwards the budget per annum is bigger than before.

The budgets span a wide range of research topics: agriculture, health, ICT, nanotechnology, energy, environment, traffic, social sciences, space, and security. ICT's share in FP7 was €9 billion for a 7-year period, which was about 18% of the frame program's total budget [2].

Table 3.1 History of the frame programmes.

FP	Start	End
1	1984	1987
2	1987	1991
3	1991	1994
4	1994	1998
5	1998	2002
6	2002	2007
7	2007	2013
8	2014	2020

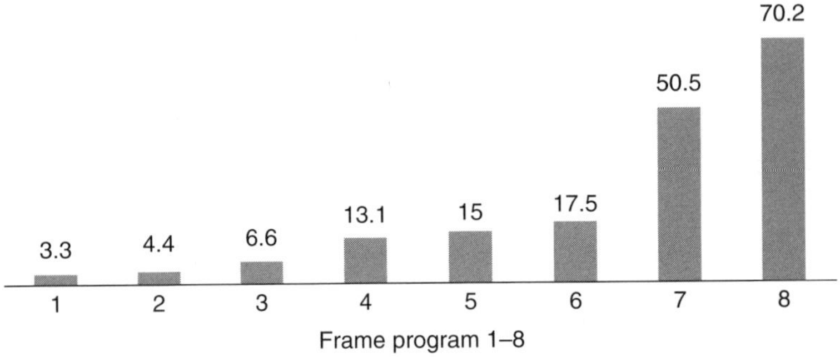

Figure 3.1 EU funding budget per programme (€ billions).

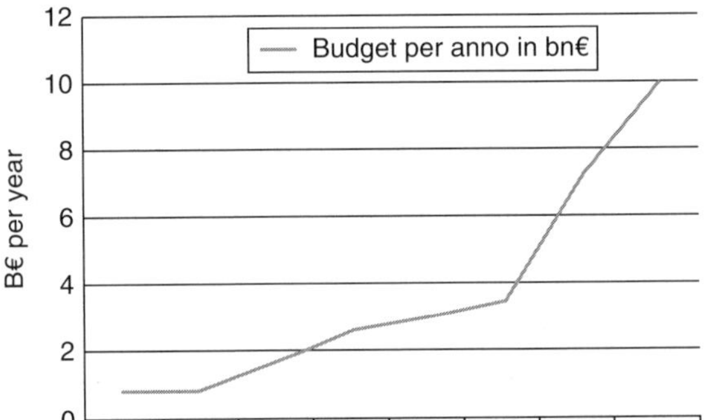

1	2	3	4	5	6	7	8
0.852	0.88	1.65	2.62	3	3.5	7.2	10

Figure 3.2 Budget per annum in relation to the frame programmes (€ billions).

Within these programmes some of the funded projects had a flagship character for the ICT sector, especially for the mobile industry. In the following, some examples are listed for projects that left significant traces in the field of mobile technologies.

ATDMA was a project research on concepts for beyond-GSM technology; baseline work for the following FRAMES project. The FRAMES project worked on the wideband-CDMA and TD-CDMA concepts. The organizations involved finally ended up as the Alpha and Delta proposal groups in the ETSI beauty contest 1997 for the UMTS decision. These concepts delivered the fundamentals of WCDMA and TDS-CDMA in the later 3GPP releases. Winner I and II have been important for the ITU spectrum requirements study and in the Super-3G requirements discussion, which led to the LTE requirements. The Winner channel models and simulation guidelines have been adopted by 3GPP. The project also did the first research in areas like heterogeneous networks (HetNets), relays, and D2D.

ARTIST4G had the research objective of improving fairness between users located at the cell centre and those at the cell border [3]. The research was based on known LTE-A features, and studied enhancements for CoMP, IF cancellation, advanced and moving relays, and the network architecture impact. It also looked at the benefit of the combination of LTE-A features. Many company contributions had notable indirect impacts on 3GPP.

EARTH was a research project to study energy efficiency (EE) in networks [4]. It received high recognition by the EU and several research awards. The studies examined EE deployment strategies and network architectures. New network management solutions and innovative EE component designs were formulated, for example adaptive operating points and resource-management protocols for multicell cooperation.

METIS2020, which started in November 2012, had research objectives such as air-interface advances for machine-type (MTC) and device-to-device (D2D) communication, D2D evolution in networks, exploitation of massive antenna arrays, new modulation schemes, HetNet aspects and multiradio–common radio resource management [5]. The METIS2020 title is often confused with the EU Horizon 2020 program, which is the nickname for FP8. METIS2020 is part of FP7 and not FP8 activity. However, the research topics mentioned here have a clear impact on 5G.

There are many other projects that may also be worth mentioning, but we may look especially to the final phase of the FP7, where many projects have already picked up crucial 5G design elements. The 5GNOW project [6] focused on new waveforms, and mmWave technology was addressed by projects such as MiWeba [7] and MiWaves [8]. There are many more projects worthy of mention, but all these are public and details can be found on the web pages of the EU [9].

3.2.2 EU Bodies, Structure, Roles, and Project Creation

The EU has conducted long-term research programmes for many decades, and has developed and refined practices and structures over these years. The multitude of names, definitions, and organizations in the field of EU-funded research can cause confusion. This chapter will help readers to understand the terminology used and will explain how these bodies interact.

A frame programme (FP) defines a specific EU research programme spanning 6–7 years. The last programme concluded was FP7, which has still projects active. The most recent FP, started in 2014, is FP8 [10].

The content, rules, and budget of an FP are drafted and proposed by the European Commission, and have to be agreed by the European Parliament (EP) and the Council of Ministers.

The Commission is the executive body of the EU and is responsible for proposing legislation, implementing decisions, upholding the EU's treaties and managing its day-to-day running. The EP is the directly elected parliamentary institution of the EU. The Council of the EU, also called the "Council" or more accurately the "Council of Ministers", is part of the bicameral EU legislature, representing the executives of EU member states. The other legislative body is the EP. The Council is composed of the 28 national (prime-) ministers, one per state. Council leadership changes every half year from state to state.

The Commission is in charge of developing and planning FPs, and has to develop a proposal. This proposal then goes through several review and discussion cycles with the EP and the Council before it gains final legal approval.

For the FP8 proposal, this was about 2 years of work and involved several revised versions, after the commission issued its first proposed version to the legislative bodies. If an FP and its budget are approved, the EU will issue a call for proposals (a "call"). Within a program period, the EU issues biannual calls for proposals related to EU work programmes for specific fields and phases, substructured in so called objectives. Calls often have a focus on particular sectors, so any call will not necessarily be focused on ICT, but some are.

Several calls are made against an FP, seeking project proposals by proposal consortia. Each call has specific objectives and these are the targets for the projects submitted and granted for this call. In FP 7 we have seen a total of 11 calls.

In response to a call, a project proposal is submitted to the EU. A proposal is compiled by a group of partners (proposal consortia) with the intention of jointly conducting this project. The project proposal is described in a response to the EU aiming to fulfil the objectives. These are split into two parts. Part A describes the planned efforts, the organizations involved, and the funding and re-imbursement plan. Part B contains the description of work (DoW), which describes the planned research activities, and how these fulfil the objectives of the call.

The EU Commission has project officers (POs) dealing with these proposals in the various sectors. The PO organizes a review of the proposals, using a group of external and internal auditors, who have to evaluate and rank the proposals. In past programmes, proposals with an evaluation grade above a certain threshold were invited to hearings and negotiation meetings. During this phase the EU made adjustments of the proposed budgets organizational, structural, or technical gaps were pointed out. This led usually to revisions of Parts A and B before the project proposal was finally accepted. For FP8 the rules have been changed, and there will no longer be hearings and adjustments in which the POs ask for changes to the proposals regarding drafting, the match to the budget, and the organizational requirements and objectives. If the reviewer's assessment is now negative due to a failure in one of these aspects, the proposal is rejected.

A project lasts typically 24–36 months. After receiving a grant for a proposal, a project consortium is formed out of the project proposal group. To build the consortium, an agreement is made between the partners, which clarifies many aspects of the collaboration, such as how intellectual property rights will be handled. The DoW given to the EU is the consortium's commitment about the intended research activities, the resources, and the deliverables. Finally the consortium will sign a grant agreement with EU, which is a legally binding contract between the partners and the EU.

3.2.3 Project Creation and Operation

Research projects in the EU framework are intended to create a pre-competitive environment. The research objectives of a proposal should be so far ahead that the features and technologies studied do not show up on company roadmaps: the different partners will only have an interest in working jointly on research that is sufficiently forward-looking. Nevertheless, the EU asks about implementation plans, to understand how and when these ideas might appear in standards or in products. The EU does not subsidise product development activities. Their aim is not to support organizations in things they would do anyhow; the aim is to drive and support innovation activity in the participating organizations.

In relation to standardization roadmaps, a significant step forward should be provided, and the topics should not match existing study and work items (SI/WI) or items already planned for future release. If a topic already has the status of a WI or SI, it is very likely that the research results will not seriously impact work in standardization. Therefore, each project proposal creation has to analyze the state of the art, and the research targets have to be sufficiently future-oriented to maintain the ability to be exploited later in standardization.

Especially in large EU projects, formerly called integrated projects (IP), many of the partners in a collaborative project are also strongly engaged in standardization. These are the network and UE vendors, chipset suppliers, and network operators. There are also academic institutes and universities. Although the academic partners are good scientific contributors, usually they do not play much of a role in the later standardization work, therefore joint pre-competitive work on potential new features can serve as a proof of concept, promoting new ideas and finding supporters for the idea for later stages when standardization starts.

3.2.3.1 Project Creation

In the previous section, a generic description was given of what a proposal consortium has to compile for a proposal submission. The EU has practical tools and templates to support proposal creation. However, it does not involve itself in project creation actions. This raises the question of how such proposal consortia are created.

The EU organizes several conferences and events every year related to the current FP. Those working on EU projects use these events as a platform to show their research results to the wider community. These conferences are also a good place for networking around future projects: the EU fosters them to help organizations to advertise themselves as potential partners, and organizations can look for additional partners who might provide a required competence for a project.

For 5G research the 5G PPP Association has established a specific brokerage site on which consortia in creation can register required competences, and individual partners can register profiles to get in contact with groups. So far this is the only active support the 5G PPP Association provides in project proposal consortia building, besides providing the non-binding pre-structuring model. In principle it is important that an organization eager to be part of a proposal starts networking and looking for contacts as early as possible. This might even start prior to the official release of a call. Engagement in European Technology Platform (ETP) activities like Networld2020 is a good way to start to get in touch with potential partners and to exchange ideas. It also provides opportunities to contribute to white papers, consultations, and workplan proposals. Such voluntary engagement helps to improve an organization's visibility and to display its competences and fields of interest.

Typically when a number of partners for a proposal have found a common interest and understanding of what they want to do, they start to prepare some boilerplate material to describe the project, and in particular they identify missing competences and potential partners who might fill the gaps. If there is agreement among the initiating group, specific partners then get invited to join. Organizations can also advertise themselves to proposals already in creation. If the core group think it fit, they can invite these partners to join. The creation of a project has to consider multiple vectors: budget constraints, budgetary balance between industry, small- and medium enterprises (SMEs), and academia, adequacy of coverage of intended research objectives, a balanced work-package structure in terms of effort, how many partners are covering specific tasks, regional diversity, the total number of partners, and other aspects too. In the end, the direction of the research in the planned project has to match the specific objectives of the call.

3.2.3.2 Project Operation

The EU reimburses costs for the research activities in granted projects. In past programmes the level of reimbursement depended on the size and role of the organization. In previous programmes industry partners got 50% of their total costs reimbursed. SMEs received 75% re-imbursement. On the academic side – universities and research institutes –100% funding was received in the past. These funding rules have been simplified for call 8, and all partners are now going to receive 100% reimbursement of the direct costs (labour, material, travel). For indirect costs (overhead), a 25% flat rate is now given, based on the direct cost claims. This may imply that SME and academia now have a disadvantage, but we have also to note that smaller organizations typically have less overhead cost.

An organization has to be eligible for funding: the organization as a legal entity and the staff executing the work must reside in the EU, or in a country that is eligible by special regulations, as is the case for countries that are part of EFTA (the European Free Trade Association). However, this does not prevent organizations outside the EU from participating in EU projects. There is the instrument of cross-region joint research actions, like the EU/JP calls in FP7. Furthermore, partners outside the EU can be part of a proposal, but without getting funding from the EU. The consortium has to explain why this partner and their competences are needed in the project. The EU can also decide to support partners.

For running project, the EU commission conducts intermediate audits, as well as a final audit. The audits have a technical and a financial part. Intermediate audits are important milestones, and allow the EU to have a progress report and quality control. These audits are done by the POs and external reviewers, often university professors, so as to have a wider and balanced view.

Over the course of the project, several deliverables have to be published. Deliverables are public and provide detailed insight to the research objectives and the results. Internal reports and confidential deliverables are accessible to the EU and the auditors only, and can contain results and material that the partners have limited interest to publish.

The EU also monitors the number of patent filings for the project at the European Patent Office. The filings have to be marked if they are a result of EU-funded research.

Another performance parameter of a project is the output of conference papers, presentations, journal papers, or other forms of publication. Many projects have specific work packages, with the main objective being to organize and foster publications.

3.2.4 Details of the FP8 Program

> Horizon 2020 should be implemented with a view to contributing directly to creating industrial leadership, growth and employment as well as citizens' welfare in Europe and should reflect the strategic vision of the Commission Communication of 6 October 2010 to the European Parliament, the Council, the European Economic and Social Committee and the Committee of the Region "Europe 2020 Flagship Initiative Innovation".
>
> Horizon 2020 should support the achievement and functioning of the European Research Area in which researchers, scientific knowledge and technology circulate freely, by strengthening cooperation both between the Union and the Member States, and among the Member States, notably through application of a coherent set of rules.
>
> *The EU's motivation for FP8 "Horizon 2020"*

The EU FP8 program, nicknamed "Horizon 2020", has three important segments:

1. excellent science
2. societal challenges
3. industrial leadership (which is the target segment).

Industrial leadership is defined as enabling and industrial technologies, including:

- ICT (which is the target sector)
- nanotechnologies
- materials
- biotechnology
- manufacturing
- space [10].

The ICT sector will issue different calls over the programme period. The work plan for the first 2 years of the programme contains three calls: call 1, a delayed call 1 for 5G PPP coordinated projects and call 2.

The Commission's plan places strategic objectives (SO) and research technology development (RTD) actions in the different calls. The LEIT (Leadership in Enabling and Industrial Technologies) work has restricted objectives and budget for 5G research in call 1 and a large RTD action in the delayed call 1. This is the objective for 5G research coordinated by the 5G PPP association [10]. In a following section the background of the 5G PPP association and its creation is explained in detail.

After passing the deadline for proposal submission in November 2014, the EU commission received 75 proposals, which have been evaluated. Seventeen projects got a positive grant and now represent the projects in the first phase of 5G PPP coordination. Similar calls for 5G PPP coordinated objectives will follow. Current plans will consider the four Phases until 2020. About 50% of the submitted proposals took orientation from the 5G PPP pre-structuring model. The other 50% now overlaps and has different potential interfaces. The main task of the 5G PPP will be to create a coherent way forward and define the interfaces for information exchange.

3.2.5 European Technology Platforms and Public–Private Partnerships

ETPs are market representation partners of different sectors of an EU programme. For ICT there was the ETP Net!works, a lobby organization with almost 1000 partners. In October 2013 the ETP Net!Works and the ETP ISI (dealing with satellite communication), merged to a new ICT-centric ETP, known as Networld2020. All the activities described in this chapter are related to ETP Net!works or Networld2020 [11].

In FP7, the EU used the PPP concept in the field of collaborative research for the first time. PPP stands for "public–private partnership", and is a formal structure that can be implemented to provide input for work programmes through Joint Technology Initiatives or other formal structures as defined in Article 187 of EU law.

In FP7 the EU used the creation of PPPs to get coordinated industry engagement, fighting against the global economic downturn in the first decade of the millennium. The difference to previous activities is that the industry partners take a role and make efforts to coordinate research activities over several concurrent research projects. This concept has been seen as a good approach for more targeted research activity.

For FP8, the EU was planning right away to install several PPPs for coordinated research. With this knowledge an initiative was formed by about 30 industry partners, all part of the ETP Net!works group. The target was to create an industry proposal for a PPP called "Advanced 5G Network Infrastructure for the Future Internet", which would address 5G research needs. This group created a proposal for the EU on how to structure, organize, and support FP8 over its lifetime, in the field of ICT 5G research, by coordinating the various projects and the research objectives.

This was finally accepted by the EU Commission, and the initiative was converted to an association, as a legal partner for the EU. The official name is the "5G PPP Infrastructure Association", often just called the 5G PPP. Prior to the official creation of the 5G PPP, a drafting group started to work on a first proposal for different research streams, and initially described research actions for 12 different fields. After the creation of the 5G PPP association, the association continued this work, which is called now the 5G PPP pre-structuring model [12].

The 5G PPP association was created as a non-profit company under Belgian law. A contract with the EU was signed with the start of call 1 of Horizon 2020. Call 1 was released the 11 December. The board members of ETP Networld2020 are the kernel of the 5G PPP association, with the addition of six extra members representing specific industry sectors. The 5G PPP infrastructure association will be the coordinative instrument of the EU, aiming to create a research ecosystem that overcomes the concept of isolated research projects producing integrated research activity, with the aim of an intense exchange between projects and a faster track towards the market.

3.2.6 Other Funded Research

Many national governments in Europe and across the world have their own research programmes aiming to foster innovation locally. Local programmes have constraints for participation, such as having a legal entity in that country, and national programme rules are often difficult from an administrative point of view. Proposal descriptions and reports sometimes have to be in the local language, or at least partly so. The re-imbursement rate is also often lower than EU rates, whereas the administrative overhead often tends to be higher than in EU projects.

There are other forms of EU research, which combine national and EU research, such as the CELTIC/CELTICPLUS [13] programme within the EU EUREKA framework. A consortium for a research proposal can deliver the proposal to CELTIC. If the proposal is found to be sound and fulfils a number of standards and quality criteria, the CELTIC organization issues a CELTIC label to the group. The individual partners can then apply at their national research funding organizations and, with their CELTIC label, have a better chance of receiving some funding. However, there is no guarantee, therefore a project can be an undertaking of different partners from different EU countries, but the individual funding remains a national issue.

References

[1] S. Borras, Science, technology and innovation in European Politics, Roskilde University, Denmark.
[2] EU, Research and Innovation, FP7. Available online at: http://ec.europa.eu/research/fp7/index_en.cfm (accessed July 27, 2016).
[3] http://cordis.europa.eu/project/rcn/93721_en.html.
[4] https://www.ict-earth.eu/.
[5] METIS project, https://www.metis2020.com/.
[6] MWC (2015) 5GNOW – 5th Generation Non-Orthogonal Waveforms for Asynchronous Signalling. Available online at: www.5gnow.eu/ (accessed July 27, 2016).
[7] MiWEBA Successful Final Review (2016). Available at: www.miweba.eu/ (accessed July 27, 2016).
[8] MiWaves (2016). Website, available at: www.miwaves.eu/ (accessed July 27, 2016).
[9] CORDIS (2016) Projects and results. Available online at: cordis.europa.eu/projects/home_en.html (accessed July 27, 2016).
[10] Horizon 2020 (n.d.) The EU Framework Programme for Research and Innovation. Available online at: ec.europa.eu/programmes/horizon2020/en/ (accessed July 27, 2016).
[11] NETWORLD2020 ETP (2012). Available online at: networld2020.eu/ (accessed July 27, 2016).
[12] https://5g-ppp.eu/.
[13] Celtic-Plus (n.d.) website. Available online at: celticplus.eu/ (accessed July 27, 2016).

4

5G: Transforming the User Wireless Experience

David Ott, Nageen Himayat, and Shilpa Talwar
Intel Corporation, Santa Clara, CA, USA

4.1 Introduction

In this chapter we present the Intel Strategic Research Alliance (ISRA) on 5G technologies, also known as "5G: Transforming the User Experience". ISRA 5G is a multi-university, multi-million-dollar university research program exploring the viability of new candidate technologies for tomorrow's wireless networks (also known as 5G). A major contribution of the program is its statement of vision for 5G, its goals of research and success criteria, and its scoping of the research areas needed to realize the next generation of wireless technologies. As such, the first half of the chapter focuses on Intel's perspectives and insights on the broader challenge of candidate 5G technologies. A summary of sponsored research projects is then presented in the second half of the chapter, including information on the researchers and their proposed approaches for exploration in the program.

4.2 Intel's Vision of 5G Technologies

Intel's stated mission for this decade is "to create and extend computing technology to connect and enrich the lives of every person on earth". As such, Intel believes that wireless communication technologies, with their unconstrained ability to seamlessly connect people to one another and to a vast world of information and services, are essential building blocks to realizing this mission. Intel also believes, as noted by former Intel fellow Justin Rattner, that users value technology not merely because it is "better" by some technical definition, but because it offers a more deeply personal "information experience". Our key goal for 5G wireless technologies, therefore, is to deliver a vastly improved, rich, and personalized user experience.

Towards 5G: Applications, Requirements and Candidate Technologies, First Edition.
Edited by Rath Vannithamby and Shilpa Talwar.

Companion website: www.wiley.com/go/vannithamby/towards5g

As a manufacturer of client-device silicon, Intel is keenly aware of the importance of future device intelligence and innovation in controlling the quality of user experience. However, it is also clear to Intel that devices operate in the context of the network and the service infrastructure they connect with. Even the most intelligent devices cannot solve the challenges faced by 5G networks without fully comprehending the role of the broader wireless network context. This includes the issues of spectrum availability, network infrastructure, service architecture, and application requirements. Therefore, Intel shares the industry view that tomorrow's 5G vision is one in which devices are jointly optimized with the surrounding network context to achieve dramatic improvements in user experience. While significant strides have been made in the technological advancements of 3G and 4G, Intel believes there is much more to be done to achieve this vision in tomorrow's 5G wireless networks.

In particular, Intel envisions key technological improvements in the areas of research set out in the following subsections.

4.2.1 Enabling New Spectrum

Increasing network capacity is a key challenge facing 5G networks, and adding additional spectrum, if available, can be a straightforward method to improve peak capacity. Low-frequency bands below 6 GHz are typically premium due to their wide-area coverage, but are heavily allocated. Allocating new spectrum below 6 GHz for mobile broadband use is challenging, and the associated regulatory process is slow and difficult. Several of the incumbents in these spectrum bands have critical uses for their spectrum, such as public safety, and this precludes the spectrum from being re-farmed. Other incumbents might use their spectrum during certain parts of the day or in certain regions, which also precludes re-farming. Nevertheless, data shows that these spectrum bands are relatively underutilized [1].

Regulators around the world are trying to enable device spectrum-sharing schemes that are mutually beneficial to the mobile operator as well as the primary holder of the underutilized spectrum. NTIA's recent report says "NTIA also believes that spectrum sharing is a vital component of satisfying the growing demand for access to spectrum and that both federal and non-federal users will need to adopt innovative sharing techniques to accommodate this demand" [2]. The European Commission's recent mandate on shared spectrum for "licensed shared access" allows a spectrum owner to grant mobile carriers access to their spectrum for a monetary benefit [3]. As regulatory bodies examine new means of sharing spectrum between commercial mobile licenses and other wireless licenses like government, public safety, or satellite services, new technologies are needed for the primary owners to share spectral resources with mobile operators or allow for its opportunistic/secondary use in order to free up this desirable spectrum.

Alternatively, high-frequency bands in the giga- to terahertz frequency range, if enabled, can potentially offer several hundred megahertz of contiguous bandwidth to drive peak data rates beyond 10 Gbits/s [4–6]. It is feasible that traditional propagation challenges associated with the use of such spectrum may be overcome if wireless and optical fiber infrastructure can be combined to support a smaller cell footprint that is matched to the range achievable with higher frequencies. Moreover, the combination of large MIMO arrays with high-frequency transmission can be exploited to yield significantly higher data rates, as well as to extend propagation range.

4.2.2 *Increasing Spectrum Efficiency*

Network capacity can also be improved by increasing the efficiency with which spectrum is used in the network. The challenge of improving spectral efficiency for scarce spectral resources has been around since the inception of the cellular industry. It is therefore not surprising that there is a limited set of technologies available to be harvested for further improvements. Recently, the focus of the research community has shifted toward technologies that aim to improve spectral reuse in cellular networks by eliminating inter-cell interference, as well as toward network topologies that aggressively reuse spectrum. Next, we list some directions that show promise for future wireless networks.

Large-scale MIMO. Advanced MIMO techniques are at the heart of achieving higher capacity for cellular systems. Multi-user MIMO (MU-MIMO) offers increased multiplexing gains [7] and, even though it has been included in the 3GPP LTE-Advanced standard, its full potential is yet to be unlocked. Drastically higher capacity can be obtained by very large MIMO (VLM) arrays [8] employed at the base-station. Increasing transmit array size has desirable implications for propagation conditions, channel-estimation errors, inter-symbol and intra-cell interference, and transmit power budget [9, 10]. Fortunately, most of the gains can be realized even at manageable antenna dimensions [11]. We believe VLM will be a core technology to create significantly higher capacity, either in the form of distributed radio heads with centralized processing [12] or in deploying hundreds of antenna elements but in higher frequency bands, such as mmWave, where antenna dimensions become more practical.

Efficient Modulation Schemes. Desirable properties of cyclic-prefix (CP) OFDM have made it the main transmission technique for advanced wireless broadband systems. There are, however, alternative modulation schemes that seek to either reduce the CP overhead [13] or take advantage of it for transmission to another user without causing interference to a primary one [14]. In the context of VLM, research results even point to the optimality of single-carrier transmission [10]. CP-OFDM itself has the potential for improved efficiency through higher-order modulation, non-orthogonal access, and superposition coding. The capability and applicability of the alternative transmission and modulation techniques for 5G systems should be analyzed with the ultimate goal of higher network capacity.

Advanced Interference Mitigation. The main bottleneck of the cellular system is high inter-cell interference levels at the coverage edge. With heterogeneous architectures becoming an integral part of future cellular networks, interference mitigation becomes even more critical and can greatly boost throughput, particularly for interference-limited edge users. Transmitter coordination is one attempt to *exploit* interference instead of *avoiding* it [7, 15–17]. Interference alignment tries to align multiple interferers in a sub-space with smaller dimension than the number of interferers [8]. In practice, both availability of channel-state information and backhaul capacity/latency have detrimental effects on the achievable gains. Robust interference-cancelling receivers are also desirable depending on the complexity limitations and the availability of the information about the interferer. We seek to further explore and expand on a variety of fundamental ideas from the last decade of developments in this area, but with a special emphasis on practical limitations such as channel feedback and backhaul imperfections.

Spectrum Reuse through Multi-cell Cooperation. The level of coordination between network nodes is a function of their backhaul characteristics (latency and bandwidth). It is expected that increasing degrees of network cooperation, from loose network-node coordination to completely centralized control (such as cloud RAN), will provide increasing levels of network capacity. For example, linear scaling of capacity with the number of infrastructure nodes deployed has been shown in multi-tier networks with the use of interference coordination techniques amongst the nodes [18, 19]. We expect this capacity to further scale with cluster size if clusters of cells co-operate to simultaneously support multiple users within the cluster. While these technologies are being investigated currently, the theoretical performance achievable with varying levels of multi-cell/tier cooperation in network and the impact of practical limitations arising from the high expense of deploying optical backhaul widely are not well understood. Intel believes a general framework for network-node cooperation based on awareness of backhaul constraints and network topology (single- or multi-tier) is needed. The ability of such a framework to enable flexible deployments in which nodes adapt their signal processing to deliver the projected capacity demand in a cost-effective manner must also be understood.

Spectrum Reuse through Device-to-Device Cooperation. Intel believes that the next set of substantial gains in spectral efficiency can be achieved through enabling direct communication between client devices. Such communication will allow for a dramatic increase in the level of spectral reuse and cooperation within the network. Here, clients may communicate locally without involving network infrastructure, helping to increase spectrum reuse with only a limited increase in network interference. Client devices may also serve as infrastructure nodes, cooperatively assisting other devices in improving the overall quality of communication. Intel believes that leveraging client devices as an integral part of the network infrastructure is the logical next dimension to explore for improving spectral efficiency. Indeed, recent developments in network theory suggest that enabling rich cooperation across devices can provide linear growth in system capacity with increasing number of devices [20, 21]. This can vastly improve upon the capacity scaling available through infrastructure nodes alone.

4.2.3 Exploiting Multiple Radio Access Technologies

The increasing availability of multi-radio access technologies (RATs) and mobile devices has fueled a growing trend towards exploiting multiple networks to address capacity as well as connectivity limitations. For example, a number of operators currently rely on spectrum aggregation across unlicensed WiFi networks as a readily available means to offload traffic demand. Another example of multi-RAT communication is the use of a smartphone device with cellular and WiFi radios to provide cellular "hotspot" connectivity for WiFi-equipped peripherals such as tablets, cameras, and so on. Intel expects the trend in exploiting multiple radio networks to accelerate for several reasons:

- Given the limited licensed spectrum, there will be ongoing momentum to access capacity across multiple RATs. In particular, we expect WiFi to continue to play an integral role as the WiFi air interface evolves towards higher data rates at additional frequency bands.
- Future spectrum allocations are likely to be fragmented and could require different transmission technologies (air interface descriptions) for use.

- As cell-sizes shrink, the footprints of cellular wide-area networks (WANs), local-area networks (LANs), and personal-area networks (PANs) will increasingly overlap. This creates an opportunity to simultaneously utilize multiple RATs for improving wireless connectivity.
- The emerging Internet of Things (IoT) will require networks to connect a diverse range of devices requiring connectivity at different scales. Interworking across the multiple RATs that are optimized for each scale of connectivity will therefore become essential.
- Network theory suggests that both short- and long-range technologies may need to work cooperatively to realize capacity gains that scale linearly with the density of devices.

Intel therefore believes that multiple RATs and the associated device/system intelligence for their use will be a fundamental characteristic of 5G networks. In particular, we expect that joint use of multiple RATs can leverage the rich multi-dimensional diversity (e.g., spatial, temporal, frequency, interference, load, etc.) across multiple radio networks (WAN, LAN, and PAN) to provide beyond-additive gains in the user-connectivity experience.

While there has been a great deal of effort in academia and industry to improve the performance of individual RATs, research on joint use of multiple networks has largely focused on a "one network at a time" model, and with a primary concern for inter-RAT hand-off technologies [22]. Moving forward, we believe that there must be a concerted effort to characterize and realize the efficiencies achievable through the simultaneous integration of multiple radio networks.

4.2.4 *Awareness of Application-specific Service Quality*

Intel believes that an application-aware wireless-system design is essential to delivering a good user experience. For example, video quality of experience (QoE) can be preserved if the wireless network is capable of prioritizing the delivery of the most perceptually important video bits in case of network congestion. Similarly, if the application is aware of the wireless link quality, it can adjust its transmission rate to match link quality and send only the most relevant content over the network. In another example, it is likely that the burstiness in throughput resulting from opportunistic methods of spectrum access may adversely impact the QoE for real-time services. Data for such services may require a more predictable spectral resource model while the data for other service types focusing on bulk data transfer may be well-suited.

Clearly, cross-layer design that jointly optimizes application content and wireless system behavior can significantly improve the overall application quality of service (QoS). While the industry has initiated some work in developing service quality metrics and service awareness within the network (e.g., 3GPP SA4 [23]), Intel believes that much more can be done to fully address application awareness in wireless network design. Of particular concern are QoE metrics in video data, which is projected to become a dominant traffic type within future wireless networks [24–26].

4.2.5 *Exploiting Context Awareness*

Application content awareness is only part of the overall context surrounding a user's QoE. Additional information, including device capabilities and configuration, user location, peer devices, physical environment features, and so on, can also be exploited to improve a user's service quality. For example, a user may significantly reduce the latency of downloading a video clip if a nearby user can provide this content locally, or a device with small display capability may share content with a nearby device with a bigger display to enhance the user viewing experience. The video user's physical environment (lighting conditions, viewing distance from the TV display, and so on) may furthermore factor into the data transmission to help reduce or adjust the amount of information transferred.

The sharing of information amongst proximate devices is another approach to exploiting context-awareness. In general, devices can augment their capabilities through the systematic use of complementary functionalities in proximate devices, as well as offloading the wide-area network by exploiting local communications. It is anticipated that heterogeneous computing devices of the future will act as a group and in cooperation with each other, rather than operating independently.

Intel believes that incorporating contextual information into the design of 5G systems can provide significant gains in minimizing the network resources needed to meet service quality, thereby maximizing the efficiency of information transfer. While some context information is currently used in today's networks in limited ways, much more work is needed to systematically consider context information within broader issues of wireless system design.

At a more theoretical level, Intel wonders whether an abstract (application-agnostic) framework for defining the efficiency of information transfer is possible. If so, could this framework capture essential features of the user context? Could such frameworks address the complexity of defining effective quantitative metrics to measure a user's application-specific service quality in general? We note that some general frameworks for addressing QoS by measuring "timely" or delay-sensitive throughput have recently appeared in the academic literature [27, 28]. We believe that further investigation is needed.

4.2.6 Improving Device Power Efficiency

The number of rich multimedia applications and services provided by cellular devices continues to grow at a high rate. This factor, coupled with an increase in device complexity due to the addition of multiple radios for different wireless networks, results in a significant increase in device energy requirements. Furthermore, the IoT is driving new usages for wireless connectivity that require a power-efficient design for unattended devices [29, 30]. Given limited expectations for future battery capacity improvements, device power efficiency remains an important design consideration for any future wireless network technology.

Intel has been actively looking at technology areas that improve power efficiency, not just to satisfy user expectations, but also to minimize our overall impact on the environment. The goal is to not just extend device battery life in order to provide a better user experience, but also to be efficient with energy as a resource within a broader system and environmental context. We expect new approaches to wireless network design to jointly consider user context, platform-implementation considerations, and modes of wireless connectivity in addressing the issue of device power consumption. This could mean exploring various techniques of exploiting context information to provide more energy-efficient performance. For example, scheduling approaches may be used for a user who is traveling and prefers to conserve energy rather than demand peak device performance, or a user may choose to select a particular RAT based on energy-efficiency considerations. Energy efficiency may also affect the choice of using cloud or local services depending upon the power cost of computation versus communication. More generally, cellular networks may be designed to minimize the duration of active communication for devices, reducing the amount of maintenance signaling for connectivity or application needs and accommodating the special needs of power-constrained devices.

A significant hurdle towards a systematic approach to energy-efficient wireless system design is its dependency upon device and infrastructure implementation specifics. In particular, it is difficult to define implementation-agnostic metrics for energy efficiency and some industry efforts have focused on targeting improvements using theoretical absolute energy efficiency metrics based on Shannon's law [31, 32]. Intel believes that there is a need to define frameworks and associated metrics for energy-efficient wireless system design that go beyond ad-hoc approaches to power reduction.

4.3 Intel Strategic Research Alliance on 5G

As a leader in technology development, Intel is a strong supporter and sponsor of research over the entire continuum between fundamental and applied technology. Intel takes the long-term view that technology advancements create a richer ecosystem that simultaneously benefits consumers and the companies who serve them. New technologies start as research on experimental approaches to difficult issues limiting the state-of-the-art in the current world of technology practice. Promising technologies are eventually proven viable using a combination of theoretical analysis, simulation, more realistic emulation, prototypes, and experimental deployments. Such technologies impact a complex ecosystem through a further process of product development, ecosystem agreement on industry standards, and changing infrastructure investments by service providers, network carriers, government organizations, and private enterprises.

Intel sees university research, in particular, as an important arena for generating new ideas, exploring high-risk approaches, and proving candidate technologies that show promise, at least in theory. To promote focused university research, Intel has recently developed an organizational framework it refers to as an Intel Strategic Research Alliance (ISRA). Typically 3 years in duration, ISRAs bring together well-known researchers from top universities worldwide to look at a challenging technical domain in need of better understanding and fundamental technology development. In 2012, Intel identified 5G as one such domain and began the process of program development in collaboration with technical leaders from Intel labs and across the company. Using a request-for-proposal (RFP) process and seedling funding of more than one million dollars, Intel assembled a collection of university research teams to explore the research vectors described in the following sections.

4.4 ISRA 5G Technical Objectives and Goals

The goal of ISRA 5G is to develop and test the viability of new candidate technologies for tomorrow's wireless networks (5G). Such technologies should dramatically improve user experience in the face of rapidly growing data volumes in wireless networks, rapidly growing numbers of wireless devices, and increasingly rich and varied requirements of application services.

Intel believes that such technologies should address the following key challenges:

Dramatic Increases in Network Capacity. 5G technologies must provide dramatically increased capacity to address a projected 100-fold increase in mobile traffic over the next 10 years. In particular, this increased capacity is critical in addressing the rising demand for rich multimedia content and the expectation that billions of diverse, machine-type devices will require network connectivity.

Uniform Connectivity Experience Throughout the Network. If 5G technologies are to support a uniform service experience, they must address the inconsistent quality of user connectivity as evident in the high degree of variability in data rates. Such variability follows from temporal variations in channel conditions, user location, the interference environment, network congestion, and the type of service desired.

High Level of Service Quality and User Experience. Beyond communication links, 5G technologies must also account for application data characteristics and user context to guarantee an overall high service quality. Here, our goal is to deliver good application-specific service quality while maximizing the efficiency of information transfer based on available context information.

Toward this end, the metrics outlined in the following subsections are used in evaluating success. It is not necessarily expected that a single technology will meet the entire performance goal for a metric. Rather, a combination of technologies will be needed to provide additive gains. Researchers are encouraged to provide an analysis of the promised gains for their proposed technologies, and to consider the aggregate contribution toward meeting the target goals when more than one technology is being proposed.

4.4.1 Goal 1: Network Capacity

Capacity-focused techniques and technologies are expected to achieve the following overall performance and capacity goals for the program (3GPP Release 11 may be used as a baseline to assess performance gains [15, 16]):

- more than 10-fold enhancement in peak data rates (bits/s) for new spectrum technologies
- more than 10-fold enhancement in average user rates (bits/s/Hz) for technologies aimed at improving spectral efficiency
- overall, more than 100-fold improvement in network capacity (bits/s/meters^2).

While capacity improvements are the main focus of this goal, a tradeoff analysis between energy efficiency (bits/Joule) and spectral efficiency for the proposed techniques is also of interest. Energy efficiency considerations, however, should not inhibit new research in this domain.

4.4.2 Goal 2: Uniform Connectivity Experience

Techniques to improve the uniformity of user-connectivity experience should meet the following program goals (3GPP Release 11 may be used as a baseline to assess performance gains, [15, 16]):

- *Greater than 10-fold reduction in data-rate variability.* Data-rate variability is calculated as the ratio between cell edge and average cell-center user data rates. The definition of

cell-edge and cell-center users is more stringent than typical 3GPP baseline: cell-edge users are defined to be the first percentile lowest data rate users; cell-center users are defined as 99th percentile highest data rate users.
- *Greater or equal cell-average spectral efficiency.* Reduction in data rate variability should not compromise cell average spectral efficiency. In particular, we expect that capacity gains achieved by Goal 1 are maintained or improved.
- *Greater or equal energy efficiency.* Energy efficiency is measured as the ratio of data rate to power (bits/Joule). Essentially, this metric requires that any additional power is consumed efficiently (i.e., data rate improvements at least offset the increase in power consumption).

4.4.3 Goal 3: Service Quality and User Experience

Techniques that utilize context information to maintain service quality while maximizing the efficiency of information transfer should target the following program goals:

- more than 10-fold reduction in the overall information rate (bits/s) required to satisfy target service quality
- more than 10-fold increase in number of users in the network achieving their desired target service quality, for example more than a 10-fold increase in the number of "satisfied" OTT application users
- more than 10-fold reduction in device power consumption.

Metrics for monitoring service quality may include application-specific metrics such as PSNR-based assessments, probability of rebuffering, mean rebuffering duration, and so on, or alternate user satisfaction metrics such as perceived battery life, mean opinion scores, and so on. Metrics that only focus on network QoS measures such as end-to-end delay, jitter, and so on, may also be used in certain cases.

4.5 ISRA 5G Project Summaries

In this section we summarize research projects from the ISRA 5G program that is currently in progress at the time of this writing.

4.5.1 Higher, Denser, Wilder: Massively Broadband and Adaptive Wireless for 5th Generation Wireless Communications

University investigators: Andreas Molisch (USC), Ted Rappaport (NYU Poly, NYU Wireless), Giuseppe Caire (TU Berlin), Alan Willner (USC), and Mung Chiang (Princeton)

Higher, Denser, Wilder explores future wireless technologies along three axes: higher carrier frequencies, increased node density, and physical layer heterogeneity.

Higher carrier frequencies refers to the potential use of electromagnetic spectrum in the 30–300 GHz range, or millimeter-wave bands. Millimeter-wave signals (say, 28, 38, 60, or 72 GHz) have a wavelength in the 1–10 mm range, creating potential for small and highly directional antennas and high data rates, say of over 10 GBps. Radio frequencies in the millimeter-wave range have been used, for example, in short-range applications (less than 10 m) but are traditionally thought to be of limited utility for access networks due to line-of-sight limitations, that is, radio signals are presumably blocked by building walls and other physical objects, and significantly attenuated by foliage. Nevertheless, millimeter-wave frequencies represent an untapped spectrum resource that could help to solve the problem of finding additional spectrum in a world where conventional radio frequencies have already been fully allocated.

A key challenge in the millimeter-wave research space is understanding the channel model for outdoor communications systems. Researchers address this problem with a real-world measurement methodology that studies the propagation environment in New York City. Key issues are how the environment impacts the availability of the channel, and how abruptly the channel changes under various scenarios. Multipath is also a focus area, including time-dispersion characteristics. An understanding of these issues could help to determine optimal packet sizing, to develop efficient rake receivers, to understand the potential for beamforming, and to study the applicability of compressive sensing which relies on signal sparsity.

Another key challenge is that of developing low-power millimeter-wave antenna arrays optimized using statistical channel models. Researchers look at the use of pilot signals to obtain antenna pointing locations that lead to strong multipath characteristics. Low-power architectures for beamforming might be possible given the right array of horn antennas and accompanying RF circuitry. What the maximum data rates are for MIMO and antenna-combining system is an open question. Researchers look for a better theoretical understanding of the multipath shape factors that can help to guide the physical direction of antennas, and the angular spread of multipath. This could, in turn, provide a basis for low-overhead signals to rapidly form beams and would be generally be useful in understanding multipath energy.

Higher, Denser, Wilder researchers also examine base-station controlled device-to-device (D2D) communications as a way to exploit the advantages of increased node density. D2D explores the notion that user devices within range of one another can communication directly; in other words, without base station intervention. Such schemes, however, can often benefit from the presence of a base station, which can enforce efficient link scheduling to maximize spectrum spatial reuse. Overall, D2D schemes could increase spectral efficiency by potentially scaling network capacity with the number of antenna elements.

Researchers address two problems in this arena. First, they look at how D2D network state can be efficiently gathered at the base station, including data waiting to be transmitted by devices, the readiness of receive buffers, and the network interference graph. Researchers look at how such state can be optimally represented at the base station and how it can be gathered in a timely way to facilitate efficient control. Second, researchers look at efficient D2D link scheduling by the base station. In contrast to maximal independent set scheduling, researchers consider coded set scheduling, an approach that uses coding across multiple slots to generate interference neutralization opportunities. Using this scheme, a device is able to cancel the interference from another device without signal decoding or interfering channel coefficients, provided it observes a clean version of the signal repeated in another slot.

Higher, Denser, Wilder work on massive MIMO first looks at approaches to beamforming processing. Conjugate beamforming allows antenna nodes to calculate precoded signals in a fully distributed manner, assuming that downlink data signals are shared across all nodes. Zero-forcing and non-linear (dirty-paper coding) precoding schemes, on the other hand, necessitate processing in a centralized manner at the backhaul, which is a demanding requirement. Researchers investigate whether intermediate solutions are possible, especially for millimeter-wave communications where opportunities for channel gains depend on only a small number of neighboring antenna nodes. Second, researchers look at whether dimensional multiuser precoding can be reduced by exploiting directional channel correlation. Results from millimeter-wave channel measurements may be combined with work on channel vector correlation analysis to allow pre-beamforming.

In another area of work, *Higher, Denser, Wilder* researchers use non-cooperative game theory to analyze communication-mode selection in heterogeneous wireless networks where mobile devices can select from a variety of RATs. While a central controller may be used to optimize a muti-RAT system for performance, this approach may not be feasible when networks are operated by different entities. Instead, a user-centric and fully distributed approach exploring autonomous decisions by devices that are simply trying to maximize their throughput may be a better alternative if convergence and efficiency can be demonstrated. How to extend the framework to incorporate a broad range of data rates, interference considerations, massive distributed MIMO, device power considerations, and so on are related challenges. Another goal is to match the heterogeneity of user application traffic with the heterogeneity of wireless access networks. Through context awareness, mobile applications with specific data-communication patterns (e.g., elephants vs mice or intermittent vs continuous data flows, could be matched to the right physical-layer transmission technology). This includes pricing control, which must consider profiling, optimal pricing approaches, and congestion-dependent pricing at fine time granularities.

4.5.2 Fundamental Limits, Self-organization, and Context Awareness for Integrated Cellular and D2D Architectures

University investigators: Gustavo de Veciana (UT Austin), Robert Heath (UT Austin), and Angel Lozano (Universitat Pompeu Fabra)

The objective of research in this project is to develop the fundamental theory and design principles for integrating cellular and D2D networks. Augmenting cellular networks with D2D communication could lead to significant increases in network capacity with no additional requirements for infrastructure or spectrum. In fact, leveraging additional D2D links for such applications as data sharing, social networking, and high-density user events could potentially scale network with the number of user devices, contributing improvements in both capacity and QoS. The problem of interference management, however, will be key.

Researchers address the D2D integration challenge in three ways. First, they use information-theoretic analysis to better understand the fundamental limits of integrated cellular and D2D communications given the potential for interference. This analysis includes both a better understanding of the interference a cellular link experiences from a D2D communication, and the interference a D2D link experiences from a cellular link. What the fundamental performance limits are for realistic two-way traffic and what the tradeoffs are for uplink/downlink resource sharing and power-control policy are open questions. Analysis also considers the overhead generated by D2D communications from, for example, link adaptation feedback, power control, exchange of channel-state information, and hybrid automatic repeat request (ARQ). The work addresses the characterization and tradeoffs between coordination costs and capacity benefits.

Second, researchers propose a suite of algorithms for managing interference at the PHY layer in order to enable increases in sum rate within the network. One focus for this work is interference pricing, which considers a sum-utility function and marginal decreases per unit as interference power increases. Interference pricing can be used to optimize sum rates for D2D links and to mitigate intra-D2D interference in a distributed manner, although techniques are needed to convey prices with minimal overhead. Another research focus is multi-hop D2D. Multi-hop can effectively extend the reach of a D2D network, leveraging devices or dedicated relays to facilitate communication beyond immediate peers. Various interference management techniques, such as interference-shaping, interference neutralization, and aligned network coding, can be applied for multi-directional communication between D2D users and leveraging multiple antenna infrastructure relays, pico base stations, and other D2D users.

Finally, researchers consider the use of context awareness to enable dynamic self-organization in D2D networks, and the implications of context awareness for architectures that manage resource allocation in integrated cellular and D2D. A key issue is that of admission control, which investigates how to proactively evaluate the interaction between cellular and D2D links. Researchers consider how prior work on interference pricing may be useful in identifying the interference outlook and shaping link-admission policy. To facilitate self-organization in cellular and D2D transmissions, researchers consider soft fractional frequency reuse patterns in OFDM-based systems, and how they can be dynamically created and adapted to traffic requirements. While base-station mediation would appear to be a solution, the measurements needed to coordinate fine-grained interactions are not available in practice. Researchers also consider new application models that leverage context and spatial locality within integrated cellular and D2D networks. How to increase such locality and the impact on capacity gains is a broad area of investigation.

4.5.3 LAWS: Large Arrays and Wide Spectrum

University investigators: Suhas Diggavi (UCLA), Ashutosh Sabharwal (Rice), and Lin Zhong (Rice)

Researchers in this effort propose a new wireless architecture that they refer to as LAWS, or Large Antennas and Wide Spectrum. Goals include significantly increasing spectral efficiency through use of a large number of base station antennas (tens to hundreds), increasing cell capacity through multi-user beamforming and full-duplex operation, and reducing link variability through intra-cell and inter-cell interference management.

The use of large antenna arrays at base stations, researchers argue, would bring with it a host of benefits, including reducing inter-cell interference, increasing user capacity, lessening inter-base station coordination requirements, increasing spectral and power efficiency in mobile terminal uplinks, and support for both beamforming and full-duplex operation. A major problem, however, is the large number of channel coefficients needed by the base station before transmission. Channel information can be imprecise when multiple base stations send pilots and pilot contamination occurs. Approaches to this problem leverage additional base-station coordination, which is counter to the objective of decoupling the channels of different interfering users.

Today's smartphone devices are capable of communicating with more than one access network, including both cellular and ISM bands. This broad (or wide) spectrum range potentially creates an opportunity for establishing side channels between users that increase cooperative communication when interference-free transmissions are not possible. Side channels could be used to support distributed full-duplex communications, scheduling schemes that increase density, and coordinated up/downlinks. A key challenge, however, is identifying the availability of such channels since they can be location dependent and unreliable. WiFi is most often not an advisable choice since access points are hardly ubiquitous, and because WiFi is heavily involved in offloading when it is available. How to discover and use side channels is an open research question.

A key tenet of the LAWS architecture is that it should be entirely on-demand. In other words, inter-cell coordination and inter-device side channels should be triggered on demand when performance destabilizes due to inadequate network capacity or interference. The problem of discovery includes understanding when mobile devices should initiate the discovery of available spectral resource, how the side channels will be managed, and how we ensure that side-channel communications do not interfere with WiFi deployments involved in cellular offloading. How resources should be divided among devices and cells is another challenge. What bandwidth should be used a given side channel, when should a resource-allocation scheme be created or updated, and how do base stations and mobile nodes coordinate on resource sharing are all key questions. Finally, how to deploy a resource-sharing scheme on demand is another significant challenge. This includes a scheme for signaling using both cellular and side channels, understanding the cost-versus-benefit tradeoffs for side-channel cooperation, and developing associated PHY-layer strategies.

4.5.4 A System View of Interference Management: Radio Circuits, PHY Mechanisms, and Protocol Designs

University investigators: Sachin Katti (Stanford) and Pramod Viswanath (UIUC)

Researchers study interference-management techniques that bring together and extend two recent advancements in wireless research. The first is full duplex radios [33], which allow a device to send and receive on the same frequency simultaneously. This capability leverages novel circuit designs and algorithms for canceling self-interference to the noise floor. The second is a positional coding technique for the PHY layer that enables the concurrent transmission of control messages without introducing overhead [34]. This is done by transmitting short, high-powered flashes. Flashes are localized in frequency and time, and their location can be used to modulate control-signal bits.

Using these advances, researchers first look at whether multiple wireless devices can concurrently send and receive interference-management feedback messages as they receive data transmissions, a feature they refer to as "almost free feedback". One aspect of the problem is avoiding a dedicated spectrum requirement. In other words, researchers wish to avoid the need for a dedicated frequency spectrum for control messages since provisioning such spectrum is both impractical and inefficient. Researchers also wish to decouple the control plane from the data plane entirely. Coordination across device radios should be achievable with or without the device's orthogonal need to transmit or receive data.

The second area of proposed work is interactive full-duplex interference alignment. Researchers would like to demonstrate that nearly instantaneous control feedback can be used to build a practical interference-alignment system. A key problem in this challenge is understanding for a fixed deterministic channel the fundamental degrees of freedom. In prior work [35], researchers have shown that full degrees of freedom are achievable when the number of transmit-receive pairs is limited. To further this work, however, distributed algorithms are needed to find interactive communication coefficients, analysis is needed of the gains for a larger number of transmit-receive systems, and an actual prototype should be developed that demonstrates the interactive-alignment scheme using full duplex.

4.5.5 *Dynamic Cloud Services Spectrum Sharing Algorithms and Mechanisms for B4G Networks*

University investigators: Eryk Dutkiewicz (Macquarie), Michael Heimlich (Macquarie), Stephen Hanly (Macquarie), and Karu Esselle (Macquarie)

Dynamic spectrum sharing addresses the challenge of wireless network capacity growth in the face of scarce spectrum availability. With licensed spectrum no longer available for auction, the scheme allows existing incumbents to make their allocations available for licensing arrangements at specific geographical locations and for specific time periods. This opportunistic model promotes a more effective utilization of existing spectrum allocations.

Researchers focus on the licensed shared access (LSA) architecture [36], which includes:

- LSA end users who dynamically enter/leave the system
- an LSA controller that manages spectrum scheduling and handover
- an LSA repository that contains information on spectrum available for leasing.

A broker may handle dynamic licensing arrangements.

Researchers focus on four areas of work. The first is determining the optimal timescale for dynamic spectrum sharing. One might imagine a continuum of options that ranges from a long-duty cycle to a short-duty cycle. What the options mean for different network scenarios and applications is an open question, as is the suitability of different options for the stakeholders involved: network operators, spectrum incumbents, end users, and device manufacturers. A related challenge is characterizing the optimal measurement timescale. Tradeoffs may exist between channel-estimate accuracy and the cost of channel feedback. Given the timescale, algorithms for dynamic spectrum licensing, spectrum scheduling, spectrum handover, and efficiency analysis may differ.

A second area of work is accurate radio environment maps (REMs). REMS are created using spectrum sensing and localization data gathered from mobile handsets. The challenge here is to understand the minimum amount of data needed for dynamic spectrum sharing given a particular timescale. REMs may include data from heterogeneous wireless network environments and be used for more complex handover schemes that consider additional options like WiFi off-loading. Compressive sensing approaches may be used to improve accuracy and reduce the amount of sensing data required. Understanding REM accuracy requirements is an important issue, as is understanding the signaling overhead tradeoffs for more and less accurate REMs.

Third, researchers look at the development of an RF front end for low-power, multi-mode devices. The front end must enable wideband spectrum sensing and fast switching between spectrum channels. Antennas must be adaptive and capable of covering narrow and broadband wireless protocols. Modulation and demodulation must be controllable in temporal and spectral dimensions, and low-power. An analog sensing RF processing module must be able to quickly sense whitespace opportunities at a range of spectrum bands while still being low-power.

Finally, researchers look at implementing the LSA architecture for a long-duty cycle case study. The proof of concept looks at various types of network architectures, such as small-cell or macro-cell, and demonstrates various LSA controller algorithms for selected applications.

4.5.6 Fundamentals of Spectrum Sharing in Device-to-Device and Heterogeneous Communication Networks

University investigator: Salman Avestimehr (USC)

Researchers consider information theoretic approaches for spectrum sharing and interference management in D2D networks. In particular, the project builds upon prior work on information theoretic link scheduling (ITLinQ) [37]. ITLinQ makes use of the notion that treating interference as noise may be information-theoretically optimal [38]. Information-theoretic independent sets are used to schedule links and can be implemented in a distributed manner. Related results show that significant efficiency is achievable, even with limited network knowledge, by coding over time and suppressing interference using its temporal structure.

Researchers divide their work into three themes. The first is advanced PHY-layer design for spectrum sharing, which asks how much successive interference cancellation and temporal interference neutralization can be used to achieve spectral efficiency. Work considers the use of powerful pico base stations in heterogeneous network contexts. Such nodes might leverage more channel information to help establish cooperation between user devices, or more transmit power and sophisticated power control schemes.

The second theme is network utility maximization for ITLinQ. Here, the goal is to generalize ITLinQ to a utility-maximization function for the entire network. In other words, the problem of maximizing user rates can be cast as a utility maximization problem for D2D distributed interference networks. There is a variety of alternatives for the utility function: weighted sum rates, min-max fairness, or QoS-based approaches. Scheduling over information-theoretic independent sets may be a promising approach.

Finally, researchers consider cached ITLinQ, where multiple network sources hold copies of data files. Work in this area explores the fundamental gains that are possible from having multiple sources available to serve a given destination node. Essentially, caching has the effect of extending information-theoretic independent sets, and thus potentially increasing spectral efficiency. Also of interest are optimal caching schemes that maximize efficiency of delivery.

4.5.7 Structured Sharing of Network and Compute Resources in a Community of Devices

University investigators: Huzur Saran (IIT Delhi), Vinay Ribeiro (IIT Delhi), and Smruti R. Sarangi (IIT Delhi)

Researchers in this project consider the technology context of emerging world markets, such as rural India, where the need for resource sharing is acutely felt. Such markets are often characterized by a relative dearth of computational resources and high-bandwidth networks.

Extremely low-cost handheld devices are common but have limited memory and computational power. PC-class hardware is increasingly available and provides a potential resource for sharing arrangements with mobile devices.

Researchers first look at network connectivity sharing in D2D networks. Network service in emerging markets is often low in QoS, restrictive in data-download policies, and generally unreliable. A common arrangement is prepaid phones that offer 2G/3G connectivity for a fixed amount of data (say, 1 GB). To address these limitations, users would benefit from flexible D2D sharing arrangements that exploit the Internet connectivity and data-transmission resources of nearby PCs or mobile devices with a more premium service arrangement.

Researchers consider three phases in the development of a connectivity sharing framework. The first is connection discovery and QoS estimation, where users with computational resources can advertise their connections to users in need of them, along with information on data and energy budgets, expected throughput and latency, and QoS estimation parameters. The latter can be derived from both active (e.g., probing) and passive measurements. The second phase is optimal connection selection with fairness, which supports the matching of user connectivity requests to available connections in a supervised manner. Fairness is a key issue and scheduling mechanisms are deployed within the D2D arrangement and at Internet gateways to enforce sharing disciplines. The third phase is content sharing and disconnection tolerance. Content sharing exploits the opportunity for devices on the same D2D network to cache content and share it with peer devices rather than waste network resources in retrieving duplicate copies. Delay-tolerant networking [39] can be used to address events like interruptions in Internet gateway service or energy/data limitations being reached.

Theme two of the research looks at automated computation-sharing arrangements within D2D networks. Researchers consider a form of device virtualization that allows users to use the capabilities of a nearby device in a seamless manner; that is, without user management. They also consider sharing arrangements that support computation offloading to more capable devices nearby. Challenges in this domain include identifying program segments with potential for offloading, understanding the computation, memory, and power requirements for execution, and evaluating the cost and benefit tradeoffs in migration communications. Researchers propose an objective function approach that weighs the power requirements of D2D communication against the cost of performing the computation locally, and takes into account the delay in execution if computation were to be migrated.

4.5.8 A Unified Framework for Enabling Energy-efficient Mobile Internet Apps and Energy-efficient Cloud Offloading

University investigator: Y. Charlie Hu (Purdue)

User QoE in 5G networks will be strongly impacted by the battery life of tomorrow's devices. This project explores our understanding of how energy is currently being used on today's smartphone platforms, and how this understanding could be used for effective offloading to the cloud within future wireless networks.

Researchers divide their focus into three areas. First, they consider the problem of profiling the energy footprint of background applications and services running on today's smartphone platforms. Work in this area builds upon prior work using *eprof* [40], a fine-grained energy profiling tool for smartphone applications ("apps"). *Eprof* leverages a variety of techniques to instrument app program code and system calls and generate traces that then drive finite state machine power models to analyze energy usage in a component-wise manner. To apply this methodology to long-running background applications, a sampling approach is needed (following the approach of the well-known GNU utility *gprof*) that gives the approximate time spent by various routines on the execution stack. Trace compression and upload management are also needed, but must be done carefully in order to avoid adding overhead.

Second, researchers consider context-aware energy analysis and optimization. For this, an understanding of user behavior is needed in order to assess which background applications and services are useful and which aren't. Researchers consider methods for logging user interactions, such as touchscreen events, and then analyzing their correlation with executing apps and background processes. This also requires an analysis of background tasks, such as Twitter feeds, and whether the user made use of the results of various energy-consuming communication and computation events. The results of this analysis can potentially be used to adapt background task behavior in a way that has little or no impact on user experience.

Finally, researchers consider the problem of using energy-profiling information to explore the benefits of cloud offloading. Their approach considers offloading at the granularity of methods, first looking at which method calls could be "remotable" – offloaded to a remote server – and for what resource types. Then, researchers consider the problem of computing the offloading energy tradeoff. Energy cost can be estimated using a combination of energy-profiling information, a model for computation offloading to cloud, and network bandwidth estimation. Additional measurements are taken to understand variations in input to a given method call, for example, the size of a key data structure.

References

[1] S. Forge, R. Horvitz, and C. Blackman (2012) Perspectives on the value of shared spectrum, Final report for the European Commission. URL: http://ec.europa.eu/information_society/policy/ecomm/radio_spectrum/_document_storage/studies/shared_use_2012/scf_study_shared_spectrum_access_20120210.pdf (accessed August 8, 2010).

[2] NTIA (2012) An assessment of the viability of accommodating wireless broadband in the 1755–1850 MHz band. URL: www.ntia.doc.gov/files/ntia/publications/ntia_1755_1850_mhz_report_march2012.pdf (accessed August 8, 2010).

[3] M. Mueck (2012) Current ETSI RRS activities of relevance to the revision of the R&TTE directive and the future EC RRS standardization mandate. *Workshop on Cognitive Radio and Software Defined Radio: Policy And Regulations* organized jointly by ECC/COST-TERRA. Mainz, Germany. May 2–3, 2012.

[4] J. Pi and F. Khan (2011) An introduction to mm-wave mobile broadband systems, *IEEE Communications Magazine*, **49**, 101–107.

[5] T.S. Rappaport (2012) The massively broadband future. In: *Spectrum 20/20 Conference*. URL: www.spectrum2020.ca/presentations/Rappaport.pdf.

[6] G.K. Chang, J. Long, S-H Fan, et al. (2011) Emerging heterogeneous optical wireless access networks for next generation telemedicine and telehealth applications. *Opto-Electronics for Communications Conference*, Kaohsiung, Taiwan.

[7] D. Gesbert, M. Kountouris, R.W. Heath, Jr. et al. (2007) From single user to multiuser communications: shifting the MIMO paradigm. *IEEE Signal Processing Magazine*, **24** (5), 36–46.

[8] F. Rusek, D. Persson, B.K. Lau, et al. (2012) Scaling up MIMO: opportunities and challenges with very large arrays, *IEEE Signal Processing Magazine*, **30**, 40–60.

[9] T.L. Marzetta (2010) Noncooperative cellular wireless with unlimited numbers of base station antennas. *IEEE Transactions on Wireless Communications*, **9** (11), 3590–3600.

[10] A. Pitarokoilis, S.K. Mohammed, and E.G. Larsson (2012) On the optimality of single-carrier transmission in large-scale antenna systems, *IEEE Wireless Communications Letters*, **1** (4), 276–279.

[11] J. Hoydis, S. ten Brink, and M. Debbah (2011) Massive MIMO: How many antennas do we need? *Proc. IEEE Allerton Conference on Communication, Control, and Computing*, Urbana-Champaign, IL.
[12] Y. Lin, L. Shao, Z. Zhu, et al. (2010) Wireless network cloud: architecture and system requirements, *IBM Journal of Research and Development*, **54** (1), 4:1–4:12.
[13] M. Muck and J.-P. Javaudin (2005) Advanced OFDM modulators considered in the IST-WINNER framework for future wireless systems. In: *Proc. 14th IST Mobile Wireless Commun. Summit.*
[14] L.S. Cardoso, M. Kobayashi, O. Ryan, and M. Debbah (2008) Vandermonde frequency division multiplexing for cognitive radio. In: *Proc. IEEE 9th Workshop on Signal Processing Advances in Wireless Communications.*
[15] 3GPP, TR 36.814: Further advancements for EUTRA physical layer aspects, V. 9.0.0, 2010.
[16] 3GPP TR 36.819: Coordinated multi-point transmission for LTE physical layer aspects, V11.1.0, 2011.
[17] Alcatel-Lucent, White paper: Light-radio portfolio: technical overview, 2011.
[18] J. Andrews (2011) Can cellular networks handle 1000x the data? URL: http://users.ece.utexas.edu/~bevans/courses/realtime/lectures/Andrews_Cellular1000x_Nov2011.pdf (accessed August 8, 2016).
[19] A. Zemlianov and G. de Veciana (2005) Capacity of ad-hoc wireless networks with infrastructure support, *IEEE Journal on Selected Areas in Communications*, **23**, 657–667.
[20] A. Ozgur, O. Leveque, and D.N.C. Tse (2007) Hierarchical cooperation achieves optimal capacity scaling in ad-hoc networks, *IEEE Transactions on Information Theory*, **53**, 3549–3572.
[21] A. Ozgur, O. Leveque, and D.N. Tse (2011) Operating regimes of large wireless network, *Foundations and Trends in Networking*, **5** (1), 1–107.
[22] L. Wang and G.S. Kuo (2012) Mathematical modeling for network selection in heterogeneous wireless networks – a tutorial, accepted for publication in *IEEE Communications Survey and Tutorials*, **15**, 271–292.
[23] 3GPP (2012) TS 26.247 v10.2.0: Transparent end-to-end packet switched streaming service (PSS); Progressive download and dynamic adaptive streaming over HTTP (3GP-DASH), Release 10.
[24] K. Piamrat, C. Viho, A. Ksentini, and J. Bonnin (2009) Quality of experience measurements for video streaming over wireless networks. In: *Sixth International Conference on Information Technology.*
[25] R.K.P. Mok, E.W.W. Chan, and R.K.C. Chang (2011) Measuring quality of experience in HTTP video streaming. In: *IEEE International Conference on Integrated Network Management.*
[26] H. Zepernick and U. Engelke (2011) Quality of experience of multi-media services: Past present and future, QoEMCS: Quality of experience for multimedia content sharing, EuroITV. URL: http://dcti.iscte.pt/events/qoemcs/papers/2011/Hans-juergen_QoEMCS.pdf (accessed August 8, 2016).
[27] I.-H. Hou, V. Borkar, and P.R. Kumar (2009) A theory of QoS for wireless. In: *INFOCOM*, 2009.
[28] S. Lashgari and A.S. Avestimehr, "Approximating the Timely Throughput of Heterogeneous Wireless Networks," *Proceedings of IEEE International Symposium on Information Theory (ISIT)*, 2012.
[29] IEEE (2010) 802.16p-10/0005: Machine-to-machine (M2M) communications technical report.
[30] 3GPP (2012) TS 22.368: Service requirements for machine type communications, v. 11.5.0.
[31] Green Touch Consortium, webpage. URL: www.greentouch.org/ (accessed August 8, 2016).
[32] IEEE (2010) C80216-10_0008r1: Energy efficiency for future wireless broadband networks.
[33] D. Bharadia, E. McMilin, and S. Katti (2013) Full duplex radios, In: *ACM SIGCOMM Computer Communication Review*, 2013.
[34] A. Cidon, K. Nagaraj, S. Katti, and Viswanath, P. (2012) Flashback: Decoupled lightweight wireless control. In: *ACM SIGCOMM*, 2012, Helsinki, Finland.
[35] Q. Geng, and P. Viswanath (2012) Interactive interference alignment. In: *IEEE Information Theory Workshop (Lausanne).*
[36] H. Okkonen, M. Palola, S. Yrjola, P. Ahokangas, and M. Mustonen (2014) Spectrum sharing using licensed shared access: the concept and its workflow for LTE-advanced networks, *IEEE Wireless Communications*, **21** (2), 72–79.
[37] N. Naderializadeh and A.S. Avestimehr (2014) ITLinQ: A new approach for spectrum sharing. In: *Device-to-Device Communication Systems.* In: *Proceedings of IEEE International Symposium on Information Theory.*
[38] C. Geng, N. Naderializadeh, A.S. Avestimehr, and S. Jafar (2014) On the optimality of treating interference as noise. In: *IEEE International Symposium on Information Theory (ISIT).*
[39] A. Mahla, D. Martin, I. Ahuja, Q. Niyaz, and A. Seth (2012) Motivation and design of a content distribution architecture for rural areas. In: *ACM DEV*, 2012.
[40] A. Pathak, Y.C. Hu, and M. Zhang (2012) Where is the energy spent inside my app? Fine grained energy accounting on smartphones with eprof. In: *Proceedings of EuroSys: European Conference on Computer Systems.*

Part II

Candidate Technologies – Evolutionary

5

Towards Green and Soft

Chih-Lin I and Shuangfeng Han
Green Communication Research Center, China Mobile Research Institute, China

5.1 Chapter Overview

Wireless communication systems have developed from the first generation (1G) to the latest fourth generation (4G) to satisfy ever-increasing and more diversified mobile-traffic demand. The anticipated 1000-fold wireless traffic increase by 2020 and the global recognition of the importance of green communications, however, pose very tough challenges for the fifth-generation (5G) design. The design metric of spectrum efficiency (SE) maximization, which was much emphasized in previous communication systems, may well be jointly pursued with, amongst other metrics, energy efficiency (EE). In this chapter, we will present green network design considerations from the perspective of network architecture, signaling and control, and high SE and EE transmission techniques. Specifically, in Section 5.2, the motivations for, requirements of, and worldwide activities regarding green radio are introduced. In Section 5.3, one EE and SE co-design framework is presented for link-level optimizations. The impact of system parameters such as frequency bandwidth, total power, antenna number, transceiver number, and so forth, on EE–SE performance is analyzed. Some promising 5G technologies, such as large-scale antenna systems (LSAS) and non-orthogonal multiple access (NOMA), are also described in terms of EE–SE co-design.

To satisfy various user requirements, future networks should be user-centric rather than, as they traditionally have been, cell centric, thus leading to 5G networks with "no more cells" (NMC). In Section 5.4 the concept of NMC is discussed. Firstly, C-RAN, as a soft radio access architecture and an enabling element for key 5G technologies, is presented. In this technology, the baseband is centralized and virtualized to flexibly allocate resource and manage interference to save both radiated power and circuit power. Since the future network will be more heterogeneous, with fluctuating traffic volume and diversified services, a fundamental

Towards 5G: Applications, Requirements and Candidate Technologies, First Edition.
Edited by Rath Vannithamby and Shilpa Talwar.

Companion website: www.wiley.com/go/vannithamby/towards5g

rethinking of signaling and control design in 5G is necessary to facilitate energy saving. Finally, an aggregator-based signaling optimization scheme is proposed to accommodate trillions of wireless nodes in the Internet of Things (IoT) in the 5G era. It is anticipated that 5G networks and beyond will be softer and greener.

5.2 Efforts on Green and Soft 5G Networks

With the maturing of 4G standards and the ongoing worldwide deployment of 4G cellular networks, research activities on 5G communication technologies have emerged in both the academic and industrial communities. Various organizations from different countries and regions have taken initiatives and launched programs aimed at potential key technologies of 5G:

- 5GNOW and METIS were launched under the European Telecommunications Standards Institute's (ETSI's) Framework 7, and study new waveforms and the fundamentals of 5G to meet the requirements proposed for 2020;
- the 5G Innovation Center was established in the United Kingdom to develop a world-class testbed for 5G technologies
- the Third Generation Partnership Project (3GPP) has drawn up its draft evolution roadmap out to 2020
- China has kicked off its IMT-2020 Promotion Group to start the study of user demands, spectrum characteristics, and technology trends [1].

Since research on 5G networks is still in an embryonic stage, there is no clear answer as to what is 5G is and what the enabling technologies are. However, the broad consensus on 5G requirement is that it includes higher SE and EE, lower end-to-end latency, and more connection nodes. From the perspective of the China Mobile Research Institute, the world of 5G should reflect two major themes: *green* and *soft*.

Global carbon emissions are increasing and sea-levels rising; global weather and air pollution in many large cities across the world are becoming more severe. Consequently, energy saving has been recognized as an urgent issue worldwide. In 2012, the annual average power consumption by information and communications technologies (ICT) was over 200 GW, of which telecoms infrastructure and devices accounted for 25% [2]. In the 5G era, mobile traffic in wireless communications networks is no longer solely generated by traditional mobile terminals. Increasingly, driven by the widespread applications of the IoT, mobile networks are being used to gather data from or transfer it between machine-type devices. These devices include, for example, intelligent wearable devices, environmental sensors for temperature and humidity, metering devices, and vehicular sensors for navigation, safety, and traffic management. Machine-type communications (MTC) are typically characterized by small packet size, low tolerance of transmission delays, low mobility, and low frequency of use. It is forecast that the number of connected MTC devices will reach tens of billions. The massive connectivity involved in MTC may not be handled efficiently by the current wireless communication networks, such as 3G and 4G. Meanwhile, the trend in networks is to become denser and denser in order to provide high data rates. Dramatic improvements in EE will be needed and new tools for jointly optimizing network SE and EE will be essential.

Several research groups and consortia have been investigating the EE of cellular networks, including Mobile VCE, EARTH, and GreenTouch. Mobile VCE has focused on the BS hardware, architecture, and operation, realizing energy saving gains of 75–92% in simulations [3]. EARTH has devised an array of new technologies, including low-loss antennas, micro discontinuous transmission (DTX), antenna muting, and adaptive sectorization according to traffic fluctuations, resulting in energy savings of 60–70%, with less than 5% throughput degradation [4]. GreenTouch has set up a much more ambitious goal of improving EE 1000 times by 2020 and proposed multiple green transmission technologies [5]. Several operators have been actively developing and deploying green technologies, including green BSs powered solely by renewable energy, and green access infrastructure such as cloud/collaborative/clean radio access network (C-RAN) [6].

Soft is the other critical characteristic of 5G networks. Carrier-grade networks are complex and composed of special-purpose nodes and hardware. New standards and features often require a variety of equipment to be developed and integrated, thus leading to very long launch cycles. In order to accommodate explosive mobile Internet traffic growth and a large number of new applications/services demanding much shorter times to market, much faster turnarounds of new network capabilities is required. Dynamic RAN reconfiguration can handle both temporal- and spatial-domain variations in mobile traffic without over-provisioning homogeneously. Soft technologies are the key to resolving these issues.

By separating software and hardware, control plane and data plane, and building software over general-purpose processors (GPPs) via programming interfaces and virtualization technology, it is possible to achieve lower costs and higher efficiency using software defined networks and network functions virtualization [7]. The OpenRoad project at Stanford University introduced Open-flow, FlowVisor, and SNMPVisor to wireless networks to enhance the control plane. Base-station virtualization from NEC concentrated on slicing radio resources at the medium access control layer. CloudEPC from Ericsson modified the Long-term Evolution (LTE) control plane to control open-flow switches. CellSDN from Alcatel-Lucent considered a logically centralized control plane, and scalable distributed enforcement of quality of service (QoS) and firewall policies in the data plane. C-RAN implements a soft and virtualized BS with multiple baseband units (BBUs) integrated as virtual machines on the same server, supporting multiple radio access technologies (RATs). A soft end-to-end solution from the core network to the RAN can enable the 5G goals of spectral and energy efficiency.

5.3 Rethink Shannon: EE and SE Co-design for a Green Network

5.3.1 EE and SE Co-design Fundamentals

Given limited spectrum and ever-increasing capacity demand, SE has been pursued for decades as the top design priority of all major wireless standards, ranging from cellular networks to local- and personal-area networks. The cellular data rate has improved from kilobits per second in 2G to gigabits per second in 4G. SE-oriented designs, however, have overlooked the issues of infrastructure power consumption. Currently, RANs consume 70% of the total power. In contrast to the exponential growth of traffic volume on the mobile Internet, both the associated revenue growth and the network EE improvement lag by orders of magnitude. A sustainable future wireless network must therefore be not only spectrum-efficient but also energy-efficient. Therefore EE and SE joint optimization is a critical part of 5G research. Looking at traditional cellular systems, there are many opportunities to become greener, from the equipment level through, say, more efficient power amplifiers using envelope tracking, to the network level, for example through dynamic operation in line with traffic variations, both

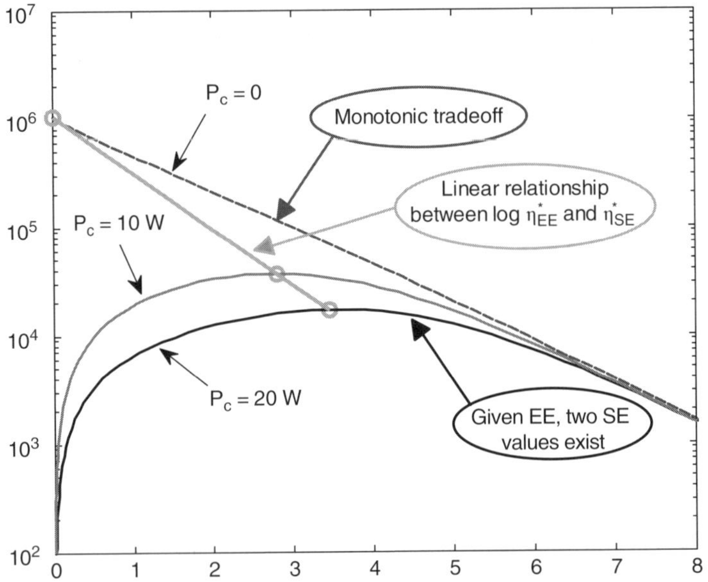

Figure 5.1 SE and EE relationship for different circuit powers.

in time and space. For the fundamental principles of EE and SE co-design, one must first revisit the classic Shannon theory and reformulate it in terms of EE and SE. In classic Shannon theory, the channel capacity is a function of the log of the transmit power (P_t) noise power spectral density (N_0), and system bandwidth (W). The total system power consumption is a sum of _P_t and the circuit power P_c, that is,

$$P_{tot} = P_t / \rho + P_c \tag{5.1}$$

where ρ is power amplifier (PA) efficiency defined as the ratio of the input of the PA to the output of the PA. EE is, by definition, equal to the channel capacity normalized by the system power consumption. SE is the channel capacity normalized by system bandwidth. The relationship of EE and SE can be shown as a function of PA efficiency and P_c (see Figure 5.1). It can be seen that when P_c is zero, there is a monotonic trade-off between η_{EE} and η_{SE} as predicted by the classic Shannon theory. For non-zero P_c, however, η_{EE} increases in the low-SE region and decreases in the high-SE region with η_{SE}; for a given η_{EE} there are two values of η_{SE} [8]. As P_c increases, the EE–SE curve appears flatter. Furthermore, when taking the derivative of ηEE over η_{SE}, the maximum EE (η^*_{EE}) and its corresponding SE (η^*_{EE}) then satisfy:

$$\log_2 \eta^*_{EE} = \frac{\log_2 \rho}{N_0 \ln 2} - \eta^*_{SE} \tag{5.2}$$

This means there is a linear relationship between $\log_2 \eta^*_{EE}$ and η^*_{SE}, and the EE–SE relationship at the EE optimal points is independent of P_c. This observation implies that as P_c decreases, an exponential EE gain may be obtained at the cost of linear SE loss. Figure 5.2 compares the

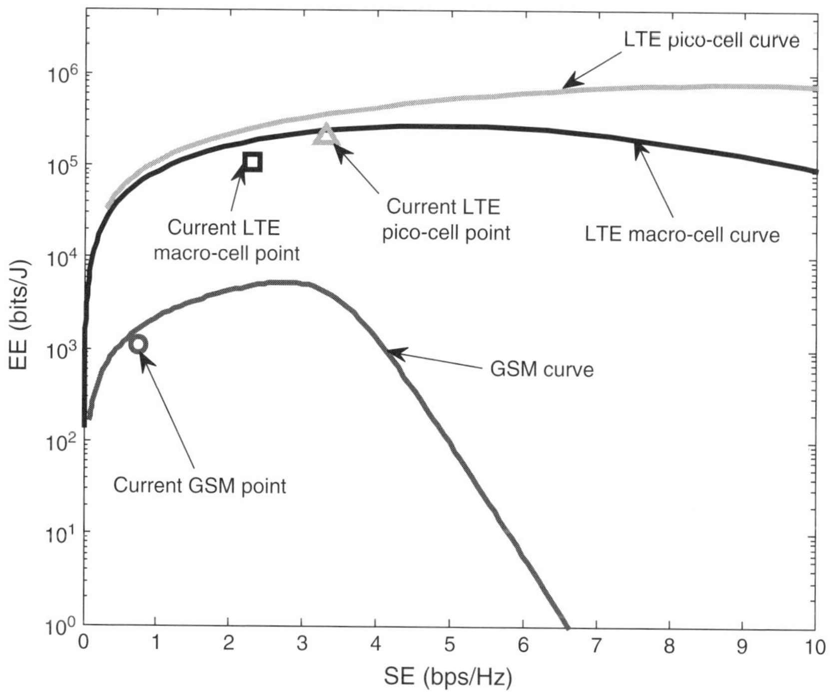

Figure 5.2 SE and EE relationship for current networks.

EE–SE performance of current Global System for Mobile Communications (GSM) and LTE BSs. LTE performs better than GSM in terms of both SE and EE; both, however, are working in a low-SE region, indicating room for improvement.

Antenna muting is proposed in EARTH to improve EE, while LSAS stipulates EE improvement by increasing the number of antennas. These seemingly contradicting conclusions are actually consistent with the analysis presented previously; the difference is that the former operates in a low-SE region, whereas the latter operates in a high-SE region. While some progress has been made in EE and SE co-design investigation, there is still a long way to go to develop a unified framework and comprehensive understanding in this area. In particular, new techniques, such as spatial modulation, can utilize the antenna resource more efficiently. Ideally, the EE–SE curve in future systems should achieve the following criteria:

- The EE value should be improved for each SE operation point.
- The EE–SE win-win region should be enlarged and the EE–SE trade-off region should be reduced.
- The slope of the EE–SE curve in the tradeoff region should be reduced.

MIMO systems have been studied under these criteria. As shown in Figure 5.3, besides the linear relationship between $\log_2 \eta^*_{\mathrm{EE}}$ and η^*_{SE}, the slopes of the EE–SE curve are also affected by different parameters. We can see that when SE approaches zero, the slope depends only on the bandwidth and circuit power, and is independent of the antenna configuration and the

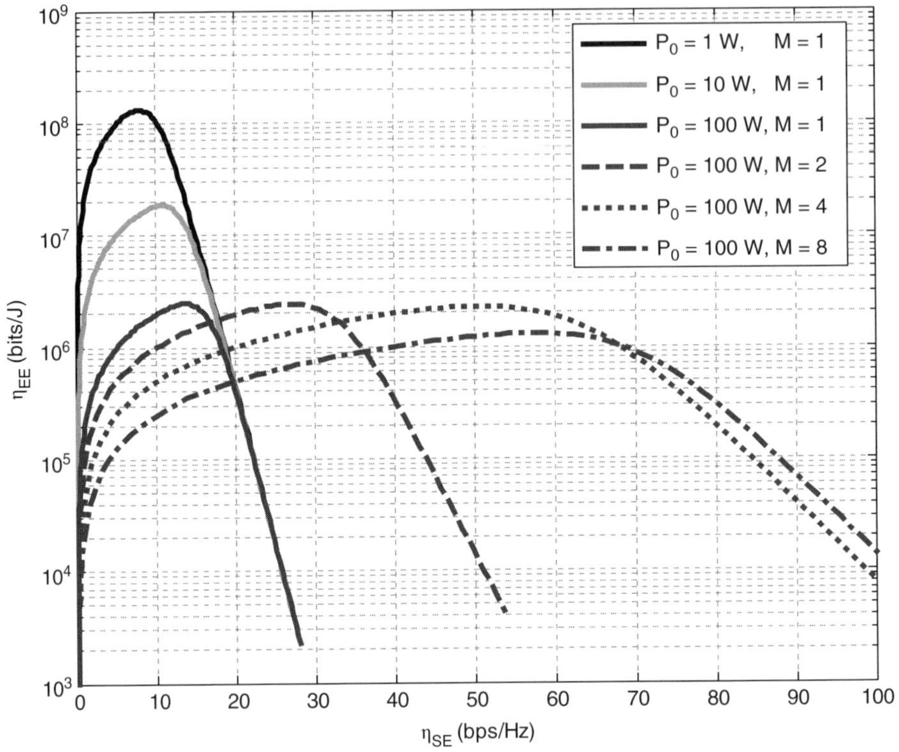

Figure 5.3 The impact of the circuit power and the number of antennas on the EE–SE curves.

knowledge of channel-state information (CSI) at the transmitter. From the solid curves, we can see that reducing the circuit power can improve EE in the low-SE region significantly. When SE approaches infinity, the slope is only related to the number of data streams – the multiplexing gain – and independent of circuit power. From the dashed curves, it can be observed that when increasing the number of antennas, the EE improvement mainly happens in the high-SE region. This observation implies that reducing circuit power has advantages in the low-SE region while developing transmission techniques can improve EE in the high-SE region.

When considering large-scale MIMO systems, we can derive the EE–SE relationship with different kinds of CSI. Consider a transmitter with M antennas and a receiver with N antennas. The propagation channel coefficients between transmit and receive antennas are assumed to be independent and identically distributed random variables with zero mean and variance ψ. When the CSI is not known at the transmitter, the EE–SE relationship satisfies the following expression:

$$\log_2\left(\frac{\rho\psi N}{N_0 \ln 2}\right) - \frac{1}{N}\eta_{SE}^* = \log_2 \eta_{EE}^* \tag{5.3}$$

When the CSI is known at the transmitter, the EE–SE relationship is

$$\log_2\left(\frac{\rho\psi M}{N_0 \ln 2}\right) - \frac{1}{N}\eta_{SE}^{*} = \log_2 \eta_{EE}^{*} \tag{5.4}$$

From Eqs (5.3) and (5.4), it can be found that the linear relationship depends on the number of antennas and CSI knowledge. We can find that the intercept in Eq. (5.4) is larger than that in Eq. (5.3) for a large-scale antenna array at the transmitter, which infers the EE–SE gain from CSI knowledge. With K single-antenna users, the EE–SE relationship is highly dependent on the number of users as shown in Eq. (5.5). When increasing K, the slope, $-1/k$, increases, which indicates the decrease in speed of EE with SE reduces.

$$\frac{1}{K}\sum_{k=1}^{K}\log_2\left(\frac{\rho B}{\ln 2}\frac{\psi_k M}{\sigma^2}\right) - \frac{1}{K}\eta_{SE}^{*} = \log_2 \eta_{EE}^{*} \tag{5.5}$$

5.3.2 5G Candidate Technologies with EE–SE Co-design

5.3.2.1 Hybrid BF for LSAS

LSAS is considered to be one of several promising technologies in 5G. Theoretically, LSAS with a full digital beamforming (BF) can yield optimal performance, significantly increasing the system EE and SE via multiuser BF. When a large number of antennas are implemented to achieve better BF gains, however, implementing the same number of transceivers may not be feasible due to excessive demand on real-time signal processing for high BF gains, high power consumption and cost (especially the high cost and power consumption of mixed-signal devices in mmWave systems). A BF structure with much smaller number of digital transceivers than the total antenna number will therefore be more practical and cost-effective to deploy.

An N by M hybrid BF structure is shown in Figure 5.4, where each of the N transceivers is connected with M antennas. Analog BF is performed over only M RF paths in each transceiver and digital BF is performed over N transceivers [9–11]. This structure is more practical for base station antenna deployment in current cellular systems, where each transceiver is generally connected to a column of antennas.

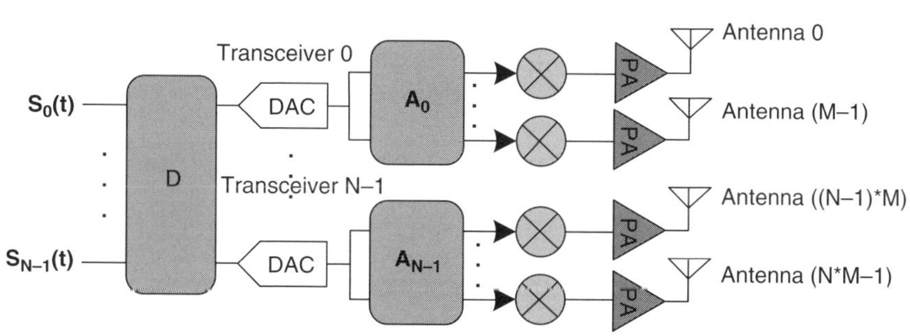

Figure 5.4 *N* by *M* beamforming structure.

There are still many challenging issues regarding the hybrid BF structures, including the joint digital and analog BF design for multi-user scenarios, EE–SE optimized BF structure design, and efficient reference signal design.

EE–SE Relationship

Consider the N by M BF structure: perfect analog BF is assumed between M antennas per transceiver for one user (in total there are N users). There is zero inter-user interference with the assumption of a large enough M or via proper user scheduling, even when M is not so large. For example, the beamwidth of the analog beam generated by a 50-element linear antenna array with half-wavelength spacing can be as small as 2°. It would not be difficult to schedule, say, $N=10$ users with negligible inter-user interference in one cell. The N user sum capacity of this structure is derived as $C=WN\log_2(1+MP\eta_{SE}/WN_0)$, where W is the system bandwidth, P is the total power of M PAs per transceiver, η_{PA} is the PA efficiency, and N_0 is the thermal noise density. Channel gain is assumed to be 1. The SE of this structure is written as $\eta_{SE}=C/W$.

An accurate power model is needed to calculate the EE. However, this is not straightforward, since base stations have different types (macro, pico, femto) and are generally produced by different vendors with different implementation technologies. In this section, the following simple power model is used: $P_{total}=NP+P_{static}=NP+NP_0+P_{common}+NMP_{rf\_circuit}$, where P_{total} is the total power, NP is the RF power of total N transceivers, P_{static} is the static power of the base station, including the part of the power NP_0 that scales with the number of transceivers N, P_{common}, which is common for any transceiver number, and $NMP_{rf\_circuit}$, which scales with total antenna number NM. The EE–SE relationship can be written as,

$$\eta_{EE}=C/P_{total}=\frac{\eta_{SE}}{\left(2^{\frac{\eta_{SE}}{N}}-1\right)\frac{N_0}{\eta_{PA}}\frac{N}{M}+\frac{NP_0+P_{common}+NMP_{rf\_circuit}}{W}} \tag{5.6}$$

EE–SE Relationship at the Green Points

As shown by Li et al. [8], the EE–SE relationship based on Shannon's theory is monotonic, and a higher SE will always lead to a lower EE. When the circuit power is considered, however, there is a green point on the EE–SE curve where maximum energy efficiency η^*_{EE} is achieved. Two cases are discussed here for the N by M hybrid BF structure: the case when $NM=L$ (i.e., the total antenna number is fixed to be L, but N and M are variable) and the case when N and M are independent. The analytical goal of the first case is to find the optimal number of transceivers and, correspondingly, the optimal number of antennas per transceiver given the total number of antennas. The investigation of the second case involves exploring how the independent N and M should be jointly optimized.

It can be shown based on Eq. (5.6) that there exists only one green point on the EE–SE curve for each case [10], in other words there is only one η^*_{SE} that maximizes EE performance. The relationship between η^*_{EE} and η^*_{SE} is further given as $lg(\eta^*_{EE})=-\eta^*_{SE}\,lg2/N+lg(M\eta_{PA}/N_0\ln 2)$, in other words $lg(\eta^*_{EE})$ scales with η^*_{SE} linearly with a slope of $-lg2/N$. Similar to the EE–SE relationship in classic Shannon theory, a higher η^*_{SE} will always lead to a lower η^*_{EE}. Interestingly, the relationship between η^*_{EE} and η^*_{SE} is independent of P_0, P_{common}, $P_{rf\_circuit}$ and W, though, as can be seen from Eq. (5.6), η^*_{SE} and η^*_{EE} are determined based on all the other parameters.

Assume $W=2\times10^7$ Hz, $N_0=10^{-17}$ dBm/Hz and a channel gain of −100 dB. The EE–SE relationship is depicted for two scenarios: the scenario with $N=4$ and $M=100$, and the scenario with $N=8$ and $M=50$. For each scenario, four ($P_{rf\_circuit}$, P_{common}, P_0) combinations are simulated, including:

1 W, 50 W, 1 W
5 W, 100 W, 5 W
5 W, 100 W, 2 W
1 W, 50 W, 3 W.

EE (bits/Joule) is depicted on a log-scale and SE (bits/s/Hz) is depicted on a linear scale. As shown in Figure 5.5, the EE–SE curve is different with different ($P_{rf\_circuit}$, P_{common}, P_0) combinations. On each curve, there exists only one green point. Also, the green points are actually in a straight line for both scenarios, with a large slope in the first scenario.

With this analysis, therefore, it is expected that the system operates at the green point. Also it is important that η^*_{SE} satisfies the system SE requirement. Moreover, η^*_{EE} should be high enough. These require careful design of the following parameters: P_0, P_{common}, $P_{rf\_circuit}$, W, η_{PA}, N and M. For example, when other parameters are given, N can be designed to maximize η^*_{EE}.

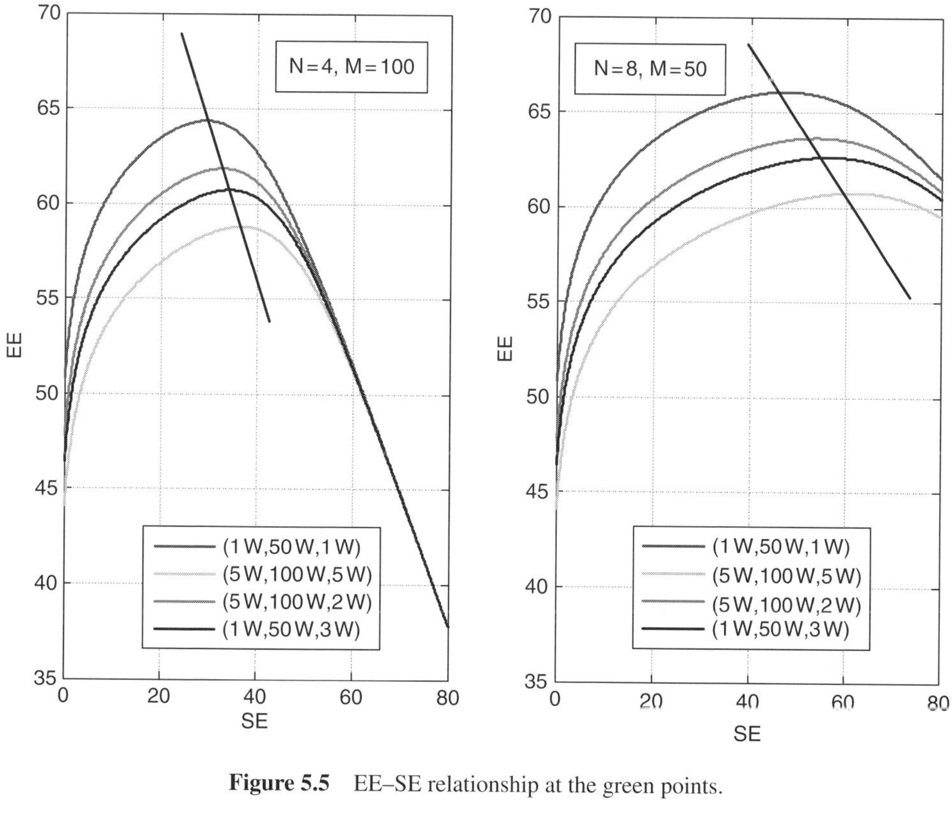

Figure 5.5 EE–SE relationship at the green points.

Impact of N on EE-SE

When the required SE is pre-determined, it is desirable that the transceiver number N be optimized, yielding the highest EE performance with the minimum transceiver number. Based on Eq. (5.6), it is found that in the cases $NM=L$ and independent N and M, for any given SE, there exists only one optimal N to yield the best EE. A detailed proof can be found in the literature [10]. The practical meaning of the existence of the optimal N is that with a given SE, a system designer does not need to implement too many transceivers to achieve the best EE performance.

Assume $P_{rf\_circuit}=1$ W, $P_{common}=50$ W, $P_0=1$ W, $W=2\times107$ Hz, $N_0=10$–17 dBm/Hz, and a channel gain of −100 dB. Considering the $NM=500$ case, the impact of N (from 1 to 10) on EE performance is shown in Subplot 1 of Figure 5.6, where five SE values are simulated. Note that since N and M are integers, there are only five valid (N, M) combinations for $NM=500$; that is, (1, 500), (2, 250), (4, 125), (5, 100), and (10, 50). It can be observed that on each curve there exists one optimal N that yields the highest EE. For example, when SE is 20 bps/Hz, the optimal N is 4. When SE is 8 bps/Hz, the optimal N is just 1. The case when N is larger than 10 is not shown in the figure, since it may be difficult to schedule users with negligible inter-user interference when M is very small.

When N and M are independent, the impact of N on EE performance is shown in Subplot 2 in Figure 5.6, where $M=50$ and the other parameters are the same as in Subplot 1. Similar to

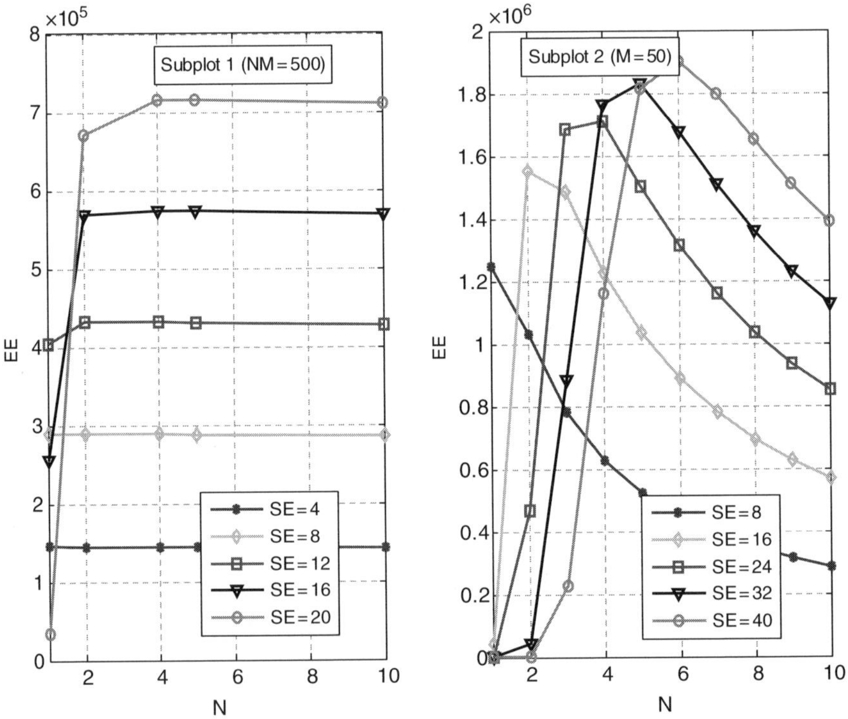

Figure 5.6 N vs EE with different SE values.

the fixed *NM* case, on each curve there exists one optimal *N*. For example, when SE is 40 bps/Hz, the optimal *N* is 6. In contrast to the fixed *NM* case, EE performance is very sensitive to *N* because the total antenna number scales with *N*.

In a practical system, the SE requirement may vary according to the traffic load and service types. For example, as shown in Subplot 2 of Figure 5.6, with the maximum 40 bps/Hz SE requirement, the optimum *N* should be designed to be 6. But when the SE requirement is reduced to 8 bps/Hz, the optimal *N* should be 1. Therefore, according to the possible SE range, the system is designed with the largest optimal *N*, and selects the best *N* according to the SE requirement via transceiver on/off. This can help to further enhance the EE performance according to the system traffic load.

5.3.2.2 NOMA with EE–SE Co-design

NOMA is one of several promising technologies that could improve the sum capacity [12–14]. It is well known that NOMA, using superposition coding [15] at the transmitter and successive interference cancellation at the receiver, achieves the capacity region of the downlink broadcast channel and outperforms orthogonal multiple access [16]. Originally, NOMA was designed to maximize system SE, with little attention given to the EE–SE co-design. In this section, an energy-efficient two-user NOMA design is presented. It is shown that there exists one optimal SE to maximize the overall EE given another user's SE requirement.

Consider a cellular system with two cells and four users (each user has one antenna). In each cell a two-user NOMA scheme with super position coding is adopted (say, user 0 and user 1 are served by a single antenna BS0 and user 1 and user 2 are served by a single antenna BS1). As shown in Figure 5.7, the large-scale channel gain of the *i*th ($i=0, 1, 2, 3$) user to BS0 are G_0, G_1, g_2, and g_3. The large-scale channel gain of the *i*th ($i=0, 1, 2, 3$) user to BS1 are g_0, g_1, G_2 and G_3. We assume $G_0 > G_1$ and $G_2 > G_3$. For each base station, the total power consumption P_{total} is modeled as

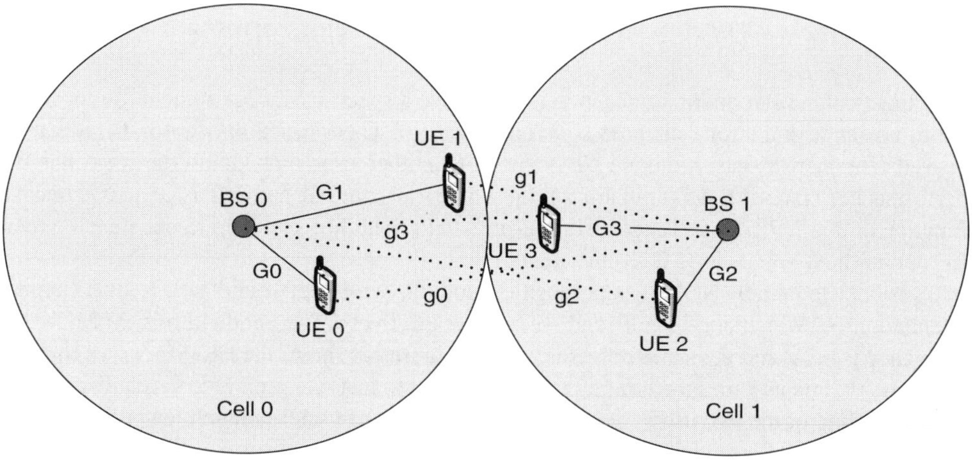

Figure 5.7 Network NOMA with two cells and four users.

$$P_{total} = \sum_{k=0}^{1} P_k + P_{static} \tag{5.7}$$

where P_k is the total power of the PA for the kth user, and P_{static} is the static power, which includes circuit power and processing power. Consider the single-cell NOMA of cell 0; the sum capacity of the two-user NOMA is expressed as

$$C_{cell0} = W\log\left(1+\frac{\eta_{PA}P_0L_0}{WN_0+\eta_{PA}L_0\left(P-P_0\right)}\right)+W\log\left(1+\frac{\eta_{PA}\left(P-P_0\right)L_1}{WN_0}\right) \tag{5.8}$$

where W is the system bandwidth, η_{PA} is PA efficiency, and N_0 is the thermal noise density. L_0 and L_1 are the channel gains of user 0 and user 1, respectively. The relationship between the total EE η_{EE} and the total SE η_{SE} of this NOMA is

$$\eta_{EE} = C_{cell0} / P_{total} = \eta_{SE} W / \left(\sum_{k=0}^{1} P_k + P_{static}\right) \tag{5.9}$$

Recall that the EE–SE relationship based on Shannon theory is monotonic, where a higher SE will always lead to a lower EE. When the circuit power is considered, however, there exists a green point on the EE–SE curve where the maximum energy efficiency is achieved. Interestingly, with a fixed P_{total} the EE–SE relationship as shown in Eq. (5.9) is linear, with a positive slope of $W/(P_0+P_1+P_{static})$. This indicates that an increase in SE will simultaneously bring an increase in EE. Any point in the EE–SE curve can be realized via proper power allocation between user 0 and user 1. The left-hand end point of the curve is achieved when all the power is allocated to user 1, while the right-hand end point is achieved when all the power is allocated to user 0. When the total power is adjustable, say via downlink power control, the slope of this curve is also adjustable. Suppose the SEs of user 0 and user 1 are η_{SE0} and η_{SE1}, respectively. Then the EE–SE relationship can be derived as

$$\eta_{EE} = \frac{\left(\eta_{SE0}+\eta_{SE1}\right)\eta_{PA}L_1W}{\left(\left(2^{\eta_{SE0}}-1\right)\eta_{PA}L_1+\left(2^{\eta_{SE1}}-1\right)+\left(2^{\eta_{SE0}}-1\right)\left(2^{\eta_{SE1}}-1\right)\eta_{PA}L_0\right)WN_0+L_1\eta_{PA}P_{static}} \tag{5.10}$$

In practical system operation, each user may have a specific SE requirement. Suppose the SE of a user with a strong channel is given (say, η_{SE0}). Then, based on Eq. (5.10), it can be proved that there exists a unique SE value for another user that yields the maximal EE performance, since the denominator of Eq. (5.10) is concave and EE is a quasi-concave function of η_{SE1} [17]. This EE optimization-oriented scheduling can help to enhance the total EE, while satisfying users' SE requirements.

Consider single-cell NOMA superposition coding, with user 0 and user 1, the channel attenuations of which are −77 dB and −97 dB, respectively. The bandwidth is 20 MHz, PA efficiency is 0.24, and the static power is 100 W. Figure 5.8 shows the EE–SE curves for user 0 and user 1, and assumes the total power is allocated to just one user. Due to the disparity of channel fading of the two users, the EE–SE performance of user 1 is much lower than that of user 0. Three solid straight lines are depicted for the EE–SE performance of the two-user NOMA scheme, with total transmit power being 5 W, 20 W, and 40 W (maximal power) respectively. Compared with single-user (user 1) transmission, two-user NOMA achieves

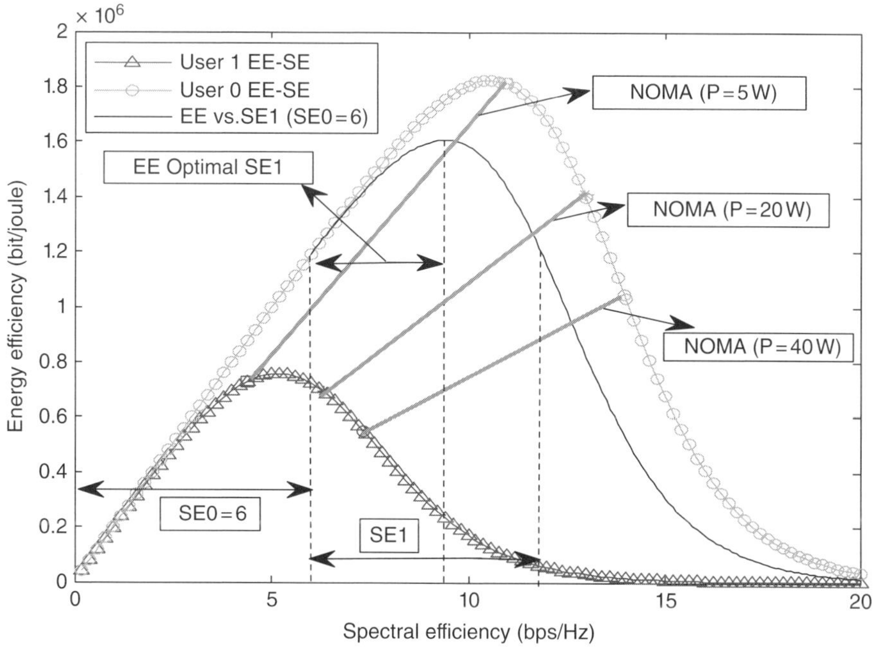

Figure 5.8 EE–SE analysis of single-cell NOMA.

much better performance. For each point on the lines, the corresponding EE–SE performance can be achieved via proper power allocation between the two users.

Given the SE of user 0 (say, 6 bps/Hz), the EE optimization can be achieved via selection of user 1s SE (SE1). As shown by the graph of total EE vs SE1, as SE1 increases from 0 (which is actually 6 bps/Hz at the x-axis) to 3.4 bps/Hz (9.4 bps/Hz at the x-axis), the EE also increases and reaches its maximal value at 3.4 bps/Hz for SE1. As SE1 continues to increase, the EE performance will degrade. Therefore, if the minimum SE requirement of user 1 is not larger than 3.4 bps/Hz, the base station can schedule user 0 and user 1 with 6 bps/Hz and 3.4 bps/Hz SE, respectively, giving optimal EE performance.

5.4 "No More Cell" for a Green and Soft Network

5.4.1 C-RAN: An Enabling Element for 5G

In the era of the mobile Internet, operators are surprised to find that their revenue is not growing in proportion to the increasing data traffic in their networks. Instead, the mobile Internet is imposing lots of challenges on them: the decoupling of revenue growth and data traffic, and low utilization efficiency of traditional telecom equipment due to large traffic variance, to name just two. In addition, it is now recognized that interference is becoming the major issue for LTE networks, which suffer much more severely than 2G and 3G systems. Unfortunately, CoMP algorithms, which are supposed to address this issue, are either too simple to achieve the desired gains or too complicated to be implemented on the current X2 interface. Finally, operators have to pay large power bills to electricity companies for power consumption from networks, especially for RANs, which use most of the energy.

A C-RAN (Centralized, Clean, Cooperative, and Cloud-RAN) system centralizes different processing resources together to form a pool, so that the resource can be managed and dynamically allocated at pool level [6]. The key enabler towards C-RAN is virtualization technology, which is widely used in modern data centers. With virtualization, standard IT servers are used as the general platform, with computation and storage as the common resources. As shown in Figure 5.9, on top of the servers run different applications in form of virtual machines (VMs). The indispensable applications in C-RAN are those to realize different RAT technologies, including 2G, 3G, 4G, and future 5G. Additional user applications such as content delivery network and web caches can also be deployed on the open virtualized platform. In addition, the C-RAN platform provides a set of standard application program interfaces, which opens the possibility of new service provision and deployment. In this way, C-RAN is no longer a single RAT processing entity but rather a platform for co-existence of diverse services.

Unlike traditional RAN systems, in which computation resources are limited within one BBU and therefore cannot be shared with other nodes, in C-RAN these resources are aggregated at a pool level and can be flexibly allocated on demand. This feature, very similar to the cloud and virtualization concepts in data centers, is called resource cloudification.

There are many advantages to C-RAN.

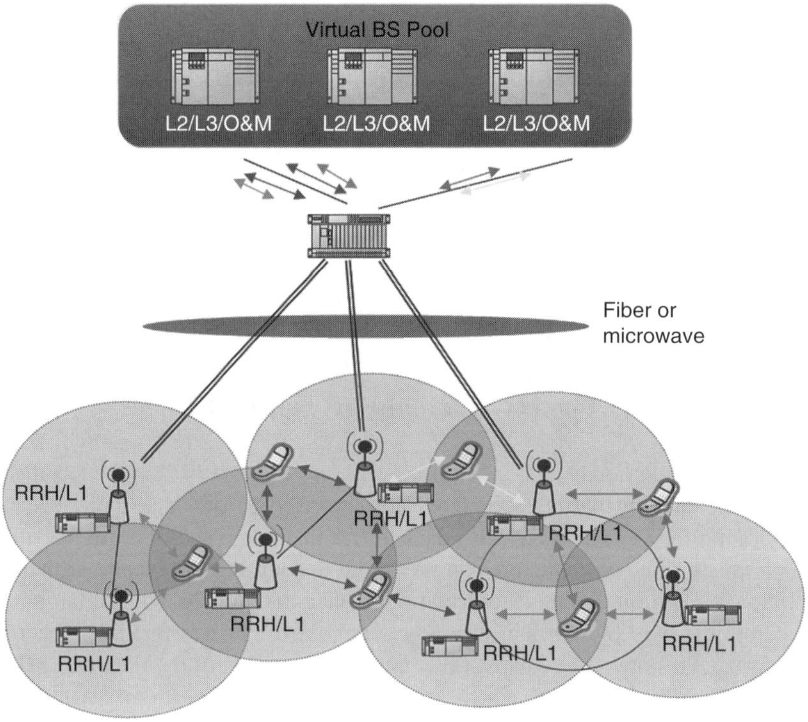

Figure 5.9 C-RAN architecture.

Total Cost of Operation Reduction. Total cost of operation (TCO) reduction mainly comes from two sources. First, centralization allows the aggregated computation nodes to share the same facilities, such as air-conditioning, which reduces the power consumption and therefore, the operating expense. Power consumption by air-conditioning usually accounts for over half of the total power consumption for operators. Second, centralization makes it easier to find a smaller number of central offices to accommodate the BBU node pool, which in turn can speed up network construction. Improved resource utilization efficiency due to resource cloudification also contributes to TCO reduction.

Improved System Performance. Internal high-bandwidth low-latency switching networks enable implementation of advanced joint processing techniques, which leads to system performance improvement.

Energy Saving. Similar to TCO reduction, energy saving is attributed to two factors: centralization and resource cloudification. First, facility sharing in the same central office helps to reduce energy consumption. Second, reduction in equipment usage due to improved resource utilization efficiency by cloudification reduces the total power consumption.

Improved Resource Utility Efficiency. This benefit is mainly from resource cloudification.

Facilitation of Service Deployment on the Edge. A C-RAN network covers a larger area and serves more users than a traditional single base station. Making use of this, it is possible to move services to or directly deploy new services on the RAN side. In this way, user experience can be improved and backhaul pressure can be relieved.

C-RAN's core feature is resource virtualization, in which processing resources can be dynamically allocated to form a soft BBU entity. Given current vendors' proprietary and closed platforms, it is advantageous to develop a new BBU platform based on virtualization technology found in modern data centers. One suitable method of network virtualization is to use network functions virtualization (NFV) that "consolidate many network equipment types onto industry standard high volume servers, switches and storage, which could be located in data centers, network nodes and in the end user premises".

Despite the simplicity of the idea of virtualization, the implementation is more difficult. Wireless communication is distinct from IT data centers in that wireless communication has extremely strict requirements for real-time processing. For example, for TDD-LTE systems it is required that an ACK/NACK must be produced and sent back to the UE/eNB within 3 ms after a frame is received. Traditional data center virtualization technologies cannot meet this requirement. Therefore, applying virtualization to base stations requires careful design and special optimization on key function blocks. In particular, some challenges are identified next.

- optimization of operating systems in order to meet the requirements of real-time mobile-signal processing, for example, to reduce the latency and virtualization overhead
- optimization of virtualization management functions to fulfill the real-time constraint: VM live migration and dynamic resource orchestration with real time signal processing
- I/O virtualization to improve the VM's I/O performance and its compatibility with live migration

- design of virtualization granularity, taking into account factors such as the correspondence between the VM and carrier and the requirements on carrier cooperation.

In the 5G vision, the network should be more user-centric from less the traditional cell-centric. Also called "no-more cell", this idea means that users should be provided with not only much higher data rate services but also that they should experience less of a difference between the cell-center and cell-edge regions. To achieve this goal, the severe interference suffered by cell-edge users should be alleviated and thus coordination among multiple cells is necessary. In 5G networks, various coordination techniques have been proposed to solve the interference problem. However, those algorithms, such as joint transmission, cannot achieve maximum performance gains under traditional architectures, such as LTE with X2 interface, which is of high latency and low bandwidth. C-RAN, on the other hand, thanks to its strong inherent central processing capability, provides an ideal structure to facilitate the implementation of coordination technologies with full or partial CSI of all users available. In fact, it has been demonstrated that C-RAN can improve CoMP performance in terms of cell-edge SE by 119% compared to non-cooperative transmission mechanisms.

C-RAN inherently supports efficiently the deployment of ultra-dense networks (UDNs), which are deemed a promising solution to the problem of absorbing highly dense user traffic. The design of UDNs involves joint consideration of many issues, including control- and user-plane decoupling, inter-site carrier aggregation and coordination, as well as interference mitigation in a heterogeneous network. In this case, C-RAN comes to play an important role, with its internal high-speed low-latency switching mechanisms and the central processing capability to implement these key technologies.

Finally, C-RAN provides a unique opportunity to support multi-RAT with the adoption of GPP and virtualization technology. In C-RAN, different RATs can be virtualized in forms of VM, and can operate separately and independently on the same platform. Thanks to highly efficient VM communication, C-RAN can also help with the multi-RAN coordination.

5.4.2 Rethink Signaling and Control for "No More Cell"

The concept of cellular systems was proposed in 1947 by two researchers from Bell Labs, Douglas H. Ring and W. Rae Young. Since the first generation of cellular standards, this cell-centric design has been maintained through every new generation of standards including 4G. The nature of a homogeneous cell-centric design is that cell planning and optimization, mobility handling, resource management, signaling and control, coverage, and signal processing are all assumed to be done either for or by each base station uniformly. Relay, distributed antenna systems (DASs), and CoMP have been implemented as short-term solutions to amend these issues. While relay and DAS are mainly used for coverage extension, CoMP aims for capacity improvement for cell-edge users. CoMP has been intensively investigated by academia, industry, and standards bodies such as 3GPP and WiMax. Inter-BS joint processing and/or coordination is sought for enhanced cell-edge and cell-average performance. Note that in the initial efforts, the cell centric network operation was hardly changed.

Network densification is viewed as a major way to meet rapid future traffic growth (potentially 1,000-fold by 2020). With the introduction of HetNets (heterogeneous networks) and UDNs, multiple layers of radio network have come into being and energy consumption, interference, and mobility issues are becoming more serious due to smaller inter-cell distances. From the network's perspective, diverse types of base stations with different coverage, transmit power, frequency bands, and so on are being introduced: macro, micro, pico, femto, and so on. From the user's perspective, traffic fluctuations are more significant than before, taking into account the emerging millions of mobile data applications. Therefore, in practical deployment, it is clear that the current system, designed for a homogeneous network, does not match with current levels of traffic variation and the diverse radio environments of today. Conventionally, radio resources are allocated semi-statically from the standpoint of the network, not that of the user equipment (UE). This causes low resource utilization and power waste. Poor performance at cell edges severely influences the consistency of user experience. User mobility makes handover happen frequently, especially in dense small-cell deployments [18]. Consequently, radio resource must be frequently reassigned, resulting in complex neighboring-cell monitoring algorithms and high signaling overhead. Moreover, frequent handover failure degrades the user experience.

Recently, in Beyond Cellular Green Generation (BCG2) [5], liquid cells, soft cells, and phantom cells have surfaced as potential radio access architectures [19]. These depart from the paradigm of cell-based coverage, resource management, and signal processing and take up the principle of NMC [20]. User demand, rather than the cell's requirements, should be the centre of network radio resource assignment and processing. All nearby radio access points, with diverse frequency bands, transmit power, and coverage, should serve one user. And the available radio resources from multiple access points should be dynamically scheduled for coordinated multipoint transmission, and the selection of control- and user-plane, and selection of uplink (UL) and downlink (DL).

The NMC concept is user-centric, with amorphous cells, decoupled signaling and data, and decoupled downlink (DL) and uplink (UL). For example, a macro BS with lower frequency and wider coverage would become a signaling BS, while small cells with higher frequency and overlapped coverage would be data-only BSs. In a HetNet scenario, the small cell is within the coverage of a macro cell. In the traditional cell paradigm, even if the small cell has no traffic, it cannot be turned off. But with a control and data decoupling scheme, the macro cell is responsible for control and the small cell only for data. Thus, when there is no data traffic in the small cell, it can be completely turned off to save energy. New users can access the macro cell, which can then coordinate with the small cell for possible data transmission. With decoupled signaling and data, mobility robustness can be improved, since handover signaling overhead is reduced – with macro signaling BS there is a more stable signaling connection – and small-cell deployment becomes much easier since careful cell planning is no longer needed. Moreover, spectrum usage in small cells will be significantly enhanced, due to the much relaxed need for control information and reference-signal transmission.

NMC is also user-centric with UL and DL decoupled. The decoupling of DL and UL is deemed to facilitate better resource allocation between cells. This can be illustrated with an example. Consider two cells: cell 1 is heavily loaded in the downlink and cell 2 overloaded in the uplink. In the traditional cell concept, if a UE device is located at the cell boundary with symmetric DL and UL data requirements, and the serving cell is cell 1, its DL requirement may not be satisfied. If the UE device's serving cell is cell 2, its UL requirement may not be satisfied.

This situation of UL/DL asymmetry is becoming even worse under HetNet deployment, since UL is always associated with DL based on best DL reference signal measurements.

However, nearby small cells with less reference signal power may provide a better UL connection. If there is a user-centric network design, the UE device's DL can be from one radio access point and its UL from another, thus meeting the UE device's requirement for asymmetric DL and UL. Moreover, CoMP with coordinated scheduling and multipoint transmission can be utilized to improve the split of the control and user planes and the UL and DL channels. For example, in dense HetNet deployments, the UE might anchor one macro/signaling base station to establish a stable signaling connection and attach to small cells for data connections. Furthermore, the UE may make UL and DL data connections from different radio access points and the DL data connection may be served dynamically by multiple radio access points.

With the introduction of UDNs in 5G, there is a great deal of overlapped coverage and the radio environment is becoming more complicated than ever. Channel information between radio access points is increasingly required to be shared in real-time and cooperation between neighboring access points is also required to enable selection of the control and user planes and the UL/DL channels from one or more points. This is not feasible in traditional RANs because too much inter-BS information sharing occurs, including dynamic user CSI and scheduling information. Fortunately, with the emergence of C-RAN, many technologies leading toward realization of the concept of NMC have been facilitated. In addition, system level and even multi-RAT optimization is also made possible.

Due to the potentially huge number of small cells and thus great cost pressures for operators, it is critical to investigate the simplification of small cells. With smaller coverage and supporting fewer users with low mobility, the time and frequency selective fading is smaller. It is then possible to reduce the overhead of broadcast signaling. Currently, the design principle of 3GPP is to apply carrier design from macro to small cells, and small-cell design tends to keep most of functionalities from macro, such as measurement, synchronization, access, handover, and so on. It is indispensable to investigate how to optimize small cell with the assistance of macro cells under the C-RAN architecture.

As shown in Figure 5.10, data-only carriers target support of the minimum control overhead without common broadcast signaling [21]. Users do not rely on a traditional broadcasting signal to implement discovery, selection, synchronization, handover, and so on. The signals and channels, therefore, need to be redesigned and also the interactions between small cells and macro cells must be enhanced.

Existing mobile networks are designed for conventional and streaming applications such as voice and video. As mobile data traffic grows exponentially, more diversified traffic profiles have emerged. They have brought new challenges to mobile networks, especially small-sized persistent bursty traffic types such as MTC and instant messaging, which contain frequent texts and photos, and periodic pings. These mobile applications would cause frequent transitions between connected and idle states. As a consequence, these transitions not only increase the battery drainage in devices, but also cause excessive signaling overhead and radio-resource fragmentation in mobile networks.

In current networks, there is only one kind of signaling/control mechanism for traffic with a variety of different profiles, in terms of rate, inter-arrival time, and tolerable delay. Signaling and control over the air is connection-oriented, resulting in relatively high overheads for bursty traffic. Each connection over the air requires several signaling and data-bearer connections, and involves more than 10 interactions. In addition, in a connection-oriented system the network needs to maintain connections during the connected state and update these connections when moving between cells.

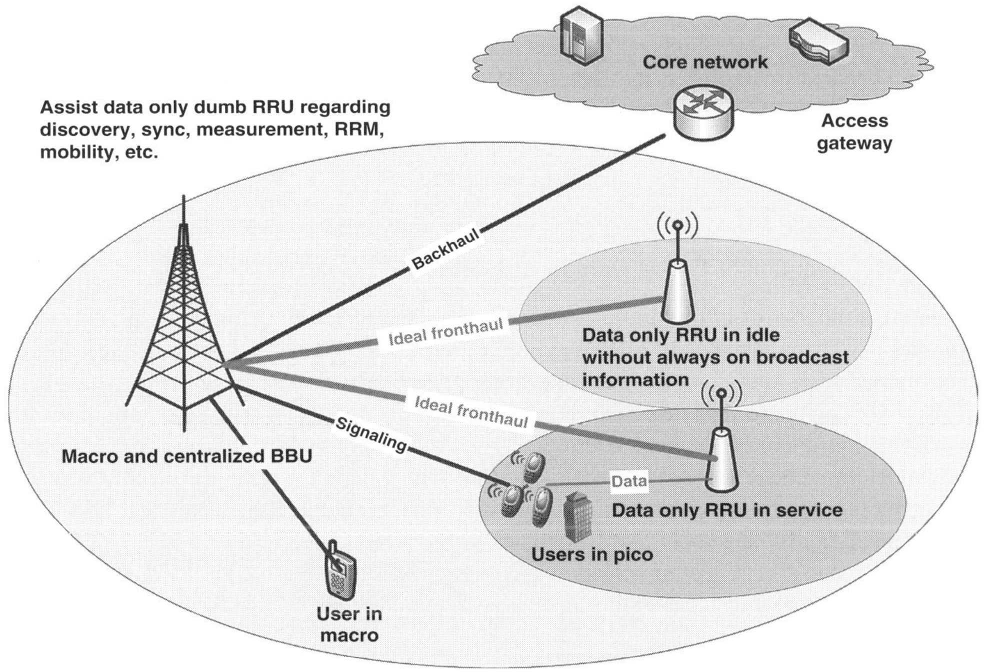

Figure 5.10 Data-only carrier in a UDN.

To fundamentally resolve the problems of excessive radio signaling overhead, the signaling and control mechanism needs to be rethought to adapted it to these diversified traffic profiles. For the 5G radio signaling and control design, significant savings in the radio connection and release might be achieved by moving away from a pure connection-oriented mechanism. Instead an adaptive signaling and control approach combining both connection-oriented and connectionless mechanisms should be devised [22]. Depending on the traffic types and network loading, an appropriate mechanism would be dynamically applied. On one hand, the current access mechanism can be simplified to piggyback small packets. On the other hand, contention-based access can be utilized, especially when the network has only a light loading and thus potential collision probabilities are under control.

5.4.3 Service Aggregator: to Accommodate Trillions of Nodes in 5G

Increasingly, driven by the widespread applications of the IoT, mobile networks are being used to gather data from or transfer data between machine-type devices. These devices include, for example, intelligent wearable devices, environmental sensors for temperature, humidity, and so on, metering devices, and vehicular sensors for navigation, safety, and traffic management. MTC has particular characteristics, such as small packet size, tolerance of transmission delay, low mobility, and low frequency of use. According to some forecasts, the number of MTC devices required in future will reach tens of billions [23].

Analogous to small-size traffic types, the massive connectivity involved in MTC may not be handled efficiently by current wireless communication networks, like 3G and 4G. The most challenging issue is not that the aggregated MTC data traffic demand is beyond the system capacity but that a large number of devices may access the network simultaneously, or even larger numbers of small packets may be generated sporadically and sent to the network independently. More importantly, it is estimated that only 7% of the total wireless traffic will be from IoT nodes in 2020. Therefore, it is important to design a new signaling mechanism that is optimized to handle the traffic profiles of the IoT, with tens of billions nodes or even trillions of nodes.

5.4.3.1 Aggregation of Packet Data from Multiple MTC Devices

There are a number of efforts in this direction. For example, various signaling optimization schemes have been discussed [24]. These aim to reduce the signaling load generated during certain procedures and scenarios, such as random access. Also, considering the fact that many MTC applications are delay-tolerant, low-access priority indication has been introduced to alleviate the problem of networks overloaded with MTC devices [25]. Elsewhere [26], a new network convergence approach has been introduced, in which the converged cellular network and wireless sensor network architecture evolves from layered to flat to decrease the hierarchical signaling exchange between the two networks.

One interesting MTC communication mode is shown in Figure 5.11. MTC devices are controlled by an aggregator, which functions as a wireless gateway to the cellular network.

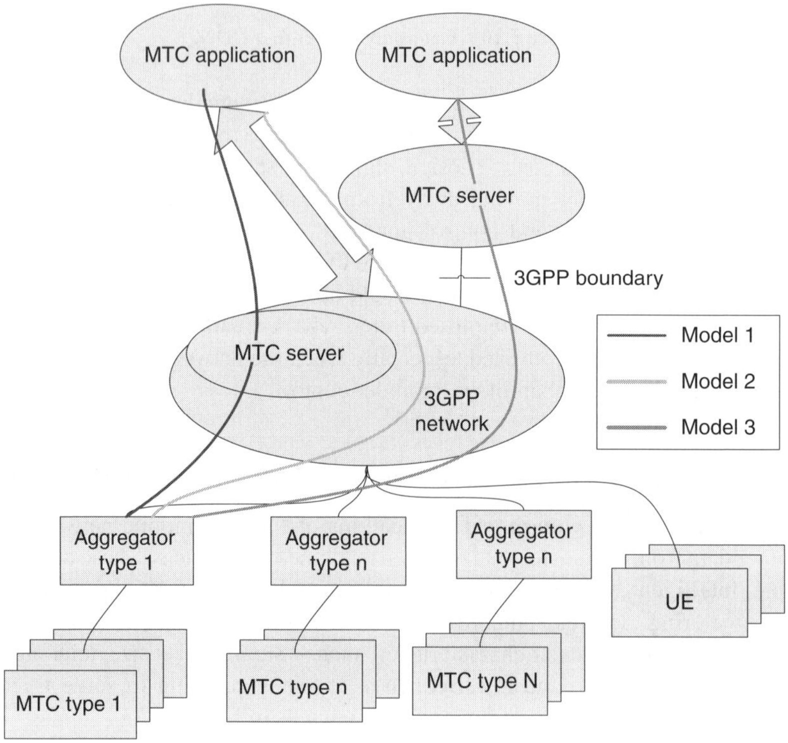

Figure 5.11 Aggregation of packet data from multiple MTC devices.

The services aggregated can be homogenous or heterogeneous, and sent to the aggregator in a scheduled manner or sporadically. The aggregator will then relay the data packets to the cellular network in an aggregated manner. Correspondingly, based on the aggregated packet types and data relay modes, the aggregators will request for an aggregator-specific RRC mechanism. The signaling overhead involved is significantly reduced and the RRC signaling efficiency is dramatically improved. Note that the service aggregation of MTC devices is analogous to the service aggregation of small-sized packets from applications running on multiple mobile devices.

Multiple MTC devices are able to connect to each specific aggregator via local transmission technologies, which could be cable, fiber, WiFi, Bluetooth, ZigBee, and even 3GPP cellular access. It is assumed that the connection between MTC devices and aggregators does not have a significant influence on the operator's mobile network and thus the impact of MTC device access to aggregators is not considered here.

5.4.3.2 Two Relay Modes of the Aggregators

Two relay modes of the aggregators are now described.

Mode 1: No RRC State Transition (Always Connected). Suppose in the network – say within one base station – there are N aggregators with the same MTC traffic and relay behavior (mode 1) on orthogonal resources, say on different frequency bands, but having the same bandwidth. They are simultaneously transmitting to the network in a frequency division multiple access (FDMA) manner. Before relaying the aggregated k packets in some time window T, the aggregator is already in connected mode. After packet transmission, the connected mode will be maintained; there is no need for RRC connection setup in the following windows.

Mode 2: RRC Connection Mode Switched to Idle Mode within Window *T*. In mode 2, we assume there are N simultaneous aggregators that equally share the time resource in each window T. The base station allocates all frequency resources to each aggregator; in other words, in a time division multiple access (TDMA) mode. Before transmitting the aggregated packets to the network, each aggregator will enter into connected mode in the time window T. The k packets will be transmitted to the cellular network in window T_{tx} (which is smaller than T) within each window T. Then, after T_{tx}, the aggregator enters into RRC idle mode directly.

For these two relay modes, same traffic is relayed in the same time window T. However, the signaling overhead of these two modes can be different. An analysis can be found in the literature [27].

5.5 Summary

Green and *soft* have been recognized as key features of future 5G wireless systems. In this chapter, the design of green and soft 5G networks has been described. The fundamental differences between classic Shannon theory and practical systems were first identified and then harmonized into a framework for EE–SE co-design. Some promising 5G technologies were re-visited based on EE–SE co-design. The characteristics of the NMC concept were described from the perspective of the infrastructure and architecture variations, with particular emphasis on C-RAN as a typical realization that will enable various soft technologies. Rethinking signaling/control for MTC communication was then explored and initial redesign mechanisms were discussed. Finally, the concept of a service aggregator was presented. This combines diversified traffic from multiple wireless devices and relays the aggregated traffic to the base station efficiently, significantly reducing the signaling overhead. Energy saving has been widely recognized as an urgent issue worldwide. 5G networks, with *green* and *soft* as key design methodologies, are expected to take on the social responsibility of energy saving, while satisfying the ever-increasing mobile traffic requirements.

Acknowledgments

The authors would like to express their gratitude to the team members in the Green Communication Research Center of the China Mobile Research Institute for their contributions to the preparation of this chapter. In particular, Dr. Zhikun Xu established a solid framework for the EE–SE co-design, and Mr. Gang Li devised the "data only carrier" concept in UDNs. Our sincere gratitude also goes to Drs Sen Wang, Qi Sun, and Zhengang Pan for their efforts in the manuscript preparation.

References

[1] MIIT, webpage. URL: www.miit.gov.cn (accessed August 3, 2010).
[2] TC Group (2008) Smart 2020: Enabling the Low Carbon Economy in the Information Age.
[3] Mobile VCE, webpage. URL: www.mobilevce.com/.
[4] P. Skillermark and P. Frenger (2012) Enhancing energy efficiency in LTE with antenna muting. *IEEE VTC Spring '12*, pp. 1–9.
[5] Y. Wu, et al. (2014) Green transmission technologies for balancing the energy efficiency and spectrum efficiency trade-off. *IEEE Communication Magazine*, **52** (11), 112–120.
[6] CMRI (2011) C-RAN: The Road Towards Green RAN. URL: http://labs.chinamobile.com/cran (accessed August 3, 2016).
[7] M. Chiosi, D. Clarke, and P. Willis (2012) Network Functions Virtualization. White paper, available at https://portal.etsi.org/Portals/0/TBpages/NFV/Docs/NFV_White_Paper3.pdf (accessed August 3, 2016).
[8] G.Y. Li., Z. Xu, C. Xiong, et al. (2011) Energy-efficient wireless communications: tutorial, survey, and open issues, *IEEE Wireless Communications*, **18** (6), 28–35.
[9] X. Huang, Y. Jay Guo, and J. Bunton (2010), A hybrid adaptive antenna array. *IEEE Transactions on Wireless Communications*, **9** (5), 1770–1779.
[10] S. Han, C.L. I, C. Rowell, et al. (2014) Large scale antenna system with hybrid digital and analog beamforming structure. *ICC Workshop 2014*, 11. pp. 842–847.
[11] S. Han, C. I, Z. Xu, et al. (2014) Reference signals design for hybrid analog and digital beamforming. *IEEE Communications Letters*, **18** (7), 1191–1193.
[12] Y. Saito, Y. Kishiyama, A. Benjebbour, et al. (2013) Non-orthogonal multiple access (NOMA) for cellular future radio access, *VTC 2013*, pp. 1–5.
[13] B. Kim, S. Lim, H. Kim, et al. (2013) Non-orthogonal multiple access in a downlink multiuser beamforming system. *MILCOM 2013*, pp. 1278–1283.
[14] J. Choi (2014) Non-orthogonal multiple access in downlink coordinated two-point systems. *IEEE Communications Letters*, **18** (2), pp. 313–316.
[15] S. Vanka, S. Srinivasa, Z. Gong, et al. (2012) Superposition coding strategies: design and experimental evaluation. *IEEE Transactions on Wireless Communications*, **11** (7), 2628–2639.
[16] D. Tse and P. Viswanath (2005) *Fundamentals of Wireless Communication*, Cambridge University Press.

[17] J. Mao, G. Xie, J. Gao, et al. (2013) Energy efficiency optimization for cognitive radio MIMO broadcast channels. *IEEE Communications Letters*, **17** (2), 337–340.
[18] T. Nakamura, S. Nagata, A. Benjebbour, et al. (2013) Trends in small cell enhancements in LTE advanced. *IEEE Communications Magazine*, **51** (2), 98–105.
[19] Y. Kishiyama, A. Benjebbour, T. Nakamura, and H. Ishii (2013) Future steps of LTE-A: evolution toward integration of local area and wide area systems. *IEEE Wireless Communications*, **20** (1), 12–18.
[20] C. I, C. Rowell, S. Han, et al. (2014) Toward green and soft: a 5G perspective. *IEEE Communications Magazine*, **52** (2), 66–73.
[21] Y. Chen, G. Li, and C. Cui (2014) Macro assisted ultra-lean data carrier and architectural design. *ZTE Technology Journal*, **2**, 17–21.
[22] Y. Chen, G. Li, Z. Pan, and C. I (2014) Small data optimized radio access network signaling/control design, *ICC WS, 2014*, pp. 49–54.
[23] Cisco (2016) White paper: Cisco visual networking index: global mobile data traffic forecast update, 2015–2020. URL: www.cisco.com/c/en/us/solutions/collateral/service-provider/visual-networking-index-vni/vni-forecast-qa.pdf.
[24] T. Taleb and A. Kunz (2012) Machine type communications in 3GPP networks: potential, challenges, and solutions. *IEEE Communications Magazine*, **50**,178–184.
[25] 3GPP (2011) TS 37.868 v11.0.0: Study on RAN improvements for machine-type communications, Release 11.
[26] J. Zhang, L. Shan, H. Hu, and Y. Yang (2012), Mobile cellular networks and wireless sensor networks: toward convergence. *IEEE Communication Magazine*, **50** (3), 164–169.
[27] C. I, S. Han, Y. Chen, and G. Li (2014), Trillions of nodes for 5G!?, *2014 IEEE/CIC International Conference on Communications in China (ICCC)*, pp. 246–250.

6

Proactive Caching in 5G Small Cell Networks

Ejder Baştuğ,[1] Mehdi Bennis,[2] and Mérouane Debbah[1]
[1] *CentraleSupélec, France*
[2] *Centre for Wireless Communications, University of Oulu, Finland*

6.1 Small Cell Networks: Past, Present, and Future Trends

Smartphones have exponentially increased the traffic load in current cellular networks, a trend that is showing no signs of slowing down [1, 2]. It is now well understood that a very effective way to increase network capacity is making cells smaller, reducing the distance to the users [3]. Indeed, cell densification has gone from the order of hundreds of square kilometers back in the 1980s to a fraction of a square meter or less with the advent of hotspots. There has recently been a great interest in deploying relays, distributed antennas, and small cellular access points, such as micro, pico, and femto cells, in residential homes, subways, enterprises, and hotspot areas. These network architectures, which are either operator-deployed or user-deployed, are referred to heterogeneous networks (HetNets) or small cell networks (SCNs) [3, 4]. By deploying additional network nodes within local-area range and making the network closer to end-users, small cells can significantly improve spatial reuse and coverage, boost capacity, and offload traffic more efficiently [4].

There is a comprehensive literature on the topic of HetNets and SCNs, tackling aspects such as interference management, cell association, stochastic network modeling, intercell interference coordination, energy-efficiency, self-organizing networks, mobility management, and LTE/WiFi interworking; see the paper by Andrews et al. for a comprehensive survey [4].

One of the key take-away points from these studies is that tight interference coordination among the macro and femto/picocell tiers is necessary for achieving cell splitting gains. This hinges on the availability of low-latency and high-capacity backhauls [5]. Network modeling approaches based on stochastic geometric tools have shown performance gains reasonably

Towards 5G: Applications, Requirements and Candidate Technologies, First Edition.
Edited by Rath Vannithamby and Shilpa Talwar.

Companion website: www.wiley.com/go/vannithamby/towards5g

close to reality in terms of system-wide and per-user capacities. Unlike time-consuming system-level simulations, fundamental insights can be gleaned from these tools, some of which have been corroborated by industry field trials and observations from detailed simulations [6].

In parallel, mobility management has received significant attention from the wireless industry, research community, and standardization bodies [7]. In conventional homogeneous networks, user terminals (UTs) typically use the same set of handover parameters – hysteresis margin, time-to-trigger (TTT), and so on. In HetNet, using the same set of handover parameters for all cells and/or for all UTs may degrade mobility performance. This is because high-mobility macro UTs may go deep inside coverage areas of small cells before the TTT optimized for macrocells expires, thus causing handover failure due to a degraded signal-to-interference-plus-noise ratio (SINR) [8].

Decentralized interference management/mitigation strategies in co-channel interference scenarios have also been studied in detail. In these, small cells are able to self-organize and optimize their transmission strategies regarding power and frequency, based on local information and with minimum information exchange [9]. This leads to a number of tradeoffs in terms of faster/slower convergence at the cost of partial/full information.

Carrier aggregation and its single/multiflow enhancements have also been investigated as a means of boosting network capacity and per-user throughput. Here, users may be served on several bands simultaneously [10]. Furthermore, with the increasing traffic asymmetry in the uplink (UL) and downlink (DL), novel cell-association mechanisms and architectures are needed to cope with new types of inter-node interferences (DL-to-UL), thereby opening new avenues for research such as flexible DL/UL communication, massive multiple-input, multiple-output (MIMO), device-to-device (D2D), full-duplexing, and so on [3, 11].

Finally, the topic of LTE and WiFi coexistence has received tremendous attention due to the multi-mode capability of small base stations (SBSs)[1] and the possibility of using both licensed and unlicensed bands. Therein, dynamic load-balancing and traffic-steering mechanisms have been proposed, leveraging the availability of WiFi for best-effort services, traffic load, delay tolerance, and so on [12].

While small cell densification is clearly the way to go, a number of technical challenges remain unsolved. Indeed, while small cell densification has been shown to boost capacity, simply adding small cells may turn out to be energy-inefficient [13]. In addition, backhaul optimization and determining the optimal location of small cells represent the main limiting factors before a full rollout of small cells takes place. The importance of the backhaul is further underscored by the unabated proliferation of smartphones and the vast array of new wireless services, such as multimedia streaming and web-browsing applications. As a result, novel approaches to backhaul-aware small cell networking have recently been proposed in the literature [14]. Examples include:

- identifying how to optimally decouple control and data planes to make cells more adaptive to traffic dynamics and network state while having a global view of the network
- backhaul offloading via smart edge caching [15–17]
- cloud radio access network (C-RAN) [18]
- software defined networking (SDN) [19]

[1] The term "SBS" will be used interchangeably with "small cell" in this work.

- resource/network virtualization
- ultra-dense networks
- massive MIMO.

Among these approaches, in this chapter we focus on proactive edge caching as a way of dealing with backhaul offloading in SCNs, which is especially important in dense deployments.

The rest of this chapter is organized as follows. We give an introduction of cache-enabled proactive SCNs in Section 6.2. Our system model and the corresponding problem formulation is presented in Section 6.3. The details of proactive caching at the SBSs and UTs are given in Sections 6.4 and 6.5, respectively, and discussions of numerical results are carried out in the same sections. The current directions of caching in wireless networks and relevant research are discussed in Section 6.6. Finally, Section 6.7 draws some conclusions and outlines future work.

6.2 Cache-enabled Proactive Small Cell Networks

Most of the existing studies on SCNs are based on the classical networking paradigm, known as *reactive*, in which users' content requests are served immediately, yielding outages otherwise. In such a situation, sustaining peak traffic demands in these networks requires expensive high-speed backhaul, resulting in tremendous operational expenditures. Given the fact that such a cost may not be affordable, a novel networking paradigm is clearly needed for densely deployed SCNs. This can be done by exploiting recent advances in storage, context-awareness, social networking, and D2D [17].

This novel network paradigm is proactive, in the sense that the nodes at the edge of the network – the SBSs and UTs – predict users' context information and cache their strategic contents accordingly, in order offload the backhaul and satisfy users' quality-of-service (QoS) demands. This goes beyond the scope of traditional cellular networks, which have been designed assuming *dumb* UTs with limited storage and processing capabilities. Nowadays, UTs are much more sophisticated, giving an opportunity to exploit their capabilities in conjunction with SCNs by storing the predicted content at the network edge. This in turn yields significant gains in terms of network resources, minimizing operational and capital expenditures [3].

Because a huge amount of user information is often available and because human behaviour has a certain predictability [20], users' future events can be inferred. Therefore, in this chapter, we explore a proactive caching framework that leverages context-awareness and storage capabilities at the edge of the network in order to sustain peak data demands and offload the backhaul. More precisely, estimates of users' future demands and the popularity of particular content can be used to identify content that should be proactively stored before actual requests to deliver it take place. In addition, whenever D2D communication is available, the proactive caching approach exploits users' physical proximity, their social relationships (and their influence within the social community), as well as their storage, for content dissemination.

As stated before, recent results have shown that the human behaviour is correlated and predictable to a large extent [20]. Therefore, SBSs are assumed to be equipped with storage units and low-speed backhaul is used for their broadband connections. Then, as will be shown, proactively caching users' content at SBSs alleviates the backhaul load and brings higher user satisfaction. The proactive caching procedure is based on the idea of storing popular content at the SBSs. To achieve this, the popularity of the content has to be estimated. Using tools from machine learning and analyzing the infrastructure logs [21]), a trove of hidden information about users' behaviour can be revealed. Analyzing this information falls into the big data phenomenon, in which collaborative filtering (CF) methods can be successfully applied for inference.

Yet another approach for storing content at the edge of the network is via caching at the users' devices, and leveraging D2D communications for content dissemination. Online social networks, such as Facebook, Twitter, and Digg, have become instrumental in disseminating content across social communities [2]. Typically, users tend to highly value content recommended by their friends or people with similar interests. Thus, exploiting users' social relationships and proactively storing content in their devices can alleviate peak traffic demand. Notably, the strategic content storeed in the caches of popular/influential users can ease backhaul congestion and yield considerable network savings. In order to show such network savings, we first detail our system model in the following section.

6.3 System Model

Let us consider a scenario of M SBSs $\mathcal{M} = \{1,...,M\}$ and N UTs $\mathcal{N} = \{1,...,N\}$. The broadband connection of every SBS $m \in \mathcal{M}$ is provided by a central scheduler via a limited backhaul link with capacity c_m.[2] We suppose that the capacity of the wireless small cell link between SBS m and UT n is given by $c_{m,n}$. Depending on the content availability and users' proximity, the SBSs can establish D2D communications between users n and n'; the corresponding D2D link capacity is denoted by $\ddot{c}_{n,n'}$. This scenario is illustrated in Figure 6.1. Suppose that user n requests some content from a library of F files, represented by $\mathcal{F} = \{1,...,F\}$, according to probabilities $\mathcal{P}_n = \{p_{n,1},...,p_{n,F}\}$. In this library, the length of the contents are $\mathcal{L} = \{l_1,...,l_F\}$ and the bitrates are given by the set of $\mathcal{B} = \{b_1,...,b_F\}$. Now, suppose that a number R of content

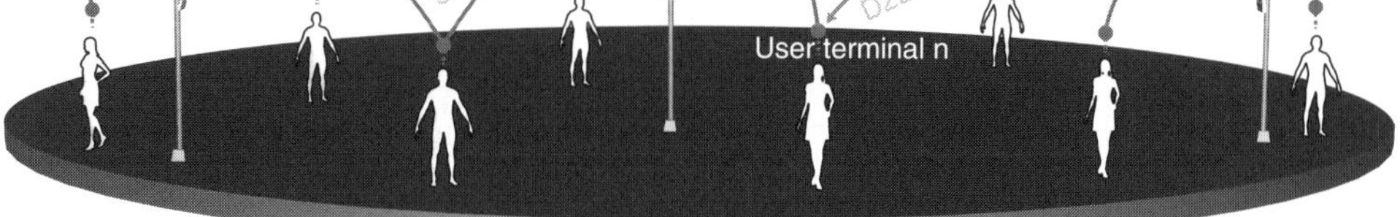

Figure 6.1 A sketch of the scenario in the system model.

[2] This controller is typically a network entity located at the evolved packet core (EPC) or at the network edge (small cell gateway).

requests are drawn by users randomly during T time slots. Then, we say that a request $r \in \mathcal{R} = \{1,...,R\}$ is satisfied if the rate of delivery is equal or greater than the bitrate of the requested content as follows:

$$\frac{l_r}{t'_r - t_r} \geq b_r \tag{6.1}$$

where $l_r \in \mathcal{L}$ represent the length of the requested content, t_r (t'_r) is the start (end) time of the delivery, and $b_r \in \mathcal{B}$ is the bitrate of the content $f_r \in \mathcal{F}$. Given this definition, the *satisfaction ratio* can be expressed as:

$$\eta(\mathcal{R}) = \frac{1}{R}\sum_{r\in\mathcal{R}} \mathbb{1}\left\{\frac{l_r}{t'_r - t_r} \geq b_r\right\} \tag{6.2}$$

where $\mathbb{1}\{\ldots\}$ is the indicator function, which yields 1 when the condition holds and 0 otherwise.

Our target as network operator is to keep the satisfaction ratio above a given threshold, while minimizing the use of the backhaul. As stated before, this can be done via proactive caching in the SBSs and UTs, an approach that we detail for two case studies separately in the following sections.

6.4 Proactive Caching at Base Stations

Research has shown that the backhaul constitutes one of the most important challenges for SCN deployments and that this problem is going to increase dramatically due to the densely deployed SBSs expected in the future. Suppose that the total capacity of the backhaul is lower than the available wireless link capacity between SBSs and UTs: $\sum_{m\in\mathcal{M}} c_m \ll \sum_{m\in\mathcal{M}}\sum_{n\in\mathcal{N}} c_{m,n}$. Since in this case we suppose that the backhaul is the bottleneck, one reasonable option is to avoid its use by storing the users' content proactively at the SBSs, during off-peak hours. In other words, if the users' content can be stored at SBSs before the users' actual demands for content arrives, the backhaul may not be used some of the time, depending on on how smartly the content is placed.

Let us consider that the rate of the backhaul link during the content delivery for request r at time t is $\lambda_r(t)$. Then, the backhaul load under given these definitions can be expressed as follows:

$$\rho(\mathcal{R}) = \frac{1}{R}\sum_{r\in\mathcal{R}} \frac{1}{l_r}\sum_{t=t_r}^{t=t'_r} \lambda_r(t) \tag{6.3}$$

Additionally, suppose that the storage capacity of SBS m is given by s_m and the amount of its consumption at time t is denoted by $\kappa_m(t)$. The backhaul minimization problem subject to the link capacities, storage, and QoS constraints, can then be formulated as follows:

$$\begin{aligned}
\underset{t'_r, r\in\mathcal{R}}{\text{minimize}} \quad & \rho(\mathcal{R}) \\
\text{subject to} \quad & \lambda_r(t) \leq c_m, && \forall m \in \mathcal{M}, \\
& \kappa_m(t) \leq s_m, && \forall m \in \mathcal{M}, \\
& \eta(\mathcal{R}) \geq \eta_{min}, && \forall r \in \mathcal{R},
\end{aligned} \tag{6.4}$$

where η_{min} is the target satisfaction ratio. Since dealing with Eq. (6.4) is computationally intractable, a heuristic approach similar to the one outlined by Baştuğ et al. [22] can be used, storing popular content in the cache of SBSs. Before such a caching procedure is applied, we suppose that each SBS m has to track, learn, and build its user' content profile to infer the future demands. Assume that $\mathbf{P}_m$ are the discrete content probabilities of users in SBS m; we refer to this as a popularity matrix, each row representing the users and the columns the content popularities/ratings. A perfectly known $\mathbf{P}_m$ would allow us to store the content according to this caching procedure. Unfortunately, this situation is not the case in practice because the matrix is not perfectly known, and in fact is large and indeed sparse. Given these observations and inspired from the Netflix paradigm [23], supervised machine learning tools can be used to exploit user–content correlations. Inferring the probability that user n requests content f (namely estimating the popularity matrix), and storing the predicted content accordingly can clearly offload the backhaul.

$$\min_{\{b_n, b_f\}} \sum_{n,f} \left(r_{nf} - \hat{r}_{nf}\right)^2 + \lambda \left(\sum_n b_n^2 + \sum_f b_f^2 \right) \tag{6.5}$$

where the sum is over the (n,f) user–content pairs in the training set, containing how user n rated content f (that is, r_{nf}). The total number of users in the training set is N and F is the total number of contents, thus, the minimization is done over all the $N + F$ parameters. In this formulation, $\hat{r}_{nf} = \bar{r} + b_n + b_f$ is the baseline estimator, in which b_f is the relative quality of each content f compared to the average $\bar{r}$. The bias of each user n relative to b_n is given by $\bar{r}$. Additionally, the parameter λ is used for balancing the regularization and fitting the training data.

In the numerical setup, we use the regularized singular value decomposition (SVD) due to its numerical accuracy (see ref. [24] for a comprehensive study of CF methods including regularized SVD). Roughly speaking, since the entries of $\mathbf{P}_m$ are not fully known, the model construction is done via gradient descent using the least-squares property of the SVD. Thus, $\hat{\mathbf{P}}_m$ is constructed as the low-rank version of $\mathbf{P}_m$.

So far, we have described the first step. In the last step (namely, the placement step of the caching procedure), the content is cached proactively by storing the most popular content based on the estimation of $\hat{\mathbf{P}}_m$, until the storage capacity is full. In the following, we show the gains of proactive caching in a numerical setup and discuss the impact of several parameters of interest. A sketch of the proactive caching procedure at the BSs is shown in Figure 6.2.

6.4.1 Numerical Results and Discussions

The list of parameters used in the numerical study is provided in Table 6.1. In order to see the impact of the parameters of interest, the length and bitrate of the content, wireless small cell links and storage capacities are set to identical values. We consider three regimes of interest: low load, medium load, and high load.

In the numerical study, R requests are drawn over a time duration T; the arrival times of these requests are sampled uniformly at random. The users' content requests are drawn from the ZipF(α) distribution. Given that knowledge, at $t = 0$, the perfect popularity matrix $\mathbf{P}_m$ is constructed for each SBS m. Removing 20% of the entries of this matrix uniformly at random, the remaining entries are used for the model construction in CF. The prediction of missing

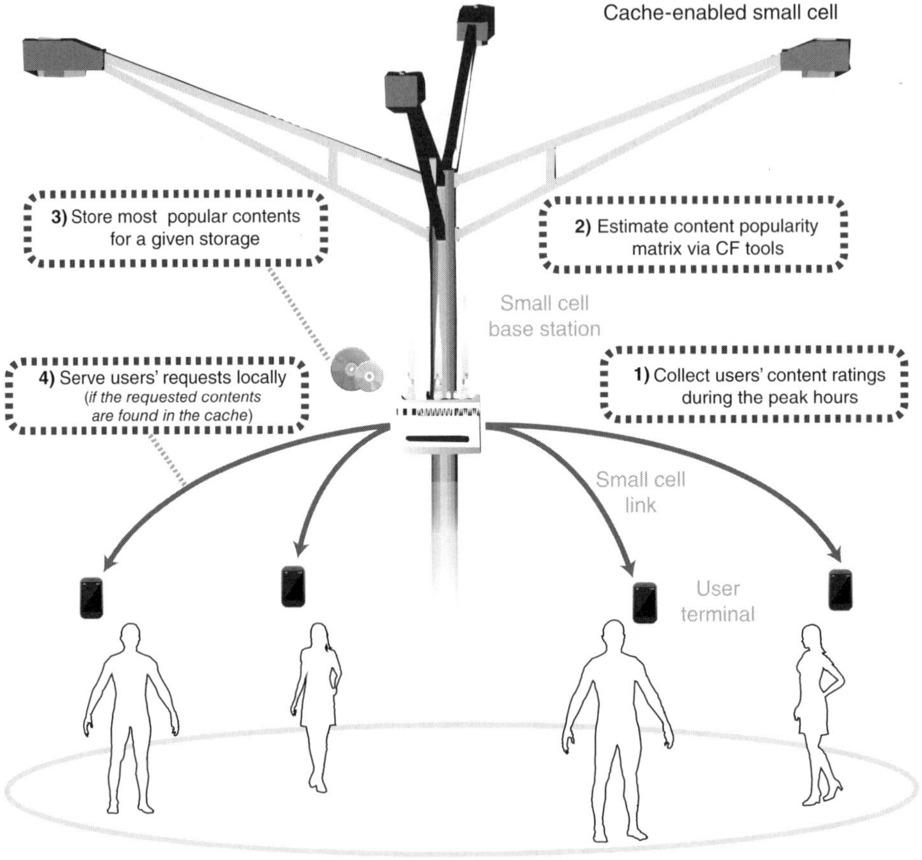

Figure 6.2 A practical procedure for proactive caching at the base stations.

Table 6.1 The numerical setup parameters for proactive caching at the SBSs.

Parameter	Description	Value
T	Time slots	1,024 s
M	Number of SBSs	4
N	Number of UTs	32
F	Number of contents	128
l_f	Length of content f	1 Mbit
b_f	Bitrate of content f	1 Mbit/s
$\sum_m c_m$	Total backhaul link capacity	2 Mbit/s
$\sum_m \sum_n c_{m,n}$	Total wireless small cell link capacity	64 Mbit/s
R	Number of requests	0 : 2,048
S	Total cache size	$0 \sim l_f \times F$
α	ZipF parameter	0 : 2

entries is then carried out by regularized SVD [25]. Once the popularity matrix is estimated, proactive caching is applied by greedily storing the most popular content subject to the constraint of the storage size of the SBS. In the numerical setup, afer completing the training and placement steps of the proactive procedure at $t = 0$, the users are served depending on their request arrival time until all content delivery processes finish. We use random caching as a baseline, referring to this as "reactive" caching.

In order to compare the benefits of caching both for the proactive and reactive cases, three parameters of interest are detailed:

- number of requests R
- total cache size S
- ZipF distribution parameter α.

The gains in the plots are normalized for ease of understanding. The evolution of the satisfaction ratios and the backhaul loads with respect to the variation of these parameters are given in Figure 6.3.

In the figure, we see that the satisfaction ratio decreases as the number of user content requests increases. The reason for this is somewhat obvious, as the capacity constraint starts to be a limiting factor for delivery of a large number of requests. Concerning the backhaul load in a very small number of requests, the reactive approach generates less load than the proactive case, something that can be explained by the "cold start" phenomenon of the CF used in the proactive case. However, as the number of requests increases, the amount of information given to the CF for the training step increases. Therefore, in the end, the proactive approach with a sufficient amount of information outperforms the reactive approach, with an almost constant gain.

One important parameter of interest in our scenario is the total storage size of the SBSs. As we increase the quantity of storage, the SBSs gains more capability to store the content from the catalog, yielding satisfaction ratios up to 1 and backhaul load as low as 0 at extreme values of the storage size. Looking at more realitic situations in which the storage size is somewhere between 0 and 1, we see that the proactive approach outperforms the reactive case in terms of the satisfaction ratio as well as the backhaul load.

The content popularity parameter α indeed has an impact on the performance metrics. At low values of α, where the distribution exhibits uniform behaviour, the proactive approach outperforms the reactive case but with a relatively low difference. However, as α increases, some content becomes more popular than the rest of the content in the catalog. Thus, the difference between the gain of proactive and reactive approaches become quite visible in terms of the satisfaction ratio and the backhaul load.

6.5 Proactive Caching at User Terminals

Another means of offloading the traffic at SBSs, and thus offloading the backhaul, can be achieved by caching user content at the UTs and exploiting D2D communications for content dissemination. For this purpose, the interplay between users' social ties and physical proximity can be taken into account for proactive caching decisions. In particular, when a content request arrives at the network, the SBS can take advantage of those influential users who already have

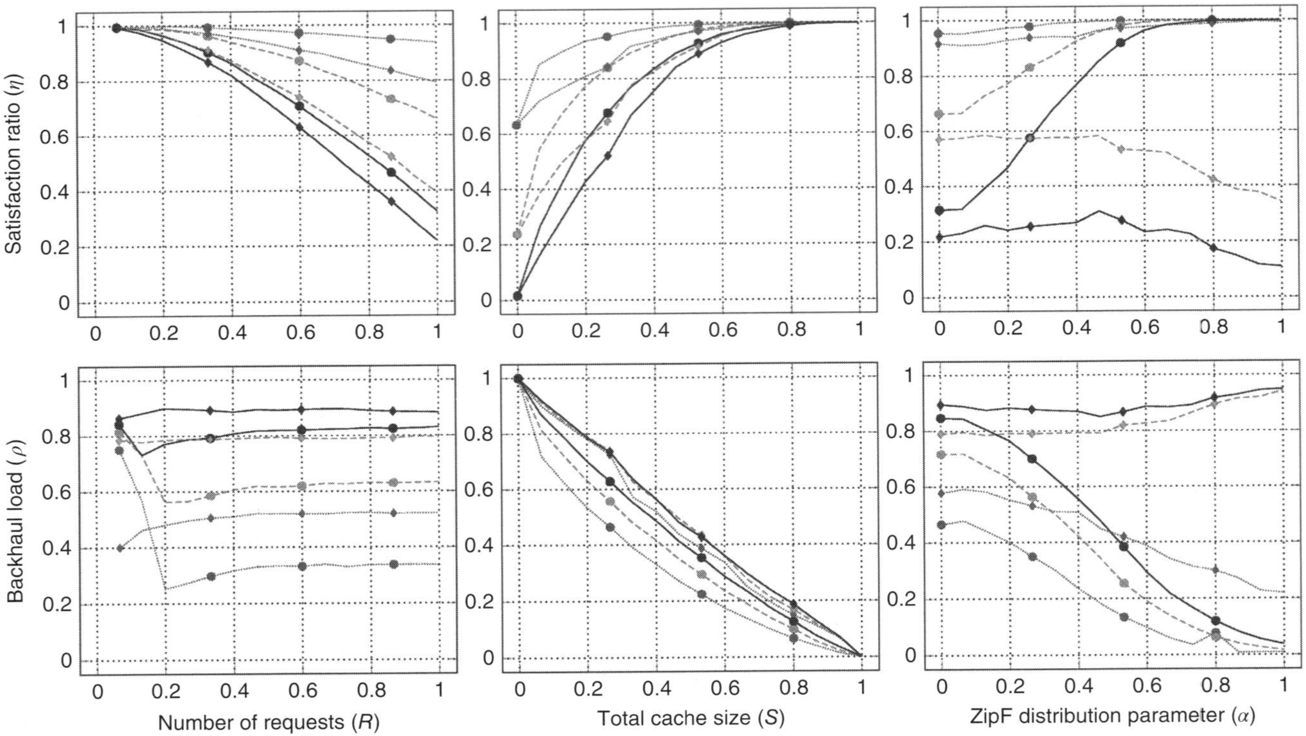

	Low load	Medium load	High load
S	0.4	0.2	0.1
α	0.3	0.2	0.1

	Low load	Medium load	High load
R	0.5	0.8	0.98
α	0.3	0.2	0.1

	Low load	Medium load	High load
R	0.5	0.8	0.98
S	0.4	0.2	0.1

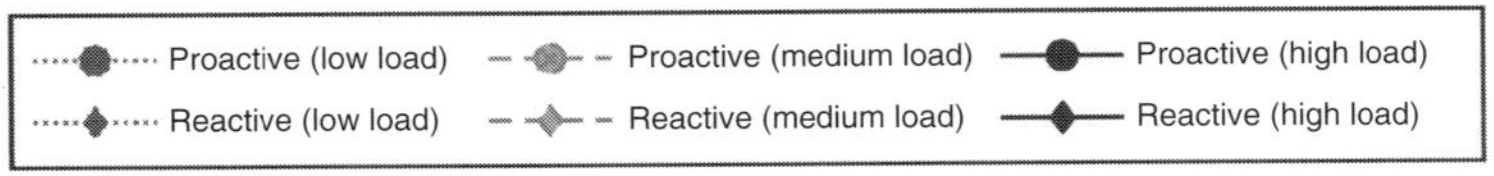

Figure 6.3 Backhaul offloading via proactive caching: Dynamics of the satisfied requests and backhaul load with respect to the number of requests, total cache size and ZipF parameter.

the content, requesting them to join the content delivery effort via D2D. If such an opportunity does not exist and the requested content is not available, as a last resort, the content can be delivered by the SBS but at the cost of using the backhaul.

Let us consider that the storage capacity of UT n is $\ddot{s}_n$ and its usage at time t is given by $\ddot{\kappa}(t)$. Also suppose that $\dot{\lambda}_r(t)$ is the total downlink rate of the SBSs during the content delivery of request r at time t, and the D2D link rate is $\ddot{\lambda}_r(t)$. Then, the small cell load can be expressed as follows:

$$\ddot{\rho}(\mathcal{R}) = \frac{1}{R}\sum_{r\in\mathcal{R}}\sum_{t=t_r}^{t=t_r'}\frac{\dot{\lambda}_r(t)}{\dot{\lambda}_r(t)+\ddot{\lambda}_r(t)} \tag{6.6}$$

Given that definition and using a formulation similar to Eq. (6.4), the D2D caching optimization problem can be written as:

$$\begin{aligned}
&\underset{t_r',r\in\mathcal{R}}{\text{minimize}} && \ddot{\rho}(\mathcal{R}) && && (6.7)\\
&\text{subject to} && \dot{\lambda}_r(t) \le c_{m,n}, && \forall m\in\mathcal{M}, \forall n\in\mathcal{N},\\
& && \ddot{\lambda}_r(t) \le \ddot{c}_{n,n'}, && \forall (n,n')\in\mathcal{N},\\
& && \ddot{\mathrm{k}}_n(t) \le \ddot{s}_n, && \forall n\in\mathcal{N},\\
& && \eta(\mathcal{R}) \ge \eta_{min} && \forall r\in\mathcal{R}.
\end{aligned}$$

According to our scenario, the first step for solving Eq. (6.7) is to infer the set of influential users. This is done via the notion of the centrality metric [26]. In general, the centrality measure is used to quantify the social influence of a node in the network and is also related to how well connected the node is. A node with a higher value of this measure means that it is more central (thus influential) than those with lower values. Several definitions of centrality metrics exist in the literature [26], but we only focus on the eigenvector centrality for exposition. Let $G = (\mathcal{N},\mathcal{E})$ be the social graph, which consists of N nodes/users, where $\mathcal{N}$ represents the set of nodes and E is set of the links between them. The graph G can be represented by its adjacency (or D2D connectivity) matrix $\mathbf{A}_{N\times N}$, where the entry $a_{n,n'}$, $n, n' = 1,..., N$ is 1 if link (or edge) $\ddot{c}_{n,n'}$ exists, or 0 otherwise. For this matrix, let the eigenvalues be represented by $\lambda_1 \ge \ldots \ge \lambda_N$ in decreasing order, and the corresponding eigenvectors of these eigenvalues be given by $\mathbf{v}_1, \ldots \mathbf{v}_N$. The eigenvector-centrality in this case is the eigenvector $\mathbf{v}_1$ that has the largest eigenvalue λ_1. Knowing the K most-influential users of the social network via the notion of centrality, a clustering method (that is, K-means [27]) can then be formed around the users for community formation.

Once the set of influential users is identified and their communities are formed, the next step is to analyze content dissemination within each social community. By doing so, the critical content of each community can be stored in the caches of influential users. To show this, suppose that there is a set number of available contents, denoted by $\mathcal{F} = \mathcal{F}_0 + \mathcal{F}_h$, where $\mathcal{F}_h$ is the set of contents with viewing history and $\mathcal{F}_0$ represents the set of contents without history. We further assume that each user is interested in only one type of available content $\mathcal{F}$. Let π_f be the probability that content f is chosen by a given user, and as a prior [28], assume that the distribution follows a beta distribution [28]. Then, the selection of user n given as the conjugate probability of the beta distribution has a Bernoulli distribution. This in turn shows that the resulting user-content partition is analogous to that of the Chinese restaurant process (CRP) [28]. The CRP is a metaphor in which the objects are customers in a restaurant, and the classes are represented by the tables at which the customers sit. More precisely, in CRP, there is a restaurant with a large number of tables, each with an infinite number of seats, and customers arrive sequentially, each of them choosing a table at random.

In the CRP, with concentration parameter β, each customer decides to occupy a table with a probability proportional to the number of occupiers of that table, and chooses the next available table in proportion to the parameter β. Being more specific, the first customer selects the first table with probability $\frac{\beta}{\beta}=1$. The second customer selects the first table with probability $\frac{1}{1+\beta}$, and the second table with probability $\frac{\beta}{1+\beta}$. Once the second customer has selected a table, the third customer selects the first table with probability $\frac{1}{2+\beta}$, the second table with probability $\frac{1}{2+\beta}$, and the third table with probability $\frac{\beta}{2+\beta}$. This selection process continues until all customers have seats, yielding a distribution of the allocation of customers to tables. In this process, the decisions of subsequent customers are affected by the feedback of previous customers; customers learn about previous customers' selections to update their beliefs and the probabilities with which they select a particular table themselves.

Content dissemination in a social network is similar to table selection in an CRP. Looking on the social network as a Chinese restaurant, and the contents as the large number of tables, and the users as the customers, we can model the content dissemination process by a CRP. Within each social community, users tend to request the sought-after content sequentially, and once a content has been downloaded a hit is recorded (that is, there is a history). This, in turn, changes the probability that this content will be requested by others within the same social community: popular content will be requested more frequently and new content less frequently. Let there be a random binary matrix $\mathbf{Z}_{N\times F}$, indicating the selection of content by users, where $z_{nf}=1$ if user n chooses content f and 0 otherwise. Then, we can show that [28]:

$$P(\mathbf{Z})=\frac{\beta^{F'}\Gamma(\beta)}{\Gamma(\beta+N)}\prod_{f=1}^{F'}(m_f-1) \tag{6.8}$$

where $\Gamma(.)$ is the Gamma function [29], m_f is the number of users already assigned to content f (from their viewing history) and F' is the number of partitions with $m_f>0$. Therefore, for a given $P(\mathbf{Z})$, the popular content of each community can be stored inside the cache of influential users. A sketch of the proactive caching procedure at the UTs is summarized in Figure 6.4.

6.5.1 Numerical Results and Discussions

In the numerical setup, for similar purposes as in the previous section, the wireless link capacities are assumed to be equal among the users. The total D2D link capacity of each user is shared among the number of social links. The list of parameters is given in Table 6.2.

Starting from $t=0$, request arrival times are drawn uniformly at random until time T. The social network is constructed using the preferential attachment model [30]. As stated before, the eigenvector centrality is used to quantize the influential users in the social network, and then the K most influential are formed into K communities via K-means clustering [27].

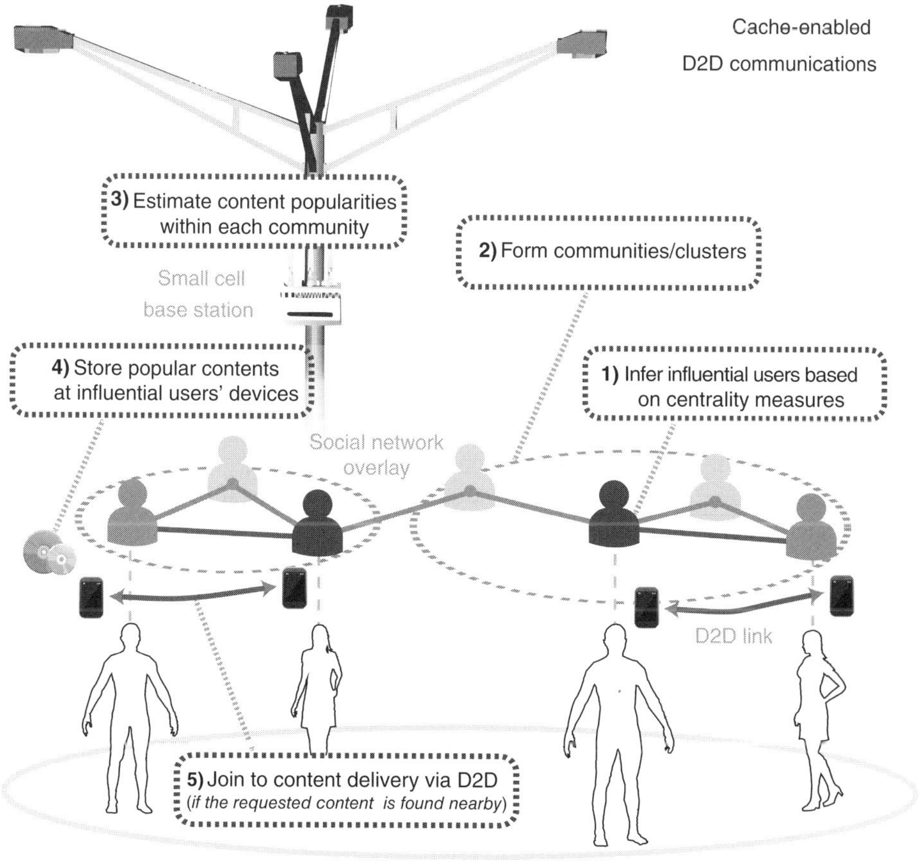

Figure 6.4 A practical procedure for proactive caching at the user terminals.

Table 6.2 The numerical setup parameters for proactive caching at the UTs.

Parameter	Description	Value
T	Time slots	1,024 s
M	Number of SBSs	4
K	Number of communities	3
N	Number of UTs	32
F	Number of contents	128
l_f	Length of content f	1 Mbit
b_f	Bitrate of content f	1 Mbit/s
$\sum_m \sum_n c_{m,n}$	Total SBSs link capacity	32 Mbit/s
$\sum_n \sum_{n',n' \neq n} \ddot{c}_{n,n'}$	Total D2D link capacity	64 Mbit/s
R	Number of requests	0 : 9,464
S	Total D2D cache size	$0 \sim l_f \times F$
β	CRP concentration parameter	0 : 100

In each community, the content popularity distribution is sampled from the CRP(β). Given the content popularity, the proactive caching is done by storing the popular files greedily inside the influential users terminals until no storage space remains. Similar to the case study in previous section, random caching is used as a baseline.

Parameters of interest in this case are:

- number of requests R
- total D2D cache size S
- CRP concentration parameter β.

The results are normalized for ease of understanding. The impact of the parameters of interest on the satisfaction ratio and small cell load are given in Figure 6.5.

In the figure, increasing the number of requests, we see that the satisfaction ratio decreases rapidly and the small cell load decreases at a low pace. The gains of the proactive caching approach are higher than the reactive approach in all regimes.

When there is an increment of D2D size, we observe an increment in the satisfaction ratio and decrement in the small cell load. Even though both proactive and reactive cases have gains, the proactive approach has better performance than the reactive approach.

The concentration parameter β has also an impact on the performance. When β increases (that is, the number of distinct contents grows), the satisfaction ratio and the small cell loads tend to be almost constant in the reactive approach. On the other hand, as β increases, the satisfaction ratio in the proactive approach decreases and the small cell load increases. The performance gap between the proactive and reactive approaches gets closer and closer as β increases. This is due to the facts that the content catalog size is growing while the UTs have a limited cache size.

6.6 Related Work and Research Directions

In this chapter, we have highlighted the proactive caching framework [17]. Indeed, the idea of caching goes back to the 1960s in the context of algorithm design in operating systems [31]. According to Belady [31], the optimal content removing strategy in the case of new content arrival is to remove from the memory any content that is not going to be requested in the near future. Beside this line of work, there has also been extensive study of web caching schemes in recent decades, aiming to improve the scalability of the World Wide Web and offloading the network by caching content in proxy servers and/or intermediate nodes of the network; see the paper by Wang for a brief literature review [32]. Numerous caching algorithms for content delivery networks have emerged in the recent years [33], allowing content providers to reduce access delays to the requested content. At the concept stage, there are also information-centric networks, which aim to change the way of accessing the content on the Internet by uniquely naming the contents and smartly storing it across the network, rather than the traditional approach of having one source for the content access [34, 35]. Besides these lines of work, studies of local caching as a way of offloading the wireless communications infrastructure are recent. Similar to what we have presented here, the growing literature is mostly based on caching at the edge of the network. In the following subsections, we summarize some of these studies based on their similarities and directions.

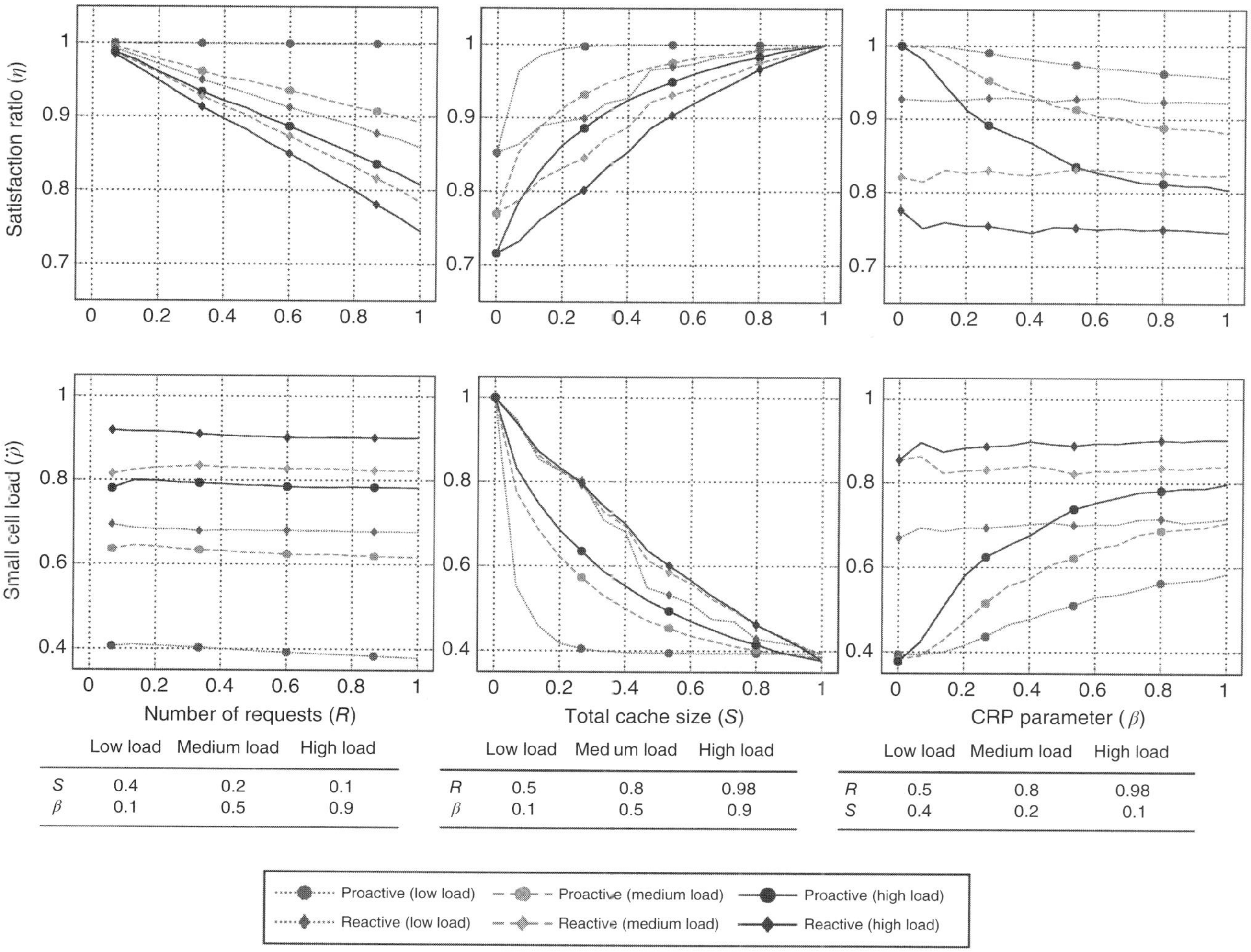

Figure 6.5 Social-aware caching via D2D: dynamics of the satisfied requests and small cell load with respect to the number of requests, total cache size, and CRP concentration parameter β.

6.6.1 Proactive Caching and Content Popularity Estimation

Proactive caching in SCNs with perfect knowledge of content popularity is described by Baştuğ et al. [22]. A team from the same group exploited context-awareness, social networks, and D2D communications, and described proactive caching approaches for SCNs both at the SBSs and UTs, showing that several gains are possible under the given numerical setup [17]. Instead of assuming perfect knowledge of the content popularity, an estimation was made via machine learning tools (the CF in particular), exploiting correlations of human behaviour and their preferences. Having such an estimation, the caching decision was more efficient, yielding better performance in terms of the user satisfaction and offloading of the network.

On the other hand, a well-known problem in the CF literature is the cold-start problem. This can occur in the case of estimation when there is very little information available. To improve the content popularity estimation, one approach harnessing the machine learning literature is transfer learning, based on the idea of smartly transferring information from a target domain to a source domain [36]. Inspired from this, Baştuğ et al. did a preliminary study on transfer learning for caching in SCNs [37]. Even though it has naturally its own challenges (i.e., negative transfer), they have shown that content popularity estimation via CF can be improved by this approach [38]. Further investigations are needed to combine this approach with proactive caching in SCNs. Additionally, in the context of proactive caching, Baştuğ et al. have exploited the centrality measures for content placement [39]. They introduced a simple content dissemination process and the preliminary performance results of this centrality-based content placement methods were given via numerical simulations. As an alternative to these proactive approaches, a game-theoretical formulation of the proactive caching problem as a many-to-many matching game was introduced by Hamidouche et al. [40]. A matching algorithm that reaches a pairwise stable outcome was provided for the caching problem, showing that the number of satisfied requests can be reach up to three times the satisfaction levels of a random caching policy.

6.6.2 Approximation Algorithms

The idea of femtocaching was described by Golrezaei et al. [16], with the SBSs (helpers) with low-rate backhaul but high-storage units in charge of delivering the content to users via short-range transmissions. The analysis was carried out both for coded and uncoded cases, showing that the optimum content assignment is NP-hard, whereas the coded case is formulated as a convex problem that can be reduced to a linear program. A greedy algorithm for the coded case and numerical results were provided, showing that video throughput can be improved by a factor of 3–5 in realistic settings. Extensions to this work, including the D2D case, have also been given [41, 42]. Alternatively, a multicast-aware caching problem was formulated by Poularakis et al. [43], who provided a heuristic algorithm, showing that servicing cost can be reduced by 52% compared to the multicast-agnostic case.

Optimal content placement in a SBS with limited backhaul capacity was studied by Blasco and Gunduz [44], showing that the problem can be reduced to a knapsack problem when the content popularity distribution is known. Assuming that the content popularity distribution is not known in advance, the problem is formulated as a multi-armed band (MAB) problem so that the content popularity distribution can be learned online and content placement can be decided. Three different caching algorithms were provided to show the exploration–versus-exploitation trade-offs of this problem. As an extension, a derivation of regret bounds and more extensive analysis of the algorithms through numerical simulations were later presented [45].

Additionally, Sengupta et al. described a distributed caching model with multiple SBS in the framework of MAB problem, showing that coded caching can outperform the uncoded case [46]. Besides MAB approaches, an approximation framework based on the facility-location problem was given by Poularakis et al. [47] and Abboud et al. presented a distributed caching algorithm based on the alternating direction method of multipliers [48].

6.6.3 Coded Caching Gains

Maddah-Ali and Niesen adopted an information-theoretic formulation of the caching problem [49]. Local and global caching gains, which depend on the available memory of each user and the cumulative memory of all users, respectively, were derived from a coded caching scheme. The proposed scheme consists of placement and delivery phases and is given for a centralized setup where the content placement is handled by a central server. It is essentially offline as there is no content placement during the delivery phase. This scheme has been shown to outperform conventional uncoded schemes under uniform content popularities, and works in a single shared link instead of more general networks. These results have been extended to:

- non-uniform content popularities [50, 51]
- non-uniform cache access [52]
- heterogeneous cache sizes [53]
- online caching systems [54]
- hierarchical caching networks [55]
- the multi-server case [56].

Moreover, improved bounds have been described [57, 58], as has then delay-sensitive content case [59], and the information-theoretic security aspects [60]. Similar to these efforts, a decentralized approach for D2D networks with random coded caching has been described in terms of scaling laws [61, 62], where a protocol channel model similar to that set out by Gupta and Kumar [63] is taken into account. In the same vein, the performance of decentralized random caching placement with a coded delivery scheme has been described [64, 65], with the expected rate characterized for random demands with the Zipf popularity distribution.

In the context of distributed storage systems and coding, the performance of simple caching, replication and regenerating codes was studied in a D2D scenario by Paakkonen et al. [66]. They derived a simple decision rule for choosing simple caching and replication so as to minimize the expected total cost in terms of energy consumption. Hollanti et al. studied the physical layer functionality of wireless distributed storage systems from the point of space-time storage codes [67]. Based on that work, Barreal et al. described a wireless storage system that communicates over a fading channel and propsed a novel protocol for the transmission based on algebraic space-time codes, in order to improve the system reliability while keeping the decoding at a feasible level [68]. It was shown that the proposed protocol performs better than the simple time-division multiple access protocol and falls behind the optimal diversity-multiplexing gain tradeoff. Alternatively, a triangular network coding approach for cache content placement was presented by Ostovari et al. [69], who compared uncoded content placement and triangular network coding strategies in a numerical setup. Additionally, a coded caching scheme over a wireless fading channel was presented by Huang et al. [70], and Wang et al. cast the caching problem into a multi-terminal source coding problem with side information [71].

6.6.4 *Joint Designs*

In terms of joint designs, a two time-scale joint optimization of power and cache control was given by Liu and Lau [72] for cache-enabled opportunistic cooperative MIMO. First, for short timescales, the closed-form expressions for the power control were derived from an approximated Bellman equation. Then, for long timescales, the caching problem was translated into a convex stochastic optimization problem and a stochastic subgradient algorithm was provided for its solution. The proposed solution was shown to be asymptotically optimal for high signal-to-noise ratios and its comparison with baseline approaches was done via simulations.

Another mixed timescale solution for cooperative MIMO was given by the same team [73]. In order to minimize the transmit power under the QoS constraint, the MIMO precoding was optimized for short timescales and cache control was done over long timescales. They have also described the joint optimization of cache control and playback buffer management for video streaming [74]. Meanwhile, Peng et al. have described joint caching and beamforming for backhaul-limited caching networks [75], and joint caching and interference alignment in MIMO interference channels under limited backhaul capacity was presented by Deghel et al. [76].

6.6.5 *Mobility*

Poularakis and Tassiulas analyzed mobility aspects of coded content delivery based on a discrete-time Markov chain model [77]. In order to minimize the probability of using the main base station in this model, a distributed approximation algorithm based on large deviation inequalities was introduced and numerical experiments on a real world dataset were conducted for the proposed algorithm. Another caching scheme that exploits users' mobility was given by Siris et al. [78], who investigated the influence of the system parameters on the delay gains via system-level simulations. Other papers also consider the impact of mobility in cache-enabled networks [79, 80].

6.6.6 *Energy Consumption*

Perabathini et al. investigated energy-consumption aspects of caching, both in terms of area power consumption and energy efficiency [81]. Cache-enabled base stations were distributed according to a homogeneous Poisson point process and the optimization was done using a detailed power model.

Meanwhile, Zhou et al. highlighted energy-harvesting aspects of proactive caching [82], and an effective push mechanism for energy harvesting powered small-cell base stations [83], and joint caching and base-station activation for green cellular networks [84] have also been proposed.

6.6.7 *Deployment Aspects*

Baştuğ et al. considered the deployment aspects of cache-enabled SBSs with limited backhaul [85]. In that study, the cache-enabled SBSs were stochastically distributed for the analysis rather than their positions being determined using traditional grid models. The expressions for the outage probability and average content delivery rate were derived as a function of the SINR, SBSs intensity, target-content bitrate, cache size, and the shape of the content popularity distribution. Following that paper, Blaszczyszyn and Giovanidis have shown that storing the most popular content is beneficial only in particular deployment scenarios [86]. On the other hand, for cache-enabled D2D communications, another stochastic framework has been described [87], that relies on two performance metrics that quantify the local and global fraction of served content requests. Yet another study for the stochastically distributed cache-enabled nodes is that of Altman et al. [88], who investigated the expected cost of obtaining the complete content under coded as well as uncoded content allocation strategies in the light of the observation that cost is a function of distance. As an extension to their work, the expected deployment cost of caches versus the expected content retrieval from the caches was analyzed by Mitici et al. [89].

6.7 Conclusions

In this chapter, we have discussed current advances in SCNs and proposed a novel proactive network paradigm based on caching at the edge of the network. Using tools from machine learning, we exploited users' predictable behaviour and their social relationships to optimize caching at the edge of the network. Our approach showed that peak mobile traffic demands can be significantly minimized, yielding backhaul offloading and resource savings. According to our findings and the growing literature, caching is now seen as a disruptive solution for 5 G SCNs. An interesting new direction of the work presented here is be the estimation of content popularity when the time and spatial dynamics of mobile users are involved. This clearly requires the development of novel algorithms and machine learning tools that can infer content popularity patterns from available data. Additionally, the benefits of caching in complex network structures – hierarchical networks, multi-hop networks, heterogeneous networks, and combinations of these – could be investigated while considering network constraints and physical-layer aspects. On the other hand, adaptive proactive caching schemes that can predict users' behaviour online and cache the contents accordingly are still in their infancy, and in this regard, establishing trade-offs between the feedback overhead and possible performance gains will be interesting. Also, joint design (say, caching and scheduling, network/index-coding-aided caching, and so on) is another direction for research. On top of these, experimental testbeds will allow network operators to see the practical gains for cache-enabled 5 G SCNs.

Acknowledgments

This research has been supported by ERC Starting Grant 305123 MORE (Advanced Mathematical Tools for Complex Network Engineering), the SHARING project under Finland grant 128010, and the BESTCOM project.

References

[1] Cisco (2014) Cisco visual networking index: global mobile data traffic forecast update, 2013–2018. URL: http://goo.gl/l77HAJ.

[2] Ericsson (2012) 5G radio access – research and vision. URL: http://goo.gl/Huf0b6.

[3] J.G. Andrews (2013) Seven ways that HetNets are a cellular paradigm shift. *IEEE Communications Magazine*, **51** (3), 136–144.

[4] J.G. Andrews, H. Claussen, M. Dohler, et al. (2012) Femtocells: past, present, and future. *IEEE Journal on Selected Areas in Communications*, **30** (3), 497–508.

[5] A. Damnjanovic, J. Montojo, Y. Wei, et al. (2011) A survey on 3GPP heterogeneous networks. *IEEE Wireless Communications*, **18** (3), 10–21.
[6] H.S. Dhillon, R.K. Ganti, F. Baccelli, and J.G. Andrews (2012) Modeling and analysis of k-tier downlink heterogeneous cellular networks. *IEEE Journal on Selected Areas in Communications*, **30** (3), 550–560.
[7] 3GPP (2014) TR 36.839 v11.0.0: Mobility enhancements in heterogeneous networks (release 11). URL: http://www.3gpp.org/DynaReport/36913.htm.
[8] M. Simsek, M. Bennis, and I. Guvenc (2015) Mobility management in hetnets: a learning-based perspective, *EURASIP Journal on Wireless Communications and Networking*, **1**, 26.
[9] M. Bennis, S.M. Perlaza, P. Blasco, et al. (2013) Self-organization in small cell networks: a reinforcement learning approach. *IEEE Transactions on Wireless Communications*, **12** (7), 3202–3212.
[10] M. Simsek, M. Bennis, and I. Guvenc (2013) Enhanced intercell interference coordination in hetnets: single vs. multiflow approach. *IEEE Globecom Workshops* (*GC Wkshps*), pp. 725–729.
[11] M.S. El Bamby, M. Bennis, W. Saad, and M. Latva-aho (2014) Dynamic uplink-downlink optimization in TDD-based small cell networks. *ArXiv preprint arXiv:1402.7292.*
[12] M. Bennis, M. Simsek, A. Czylwik, et al. (2013) When cellular meets WiFi in wireless small cell networks. *IEEE Communications Magazine*, **51** (6), 44–50.
[13] Q. Ye, M. Al-Shalash, C. Caramanis, and J.G. Andrews (2013) On/off macrocells and load balancing in heterogeneous cellular networks *ArXiv preprint arXiv:1305.5585.*
[14] F. Boccardi, R.W. Heath Jr, A. Lozano, et al. (2013) Five disruptive technology directions for 5g. *ArXiv preprint arXiv:1312.0229.*
[15] S. Göbbels (2010) Disruption tolerant networking by smart caching. *IEEE International Journal of Communication Systems*, **23** (5), 569–595.
[16] N. Golrezaei, K. Shanmugam, A.G. Dimakis, et al. (2012) Femtocaching: wireless video content delivery through distributed caching helpers IEEE INFOCOM, pp. 1107–1115.
[17] E. Baştuğ, M. Bennis, and M. Debbah (2014) Living on the edge: the role of proactive caching in 5G wireless networks. *IEEE Communications Magazine*, **52** (8), 82–89.
[18] CMRI (2011) C-RAN: The road towards green RAN. URL: http://goo.gl/jZt7TR.
[19] S. Sezer, S. Scott-Hayward, P.-K. Chouhan, et al. (2013) Are we ready for SDN? Implementation challenges for software-defined networks, *IEEE Communications Magazine*, **51** (7), 36–43.
[20] C. Song, Z. Qu, N. Blumm, and A.-L. Barabasi (2010) Limits of predictability in human mobility, *Science*, **327** (5968), 1018–1021.
[21] V. Etter, M. Kafsi, and E. Kazemi (2012) Been there, done that: what your mobility traces reveal about your behavior. *Mobile Data Challenge by Nokia Workshop, in conjunction with Int. Conf. on Pervasive Computing*, pp. 1–6.
[22] E. Baştuğ, J.-L. Guénégo, and M. Debbah (2013) Proactive small cell networks. *20th International Conference on Telecommunications (ICT)* Casablanca, Morocco, pp. 1–5.
[23] Netflix (2009) Netflix prize webpage. URL: netflixprize.com.
[24] J. Lee, M. Sun, and G. Lebanon (2012) A comparative study of collaborative filtering algorithms. *ArXiv preprint arXiv:1205.3193.*
[25] P. Arkadiusz (2007) Improving regularized singular value decomposition for collaborative filtering. *Proceedings of KDD Cup and Workshop Vol. 2007*, pp. 1–4.
[26] M. Newman (2009) *Networks: An Introduction.* Oxford University Press.
[27] A.K. Jain (2010) Data clustering: 50 years beyond k-means, *Pattern Recognition Letters*, **31** (8), 651–666.
[28] T.L. Griffiths and Z. Ghahramani (2011) The Indian buffet process: an introduction and review. *Journal of Machine Learning Research*, **12**, 1185–1224.
[29] M. Abramowitz and I.A. Stegun (1972) *Handbook of Mathematical Functions with Formulas, Graphs, and Mathematical Tables.* National Bureau of Standards Applied Mathematics Series 55.
[30] A.-L. Barabási and R. Albert (1999) Emergence of scaling in random networks, *Science*, **286** (5439), 509–512.
[31] L.A. Belady (1966) A study of replacement algorithms for a virtual-storage computer, *IBM System Journal*, **5** (2), 78–101.
[32] J. Wang (1999) A survey of web caching schemes for the internet. *ACM SIGCOMM Computer Communication Review*, **29** (5), 36–46.
[33] S. Borst, V. Gupta, and A. Walid (2010) Distributed caching algorithms for content distribution networks. *Proceedings IEEE INFOCOM, 2010,* pp. 1–9.
[34] A. Araldo, M. Mangili, F. Martignon, and D. Rossi (2014) Cost-aware caching: optimizing cache provisioning and object placement in ICN. *ArXiv preprint arXiv:1406.5935.*

[35] B. Ahlgren, C. Dannewitz, C. Imbrenda, et al. (2012) A survey of information-centric networking. *IEEE Communications Magazine*, **50** (7), 26–36.
[36] S.J. Pan and Q. Yang (2010) A survey on transfer learning. *IEEE Transactions on Knowledge and Data Engineering*, **22** (10), 1345–1359.
[37] E. Baştuğ, M. Bennis, and M. Debbah (2014) Anticipatory caching in small cell networks: a transfer learning approach. *1st KuVS Workshop on Anticipatory Networks*, Stuttgart, Germany, pp. 1–3.
[38] E. Baştuğ, M. Bennis, and M. Debbah (2015) A transfer learning approach for cache-enabled wireless networks. *ArXiv preprint arXiv:1503.05448.*
[39] E. Baştuğ, K. Hamidouche, W. Saad, and M. Debbah (2014) Centrality-based caching for mobile wireless networks. *1st KuVS Workshop on Anticipatory Networks*, Stuttgart, Germany, pp. 1–3.
[40] K. Hamidouche, W. Saad, and M. Debbah (2014) Many-to-many matching games for proactive social-caching in wireless small cell networks. *WNC3 workshop, WiOpt*, Hammamet, Tunisia, pp. 569–574.
[41] N. Golrezaei, A.F. Molisch, A.G. Dimakis, and G. Caire (2013) Femtocaching and device-to-device collaboration: A new architecture for wireless video distribution. *IEEE Communications Magazine*, **51** (4), 142–149.
[42] A.F. Molisch, G. Caire, D. Ott, et al. (2014) Caching eliminates the wireless bottleneck in video-aware wireless networks. *ArXiv preprint arXiv:1405.5864.*
[43] K. Poularakis, G. Iosifidis, V. Sourlas, and L. Tassiulas (2014) Multicast-aware caching for small cell networks. *ArXiv preprint arXiv:1402.7314.*
[44] P. Blasco and D. Gunduz (2014) Learning-based optimization of cache content in a small cell base station. *ArXiv preprint arXiv:1402.3247.*
[45] P. Blasco and D. Günduz (2014) Content-level selective offloading in heterogeneous networks: multi-armed bandit optimization and regret bounds. *ArXiv preprint arXiv:1407.6154.*
[46] A. Sengupta, S. Amuru, R. Tandon, et al. (2014) Learning distributed caching strategies in small cell networks. *International Symposium on Wireless Communication Systems (ISWCS)*, Barcelona, Spain, pp. 917–921.
[47] K. Poularakis, G. Iosifidis, and L. Tassiulas (2014) Approximation algorithms for mobile data caching in small cell networks. *IEEE Transactions on Communications*, **62** (10), 3665–3677.
[48] A. Abboud, E. Baştuğ, K. Hamidouche, and M. Debbah (2015) Distributed caching in 5G networks: An alternating direction method of multipliers approach. URL: http://goo.gl/vBdhV7.
[49] M. Maddah-Ali and U. Niesen (2014) Fundamental limits of caching. *IEEE Transactions on Information Theory*, **60** (5), 2856–2867.
[50] U. Niesen and M.A. Maddah-Ali (2013) Coded caching with nonuniform demands. *ArXiv preprint arXiv:1308.0178.*
[51] J. Hachem, N. Karamchandani, and S. Diggavi (2014) Multi-level coded caching. *ArXiv preprint arXiv:1404.6563.*
[52] J. Hachem, N. Karamchandani, and S. Diggavi (2014) Coded caching for heterogeneous wireless networks with multi-level access. *ArXiv preprint arXiv*:1404.6560.
[53] S. Wang, W. Li, X. Tian, and H. Liu (2015) Fundamental limits of heterogenous cache. *ArXiv preprint arXiv:1504.01123.*
[54] R. Pedarsani, M.A. Maddah-Ali, and U. Niesen (2013) Online coded caching. *ArXiv preprint arXiv:1311.3646.*
[55] N. Karamchandani, U. Niesen, M.A. Maddah-Ali, and S. Diggavi (2014) Hierarchical coded caching. *IEEE International Symposium on Information Theory (ISIT)*, pp. 2142–2146.
[56] S.P. Shariatpanahi, S.A. Motahari, and B.H. Khalaj (2015) Multi-server coded caching. *ArXiv preprint arXiv:1503.00265.*
[57] Z. Chen (2014) Fundamental limits of caching: improved bounds for small buffer users. *ArXiv preprint arXiv:1407.1935.*
[58] A. Krishnan, N.S. Prem, V.M. Prabhakaran, and R. Vaze (2013) Critical database size for effective caching. *ArXiv preprint arXiv:1501.02549.*
[59] U. Niesen and M.A. Maddah-Ali (2014) Coded caching for delay-sensitive content. *ArXiv preprint arXiv:1407.4489.*
[60] A. Sengupta, R. Tandon, and T.C. Clancy (2013) Fundamental limits of caching with secure delivery. *ArXiv preprint arXiv:1312.3961.*
[61] M. Ji, G. Caire, and A.F. Molisch (2014) Wireless device-to-device caching networks: Basic principles and system performance. *ArXiv preprint arXiv:1305.5216*
[62] M. Ji, G. Caire, and A.F. Molisch (2014) Fundamental limits of caching in wireless D2D networks. *ArXiv preprint arXiv:1405.5336.*
[63] P. Gupta and P.R. Kumar (2000) The capacity of wireless networks, *IEEE Transactions on Information Theory*, **46** (2), 388–404.

[64] M. Ji, A. Tulino, J. Llorca, and G. Caire (2014) On the average performance of caching and coded multicasting with random demands. *11th International Symposium on Wireless Communication Systems (ISWCS'14)*, Barcelona, Spain, pp. 922–926.
[65] M. Ji, A.M. Tulino, J. Llorca, and G. Caire (2015) Order-optimal rate of caching and coded multicasting with random demands. *ArXiv preprint arXiv:1502.03124.*
[66] J. Paakkonen, C. Hollanti, and O. Tirkkonen (2013) Device-to-device data storage for mobile cellular systems. *IEEE Globecom Workshops (GC Workshops)*, pp. 671–676.
[67] C. Hollanti, D. Karpuk, A. Barreal, and H.-F. Lu (2014) Space-time storage codes for wireless distributed storage systems. *ArXiv preprint arXiv:1404.6645.*
[68] A. Barreal, C. Hollanti, D. Karpuk, and H.-F. Lu (2014) Algebraic codes and a new physical layer transmission protocol for wireless distributed storage systems. *ArXiv preprint arXiv:1405.4375.*
[69] P. Ostovari, A. Khreishah, and J. Wu (2013) Cache content placement using triangular network coding. *IEEE Wireless Communications and Networking Conference (WCNC)*, pp. 1375–1380.
[70] W. Huang, S. Wang, L. Ding, et a;/(2015) The performance analysis of coded cache in wireless fading channel. *ArXiv preprint arXiv:1504.01452.*
[71] C.-Y. Wang, S.H. Lim, and M. Gastpar (2015) Information-theoretic caching: sequential coding for computing. *ArXiv preprint arXiv:1504.00553.*
[72] A. Liu and V. Lau (2014) Cache-enabled opportunistic cooperative MIMO for video streaming in wireless systems. *IEEE Transactions on Signal Processing*, **62** (2), 390–402.
[73] A. Liu and V. Lau (2014) Cache-induced opportunistic MIMO cooperation: a new paradigm for future wireless content access networks. *IEEE International Symposium on Information Theory (ISIT)*, pp. 46–50.
[74] A. Liu and V. Lau (2015) Exploiting base station caching in MIMO cellular networks: opportunistic cooperation for video streaming. *IEEE Transactions on Signal Processing*, **63** (1), 57–69.
[75] X. Peng, J.-C. Shen, J. Zhang, and K.B. Letaief (2014) Joint data assignment and beamforming for backhaul limited caching networks. *International Symposium on Personal, Indoor and Mobile Radio Communications (PIMRC'14)*, Washington, DC, pp. 1370–1374.
[76] M. Deghel, E. Baştuğ, M. Assaad, and M. Debbah (2015) On the benefits of edge caching for MIMO interference alignment. URL: http://goo.gl/uqp5Uj.
[77] K. Poularakis and L. Tassiulas (2013) Exploiting user mobility for wireless content delivery. *2013 IEEE International Symposium on Information Theory Proceedings (ISIT)*, pp. 1017–1021.
[78] V.A. Siris, X. Vasilakos, and G.C. Polyzos (2014) Efficient proactive caching for supporting seamless mobility. *ArXiv preprint arXiv:1404.4754.*
[79] P. Sermpezis, L. Vigneri, and T. Spyropoulos (2015) Offloading on the edge: analysis and optimization of local data storage and offloading in HetNets. *ArXiv preprint arXiv:1503.00648.*
[80] G. Alfano, M. Garetto, and E. Leonardi (2013) Content-centric wireless networks with limited buffers: when mobility hurts. *Proceedings IEEE INFOCOM, 2013*, pp. 1815–1823.
[81] B. Perabathini, E. Baştuğ, M. Kountouris, M. Debbah, and A. Conte (2015) Caching on the edge: a green perspective for 5G networks. *IEEE International Conference on Communications (ICC'15)*, London, pp. 2830–2835.
[82] S. Zhou, J. Gong, Z. Zhou, et al. (2015) Green delivery: proactive content caching and push with energy harvesting based small cells. *ArXiv preprint arXiv:1503.04254.*
[83] J. Gong, S. Zhou, Z. Zhou, and Z. Niu (2015) Proactive push with energy harvesting based small cells in heterogeneous networks. *ArXiv preprint arXiv:1501.06239.*
[84] K. Poularakis, G. Iosifidis, and L. Tassiulas (2015) Joint caching and base station activation for green heterogeneous cellular networks. *IEEE International Conference on Communications (ICC'15)*, London, pp. 3364–3369.
[85] E. Baştuğ, M. Bennis, M. Kountouris, and M. Debbah (2015) Cache-enabled small cell networks: modeling and tradeoffs, *EURASIP Journal on Wireless Communications and Networking*, **1**, 41.
[86] B. Blaszczyszyn and A. Giovanidis (2014) Optimal geographic caching in cellular networks. *ArXiv preprint: 1409.7626.*
[87] A. Altieri, P. Piantanida, L.R. Vega, and C. Galarza (2014) On fundamental trade-offs of device-to-device communications in large wireless networks. *ArXiv preprint arXiv:1405.2295.*
[88] E. Altman, K. Avrachenkov, and J. Goseling (2013) Coding for caches in the plane. *ArXiv preprint arXiv:1309.0604.*
[89] M. Mitici, J. Goseling, M. de Graaf, and R.J. Boucherie (2013) Deployment vs. data retrieval costs for caches in the plane. URL: http://doc.utwente.nl/88064/ (accessed August 3, 2016).

7

Modeling Multi-Radio Coordination and Integration in Converged Heterogeneous Networks

Olga Galinina,[1] Sergey Andreev,[1] Alexander Pyattaev,[1]
Mikhail Gerasimenko,[1] Yevgeni Koucheryavy,[1] Nageen Himayat,[2]
Kerstin Johnsson,[2] and Shu-ping Yeh[2]
[1] *Tampere University of Technology, Finland*
[2] *Intel Corporation, Santa Clara, CA, USA*

7.1 Enabling Technologies for Multi-Radio Heterogeneous Networks

7.1.1 Understanding Challenges in Mobile Wireless Networking

The rapid expansion of wireless communications over recent decades has introduced fundamental changes to "anytime, anywhere" mobile Internet access, and has also posed new challenges for the research community. In recent years, the wireless standards community has completed the fourth generation (4G) of broadband communication technology [1], which offers numerous advances in wireless system design, resulting in significant improvements in spectral and energy efficiencies as well as user quality of service (QoS). However, with the predicted explosion in both types and numbers of wireless devices it is commonly believed that despite novel 4G technologies, mobile broadband networks will face a capacity crunch in the near future.

Indeed, current multi-tier cellular technologies are already seeing high diversity in data rates, suffer from excessive time delays, or sometimes even service outages due to poor coverage and harsh interference conditions. Cellular coverage also remains unsatisfactory

Towards 5G: Applications, Requirements and Candidate Technologies, First Edition.
Edited by Rath Vannithamby and Shilpa Talwar.

Companion website: www.wiley.com/go/vannithamby/towards5g

in indoor environments despite aggressive spectrum reuse and sophisticated techniques for interference coordination [2]. To make matters worse, billions of machine-type devices connect to the network thus reshaping the Internet as we know it today. All these technological challenges accentuate the need to explore novel solutions within the context of 5G networks.

Given a historical 10-year cycle for each generation, it is expected that 5G systems will be deployed sometime around year 2020. As the dust around 5G communications technology settles, it becomes clear that it will be a synergistic integration of diverse techniques and solutions, rather than one "killer" technology, with the goal of dramatically improving the performance of next-generation wireless systems (see Figure 7.1). Many agree in that the only comprehensive solution to mitigate the increasing disproportionality between the user QoS and the available wireless resources is by deploying the higher density of femto and pico cells in the current cellular architecture. This improves network capacity by increasing the frequency reuse per unit area and the average data rate per transmission [3]; that is, smaller cells yield shorter radio links and thus improve data rates.

However, given the current levels of network loading and at least 13-fold growth over the next 5 years [4], the network may need additional small cells beyond those created by fixed access points. These should allow clients to offload their traffic onto direct device-to-device (D2D) radio links, which are generally shorter and lower-to-the-ground than standard "small cell" connections. With D2D, neighboring wireless devices can communicate without use of the network infrastructure, thus enabling a dramatic improvement in spectral reuse [5]. In

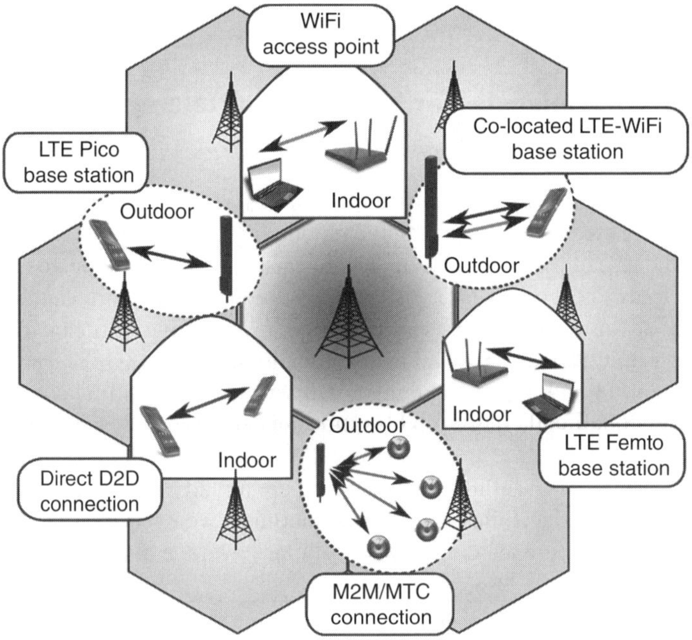

Figure 7.1 Envisioned 5G heterogeneous network.

addition, the proximity of user devices promises higher data rates, lower transfer delays, and reduced power consumption [6]. The potential applications of D2D in cellular networks are numerous [7] and include local voice services (offloading calls between proximate users), multimedia content sharing, gaming, group multicast, context-aware applications, and public safety.

Another pressing challenge is that licensed spectrum continues to be scarce and expensive, and traditional methods to improve its efficient use are approaching their theoretical limits [8]. As cell sizes shrink, the footprints of cellular, local, and personal area networks are increasingly overlapping, which creates an opportunity to simultaneously utilize multiple radio access technologies (RATs) for improved capacity and connectivity [9, 10]. Consequently, the incentive to efficiently coordinate between the alternative RATs is growing stronger [11]. To this end, the distributed unlicensed-band system – wireless local area networks (WLAN) – may take advantage of the centralized control function residing in the cellular network to effectively perform dynamic multi-RAT network association.

In summary, we firmly believe that intelligent coupling between multiple RATs may leverage several dimensions of diversity and that both short- and long-range technologies may need to work cooperatively to realize the desired improvements in capacity and service experience [12].

7.1.2 5G Technology Trends: Heterogeneous Networks

A transformation of mobile user experience requires revolutionary changes in both network infrastructure and device architecture, where the user equipment (UE) is jointly optimized with the surrounding network context [3]. Consequently, tighter interworking between various RATs has been receiving more attention in recent years. While previously cellular and WLAN technologies were developing largely independently, today WiFi is becoming an integral part of an operator's cellular network [13]. As the result, it becomes crucial to aggregate different radio technologies as part of a common *converged* radio network, in a manner transparent to the end user, and develop techniques that can efficiently utilize the radio resources available across different spectral bands potentially using various RATs [14].

In light of this, heterogeneous networks (HetNets) have recently emerged as an advanced networking architecture (see Figure 7.1) enabling aggressive capacity and coverage improvements in future 5G networks [15]. This architecture comprises hierarchical deployment of wide-area macrocells for basic connectivity and coverage, augmented with densely deployed small cells of various footprints [16] and by different RATs (femto and pico cells, WiFi access points, relay nodes, integrated WiFi-LTE small cells, etc.) to boost capacity [17]. In particular, unlicensed-band technologies are increasingly managed as part of an operator's cellular network to unlock advanced levels of interworking between cellular and WLAN RATs. This is, on the one hand, due to the fact that contemporary consumer devices massively support WiFi together with other RATs. On the other hand, mobile network operators increasingly rely on WLAN-based offloading to relieve congestion on their cellular networks [18] and hence desire more control of how WLAN is utilized and managed.

Not surprisingly, the recent literature has been rich in papers addressing the important aspects of load balancing and access network selection for multi-RAT HetNets [19]. The existing publications range from considering simpler user-centric network selection strategies (known as vertical handover) to full multi-tier and multi-radio cooperation [20, 21], for example where WiFi becomes a "virtual carrier" anchored on the cellular network. However, the focus has been mostly on centrally-managed systems with full control at the base station or totally distributed solutions, but not so much on network-assisted schemes. Most recently, the concept of LTE-unlicensed has attracted the interest of industry and academia alike with the goal of allowing LTE systems to utilize bandwidth-rich unlicensed spectrum around the 5-GHz band to augment their capacity [22]. Another emerging industry trend considered in the latest publications is multi-radio small cells with *co-located* cellular and WLAN interfaces that are able to reduce deployment costs and leverage common infrastructure across heterogeneous cells [23].

Reacting to this recent interest, the Third Generation Partnership Project (3GPP) is becoming increasingly active in developing new interworking solutions between 3GPP cellular technologies, such as UMTS or LTE, and WiFi (IEEE 802.11) [24]. However, given that co-located cellular/WLAN deployments are presently not common, current standardization efforts focus more on user-centric interworking architectures while only assuming limited degrees of cooperation/assistance across the HetNet [25]. The field of investigation spans:

- schemes for trusted access to 3GPP services with WLAN devices
- support for access network discovery and selection functions
- seamless mobility between cellular and WLAN technologies.

More recently, several new study/work items have been opened targeting the interworking solutions that involve cooperation within the radio access network (RAN) [26], in contrast to prior schemes that have loosely defined functions within the 3GPP core network, such as security and inter-RAT mobility [27]. This shift is dictated by the need to support improved QoS on WLAN networks, as prescribed by a consortium of network operators with their tighter requirements for carrier-grade WiFi. The WLAN community has also responded with their new initiatives on Hot Spot 2.0, as well as an emerging "High Efficiency WLAN" effort by the IEEE 802.11 work group. Therefore, we expect the trend for tighter integration of cellular and WLAN technologies to continue, potentially encompassing other radio technologies beyond current WiFi and additional-use cases beyond spectrum aggregation.

However, introducing an increasing number of serving stations to bridge the capacity gap incurs extra complexity, due to the need for more cumbersome interference management, higher rental fees, and increased infrastructure maintenance costs [28]. Even when additional spectrum is allocated, these new frequencies are likely to remain fragmented and could require diverse transmission technologies. We therefore expect that the majority of immediate gains will come from advanced architectures and protocols that would leverage the unlicensed spectrum. For example, mobile users with direct D2D communication capability may take advantage of their unlicensed-band radios and cooperate with other proximate users to improve local access in a cost-efficient way [29].

7.1.3 5G Technology Trends: Direct Communications

Currently, the lion's share of expected mobile traffic growth comes from peer-to-peer services that naturally involve clients in close proximity [30]. Hence, we envision that, whenever possible, neighboring client devices will use their direct connectivity capabilities, instead of cellular links. Consequently, D2D connections are expected to become an effective solution that will unlock substantial gains in capacity [12] and relieve congestion [31] in 5G mobile networks. For mobile network operators, D2D connectivity is becoming vital to enable traffic offloading from the core network and to realize efficient support of social networking through localization.

Over the last decade, 3GPP has invested much research effort into the characterization of D2D connections as part of LTE cellular technology in licensed bands [1], where a license grants a network operator the right to use spectrum exclusively. Driven by a wealth of potential practical applications, the concept of D2D communication as an underlay to a cellular network has been developed in the seminal work of Doppler et al. [32] and numerous subsequent papers. As in cognitive radio, D2D underlay is operating on the same resources as the cellular network and D2D users control their transmit power to suppress the resultant interference to cellular users [34]. Given its growing importance, licensed-bands D2D is becoming an attractive research area, where many fundamental questions still remain open including the information-theoretic capacity of the D2D underlay. However, the corresponding standardization efforts are developing slowly, and products employing the D2D underlay may not be the first to market.

For more than 5 billion cellular users, network-assisted D2D communication is becoming a natural next step to achieve enhanced resource utilization as traditional methods to improve the use of licensed spectrum approach their theoretical limits. Consequently, there has already been some coverage in the literature on direct user connectivity, with different levels of network involvement ranging from minimal degrees of assistance (as in Aura-net/FlashLinQ [34]) to fully controlled solutions (as in cellular underlay [32]). The latter is naturally more challenging and generally requires interference control to enable simultaneous direct connections [35].

For the underlay to work, the network should employ proper admission and power control on D2D transmitters as well as allocate radio resource to them. As a result, D2D links may:

- reuse resources reserved for cellular use
- use free resources not allocated for cellular use
- relay traffic through the infrastructure network avoiding direct transmissions.

The choice between these alternatives is known as transmission-mode selection [36] and has attracted many researchers, with a focus on optimization targets, from signal to interference plus noise ratio (SINR) and throughput to energy efficiency, data delay, fairness, and outage probability [37]. The main differences between existing studies are in:

- the numbers of communicating entities of each type considered (base stations, cellular, and D2D users)
- the emphasis on uplink (UL) or downlink (DL) connection and the resulting interference

- orthogonal versus non-orthogonal resource sharing
- the degree of available network assistance
- the network/D2D duplexing mode.

Unlicensed bands can be used freely, which gives opportunity to leverage D2D benefits almost immediately. While there already exists a plethora of unlicensed spectrum protocols to technically enable direct connectivity, there is neither centralized control of radio resources to manage QoS on D2D links nor is there any scalable device-discovery solution [8]. Augmenting current technology, we envision that devices be continually associated with the cellular network and use this connectivity to help manage their D2D connections in unlicensed bands. Therefore, as has been the case for HetNets, in the near-term we expect that the majority of gains will come from advanced architectures and protocols that leverage the unlicensed spectrum.

In other words, in conventional WLANs, the access point (AP) has no measures to control the resources used by ad-hoc user connections, which contend for the same channel. This is where the LTE network can be of much help. If clients are continuously connected to the LTE network, it knows which cell(s) they are associated with, which tracking area(s) they are in, and their locations to within a few meters (if location services are enabled). Therefore, the network can quickly and without significant overhead determine if/when clients are potentially within D2D range and inform them accordingly. Additionally, network assistance can help with mode selection (LTE/WiFi), power control, and selecting transmission format (modulation and coding rates, MIMO transmission mode, etc.) [7].

7.1.4 Focus and Contributions of our 5G Research

We have recently completed an advanced, in-depth characterization of multi-radio HetNets with the goal of fully understanding the performance gains and resolving any impediments to them. In this chapter, we reveal our most important findings by concentrating on network-assisted RAT assignment, which provides significant improvements in overall system performance. In what follows, our focus is on integration between multiple RATs within the envisioned 5G HetNet architecture (see Figure 7.1). As our case study, we consider convergence of WLAN- and D2D-based small "cells" with operator-managed cellular deployment. This illustrates feasible solutions for integration and their associated performance benefits. We explore the potential benefits in terms of augmented system capacity and improved user connectivity experience for a diverse range of devices that require connectivity at different scales.

We emphasize that interworking between WLAN and cellular networks has already been considered in the past, but largely from the perspective of inter-network (vertical) hand-off [13]. The cellular standards community, represented by the 3GPP, has also been involved in developing specifications that address cellular/WLAN interworking for a number of years. Several new study and work items have recently emerged that seek to develop specifications for tighter integration of WLAN with cellular networks. While much of this effort has focused on loose interworking solutions only requiring changes within the core network, there has been a recent shift in 3GPP Release 12 to address interworking within the RAN [38].

In this work, we specifically address the important problem of network selection between multi-tier cellular and WLAN RATs [39] assuming that WLAN belongs to an operator-deployed and *managed* multi-RAT HetNet. We also take a step ahead with respect to the current 3GPP efforts and consider intelligent *assistance* from the cellular network in the RAT selection process, when a new coordinating entity in the cellular RAN is made to receive relevant information from multi-radio devices, for example their position, their QoS requirements, how much interference/load they sense on the nearby WLAN networks, and so on, and then advises users on the attractiveness of the connectivity options.

For consistency with current network deployments, we concentrate on distributed small cell overlay with standalone WiFi access points as well as picocell base stations, assuming that there is no direct interface between the cellular and WLAN radio networks. However, our methodology may also characterize co-located cellular/WLAN deployments as well as more advanced technologies and scenarios to become appealing in the context of 5G networks [40]. More specifically, we focus on *uplink* performance, as it has not been fully addressed in the existing literature due to the more challenging nature of the interference-related aspects [41].

To further advance the state-of-the-art research on HetNets, which has primarily focused so far on static (full-buffer) steady-state formulations, we target flow-level performance and consider stochastic traffic loads. In particular, new data flows representing, for example, real-time data sessions with the minimum target bitrate, are arriving randomly and leaving the system after the service has been received [42]. Consequently, the number of active flows varies with time, which is often referred to as the flow-level dynamics. Analyzing dynamic setups is important to gain a better understanding of real-world systems, but it also involves extra complexity. Therefore, dynamic systems have received much less research attention than their static alternatives; that is, those with a fixed set of backlogged users.

In what follows, we outline a general methodology for modeling operation of a converged multi-radio HetNet that is capable of offloading user sessions onto small cell and D2D connections across both licensed and unlicensed spectrum. The proposed methodology is broad enough to accommodate various offloading scenarios, radio-selection algorithms, user-performance characteristics, and advanced wireless technologies such as WiFi and LTE. Here we are primarily interested in evaluating session blocking and reject probabilities. However, given the increasing importance of energy efficiency for mobile battery-driven user devices [43], we also aim at characterizing the energy expenditure of a typical user session.

7.2 Comprehensive Methodology for Space-Time Network Analysis

7.2.1 Capabilities of the Proposed Mathematical Approach

The performance of a converged HetNet depends considerably on the users' arrival rate as well as on their geographical locations and, as such, studying dynamic wireless systems together with the spatial features is becoming increasingly important. In particular, every data flow in a dynamic network may generally represent a stream of packets corresponding to a new file transfer, web-page browsing, or real-time voice/video session. To mimic the flows produced by a large population of independent users, Poisson processes have been extensively applied in the past. Originally, flow-level frameworks were helpful in investigating flexible bandwidth-allocation mechanisms in the context of wired systems. Extending their applicability to wireless networks, it was concluded that the throughput experienced by a dynamic user population can substantially differ from that received by a fixed number of users [44].

Another important aspect of HetNets is that locations of the network users relative to each other highly impact the resulting system performance [45]. Indeed, given that users are not regularly spaced, there may be a high degree of spatial randomness that needs to be considered explicitly. We thus adopt an appropriate random spatial model where user locations are drawn from a particular realization of a random process. Coupling such topological randomness with system dynamics requires a fundamental difference in characterizing user signal power and interference. Fortunately, the field of stochastic geometry provides us with a rich set of powerful results and analytical tools that can capture the network-wide performance of a random user deployment [46].

The use of stochastic geometry – that is, the statistical modeling of spatial relationships – has become increasingly popular over recent decades for analysis of network performance averaged over multiple spatial realizations. As part of a more recent surge, it has also been useful in characterizing many important aspects of current cellular technology, from conventional macrocell deployments to hyperdense heterogeneous and small cell networks [47]. The application of stochastic geometry typically features a particular spatial point process to statistically capture, for example, user locations, yielding insights into the impacts of user density, transmit power, path loss, and interference.

The application of queuing theory makes it possible to model user sessions arriving at random and leaving the system after being served. A session is a real-time data flow from one user to another; in our work, sessions are initiated according to a Poisson point process. Such spatial processes have been used extensively to:

- characterize the coexistence of cellular and mobile ad-hoc networks [37]
- study device discovery aspects of FlashLinQ [48]
- assess the performance of multi-tier heterogeneous cellular systems [41]
- capture the distributions of transmit power and SINR in D2D networks [49].

However, in most cases, the use of stochastic geometry does not directly enable system dynamics. Moreover, the existing literature fails to provide a unified framework for modeling the intricate interactions between a cellular network in the licensed bands and a D2D network in the unlicensed bands.

7.2.2 Proposed Taxonomy for HetNets

Next, we describe our proposed classification for the characteristic types of HetNets, embracing their distinctive features from the analytical perspective. For the constant bitrate traffic considered, we distinguish three main components, which primarily determine the corresponding mathematical constructs:

- interference: insignificant, so that it may be neglected or significant, which is to be accounted for explicitly
- power control/allocation: varies from fixed power allocation – that is, absence of power control – to processor sharing, while optimal power allocation may be located somewhere between these extremes [42]
- resource allocation: dedicated resources per a communication link versus shared channel access by several links.

In Figure 7.2, we display the three-dimensional space formed by these three criteria and specify various HetNet flavors as examples in the introduced space. We split the provided points (examples) into three groups and further consider these separately as individual models:

- Type I: resource is shared between several links, impact of interference may be neglected due to the technology-related features (e.g., coordination, spatial reuse)
- Type II: resource is shared between several links, interference between the neighboring cells has to be taken into account
- Type III: resource is exclusive for one link, interference has to be taken into account.

Due to a wide variety of possible power-control mechanisms, we combine all the power allocation schemes into one group vertically, and hereinafter refer to the system types as given here. As an example, the point WiFi, DCF based on the IEEE 802.11 WLAN standard reflects significant interference between the WLAN cells, while the resource of a

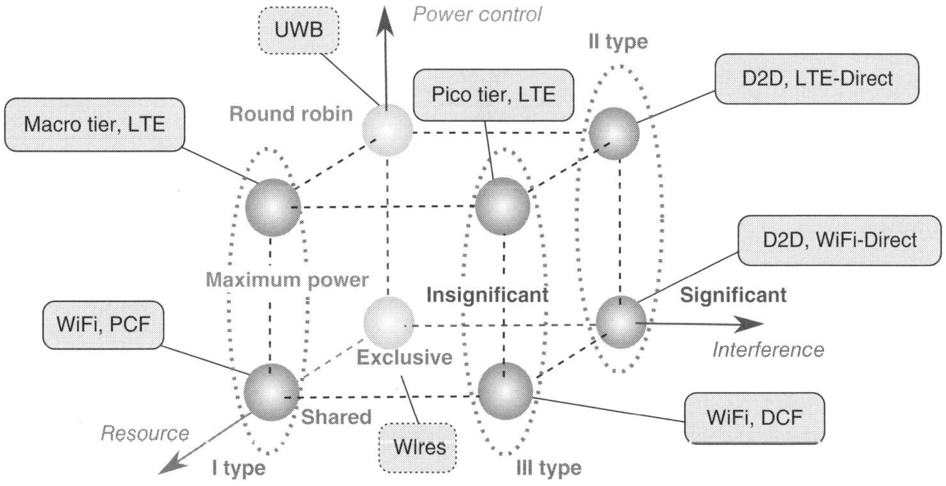

Figure 7.2 Proposed taxonomy of various HetNet types.

single AP is shared between several subscribers without the power control. Another example is the point WiFi, LTE-Direct, where interference between the neighboring links may be high enough, but up to the entire uplink resource per link may be given to one transmitter.

7.2.3 General Assumptions of the Model

In what follows, we introduce our integrated system model, comprising a number of macro and pico cellular, WLAN, as well as D2D networks, which we refer to as *tiers*. Below we summarize the core assumptions made throughout this chapter.

We study one (typical) cell of a macro network with radius R featuring a macro base station (BS) in its center. We assume that T wireless networks are available within the macro cell. Every network is an instance of one of the previously mentioned three categories (see, e.g., Figure 7.3); that is, it belongs to one of type I, II, or III. All the networks are serving UL data from their wireless users concurrently. The traffic considered is characteristic of real-time sessions, with the target bitrate of r_0.

Based on recent specifications [26], we further assume non-overlapping frequency bands for all tiers. Therefore, user transmissions on one tier do not interfere with those on the others. However, all links of types II or III (for instance, pico cells, WLAN, or D2D) share the frequency bands of their respective tiers and thus interfere, whereas the tier of type I is interference-free. Our general system model is illustrated in Figure 7.3. Areas of the macro cell are

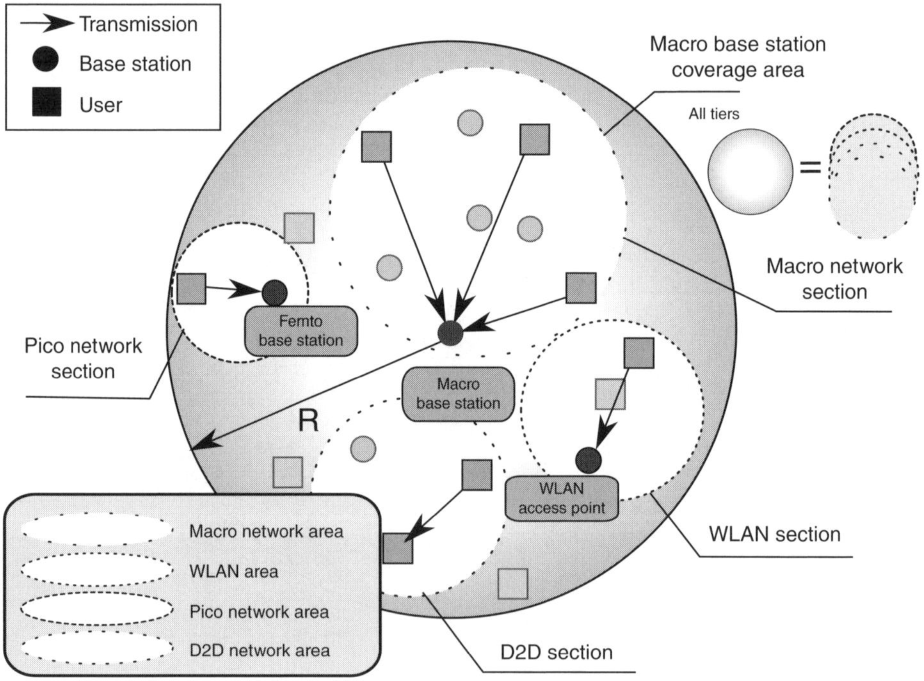

Figure 7.3 System model of a four-tier HetNet within a macro cell of radius R; the cuts demonstrate different network tiers.

represented as of tier I, pico and WLAN tiers as of type III, and D2D tier as of type II, together with the corresponding users and infrastructure nodes.

We assume that the transmitting users (or, *sessions*) with some uplink data traffic demand arrive on the joint network according to a one-dimensional Poisson process of rate λ in the time domain. We thus associate a newly arrived user with its session and its location, which is assumed fixed throughout the session's lifetime. For the sake of tractability, we also assume that the duration of a user session is exponentially distributed with a mean of μ^{-1}, which may correspond to, for example, a real-time voice/video call.

To explicitly model system spatiality, we make the several principal assumptions.

Assumption 1: Spatial Distribution of Users. The locations of arriving users follow a Poisson point process (PPP) on the two-dimensional plane. The area of our interest is limited by the considered macrocell (e.g., a circle of radius R), resulting in uniform distribution of users within the circle.

Assumption 2: Spatial Distribution of Infrastructure. Further, we adopt the following assumption about the locations of users and network infrastructure:

- Type II: For every session i of the tier type II, we differentiate between the transmitting user T_i, which is the data originator, and the receiving user R_i, which is the respective destination. We further assume that for a transmitting user T_i, the corresponding receiving user R_i arrives simultaneously with T_i, such that the location of R_i is distributed uniformly within the same circle of radius R.
- Type III: The locations of receivers (for example, APs/BSs) on the tier type III are independent and spatially distributed according to a PPP on the two-dimensional plane with the rates of L_i, where i is the number of the corresponding tier.

We note that in practice the constraint of deploying users within a particular area may be dictated by, for example, maximum transmit power restrictions and/or channel degradation factors. Moreover, uniform distribution is only assumed here as a baseline example. Generally, we may consider an arbitrary joint distribution $f(x, y)$ of user locations, which would somewhat complicate further analysis technically, but without significant impact on the derivation methodology.

Assumption 3: Signal Propagation. For all tiers, we assume for tractability that for the session i the wireless channel gain $\gamma_{k,j}$ between the user k and the receiver j depends on the distance $d_{k,j}$ separating them, and therefore it may be expressed as:

$$\gamma_{kj} = \frac{G}{d_{k,j}^{\kappa}} \tag{7.1}$$

where $d_{k,j}$ is the distance between the receiver and the transmitting user, κ is the propagation exponent, and G is the propagation constant determined by a particular RAT and accounting for the corresponding channel model.

Without the loss of generality, we further assume that the data rate is continuous and that the power-rate mapping is defined by the Shannon formula. This consideration has recently been shown in [50] to remain very accurate for current wireless networks.

Assumption 4: Power-rate Mapping. We assume that the data rate is continuous and that the power-rate mapping is given by the Shannon formula. The transmit power p_i of a user i and its data rate r_i (in [nats/s]) are therefore coupled by the generalized Shannon capacity theorem:

$$r_i = \min\left\{B \log\left(1 + Ap_i\right), r_{\lim}\right\}, \tag{7.2}$$

where p_i is the output power of the radio frequency power amplifier, and A and B are the scaling coefficients depending on the particular wireless technology used.

For the sake of an example, these are given as $A = \frac{\eta\gamma_{i,i}}{N_0 + I}$, $B = w$, where $\gamma_{i,i}$ is the path gain between the user and the receiver for session i, η is the fading margin, w is the channel bandwidth, N_0 is the noise level, and I is the interference level at the receiver. The constant $r_{\lim}$ defines the constraint of the data-rate growth based on the fixed set of modulation and coding schemes. Hence, further increase in the signal-to-noise ratio (SNR) does not yield unbounded data-rate growth after a certain value of $d_0 = \left[\frac{G \cdot p}{(N_0 + I)(e^{r_{\lim}/w} - 1)}\right]^{1/\kappa}$.

While random network topology is the primary focus of our model, we also investigate flow-level system dynamics. This involves an appropriate queuing model, where the session arrives and leaves the system after being served; the service time is determined by the random session length. When a new session arrives or a served session leaves the system, the centralized assisting entity in the cellular network performs admission and power control on all tiers by deciding whether the session should be admitted to a particular tier or not (admission control) and/or advising on the user's transmit power (power control).

For each of three considered tier types, the corresponding data link is governed by applying transmission policies. A particular policy generally determines user admission, scheduling, and transmit power. Whenever admitted, a transmitting user occupies a fraction of the time-frame resource and sets its power as commanded by the BS to achieve the data rate given by Eq. (7.2). The BS makes a new decision on scheduling allocations and transmission power for all active users at every new arrival or when an existing session is served and leaves the system.

Assumption 5: Power Control and Scheduling. The transmission policies considered are the following:

- The maximum rate (MR) policy assumes no explicit power control and sets a fixed transmit power, which is the allowable maximum for a particular radio technology. Then, admission control checks if the target bitrate can be achieved with this maximum power. Given the relationship in Eq. (7.2), the instantaneous data rate for session k is determined by the maximum transmit power $p_{\max}$ as:

$$r_k^{\max} = \min\left\{w \log\left(1 + \gamma_{k,k} p_{\max}\right), r_{\lim}\right\} \tag{7.3}$$

- The round robin (RR) policy ensures that the system resource is always shared between the users equally. Each admitted session out of n running sessions is allocated an equal portion of the total time resource: $\frac{r_0}{r_k} = \frac{1}{n}$. Then, the users adjust their transmit power to match their required target bitrate as long as it does not exceed the maximum allowed power level. Clearly, in case of n active sessions, $r_k = r_0 n, \forall k = \overline{1,n}$.

In summary, both MR and RR policies offer a flexible choice between more system capacity, resulting also in higher power consumption, and better network resource utilization, enabling some transmit power savings. By considering these next, we ensure that the HetNet offers a good balance between network- and user-side performance.

Assumption 6: Admission Control. The assumptions for admission control are as follows:

- Types I and III: Since a real-time session requires the bitrate of r_0, the system admits a newly arrived session if there is still sufficient resource to serve it. In other words, each ongoing session k has to occupy exactly r_0/r_k of the system time (where the overhead is accounted for later), while for all the active sessions it holds as the following:

$$\sum_{all\ sessions}\left(\frac{r_0}{r_k^{\max}}\right) \leq \delta \tag{7.4}$$

  where δ is the resource available for sharing at a particular receiver (AP or BS; for example, excluding resources allocated for fading compensation), $r_k \leq r_k^{max}$ is the instantaneous data rate depending on the distance between the user and the receiver, and r_k^{max} is the maximum achievable data rate at the maximum power level. Admission control of tier type III also incorporates interference control, which is detailed later (see Section 7.2.3.8).
- Type II: Since the resource is exclusive for the link of this tier, admission control focuses on ensuring that the rate r_0 is achievable for the newly arrived link, so that the interference is controlled as well.

For tier types II and III, admission control policy may determine whether current interference exceeds a particular given threshold or not.

Assumption 7: Interference Margin. We also assume that the noise plus interference has the form of $N_0 + I = KN_0$, where the value of K is a scaling factor fixed across the network. It has the meaning of interference margin per receiver. We may thus aggregate the individual interferences created by the proximate users of a particular tier into a cumulative background noise level, which in the practice of network planning is taken into account as a particular interference margin.

Assumption 8: Interference Assessment. The assumptions for interference assessment are as follows:

- Type II: We assume that the noise plus interference power does not exceed some network-wide threshold: $N_0 + I \leq KN_0$. Further, it is assumed that the tier type I with $n-1$ active users admits a new session n if for the set $\{T_k\}_{k=1}^{n}$ of transmitters the following conditions hold at each receiver R_j, $j = \overline{1,n}$:

$$\frac{p_k \gamma_{k,k}}{KN_0} \geq e^{r_0/(w\delta_{i,k})} - 1 \quad \text{and} \quad p_k \gamma_{k,j} \leq N_0, \quad \forall k, \quad j \neq k \tag{7.5}$$

  where r_0 is the target bitrate, $\delta_{i,k}$ is the actually available resource for the given link of the tier i after removing the overheads and signaling, and the value of K is fixed for this tier. These conditions imply that the required bitrate r_0 can be achieved on each link k (see the left-hand side of Eq. (7.5)) and that the interference on R_j produced by T_k does not exceed the given threshold N_0 (see the right-hand side of Eq. (7.5)).

- Type III: It is imposed that a tier with $n-1$ active users admits a new session n if for the set $\{U_k\}_{k-1}^{n}$ of all users the following conditions hold at each receiver:

$$\frac{p_k \gamma_{k,k}}{KN_0} \geq e^{r_0/w} - 1 \text{ and } p_k \gamma_{k,j} \leq N_0, \quad \forall j, k \neq j \tag{7.6}$$

where $\gamma_{k,j}$ is the path gain between the user k and its receiver j and p_k is the corresponding allocated power.

By that, the admission control function ensures that the required minimum bitrate can be achieved by a user, and that the interference at the receiver j produced by the user k does not exceed a given threshold depending on the technology-related features. We also note here that our interference and rate estimation has predictive character and assists the network in making a guided decision on whether a user should be admitted or not.

7.2.4 The HetNet Operation Considered

In this work, we begin by examining the individual operation of tiers belonging to the main types. Also, for the entire converged HetNet, we explore a particular *sequential* mechanism of user admission and network selection as a characteristic example of future 5G performance. This is illustrated in Figure 7.4, where we assume a cascade service for any new session arriving into the system. Correspondingly, the network-selection assistance entity attempts to offload the newly arrived session onto the first network according a particular RAT priority. In the case when a RAT operates over a shared resource, the network selection entity attempts to offload the user session to the nearest receiver – the closest AP/BS – by performing the corresponding admission control, which is managed centrally. We note that the nearest receiver may also be located outside of the circle R.

If the session is accepted on the current tier, it is served there without interruption until it successfully leaves the system. Otherwise, if this session cannot be admitted onto the attempted tier, the network admission function attempts the following RAT in the order of decreasing priority. Hence, either the session is accepted onto one of the $T-1$ tiers and served or the macro network T (which is always attempted last) tries to serve this session. Eventually, if the session cannot be admitted onto the macro network either, it is considered to be permanently blocked and leaves the system unserved without any impact on the new arrivals.

Whenever admitted, a transmitting user exploits a fraction of the system time resource and sets its power either as fixed or as commanded by the power-control function to achieve its required data rate. The system makes a new decision on scheduling and transmission power allocations for all active users at every new arrival or when an existing session is served and leaves the system. For each tier, we introduce the corresponding blocking probability $P_{block}^{(i)}$ and acceptance probability $P_a^{(i)} = 1 - P_{block}^{(i)}$, where i is the index corresponding to one of the tiers, respectively. Moreover, we remind ourselves that the session arrival rate on the first tier is λ (see Assumption 1).

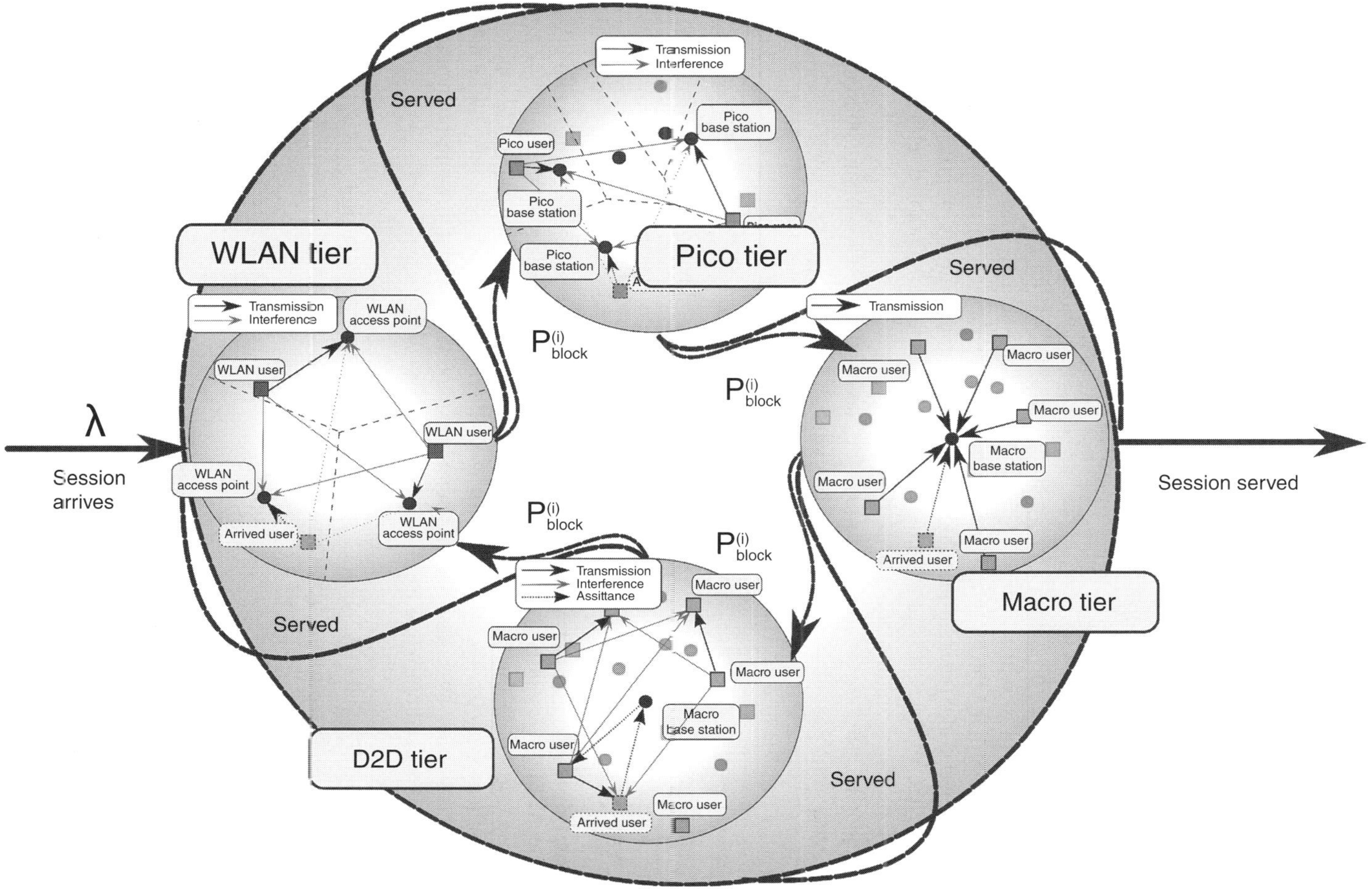

Figure 7.4 Operation of a multi-RAT HetNet.

Assumption 9: Decoupling Assumption. To preserve analytical tractability of our mathematical model, we assume that all types of network tiers serve their users independently, which results in a random thinning of the arrival process with the corresponding acceptance probabilities. This assumption is a natural methodological move to decompose the system into a set of tractable and well-defined components, which may be easily replaced and/or interchanged.

Proposition 1 Due to the Poisson property of the thinned flow, the arrivals on the following tier (those not accepted by the current tier) follow a Poisson process of density $\lambda_{i+1} = \lambda_i(1 - P_a^{(i)})$, where $P_a^{(i)}$ is the tier i accept probability.

Abstracting away the locations of users for analytical tractability, we further assume that the arrivals on the subsequent tiers are also placed uniformly within the circle of radius R. In contrast to Proposition 1, the latter consideration *does not* actually hold in reality. Instead, there is some pattern in which users are taken for service by the tiers. However, our simulation results (as reported next) reveal that the assumption of uniformity is surprisingly accurate. This makes the analysis of our system under the previously mentioned assumptions an adequate approximation for a practical HetNet operation [50].

Consequently, denoting the macro network (the "last resort" tier) accept probability as $P_a^{(T)}$, we may easily establish the overall system blocking probability P_{block} as follows:

$$P_{block} = 1 - \sum_{i=1}^{T} \prod_{j=1}^{i-1} \left(1 - P_a^{(j)}\right) P_a^{(i)} \tag{7.7}$$

7.3 Analysis of Random Dynamic HetNets

Next, we provide a summary of our rigorous analytical efforts to evaluate the important HetNet-related performance metrics. Hereinafter, we consider different tiers separately. We underline here that our system analysis is built on the decoupling principle as per Assumption 9. This technique is used widely and allows evaluation of even very complex systems by regarding them as an integrated set of tractable components. The following mathematical models and associated reasoning are divided into two major parts:

- tier types I and II, where the system is determined by the state of the links
- tier type III, where additional information on the receivers is needed.

7.3.1 Core Stochastic Model

Here, we outline our general stochastic model for the tiers based on the assumptions previously. Assume that the arrivals on all the tiers follow a Poisson process with rates $\lambda_1 = \lambda$, λ_i, λ_T. We observe tier types I and II at the particular moments t of session (user) arrivals and departures. Since the arrivals follow a Poisson process and the service (session length) is distributed exponentially, our system behavior may be represented by a stochastic Markov process $S(t)$, where the future process evolution is determined by the set of ongoing sessions that are currently being served on a given tier.

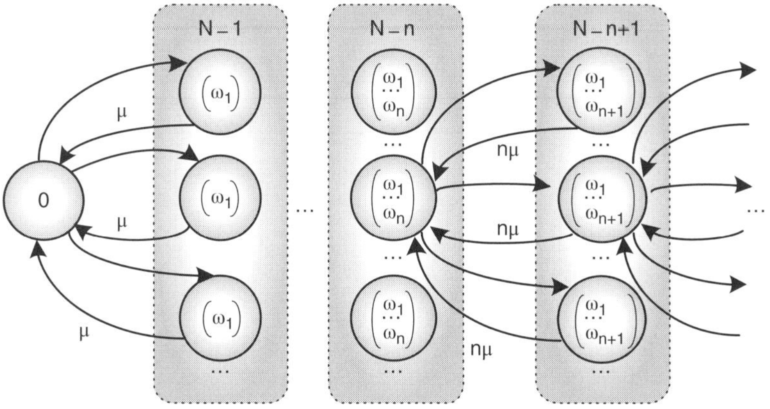

Figure 7.5 State diagram for tier types I and II.

7.3.1.1 Tier Types I and II Analysis

For tier types I and II, the state of the process *S*(*t*) is determined by the characteristics of the ongoing sessions within the target circle. For convenience, we denote these abstract characteristics as ω and note that they depend on the distance between the transmitter and the receiver. Therefore, the system state is represented by the vector $(\omega_1,\ldots,\omega_n)$, where n is the number of sessions in service (see Figure 7.5).

Let a tier have n running sessions in the state s. We denote the rejection probability at the state s for the newly arrived session as $Q_{n+1|s}$. Then, transitions from the state $s = (\omega_1,\ldots,\omega_n)$ to the state $(\omega_1,\ldots,\omega_n,\omega_{n+1})$ and backwards have the rates of $\lambda_T(1-Q_{n+1|s})$ and $(n+1)\mu$, respectively.

7.3.1.2 Tier Type III Analysis

Tier type III comprises several shared receivers (APs/BSs), which are distributed on the plane with densities of L_i. Such a tier is interference-limited, and hence the respective stochastic processes have state-dependent properties, which are different from those discussed previously. The state of the stochastic Markov process *S*(*t*) may be represented by the set of sessions with respect to the corresponding receivers. Similarly, we adopt notation ω for the session characterization. Then, the state s of a tier type III is represented by:

$$\left(\omega_1,\ldots,\omega_{n_1};\omega_{n_1+1},\ldots,\omega_{n_1+n_2};\ldots;\omega_{s_n+1},\ldots,\omega_{s_n+n_{N_i}}\right)$$

where $s_n = \sum_{i=1}^{N_i-1} n_i$, and n_1, n_2, and n_{N_i} are the numbers of users associated with the first, second, and last link, respectively. The random variable N_i corresponds to the number of receivers in a certain area and follows a Poisson distribution. The state diagram of the system is illustrated in Figure 7.6.

We consider state s, where tier type III is serving n ongoing sessions with a random number of N_i receivers. We denote the rejection probability for the newly arrived session as $Q_{n+1|s}$. Then, the transitions from the state s to the state of $n+1$ active sessions have a rate of $\lambda_i(1-Q_{n+1|s})$. The backward rate equals $(n+1)\mu$ since the service does not depend on the state, but rather on the number of simultaneously served sessions.

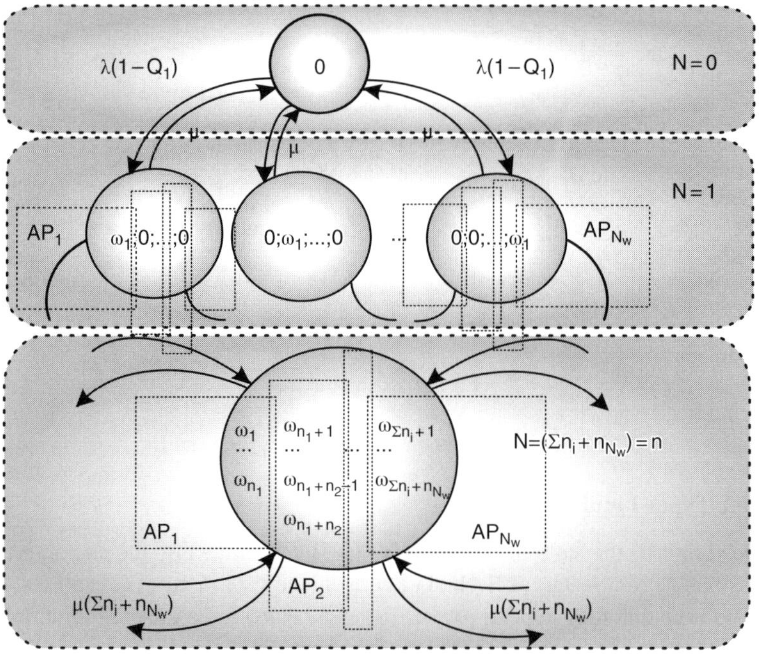

Figure 7.6 State diagram for tier type III.

7.3.2 Calculating the Steady-State Distribution

Due to the uncountable number of states in the considered system, it may be complicated to attack the steady-state distribution straightforwardly. However, the corresponding Markov process may be simplified by employing the state aggregation technique.

Hence, for tier types I and II, we aggregate the states $\{(\omega_1,...,\omega_n)\}_{\omega\in\Omega}$ by n (where Ω is the space of all possible vectors $(\omega_1,..., \omega_n)$, $n \in N$). For the more complex tier type III, we combine all possible states of the system (which contain n ongoing sessions) into the state n. This combining process is illustrated in Figure 7.6.

Assumption 10: State Aggregation. The state aggregation assumptions are as follows:

- For tier types I and II, we aggregate all the states containing n sessions into the unifying state n, regardless of the actual locations of users.
- For the more complex tier type III, we combine all possible states of the system (which contain n ongoing sessions) into the state n, regardless of the locations of the current users or their connections to a certain receiver.
- In order to keep our system memoryless, we adopt a simplification, where the sessions at the state n, while keeping all of their other properties, do not preserve their locations from state to state. For the sake of analytical tractability, these locations are assumed to be generated anew at every particular state n.

We note, however, that the system still keeps track of the previously admitted sessions owing to the probabilities Q_{n+1} to reject the session arrived at the state n conditioning on the fact that the current n -th session satisfies the admission control criteria.

As a result of the state aggregation, we arrive at the birth-death processes for all the tiers with the rates of $\lambda_i(1-Q_{n+1})$ and $(n+1)\mu$. We further denote the arrival rate λ_i into the system simply as λ and formulate the following proposition.

Proposition 2 The steady-state distribution $\{\pi_i\}_{i=0}^{\infty}$ for the considered process S(t) with the transitions $\lambda(1-Q_{n+1})$ and $(n+1)\mu$ can be closely approximated by:

$$\pi_n = \pi_0 \frac{\lambda^n}{\mu^n} \frac{\prod_{i=1}^{n}(1-Q_n)}{n!} \tag{7.8}$$

where

$$\pi_0 = \left(\sum_{i=0}^{\infty} \frac{\lambda^n}{\mu^n} \frac{\prod_{i=1}^{n}(1-Q_n)}{n!} \right)^{-1}$$

and Q_{n+1} is the reject probability on the transition from the state n to the state $n+1$.

Regarding a proof of this proposition, we refer readers to the steady-state distribution for the well-known birth–death process, the derivation of which may be found in the corresponding literature.

Based on the steady-state distribution and assuming that it exists, our approach empowers us to estimate a wide class of stationary characteristics in the system considered, such as the expected number of ongoing sessions, the probability of a session being permanently blocked, and even its energy consumption. However, the latter is left out of the scope of this work due to space limitations. The average number of active sessions and the system blocking probability are defined as:

$$E[N] = \sum_{n=0}^{\infty} n\pi_n, \quad P_{block} = \sum_{n=0}^{\infty} Q_{n+1}\pi_n \tag{7.9}$$

The latter metric may also be used as the system (area) capacity prediction for sufficiently high arrival rates.

In our analysis, we disregard the history of the system processes from the perspective of the ongoing sessions. We thus replace the initial stateful systems with memoryless processes, for which we examine the arbitrary set of respective random variables at each point t. If the reject probabilities Q_{n+1} are known for all tiers, we easily obtain the steady-state distribution by using Eq. (7.8). Therefore, in what follows we concentrate on calculating the values of Q_{n+1}. In order to take into account the memory property that we have thus omitted, we will refer to the corresponding conditional probabilities further on.

The methodology is rather generic and can be applied to a wide range of RATs, provided that they are sufficiently simplified to meet our taxonomy. The distinctive features defining the

Table 7.1 Selected systems for tier types I, II, and III.

	MP	RR
Type I		×
Type II	×	
Type III	×	

selected model behavior are defined by the particular values of parameters as well as by the calculation of the probabilities Q_{n+1}, which, in turn, depend on the power allocation policy:

$$(1-Q_{n+1}) = \Pr\{\text{new session } n+1 \text{ is admitted} \mid n \text{ sessions are already active}\} \quad (7.10)$$

where by this conditional probability we account for the previous system history, while the new system evolution process (n-based) is memoryless. In other words, we estimate the probability of sharing the resource between $n+1$ random sessions if n other stochastically different sessions have already been admitted at the previous state.

Due to the space limitations, we restrict ourselves to consideration of one system of each type by selecting a particular power policy (as noted in Table 7.1). Alternative power-allocation policies may be accounted for in a similar way.

7.3.3 Characterizing Transitions for Important HetNet Examples

We continue by evaluating the probabilities Q_{n+1} and the transition rates $\lambda_i(1-Q_{n+1})$ necessary for deriving the steady-state distribution. The rest of the text is organized in the following order. First, we illustrate our approach on the simpler tier types I and II, selecting as examples the RR (round robin) and MP (maximum power) policies, correspondingly, which may characterize LTE macro cell and D2D over WiFi-Direct. Then, we continue with the more complex (due to the presence of interference) tier type III, represented by the system with the maximum power policy and corresponding to, for example, WLAN operation.

7.3.3.1 Tier Type I Transitions

We begin with tier type I under the RR transmission policy and detail calculations that are necessary for characterizing this policy conditioning on:

- the absence of interference
- equal sharing of the resource among all the links.

Transitions from state n to state $n+1$ are therefore defined by:

$$\lambda_m(1-Q_{n+1}) = \lambda_m\left(\Pr\left\{\frac{r_0}{r_i^{max}} \le \frac{\delta_m}{n+1}, \forall i = 1, n+1 \,\middle|\, \frac{r_0}{r_i^{max}} \le \frac{\delta_m}{n}, \forall i = 1, n\right\}\right) \quad (7.11)$$

Further, we formulate the following theorem, the proof of which may be found in the literature [50].

Theorem 1 For tier type I under the RR policy, the accept probabilities $\Pr\{accepted \mid arrived\} = 1 - Q_{n+1}$ can be obtained by:

$$1 - Q_{n+1} = \Pr\left\{r_{n+1}^{\max} \geq \frac{r_0(n+1)}{\delta_m}\right\}\left(\frac{\Pr\left\{r_i^{\max} \geq \frac{r_0}{\delta_m}(n+1)\right\}}{\Pr\left\{r_i^{\max} \geq \frac{r_0 n}{\delta_m}\right\}}\right)^n \tag{7.12}$$

where

$$\Pr\{r \geq x\} = 1 - \Pr\{r < x\} = 1 - F_r(x) \tag{7.13}$$

and

$$F_r(x) = 1 - \frac{1}{R^2}\left[\frac{\eta G p}{N_0}\right]^{2/\kappa}\left(e^{r/w} - 1\right)^{-2/\kappa}, r_R \leq x < r_{lim}; \quad F_r(x) = 1, x \geq r_{lim}$$

The latter completes the formula of Eq. (7.11) and delivers the steady-state distribution for Eq. (7.8), as well as other relevant stationary metrics such as those in Eq. (7.9).

7.3.3.2 Tier Type II Transitions

We continue with tier type II, for which the resource access is exclusive, the admission control regulates interference, and the transmit power is set to its maximum. First, let n sessions already exist in the system. Hence, for all $i = 1, n$ we require the following target data rate condition to hold:

$$r \leq w \log\left(1 + \frac{p_{\max}\gamma_{i,i}}{KN_0}\right) \Leftrightarrow p_{\max}\gamma_{i,i} \geq KN_0\left(e^{\frac{r}{w}} - 1\right) \tag{7.14}$$

Therefore, the following theorem can be formulated.

Theorem 2 For tier type II under the MP policy, if admission control is performed according to (7.5) and, in particular, accounting for (7.14), then the reject probabilities Q_{n+1} can be closely approximated by:

$$Q_{n+1} = 1 - \Pr\{accepted \mid arrived\} = \left[F_\gamma\left(\frac{N_0}{p_{\max}}\right)\right]^{2n-1}\left[1 - F_\gamma\left(\frac{\theta_0}{p_{\max}}\right)\right] \tag{7.15}$$

where $\theta_0 = KN_0(e^{\frac{r}{w}} - 1)$ and the cumulative distribution function for the SNR per power unit γ is given as:

$$F_\gamma(\gamma) - 1 + \frac{G^{\frac{4}{k}}\gamma^{-\frac{4}{k}}}{8R^4} - \frac{G^{\frac{2}{k}}\gamma^{-\frac{2}{k}}}{R^2}\ln 2, \text{if } \frac{G}{\left(2R^2\right)^{\frac{k}{2}}} \le \gamma \le \gamma_{\max}, \gamma_{\max} = \frac{KN_0}{p_{\max}}\left(e^{\frac{r_{\max}}{w}} - 1\right)$$

$$F_\gamma(\gamma) = 1 - \frac{1}{R^2}\left(\frac{G^{\frac{4}{k}}\gamma^{-\frac{4}{k}}}{8R^2} + G^{\frac{2}{k}}\gamma^{-\frac{2}{k}}\ln\frac{4R^2\gamma^{\frac{2}{k}}}{G^{2k}}\right), if \frac{G}{(2R)^k} \le \gamma \le \frac{G}{\left(2R^2\right)^{\frac{k}{2}}}$$

7.3.3.3 Tier Type III Transitions

We continue by considering tier type III and detailing the calculations that are necessary for characterizing the MP transmission policy. The transitions from state n to state $n+1$ are thus defined by:

$$\lambda_w\left(1 - Q_{n+1}\right) = \lambda_w \Pr\left\{A_j^{(n+1)}, j = \overline{1, n+1} \middle| A_j^{(n)}, j = \overline{1, n}\right\} \quad (7.16)$$

where event $A_j^{(n)}$ is given as:

$$A_j^{(n)} = \left\{\frac{r_0}{r_j^{\max}} \le \delta_w - \sigma_n \text{ and } \gamma_{j,k} p_{\max} \le N_0, \forall k \ne j\right\}$$

where δ_w is a share of the available resource at the receiver (without the signaling overhead and collisions) and σ_n is a part of the resource given to other sessions at the same receiver in the current state. We further denote $r_0 / \left(\delta_w - \sigma_n\right)$ as $\tilde{r}_{0,n}$.

The calculation of σ_n is based on the following assumption.

Assumption 11: AP Link Abstraction. Here, to abstract away the session-receiver details at the state *n*, we assume that upon its arrival into the system, a session observes the average (typical) number of users at the nearest receiver (see Theorem 3). This average number depends on the number of ongoing sessions; that is, on the state index *n*, as well as on the parameter r_0.

Theorem 3 For tier type III, the average number of sessions per receiver (AP/BS) n_0 tends to $\frac{n}{L_i(\pi R^2)} = \frac{n}{E[N_i]}$ for large areas, where $E[N_i]$ is the expected number of receivers of tier *i* within the circle *R* [50].

Note that Theorem 3 is similar in its meaning to the research findings obtained previously by Foss and Zuyev [51]. Then, based on these results, we may reformulate as stated in Assumption 11. A newly arrived session observes the system, where, on average, every receiver already serves $n_0 = \frac{n}{L_i(\pi R^2)}$ sessions.

Theorem 4 For tier type III under the MP policy, the corresponding transition rates may be calculated as $\Pr\{accepted \mid arrived\} = 1 - Q_{n+1}$, accounting for the following:

$$1-Q_{n+1} = \frac{\left(1-e^{-\pi L_w d_{r,n+1}^2}\right)^{n+1}}{\left(1-e^{-\pi L_w d_{r,n}^2}\right)^{n}} \left(L_w \pi d_{thr}^2 e^{-L_w \pi d_{thr}^2} + e^{-L_w \pi d_{thr}^2}\right) \tag{7.17}$$

where $d_{thr} = \left[\frac{G p_{\max}}{N_0}\right]^{\frac{1}{\kappa}}$ and the constant value $d_{r,n}$ is defined as $\left(\frac{p_{\max}\eta G}{K N_0}\right)^{\frac{1}{\kappa}} \left(e^{\frac{\tilde{r}_{0,n}}{w}} - 1\right)^{-\frac{1}{\kappa}}$.

In summary, by introducing $d_{r,n}$, we emphasize that it depends on $\tilde{r}_{0,n} = r_0 / (\delta_w - \sigma_n)$. This, in turn, is a function of the number of sessions on tier type III via the occupied resource σ_n representing the average share of the resource exploited at the state n and is given by:

$$\sigma_n = E\left[\frac{r_0}{r_i^{\max}} \middle| \frac{r_0}{r_i^{\max}} \le \delta_w\right] \frac{n}{E[N_i]} = E[y \mid y \le \delta_i] \frac{n}{E[N_i]} \tag{7.18}$$

and $E[y \mid y \le \delta_i]$ may be found as:

$$E[y \mid y \le \delta_i] = \int_{y_0}^{\delta_i} y f_y(y \mid y \le \delta_i)\, dy = \frac{2\pi L_w}{C_3} \int_{y_0}^{\delta_i} y d(y) d'(y) e^{-\pi L_w \left(\frac{p_{\max}\eta G}{K N_0}\right)^{\frac{2}{\kappa}} \left(e^{\frac{r_0}{wy}} - 1\right)^{\frac{2}{\kappa}}} dy$$
$$+ \frac{2\pi L_i}{C_3} y_0 \int_{0}^{y_0} d(y) d'(y) e^{-\pi L_w \left(\frac{p_{\max}\eta G}{K N_0}\right)^{\frac{2}{\kappa}} \left(e^{\frac{r_0}{wy}} - 1\right)^{-\frac{2}{\kappa}}} dy \tag{7.19}$$

where $C_3 = \Pr\{y \le \delta_i\} = F_y(\delta_i)$ and y_0 is assumed to be less than δ_i.

Equation (7.17) finally enables us to derive the key performance metrics of interest, such as the expected number of ongoing sessions and the overall blocking probability of Eq. (7.9).

7.4 Quantifying Performance with System-level Evaluations

7.4.1 Features of our 5G System-level Simulator

To complement our analytical study, we have developed an advanced system-level simulator (SLS) based on:

- the up-to-date 3GPP LTE evaluation methodology (3GPP LTE Release-10 FDD)
- current IEEE 802.11 specifications (IEEE 802.11-2012 supporting WiFi-Direct features).

Presently, neither free nor commercially-available simulation platforms are readily applicable for evaluating 5G multi-RAT systems, as they lack the necessary features and scalability to adequately capture the dependencies between the variables studied. By contrast, our SLS is a flexible tool designed to support diverse deployment strategies, traffic models, channel characteristics, and wireless protocols.

Our SLS tool consists of software modules:

- to model the deployment of wireless infrastructure and user devices
- to control events related to transmission of signals between several distinct types of transmitters and receivers
- to abstract for wireless channels
- to quantify system performance by collecting measurements and statistics.

It models all of the conventional LTE/WiFi infrastructure and user deployment choices:

- arbitarily shaped cells
- environment with or without wraparound
- uniform versus clustered user distribution.

To optimize the performance of our SLS, we make several simplifications of realistic wireless systems, yet we attempt to mimic the most important mechanisms and dependencies explicitly. We are specifically interested in dense deployments so that we can analyze the conditions where intelligent RAT selection would be most needed; when the cellular network would have difficulty supporting the offered traffic load on its own.

Below we construct a multi-RAT simulation model representative of an urban deployment, where WiFi and D2D small "cells" are overlaid on top of the multi-tier 3GPP LTE network. Outdoor deployments are considered and are based on recommendations in 3GPP's TR 36.814 specification [52] combining that with varying pico BS and WLAN AP densities per TR 36.819 [53]. Hence, our scenario represents a harmonized 3GPP vision of a characteristic HetNet deployment. A part of it concentrates on an area of interest, in which co-located cellular and D2D/WLAN networks cover a limited region with many users requiring service, for example, a shopping mall or business center. For the D2D/WLAN systems, the simulation is largely based on the IEEE 802.11 medium access control procedure with carrier sensing. We also assume that all APs and their respective users run the same version of the technology as WiFi-Direct clients, namely IEEE 802.11-2012. For calibration purposes, we employ reliable results from publications on ad-hoc WLAN deployments.

For the LTE system, the simulation captures the following practical features (as opposed to this analytical methodology): data frame structure, bandwidth requests, and scheduling by the BS. Here, our example scenario comprises 19 hexagonal cells supporting 3GPP LTE Release 10 technology, and a wrap-around technique is used to improve precision of the simulation at the edges of the deployment area. The system works over two 10-MHz bands for FDD operation (for UL and DL), shared by all cells with three sectors in each, resulting in a $1 \times 3 \times 1$ reuse pattern. For more details on the configuration of the reference LTE network, the interested reader is directed to the relevant standardization documents (e.g., 3GPP TR 36.814-900 and ITU-R M.2135-1). For performance verification purposes, we also implemented a calibration scenario from 3GPP TR 36.814-900, Table A-2.1, and ran the corresponding tests. Our simulation results fall well within the required limits for both cell-center and cell-edge spectral efficiency targets.

7.4.2 Discussing Representative Numerical Results

We begin with Figure 7.7, where we detail the blocking probabilities (or the proportion of service requests that cannot be served by the network) for the integrated HetNet as well as for the three tiers individually: macrocell, picocell, and WLAN tier. Our observation is that with two additional overlay tiers, the HetNet performance improves significantly over what can be achieved in the macro-only networks (cellular baseline).

By employing our advanced SLS tool, we are able to demonstrate in detail how the components of the blocking probability P_{block} evolve with increasing load on the network

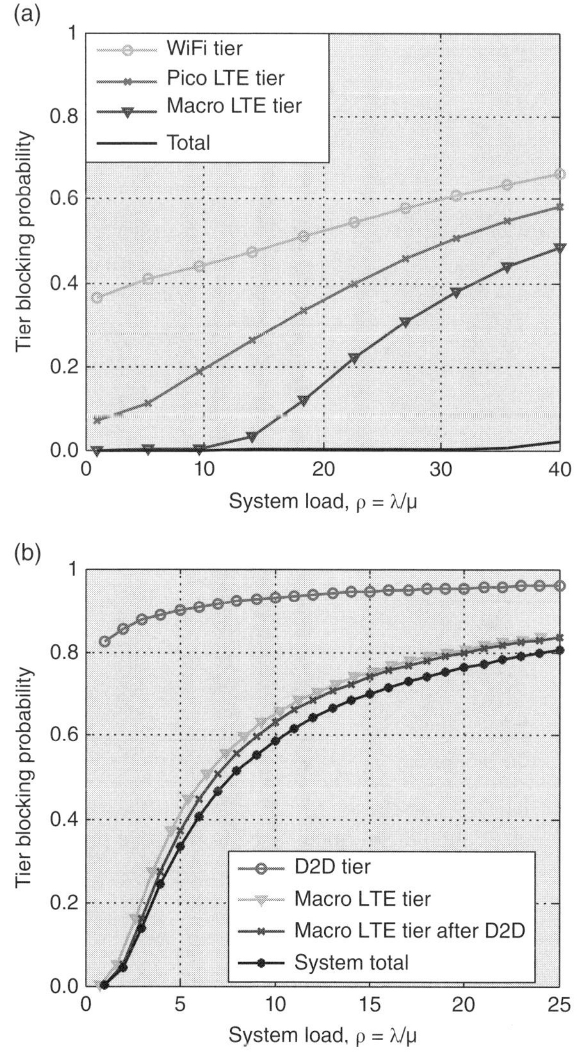

Figure 7.7 Analyzing blocking probabilities: (a) in three-tier and (b) two-tier HetNets.

(see Figure 7.8a). The session is blocked if it cannot fit into the schedule at the time of arrival, and for the D2D network we differentiate between session rejections due to:

- prohibitive interference from the existing transmissions
- excessive link length to support the required bitrate (given that the interference constraint has been satisfied).

It is important to analyze the structure of the blocking processes for both systems. For the D2D system, at low loads the blocking is primarily caused by the link length, whereas as the load increases the probability of a blocking due to interference becomes dominant.

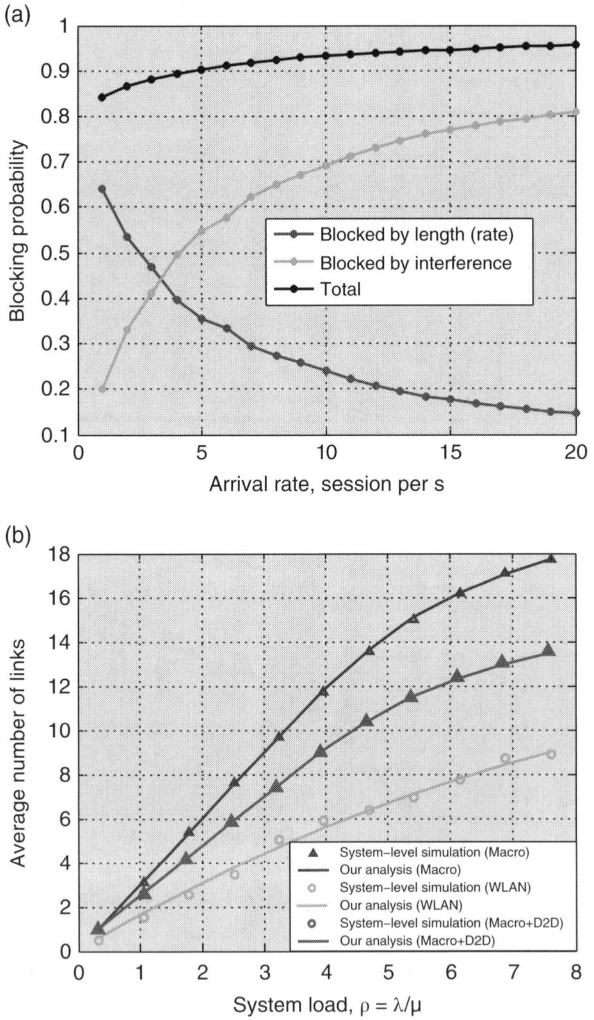

Figure 7.8 Evolution of blocking probability with network load: (a) session blocking/reject probability simulations and (b) system capacity simulation versus analytical approximations.

In addition, the Figure 7.8b contrasts the simulation results against our previous analysis for LTE, WiFi, and integrated LTE+D2D network to confirm good convergence between the analytical approximations and the SLS results. Here, continuous lines indicate analytical data, while symbols correspond to simulated values. Clearly, the overall trend is an increase in the expected number of running links, up to the saturation point, which depends on the deployment, scheduling, and multiplexing methods used.

In order to understand the more subtle effects associated with the HetNet operation, let us finally take a look at the quality of the links in our system (see Figure 7.9). When the cellular system is empty, it can afford to accept all links, no matter the quality. Under such conditions,

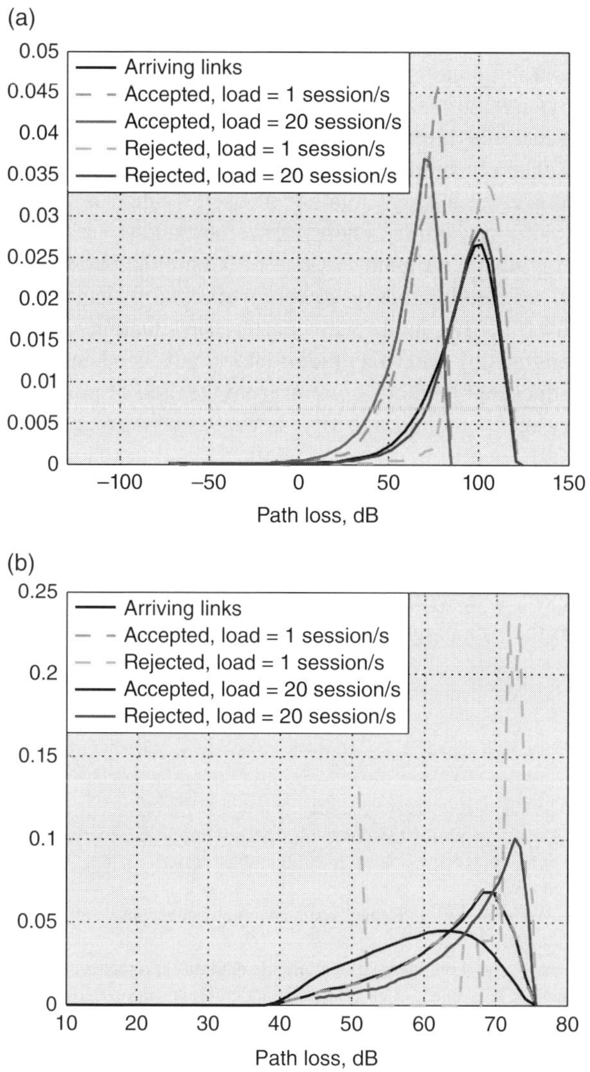

Figure 7.9 Link quality: (a) D2D tiers; (b) LTE tiers.

the link quality for arrivals and accepted links is similar, and there are almost no discards (see Figure 7.9a). As the cellular system becomes loaded, however, we see that it takes only shorter links in, as those have significantly better chances of fitting into the schedule (see Figure 7.9b).

7.5 Summary and Conclusions

In this chapter, we have provided a unified mathematical methodology, supported by advanced system-level simulations, allowing us to capture the crucial HetNet dynamics together with the geometrical randomness of realistic user deployment. Moreover, we proposed a comprehensive classification of practical HetNet examples embracing the available 5G technologies in a systematic way. Abstracting away less impactful system properties, we thus focused on the generic access technology groups within the proposed classification. To this end, we thoroughly described the analytical methods required to calculate important performance-related parameters. By doing that, we rigorously covered a broad range of attractive HetNet configurations by providing relatively simple and accurate approximations for the stationary metrics of interest. These include the average number of users (which is equivalent to the system capacity), system blocking probability, energy efficiency, and so on.

More generally, studying the ultimate capacity of 5G multi-radio HetNets remains an open problem in the field of information theory, and our methodology has the potential to shed light on it, given that it can explicitly capture new interference situations and hence the achievable data rates. This challenging objective may require novel advanced analytical tools to interconnect and apply techniques and methods coming from the area of point processes, probability theory, queuing theory, and percolation theory, as well as modern engineering insights.

Acknowledgments

This work is supported by Intel Corporation, GETA, TISE, and the IoT SRA program of Digile, funded by Tekes. The work of the second author is supported with a postdoctoral researcher grant from the Academy of Finland, as well as with a Jorma Ollila grant by the Nokia Foundation.

References

[1] 3GPP LTE Release 10 and beyond (LTE-Advanced)

[2] R. Amir, J. Martin, J. Deaton, et al. (2013) Balancing spectral efficiency, energy consumption, and fairness in future heterogeneous wireless systems with reconfigurable devices. *IEEE Journal on Selected Areas in Communications*, **31**, 969–980.

[3] S.P. Yeh, S. Talwar, G. Wu, et al. (2011) Capacity and coverage enhancement in heterogeneous networks. *IEEE Wireless Communications* **18** (3), 32–38.

[4] Cisco (2014) Cisco visual networking index: global mobile data traffic forecast update, 2013–2018

[5] J. Lehtomaki, I. Suliman, J. Vartiainen, et al. (2008) Direct communication between terminals in infrastructure based networks *Proc. of the ICT-MobileSummit*, pp. 1–8.

[6] G. Fodor, E. Dahlman, G. Mildh, et al. (2012) Design aspects of network assisted device-to-device communications. *IEEE Communications Magazine*, **50** (3), 170–177.

[7] L. Lei, Z. Zhong, C. Lin, and X. Shen (2012) Operator controlled device-to-device communications in LTE-Advanced networks. *IEEE Wireless Communications*, **19** (3), 96–104.

[8] L. Al-Kanj, Z. Dawy, and E. Yaacoub (2013) Energy-aware cooperative content distribution over wireless networks: Design alternatives and implementation aspects. *IEEE Communications Surveys & Tutorials*, **15** (4), 1736–1760.

[9] M. Gerasimenko, N. Himayat, S. Yeh, et al. (2013) Characterizing Performance of load-aware network selection in multi-radio (WiFi/LTE) heterogeneous networks. *GLOBECOM Workshops (GC Wkshps)*.

[10] A.Y. Panah, S.P. Yeh, N. Himayat, and S. Talwar (2012) Utility-based radio link assignment in multi-radio heterogeneous networks. *Proc. of International Workshop on Emerging Technologies for LTE-Advanced and Beyond-4G on IEEE Globecom.*

[11] K. Andersson and C. Ahlund (2011) Optimized access network selection in a combined WLAN/LTE environment. *Wireless Personal Communications*, **61**, 739–751.

[12] A. Ozgur, O. Leveque, and D. Tse (2007) Hierarchical cooperation achieves optimal capacity scaling in ad hoc networks. *IEEE Transactions on Information Theory*, **53**, 3549–3572.

[13] M. Bennis, M. Simsek, W. Saad, et al. (2013) When cellular meets WiFi in wireless small cell networks. *IEEE Communications Magazine*, **51**, 44–50.

[14] J.G. Andrews, S. Singh, Q. Ye, et al. (2014) An overview of load balancing in HetNets: Old myths and open problems. *IEEE Wireless Communications*, **21**(2), 18–25.

[15] S.P. Yeh, A.Y. Panah, N. Himayat, and S. Talwar (2013) QoS aware scheduling and cross-RAT coordination in multi-radio heterogeneous networks. *IEEE VTC*, pp. 1–6.

[16] S. Tombaz, A. Vastberg, and J. Zander (2011) Energy- and cost-efficient ultra-high-capacity wireless access. *IEEE Wireless Communications*, **18**, 18–24.

[17] A. Prasad, O. Tirkkonen, P. Lundén, et al. (2013) Energy-efficient inter-frequency small cell discovery techniques for LTE-Advanced heterogeneous network deployments. *IEEE Communications Magazine*, **51**, 72–81.

[18] C. Sankaran (2012) Data offloading techniques in 3GPP Rel-10 networks: A tutorial. *IEEE Communications Magazine*, **50**, 46–53.

[19] S. Singh, H. Dhillon, and J. Andrews (2013) Offloading in heterogeneous networks: modeling, analysis, and design insights. *IEEE Transactions on Wireless Communications*, **12**, 2484–2497.

[20] T. Novlan, H. Dhillon, and J. Andrews (2013) Analytical modeling of uplink cellular networks. *IEEE Transactions on Wireless Communications*, **12**, 2669–2679.

[21] S. Singh and J. Andrews (2014) Joint resource partitioning and offloading in heterogeneous cellular networks. *IEEE Transactions on Wireless Communications*, **13**, 888–901.

[22] Huawei (2014) White paper: U-LTE: unlicensed spectrum utilization of LTE.

[23] N. Himayat, S.P. Yeh, A.Y. Panah, et al. (2014) Multi-radio heterogeneous networks: architectures and performance. *Proc. of IEEE International Conference on Computing, Networking and Communications (ICNC)*, pp. 252–258.

[24] IEEE (2012) 802.11-2012, Part 11: Local and metropolitan area networks.

[25] 3GPP (2013) R2-133604: Performance benefits of RAN level enhancements for WLAN/3GPP.

[26] 3GPP (2013) 3GPP TR 37.834: Study on WLAN/3GPP radio interworking.

[27] 3GPP (2013) 3GPP TS 23.402: Architecture enhancements for non-3GPP accesses.

[28] P. Marsch, B. Raaf, A. Szufarska, et al. (2012) Future mobile communication networks: challenges in the design and operation. *IEEE Vehicular Technology Magazine*, **7** (1), 16–23.

[29] S. Andreev, A. Pyattaev, K. Johnsson, et al. (2014) Cellular traffic offloading onto network-assisted device-to-device connections. *IEEE Communications Magazine*, **52**, 20–31.

[30] F. Fitzek, M. Katz, and Q. Zhang (2009) Cellular controlled short-range communication for cooperative P2P networking. *Wireless Personal Communications*, **48** (1), 141–155.

[31] C.H. Yu, K. Doppler, C. Ribeiro, and O. Tirkkonen (2011) Resource sharing optimization for device-to-device communication underlaying cellular networks. *IEEE Transactions on Wireless Communications*, **10** (8), 2752–2763.

[32] K. Doppler, M. Rinne, C. Wijting, C. Ribeiro, and and K. Hugl (2009) Device-to-device communication as an underlay to LTE-Advanced networks. *IEEE Communications Magazine*, **47**(12), 42–49.

[33] B. Kaufman, J. Lilleberg, and B. Aazhang (2013) Spectrum sharing scheme between cellular users and ad-hoc device-to-device users. *IEEE Transactions on Wireless Communications*, **12** (3), 1038–1049.

[34] M. Corson, R. Laroia, J. Li, et al. (2010) Toward proximity-aware internetworking. *IEEE Wireless Communications*, **17** (6), 26–33.

[35] S. Xu, H. Wang, T. Chen, et al. (2012) Device-to-device communication underlaying cellular networks: connection establishment and interference avoidance. *KSII Transactions on Internet and Information Systems*, **6** (1), 203–228.

[36] H. Min, W. Seo, J. Lee, et al. (2011) Reliability improvement using receive mode selection in the device-to-device uplink period underlaying cellular networks. *IEEE Transactions on Wireless Communications*, **10** (2), 413–418.

[37] K. Huang, V. Lau, and Y. Chen (2009) Spectrum sharing between cellular and mobile ad hoc networks: transmission-capacity trade-off. *IEEE Journal on Selected Areas in Communications*, **27** (7), 1256–1267.

[38] 3GPP (2013) TR 37.842: 3GPP/WLAN RAN interworking study item report.

[39] L. Wang and G.S. Kuo (2013) Mathematical modeling for network selection in heterogeneous wireless networks – a tutorial. *IEEE Communications Surveys & Tutorials*, **15**, 271–292.

[40] M. Peng, D. Liang, Y. Wei, et al. (2013) Self-configuration and self-optimization in LTE-Advanced heterogeneous networks. *IEEE Communications Magazine*, **51**, 36–45.

[41] H. Dhillon, R. Ganti, F. Baccelli, and J. Andrews (2012) Modeling and analysis of K-tier downlink heterogeneous cellular networks. *IEEE Journal on Selected Areas in Communications*, **30** (3), 550–560.

[42] H. Kim and G. de Veciana (2010) Leveraging dynamic spare capacity in wireless systems to conserve mobile terminals' energy. *IEEE/ACM Transactions on Networking*, **18** (3), 802–815.

[43] G. Li, Z. Xu, C. Xiong, et al. (2011) Energy-efficient wireless communications: tutorial, survey, and open issues. *IEEE Wireless Communications*, **18** (6), 28–35.

[44] E. Aryafar, A. Keshavarz-Haddad, M. Wang, and M. Chiang (2013) RAT selection games in HetNets. *Proc. of IEEE INFOCOM, pp.* 998–1006.

[45] J.G. Andrews, R. Ganti, M. Haenggi, and N. Jindal (2010) A primer on spatial modeling and analysis in wireless networks. *IEEE Communications Magazine*, **48**, 156–163.

[46] H. ElSawy, E. Hossain, and M. Haenggi (2013) Stochastic geometry for modeling, analysis, and design of multi-tier and cognitive cellular wireless networks: a survey. *IEEE Communications Surveys & Tutorials*, **15**, 996–1019.

[47] J.G. Andrews (2013) Seven ways that HetNets are a cellular paradigm shift. *IEEE Communications Magazine*, **51**, 136–144.

[48] F. Baccelli, N. Khude, R. Laroia, et al. (2012) On the design of device-to-device autonomous discovery, *Proc. of the International Conference on Communication Systems and Networks (COMSNETS)*, pp. 1–9.

[49] M. Erturk, S. Mukherjee, H. Ishii, and H. Arslan (2013) Distributions of transmit power and SINR in device-to-device networks. *IEEE Communications Letters*, **17** (2), 273–276.

[50] O. Galinina, S. Andreev, M. Gerasimenko, et al. (2014) Capturing spatial randomness of heterogeneous cellular/WLAN deployments with dynamic traffic. *IEEE Journal on Selected Areas in Communications*, **32** (6), 1083–1099.

[51] S. Foss and S. Zuyev (1996) On a Voronoi aggregative process related to a bivariate poisson process. *Advances in Applied Probability*, **28**, 965–981.

[52] 3GPP (2010)TR 36.814: Further advancements for E-UTRA physical layer aspects.

[53] 3GPP (2013) TR 36.819: Coordinated multi-point operation for LTE physical layer aspects.

8

Distributed Resource Allocation in 5G Cellular Networks

Monowar Hasan and Ekram Hossain
University of Manitoba, Canada

8.1 Introduction

The fifth generation (5G) of cellular networks is expected to provide a wide variety of high rate multimedia services, with speeds of 300 Mbps and 60 Mbps in the downlink and uplink, respectively, at 95% of locations and times [1]. The 5G communication platform is seen as a global unified standard with seamless connectivity across existing standards: high speed packet access, Long-Term Evolution-Advanced (LTE-A), and wireless fidelity (WiFi). Some of the emerging features and trends of 5G networks are:

- multi-tier dense heterogeneous networks [2, 3]
- device-to-device (D2D) and machine-to-machine (M2M) communications [3, 4]
- densification of the heterogeneous base stations (e.g., extensive use of relays and small cells) [5]
- cloud-based radio access network [3]
- integrated use of multiple radio access technologies [6]
- wireless network virtualization [3]
- massive and 3D MIMO [3, 7]
- millimeter waves [8]
- full duplex [9] communications.

Conventional 3G systems are single-tier and based on code division multiple access (CDMA) technology. In CDMA systems, all network nodes use the same frequency resource and are distinguished from each other by different pseudo-random spreading codes, which are

Towards 5G: Applications, Requirements and Candidate Technologies, First Edition.
Edited by Rath Vannithamby and Shilpa Talwar.

Companion website: www.wiley.com/go/vannithamby/towards5g

not exactly orthogonal. Therefore, interference among the nodes is closely related to transmit power. Through efficient power control and spreading schemes [10, 11], interference in CDMA systems can be well-managed. 4G systems, such as LTE and LTE-A, employ orthogonal frequency division multiplexing (OFDM) to improve the spectrum efficiency. Due to unplanned deployment of small cells, resource allocation and interference management in 4G networks is quite different to those in single-tier 3G networks. In a heterogeneous 4G network, which mainly consists of macro and small cells, there is always a major interferer, leading to higher dominant-interference ratios compared to single-tier networks [12]. Therefore, advanced interference management schemes, such as almost blank subframe (ABS)[1] and coordinated multi-point transmission (CoMP)[2], have been developed for LTE-A 4G networks.

In future 5G networks, dense deployments of small cells – for example pico and femto cells – and heterogeneous nodes – relays, low power access points, autonomous M2M sensors, and so on – are envisioned to improve the overall network capacity and spectrum efficiency. 5G cellular wireless systems will have a multi-tier architecture, consisting of macrocells, different types of licensed small cells, and D2D/M2M networks, to serve users with different quality-of-service (QoS) requirements in a spectrum-efficient manner. In such ultra dense 5G systems, radio resource management, for example interference mitigation and resource allocation, will be highly challenging due to the irregular and pseudo-random network topology, and therefore, existing management schemes may not be sufficient. For example, it has been shown that in single-tier systems, less than 5% of network nodes experience interference from more than two major interferers. However, in 5G heterogeneous networks, this number is expected to be 40%: almost half of nodes will be affected by more than two interferers [12, 15].

Considering the dense deployment and large number of network nodes, resource allocation and interference management are the fundamental research challenges for such multi-tier heterogeneous networks. In this chapter, we consider the radio resource allocation problem in a multi-tier orthogonal frequency division multiple access (OFDMA)-based cellular network such as 5G LTE-A. In particular, we present three novel approaches for distributed resource allocation in such networks utilizing the concepts of stable matching, factor graph-based message passing, and distributed auction.

Matching theory, a sub-field of economics, is a promising concept for distributed resource management in wireless networks. Matching theory allows low-complexity algorithmic manipulations to provide a decentralized self-organizing solution to the resource-allocation problems. In matching-based resource allocation, each of the agents – the radio resources and transmitter nodes – ranks the other using a preference relation. The solution of the matching is able to assign the resources to the transmitters depending on their preferences.

The message-passing approach for resource allocation provides a polynomial time-complexity solution by distributing the computational load among the nodes in the network.

[1] The ABS approach coordinates the subframe utilization across different cells in the time domain. Without transmitting data signals, only necessary control signals are transmitted in the subframes that are configured as ABSs in an aggressor cell; that is, the dominant interferer cell. Therefore, the user equipments (UEs) in neighboring cells suffering strong interference can be scheduled with higher data transmission priorities. For details refer to the literature [13].

[2] The basic idea of CoMP is to avoid the interference among adjacent cells. This can be achieved by a coordinated spatial-domain inter-cell scheduling, or transforming the interfering signals to desired signals via joint transmission and reception among multiple transmission points [14].

In the radio resource allocation problems, the decision-making agents – the radio resources and the transmitters – form a virtual graphical structure. Each node computes and exchanges simple messages with neighboring nodes in order to find the solution of the resource allocation problem.

Similar to matching-based allocation, the auction method is also inherited from economics and can be used in wireless-resource allocation problems. Resource allocation algorithms based on the auction method provide polynomial-complexity solutions, which are shown to output near-optimal performance. The auction process evolves with a bidding process, in which unassigned agents – transmitters – raise the cost and bid for resources simultaneously. Once the bids from all the agents are available, the resources are assigned to the highest bidder.

We illustrate each of the modeling schemes with respect to a practical radio-resource allocation problem. In particular, we consider a multi-tier network consisting of a macro base station (MBS), a set of small-cell base stations (SBSs) and corresponding small-cell user equipments (SUEs), as well as D2D user equipments (DUEs). There is a common set of radio resources – resource blocks (RBs) – available to the network tiers: the MBS, SBSs, and DUEs. The SUEs and DUEs use the available resources – the RB and power level – in an underlay manner as long as the interference caused to the macro-tier – the macro user equipments (MUEs) – remains below a given threshold. The goal of resource allocation is twofold:

- to allocate the available RBs and transmit power levels to the SUEs and DUEs in order to maximize the spectral efficiency, which is defined by the sum data rate of the SUEs and DUEs
- to keep the interference caused to the macro-tier – the MUEs – by the underlay transmitters (SBSs and DUEs) within an acceptable limit.

We show that due to the nature of the resource allocation problem, the centralized solution is computationally expensive and also incurs a huge signaling overhead. Therefore, it may not be feasible to solve the problem by a single centralized controller node, such as MBS, especially in a dense network. Therefore, distributed solutions with low signaling overhead are desirable.

We assume that readers are familiar with the basics of OFDMA-based cellular wireless networks such as LTE-A, as well as having preliminary background in theory of computing: data structures, algorithms, and computational complexity. Following a brief theoretical overview of the modeling tools – stable matching, message passing, and the auction algorithm – we present the distributed solution approaches for the resource allocation problem in the network setup described previously. We also provide a brief qualitative comparison of performance metrics for these approaches: complexity, convergence, algorithm overhead, and so on.

The organization of the rest of the chapter is as follows: beginning with the brief overview of the 5G multi-tier network architecture in Section 8.2, we present the system model, related assumptions, and the resource allocation problem for such a multi-tier network architecture in Section 8.3. The distributed solutions for the resource allocation problem – stable matching, message passing, and the auction method – are discussed in Sections 8.4, 8.5, and 8.6, respectively. The qualitative comparisons of the resource allocation approaches are presented in Section 8.7. We conclude the chapter in Section 8.8 by highlighting the directions for future research.

8.2 Multi-tier 5G Cellular: Overview and Challenges

8.2.1 Overview

The visions and requirements of 5G networks include [2, 5, 16]:
capacity and throughput improvements:

- 1,000-fold throughput improvement over 4G, cell data rates of 10 Gb/s, and signaling loads of less than 1–100%
- reduced latency: 2–5 ms end-to-end latency
- network densification: approximately 1000 times higher mobile data per unit area, 100–10,000 times higher number of connecting devices
- improved energy efficiency: 10 times longer battery life.

A promising solution to meet the 5G performance requirements is to use multiple tiers in the network architecture with a co-channel deployment scenario [3, 17]. In addition to the conventional macrocell-tier – an MBS with corresponding MUEs – these heterogeneous network tiers may include low-power nodes such as pico and femto cells and relays, as well as wireless peer-to-peer (P2P) nodes, such as D2D and M2M UEs, sensors, and so on. The heterogeneity of base stations – macrocells and small cells – not only improves the spectral efficiency but also provides improved coverage. With the reduced cell size in pico and femto cells the area spectral efficiency is increased through higher spectrum reuse. Additionally, the coverage can be improved by deploying indoor small cells in homes, office buildings, and public vehicles. By reusing the cellular radio resources, wireless P2P communication – D2D/M2M communication among UEs and autonomous sensors – underlaying cellular architecture can significantly increase the overall spectrum and energy efficiency. In addition, network-controlled P2P communications in 5G systems will allow other nodes, such as relays or M2M gateways, rather than the MBS, to control the communications among P2P nodes [18]. It is worth mentioning that the deployments of heterogeneous nodes in 5G systems will be at significantly higher density than current single-tier networks [5]. A highly dense 5G multi-tier heterogeneous network scenario is illustrated in Figure 8.1.

8.2.2 Challenges in Radio Resource Management for Multi-tier Cellular Systems

Due to the dense deployment of heterogeneous nodes in 5G networks, one approach to improve the resource utilization is to use the available resources in a spectrum underlay manner. However, for underlay communication networks, mitigating inter-cell and inter/intra-tier interference is one of the key challenges. In addition to the heterogeneity and dense deployment of wireless devices, coverage and traffic load imbalances due to the varying transmit powers of different base stations, and different access restrictions – public, private, hybrid, and so on – in different tiers make the interference management and resource allocation problems more challenging than in conventional single-tier systems. Different channel access priorities and the provision of P2P communication also complicate the dynamics of the interference. Nevertheless, the adoption of multiple tiers in the cellular network architecture will

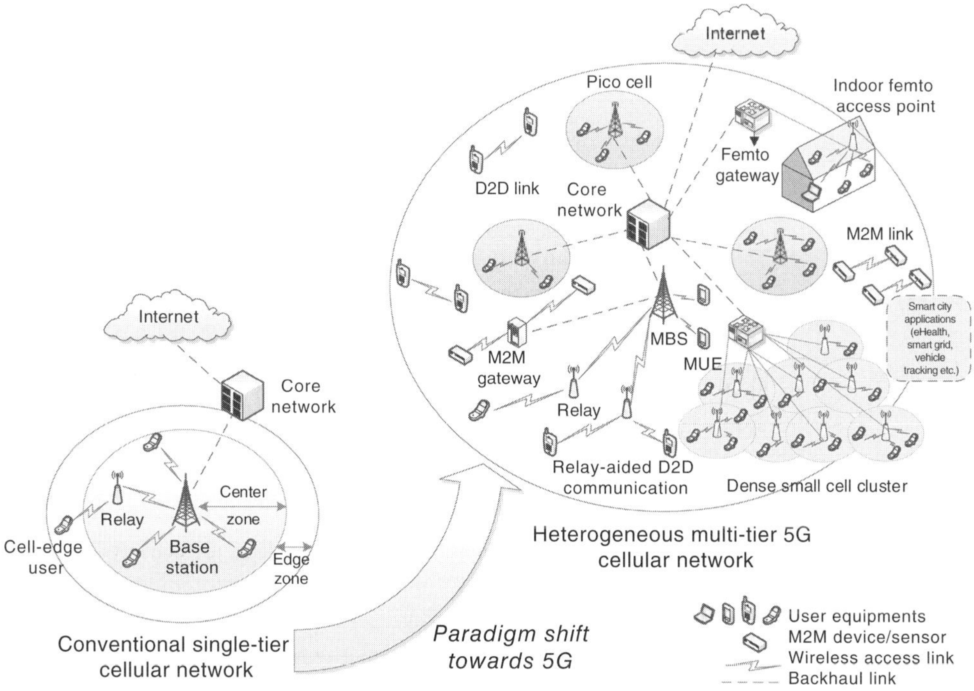

Figure 8.1 From conventional single-tier systems to future heterogeneous multi-tier 5G cellular networks.

provide better performance in terms of capacity, spectral efficiency, coverage, and power consumption; provided that there exists an efficient inter-tier and intra-tier interference management scheme [17].

Recent studies have revealed that the optimum radio resource allocation in a future OFDMA-based multi-tier network is generally an NP-hard problem, and hence computationally intractable [19, 20, 21]. Centralized methods – brute-force, sub-optimal, and heuristic-based – for solving the resource allocation problems in a dense network are not scalable. Moreover, with the centralized approaches, the controller node could be a bottleneck due to the requirement for global information to manage all the network tiers: small cells, D2D/M2M UEs, and so on. On the other hand, distributed or semi-distributed approaches with low signaling overhead are suitable for dense multi-tier networks, where the network nodes, such as SBSs and P2P nodes, perform resource allocation independently or with the minimal assistance of the central controllers: the MBSs. In addition, the distributed methods may turn out to be efficient solutions for practical implementations due to the reduced computational complexity.

This chapter aims to provide an outline of distributed resource management approaches for such multi-tier 5G network architectures. The term distributed refers to the fact that the underlay nodes in a macrocell independently determine the allocation with the minimal assistance of an MBS. The key mathematical symbols and notations used in the chapter are summarized in Table 8.1.

Table 8.1 List of major notations.

Notation	Physical interpretation
Network model	
$\mathcal{U}^{\mathrm{m}}, \mathcal{U}^{\mathrm{s}}, \mathcal{U}^{\mathrm{d}}$	Set of MUE, SUE, and D2D pairs, respectively
$\mathcal{K}^{\mathrm{T}}, \mathcal{K}^{\mathrm{R}}$	Set of underlay transmitters and receivers, respectively
$\mathcal{N}$, L	Set of RBs and power levels, respectively
K, N, L	Total number of underlay transmitters, RBs, and power levels, respectively
u_k	The UE associated with underlay transmitter k
$x_k^{(n,l)}, \mathbf{X}$	Allocation indicator, whether transmitter k using resource $\{n, l\}$ and the indicator vector, respectively
$g_{i,j}^{(n)}$	Channel gain between link i,j over RB n
$\gamma_{u_k}^{(n)}$	SINR in RB n for the UE u_k
$\Gamma_{u_k}^{(n,l)}$	Achievable SINR of the UE u_k over RB n using power level l
$p_k^{(n)}$	Transmit power of transmitter k over RB n
R_{u_k}	Achievable data rate for u_k
$I^{(n)}, I_{\max}^{(n)}$	Aggregated interference and threshold limit for the RB n, respectively
$\mathfrak{U}_k^{(n,l)}$	Utility for transmitter k using resource $\{n, l\}$
Stable matching	
μ	Matching (e.g., allocation) of transmitter to the resources
$i_1 \succeq_j i_2$	Preference relation for agent j (i.e., i_1 is more preferred than i_2)
$\mathcal{P}_k(\mathcal{N},\mathcal{L})$, $\mathcal{P}_n(\mathcal{K}^{\mathrm{T}},\mathcal{L})$	Preference profile for the transmitter k and RB n, respectively
Message passing	
$\delta_{\{n,l\}\to k}(x_k^{(n,l)})$	Message delivered by the resource $\{n, l\}$ to the transmitter k
$\delta_{k\to\{n,l\}}(x_k^{(n,l)})$	Message from transmitter k to the resource $\{n, l\}$
$\psi_{\{n,l\}\to k}$	Normalized message from the resource $\{n, l\}$ to the transmitter k
$\psi_{k\to\{n,l\}}$	Normalized message from the transmitter k to the resource $\{n, l\}$
$\tau_k^{(n,l)}$	Node marginals for the transmitter k using resource $\{n, l\}$
Auction method	
$C_k^{(n,l)}$	Cost for transmitter k using resource $\{n, l\}$
$B_k^{(n,l)}$	Data rate (multiplied by a weighting factor) achieved by transmitter k using resource $\{n, l\}$
$\mathfrak{b}_k^{(n,l)}$	Local bidding information available to transmitter k for the resource $\{n, l\}$
ϵ	Minimum bid increment parameter
$\Theta_k = \{n, l\}$	Assignment of resource $\{n, l\}$ to the transmitter k
Miscellaneous	
$\lvert\mathbf{y}\rvert$, $\lvert\mathcal{Y}\rvert$	Length (cardinality) of the vector (set) $\mathbf{y}$ ($\mathcal{Y}$)
$y(t)$	Value of variable y at any iteration t
$z := y$	Assignment of the value of variable y to the variable z
/* *comment* */	Commented text inside algorithms

8.3 System Model

8.3.1 Network Model and Assumptions

Let us consider a transmission scenario of a heterogeneous network as shown in Figure 8.2. The network consists of one MBS and a set of C cellular MUEs: $\mathcal{U}^{\mathrm{m}} = \{1,2,\cdots,C\}$. There are also D D2D pairs and a cluster of S SBSs located within the coverage area of the MBS. The set of SBSs is denoted by $\mathcal{S} = \{1,2,\cdots S\}$. For simplicity we assume that each SBS serves only one SUE for a single time instance and the set of SUE is given by $\mathcal{U}^{\mathrm{s}}$, where $|\mathcal{U}^{\mathrm{s}}| = S$. The set of D2D pairs is denoted as $\mathcal{U}^{\mathrm{d}} = \{1,2,\cdots,D\}$. In addition, the dth element of the sets $\mathcal{U}^{\mathrm{d}_T}$ and $\mathcal{U}^{\mathrm{d}_R}$ denotes the transmitter and receiver UE of the D2D pair $d \in \mathcal{U}^{\mathrm{d}}$, respectively. The set of UEs in the network is given by $\mathcal{U} = \mathcal{U}^{\mathrm{m}} \cup \mathcal{U}^{\mathrm{s}} \cup \mathcal{U}^{\mathrm{d}}$. For notational convenience, we denote $\mathcal{K}^{\mathrm{T}} = \mathcal{S} \cup \mathcal{U}^{\mathrm{d}_T}$, the set of underlay transmitters (SBSs and transmitting D2D UEs) and $\mathcal{K}^{\mathrm{R}} = \mathcal{U}^{\mathrm{s}} \cup \mathcal{U}^{\mathrm{d}_R}$, the set of underlay receivers (SUEs and receiving D2D UEs).

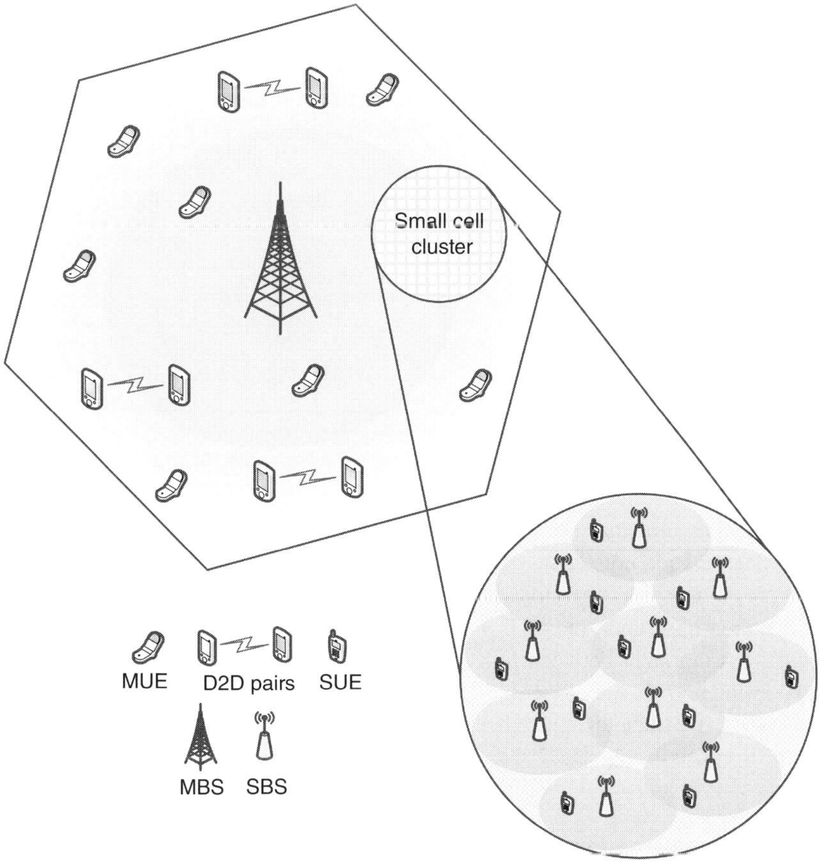

Figure 8.2 Schematic diagram of the heterogeneous network model. The D2D pairs, SBSs, and SUEs are underlaid within the macro-tier by reusing same set of radio resources.

The SBSs and DUEs are underlaid within the macro-tier (the MBS and MUEs). Both the macro-tier and the underlay-tier (the SBSs, SUEs, and D2D pairs) use the same set $\mathcal{N} = \{1, 2, \cdots N\}$ of orthogonal RBs[3]. Each transmitter node in the underlay-tier (SBS and D2D transmitter) selects one RB from the available N RBs. In addition, the underlay transmitters are capable of selecting the transmit power from a finite set of power levels: $\mathcal{L} = \{1, 2, \cdots L\}$. Each SBS and D2D transmitter should select a suitable RB-power level combination. This RB-power level combination is referred to as transmission alignment [23].[4] For each RB $n \in \mathcal{N}$, there is a predefined threshold $I_{\max}^{(n)}$ for maximum aggregated interference caused by the underlay-tier to the macro-tier. We assume that value of $I_{\max}^{(n)}$ is known to the underlay transmitters by using the feedback control channels. An underlay transmitter (SBS or transmitter DUE) is allowed to use the particular transmission alignment as long as the cross-tier interference to the MUEs is within the threshold limit.

The system model considered here is a multi-tier heterogeneous network, since each of the network tiers (macro-tier and underlay-tier consisting of small cells and D2D UEs) has a different transmit power range and coverage region, and a specific set of users with different application requirements. It is assumed that the user association to the base stations (either MBS or SBSs) is completed prior to resource allocation. In addition, the potential DUEs are discovered during the D2D session setup by transmitting a known synchronization or reference signal (beacons) [24]. According to our system model, only one MUE is served on each RB to avoid co-tier interference within the macro-tier. However multiple underlay UEs (SUEs and DUEs) can reuse the same RB to improve the spectrum utilization. This reuse causes severe cross-tier interference to the MUEs, and also co-tier interference within the underlay-tier, which leads the requirement for an efficient resource allocation scheme.

8.3.2 Achievable Data Rate

The MBS transmits to the MUEs using a fixed power $p_M^{(n)} > 0$ for $\forall n$. For each underlay transmitter $k \in \mathcal{K}^{\mathrm{T}}$, the transmit power over the RBs is determined by the vector $\mathbf{P}_k = [p_k^{(1)}, p_k^{(2)}, \cdots, p_k^{(N)}]^{\mathrm{T}}$, where $p_k^{(n)} \geq 0$ denotes the the transmit power level of the transmitter k over RB n. The transmit power $p_k^{(n)}, \forall n$ must be selected from the finite set of power levels $\mathcal{L}$. Note that if the RB n is not allocated to the transmitter k, the corresponding power variable $p_k^{(n)} = 0$. Since we assume that each underlay transmitter selects only one RB, only one element in the power vector $\mathbf{P}_k$ is non-zero.

All links are assumed to experience independent block fading. We denote by $g_{i,j}^{(n)}$ the channel gain between the links i and j over RB n and defined by $g_{i,j}^{(n)} = \beta_{i,j}^{(n)} d_{i,j}^{-\alpha}$, where $\beta_{i,j}^{(n)}$ denotes the channel fading component between link i and j over RB n, $d_{i,j}$ is the distance between node i and j, and α is the path-loss exponent.

For the SUEs, we denote u_k as the SUE associated to SBS $k \in \mathcal{S}$, and for the DUEs, u_k refer to the receiving D2D UE of the D2D transmitter $k \in \mathcal{U}^{\mathrm{d}_T}$. The received

[3] The minimum scheduling unit of LTE-A standard is referred to as an RB. One RB consists of 12 subcarriers (say, 180 kHz) in the frequency domain and one sub-frame (say, 1 ms) in the time domain. For a brief overview of heterogeneous network in the context of LTE-A standard refer to the book by Hossain [22, Chapter 1].

[4] Throughout this chapter we use the terms "resource" and "transmission alignment" interchangeably.

signal-to-interference-plus-noise ratio (SINR) for any arbitrary SUE or D2D receiver; that is, $u_k \in \mathcal{K}^{\mathrm{R}}, k \in \mathcal{K}^{\mathrm{T}}$, over RB n is given by

$$\gamma_{u_k}^{(n)} = \frac{g_{k,u_k}^{(n)} p_k^{(n)}}{\underbrace{g_{M,u_k}^{(n)} p_M^{(n)}}_{\text{interference from macro-tier}} + \underbrace{\sum_{k' \in \mathcal{K}^{\mathrm{T}}, k' \neq k} g_{k',u_k}^{(n)} p_{k'}^{(n)}}_{\text{interference from underlay-tier}} + \sigma^2} \tag{8.1}$$

where $g_{k,u_k}^{(n)}$ is the link gain between the SBS and SUE ($u_k \in \mathcal{U}^{\mathrm{s}}, k \in \mathcal{S}$) or the link gain between the D2D UEs ($u_k \in \mathcal{U}^{\mathrm{d}_R}, k \in \mathcal{U}^{\mathrm{d}_T}$), and $g_{M,u_k}^{(n)}$ is the interference gain between the MBS and the UE u_k. In Eq. (8.1), the variable $\sigma^2 = N_0 B_{\mathrm{RB}}$ where B_{RB} is the bandwidth corresponding to an RB and N_0 denotes the thermal noise. Similarly, the SINR for the MUE $m \in \mathcal{U}^{\mathrm{m}}$ over RB n can be written as follows:

$$\gamma_m^{(n)} = \frac{g_{M,m}^{(n)} p_M^{(n)}}{\sum_{k \in \mathcal{K}^{\mathrm{T}}} g_{k,m}^{(n)} p_k^{(n)} + \sigma^2} \tag{8.2}$$

Given the SINR, the data rate of the UE $u \in \mathcal{U}$ over RB n can be calculated according to the Shannon formula: $R_u^{(n)} = B_{\mathrm{RB}} \log_2(1 + \gamma_u^{(n)})$.

8.3.3 Formulation of the Resource Allocation Problem

The objective of the resource allocation problem is to obtain the assignment of RB and power level (transmission alignment) for the underlay UEs (the D2D UEs and the SUEs) that maximizes the achievable sum data rate. The RB and power-level allocation indicator for any underlay transmitter $k \in \mathcal{K}^{\mathrm{T}}$ is denoted by a binary decision variable $x_k^{(n,l)}$ where

$$x_k^{(n,l)} = \begin{cases} 1, & \text{if the transmitter } k \text{ is transmitting over RB } n \text{ with power level } l \\ 0, & \text{otherwise} \end{cases} \tag{8.3}$$

Note that the decision variable $x_k^{(n,l)} = 1$ implies that $p_k^{(n)} = l$. Let $K = S + D$ denote the total number of underlay transmitters. The achievable data rate of an underlay UE u_k with the corresponding transmitter k is written as

$$R_{u_k} = \sum_{n=1}^{N} \sum_{l=1}^{L} x_k^{(n,l)} B_{\mathrm{RB}} \log_2\left(1 + \gamma_{u_k}^{(n)}\right) \tag{8.4}$$

The aggregated interference experienced on RB n is given by $I^{(n)} = \sum_{k=1}^{K} \sum_{l=1}^{L} x_k^{(n,l)} g_{k,m_k^*}^{(n)} p_k^{(n)}$, where $m_k^* = \underset{m}{\operatorname{argmax}}\, g_{k,m}^{(n)}, \forall m \in \mathcal{U}^{\mathrm{m}}$. In order to calculate the aggregated interference $I^{(n)}$ on RB n we use the concept of the reference user [25]. For any RB n, the interference caused by the underlay transmitter k is determined by the highest gains between the transmitter k and MUEs, that is, the MUE m_k^* who is most affected by the transmitter k. Satisfying the interference constraints considering the gain with reference user will also satisfy the interference constraints for other MUEs. As mentioned in Section 8.3.1, an underlay transmitter is allowed to use a particular transmission alignment only when it does not violate the interference threshold to the MUEs: $I^{(n)} < I^{(n)}_{\max}, \forall n$. Mathematically, the resource allocation problem can be expressed by using the following optimization formulation:

(P8.1)

$$\max_{x_k^{(n,l)},\, p_k^{(n)}} \sum_{k=1}^{K}\sum_{n=1}^{N}\sum_{l=1}^{L} x_k^{(n,l)} B_{\mathrm{RB}} \log_2\left(1+\gamma_{u_k}^{(n)}\right)$$

subject to:

$$\sum_{k=1}^{K}\sum_{l=1}^{L} x_k^{(n,l)} g_{k,m_k^*}^{(n)} p_k^{(n)} < I_{\max}^{(n)}, \quad \forall n \in \mathcal{N} \tag{8.5a}$$

$$\sum_{n=1}^{N}\sum_{l=1}^{L} x_k^{(n,l)} \le 1, \qquad \forall k \in \mathcal{K}^{\mathrm{T}} \tag{8.5b}$$

$$x_k^{(n,l)} \in \{0,1\}, \ \forall k \in \mathcal{K}^{\mathrm{T}}, \forall n \in \mathcal{N}, \forall l \in \mathcal{L} \tag{8.5c}$$

where

$$\gamma_{u_k}^{(n)} = \frac{g_{k,u_k}^{(n)} p_k^{(n)}}{g_{M,u_k}^{(n)} p_M^{(n)} + \sum\limits_{\substack{k' \in \mathcal{K}^{\mathrm{T}},\\ k' \neq k}}^{K} \sum\limits_{l'=1}^{L} x_j^{(n,l')} g_{k',u_k}^{(n)} p_{k'}^{(n)} + \sigma^2} \tag{8.6}$$

The objective of the resource allocation problem **P8.1** is to maximize the data rate of the SUEs and DUEs subject to the set of constraints given by Eqs (8.5a)–(8.5c). With the constraint in Eq. (8.5a), the aggregated interference caused to the MUEs by the underlay transmitters on each RB is limited by a predefined threshold. The constraint in Eq. (8.5b) indicates that the number of RBs selected by each underlay transmitter should be at most one and each transmitter can only select one power level at each RB. The binary indicator variable for transmission alignment selection is represented by the constraint in Eq. (8.5c).

Corollary 8.3.1 The resource allocation problem **P8.1** is a combinatorial non-convex non-linear optimization problem and the centralized solution of this problem is strongly NP-hard, especially for the large set of $\mathcal{U}$, $\mathcal{N}$, and $\mathcal{L}$.

The complexity involved in solving the problem using exhaustive search is $\mathcal{O}((N\,L)^K)$. As an example, when $N = 6, L = 3$, and $K = 3 + 2 = 5$, the decision set (i.e., the search space) contains 1,889,568 possible transmission alignments.

Considering the computational overhead, it not feasible to solve the resource allocation problem by a single central controller (MBS) in a practical system; and such a centralized solution requires all the channel state information (CSI) available to the MBS.

Due to the mathematical intractability of solving the above resource allocation problem, in the following we present three distributed heuristic solution approaches: stable-matching, factor-graph-based message-passing, and distributed auction-based approaches. The distributed solutions are developed under the assumption that the system is feasible; that is, given the resources and parameters – the size of the network, the interference thresholds, and so on – it is possible to obtain an allocation that satisfies all the constraints of the original optimization problem.

8.4 Resource Allocation using Stable Matching

The resource allocation approach using stable matching involves multiple decision-making agents – the available radio resources (transmission alignments) and the underlay transmitters. The solutions – matching between transmission alignments and transmitters – are produced by individual actions of the agents themselves. Matching requests and their confirmation or rejection are determined by the given preference profiles: the agents hold lists of preferred matches in each of the opposite set. The matching outcome yields mutually beneficial assignments between the transmitters and available resources that are individually determined by such preference lists. In our model, the preference is based on CSI parameters and achievable SINR. Stability in matching implies that, with regard to their initial preferences, neither the underlay transmitters nor the MBS (transmission alignments) have an incentive to alter the allocation.

8.4.1 *Concept of Matching*

A matching (allocation) is given as an assignment of transmission alignment to the underlay transmitters forming the set $\{k,n,l\} \in \mathcal{K}^{\mathrm{T}} \times \mathcal{N} \times \mathcal{L}$. According to our system model, each underlay transmitter is assigned to only one RB, but multiple transmitters can transmit on the same RB to improve spectrum utilization. This scheme corresponds to a many-to-one matching in the theory of stable matching. More formally, the matching can be defined as follows [26]:

Definition 8.4.1 A matching μ is defined as a function, that is, $\mu : \mathcal{K}^{\mathrm{T}} \times \mathcal{N} \times \mathcal{L} \to \mathcal{K}^{\mathrm{T}} \times \mathcal{N} \times \mathcal{L}$ such that

(i) $\mu(k) \in \mathcal{N} \times \mathcal{L}$ *and* $|\mu_l(n)| \in \{0,1\}$ *and*
(ii) $\mu(n) \in \{\mathcal{K}^{\mathrm{T}} \times \mathcal{L}\} \cup \{\varnothing\}$ *and* $|\mu(n)| \in \{1,2,\ldots,K\}$

where $\mu(k) = \{n,l\} \Leftrightarrow \mu(n) = \{k,l\}$ for $\forall k \in \mathcal{K}^{\mathrm{T}}, \forall n \in \mathcal{N}, \forall l \in \mathcal{L}$, and $|\mu(\cdot)|$ denotes the cardinality of matching outcome $\mu(\cdot)$.

Definition 8.4.1 implies that μ is a one-to-one matching if the input to the function is an underlay transmitter. On the other hand, μ is a one-to-many function – $\mu_l(n)$ is not unique – if the input to the function is an RB. The interpretation of $\mu(n) = \varnothing$ implies that for some RB $n \in \mathcal{N}$ the corresponding RB is unused by any underlay transmitter under the matching μ. The outcome of the matching determines the RB allocation vector and corresponding power level, $\mu \equiv \mathbf{X}$, where

$$\mathbf{X} = \left[x_1^{(1,1)}, \cdots, x_1^{(1,L)}, \cdots, x_1^{(N,L)}, \cdots, x_K^{(N,L)} \right]^{\mathrm{T}} \tag{8.7}$$

8.4.2 Utility Function and Preference Profile

Let the parameter $\Gamma_{u_k}^{(n,l)} \triangleq \gamma_{u_k}^{(n)}|_{p_k^{(n)}=l}$ denote the achievable SINR of the UE u_k over RB n using power level l (that is, $p_k^{(n)} = l$), where $\gamma_{u_k}^{(n)}$ is given by Eq. (8.6). We express the data rate as a function of SINR. In particular, let $\mathcal{R}(\Gamma_{u_k}^{(n,l)}) = B_{\text{RB}} \log_2(1+\Gamma_{u_k}^{(n,l)})$ denote the achievable data rate for the transmitter k over RB n using power level l. The utility of an underlay transmitter for a particular transmission alignment is determined by two factors:

1. the achievable data rate for a given RB power level combination
2. an additional cost function that represents the aggregated interference caused to the MUEs on that RB.

In particular, the utility of the underlay transmitter k for a given RB n and power level l is given by

$$\mathfrak{U}_k^{(n,l)} = w_1 \mathcal{R}\left(\Gamma_{u_k}^{(n,l)}\right) - w_2\left(I^{(n)} - I_{\max}^{(n)}\right) \tag{8.8}$$

where w_1 and w_2 are the biasing factors and can be selected based on which network tier (macro-tier or underlay-tier) should be given priority for resource allocation [23]. As mentioned earlier, each underlay transmitter and RB holds a list of preferred matches. The preference profile of an underlay transmitter $k \in \mathcal{K}^{\text{T}}$ over the set of available RBs $\mathcal{N}$ and power levels $\mathcal{L}$ is defined as a vector of linear order $\mathcal{P}_k(\mathcal{N},\mathcal{L}) = [\mathfrak{U}_k^{(n,l)}]_{n\in\mathcal{N},l\in\mathcal{L}}$. We denote by $\{n_1,l_1\} \succeq_k \{n_2,l_2\}$ that the transmitter k prefers the transmission alignment $\{n_1,l_1\}$ to $\{n_2,l_2\}$, and consequently, $\mathfrak{U}_k^{(n_1,l_1)} > \mathfrak{U}_k^{(n_2,l_2)}$. Similarly, each RB holds a preference over the underlay transmitters and power levels given by $\mathcal{P}_n(\mathcal{K}^{\text{T}},\mathcal{L}) = [\mathfrak{U}_k^{(n,l)}]_{k\in\mathcal{K}^{\text{T}},l\in\mathcal{L}}$.

8.4.3 Algorithm Development

The matching between transmission alignments to the transmitters is performed in an iterative manner, as presented in Algorithm 8.1. While a transmitter is unallocated and has a non-empty preference list, the transmitter is temporarily assigned to its first preference over transmission alignments: the pair of RB and power level, $\{n,l\}$. If the allocation to the RB n does not violate the tolerable interference limit $I_{\max}^{(n)}$, the allocation will persist. Otherwise, until the aggregated interference on the RB n is below threshold, the worst preferred transmitter(s) from the preference list of RB n will be removed, even though it was allocated previously. The process terminates when no more transmitters are unallocated. Since the iterative process dynamically updates the preference lists, the procedure mentioned previously ends up with a local stable matching [27].

Algorithm 8.1 Assignment of transmission alignments using stable matching

Input: The preference profiles $\mathcal{P}_k(\mathcal{N},\mathcal{L})$, $\forall k \in \mathcal{K}^{\text{T}}$ and $\mathcal{P}_n(\mathcal{K}^{\text{T}},\mathcal{L})$, $\forall n \in \mathcal{N}$.

Output: The transmission alignment indicator $\mathbf{X} = [x_1^{(1,1)},\cdots,x_1^{(1,L)},\cdots,x_1^{(N,L)},\cdots,x_K^{(N,L)}]^{\text{T}}$.

1: Initialize $\mathbf{X} := \mathbf{0}$.
2: **while** whole some transmitter k is unassigned **and** $\mathcal{P}_k(\mathcal{N},\mathcal{L})$ is non-empty **do**
3: $\{n_{\text{mp}},l_{\text{mp}}\}$:= most preferred RB with power level l_{mp} from the profile $\mathcal{P}_k(\mathcal{N},\mathcal{L})$.
4: Set $x_k^{(n_{\text{mp}},l_{\text{mp}})} := 1$. /* *Temporarily assign the RB and power level to the transmitter k**/

$\mathfrak{I}^{(n_{\text{mp}})} :- g_{k,m_k^*}^{(n_{\text{mp}})} l_{\text{mp}} + \sum_{\substack{k' \in \mathcal{K}^{\text{T}}, \\ k' \neq k}} \sum_{l'=1}^{L} x_{k'}^{(n_{\text{mp}},l')} g_{k',m_{k'}^*}^{(n_{\text{mp}})} p_{k'}^{(n_{\text{mp}})}$. /* *Estimate interference of* n_{mp} */

if $\mathfrak{I}^{(n_{\text{mp}})} \geq I_{\max}^{(\text{mp})}$ **then**

repeat

$\{k_{\text{lp}}, l_{\text{lp}}\}$:= least preferred transmitter with power level l_{lp} assigned to n_{mp}.

Set $x_{k_{\text{lp}}}^{(n_{\text{mp}},l_{\text{lp}})} := 0$. /* *Revoke assignment due to interference threshold violation* */

$\mathfrak{I}^{(n_{\text{mp}})} := \sum_{k'=1,}^{K} \sum_{l'=1}^{L} x_{k'}^{(n_{\text{mp}},l')} g_{k',m_{k'}^*}^{(n_{\text{mp}})} p_{k'}^{(n_{\text{mp}})}$. /* *Update interference level* */

/* *Update preference profiles* */

for each successor $\{\hat{k}_{\text{lp}}, \hat{l}_{\text{lp}}\}$ of $\{k_{\text{lp}}, l_{\text{lp}}\}$ on profile $\mathcal{P}_{n_{\text{mp}}}(\mathcal{K}^{\text{T}}, \mathcal{L})$ **do**

remove $\{\hat{k}_{\text{lp}}, \hat{l}_{\text{lp}}\}$ from $\mathcal{P}_{n_{\text{mp}}}(\mathcal{K}^{\text{T}}, \mathcal{L})$

remove $\{n_{\text{mp}}, l_{\text{mp}}\}$ from $\mathcal{P}_{\hat{k}_{\text{lp}}}(\mathcal{N}, \mathcal{L})$.

end for

until $\mathfrak{I}^{(n_{\text{mp}})} < I_{\max}^{(n_{\text{mp}})}$

end if

end while

The overall stable matching-based resource allocation approach is summarized in Algorithm 8.2. Note that Algorithm 8.1 is executed repeatedly. The convergence of Algorithm 8.2 occurs when the outcome of two consecutive local matchings is similar: $\mathbf{X}(t) = \mathbf{X}(t-1)$ and as a consequence $R(t) = R(t-1)$, where $R(t) = \sum_{k=1}^{K} R_{u_k}(t)$ denotes the achievable sum rate of the underlay-tier at iteration t.

Algorithm 8.2 Stable matching-based resource allocation

Initialization:

Estimate the CSI parameters from previous time slot.

Each underlay transmitter $k \in \mathcal{K}^{\text{T}}$ randomly selects a transmission alignment and the MBS broadcasts the aggregated interference of each RB using pilot signals.

Each underlay transmitter $k \in \mathcal{K}^{\text{T}}$ builds the preference profile $\mathcal{P}_k(\mathcal{N}, \mathcal{L})$ from the CSI estimations and the utility function given by Eq. (8.8).

For each $n \in \mathcal{N}$, the MBS builds the preference profiles $\mathcal{P}_n(\mathcal{K}^{\text{T}}, \mathcal{L})$.

Initialize number of iterations $t := 1$.

Update:

while $\mathbf{X}(t) \neq \mathbf{X}(t-1)$ **and** t is less than some predefined threshold $T_{\max}$ **do**

MBS obtains a local stable matching $\mathbf{X}(t)$ using **Algorithm 8.1**, calculates the aggregated interference $I^{(n)}(t)$ for $\forall n$ and informs the transmitters.

Each underlay transmitter $k \in \mathcal{K}^{\text{T}}$ updates the preference profile $\mathcal{P}_k(\mathcal{N}, \mathcal{L})$ based on current allocation vector $\mathbf{X}(t)$ and interference level $I^{(n)}(t)$.

MBS updates the preference profile $\mathcal{P}_n(\mathcal{K}^{\text{T}}, \mathcal{L})$ for $\forall n \in \mathcal{N}$ using $\mathbf{X}(t)$ and $I^{(n)}(t)$.

Update $t := t + 1$.

end while

Transmission:

Use the resources (e.g., the RB and power levels) allocated in the final stage of update phase for data transmission.

8.4.4 Stability, Optimality, and Complexity of the Solution

In this section, we analyze the solution obtained by the stable matching approach. The stability, optimality, and the complexity of the algorithm are discussed.

8.4.4.1 Stability

The notion of stability in the matching μ means that none of the agents (underlay transmitters or resources) prefers to change the allocation obtained by μ. The matching μ is therefore stable if no transmitter and no resource that are not allocated to each other, as given in μ, prefer each other to their allocation in μ. The transmitters and resources are said to be *acceptable* if the agents prefer each other to remain unallocated. In addition, a matching μ is called *individually rational* if no agent $\tilde{\jmath}$ prefers unallocation to the matching in $\mu(\tilde{\jmath})$. Before formally defining the stability of matching, we introduce the term *blocking pair*, which is defined as follows.

Definition 8.4.2 A matching μ is blocked by a pair of agent (i,j) if they prefer each other to the matching obtain by μ; that is, $i \succeq_j \mu(j)$ and $j \succeq_i \mu(i)$.

Using this definition, the stability of the matching can be defined [28, Ch. 5]:

Definition 8.4.3 A matching μ is stable if it is individually rational and there is no tuple (k,n,l) within the set of acceptable agents such that k prefers $\{n,l\}$ to $\mu(k)$ and n prefers $\{k,l\}$ to $\mu(n)$; that is, it is not blocked by any pair of agents.

The following theorem shows that the solution obtained by the matching algorithm is stable.

Theorem 8.4.4 ***The assignment performed in Algorithm 8.1 leads to a stable allocation.***

Proof. We prove the theorem by contradiction. Let μ be a matching obtained by Algorithm 8.1. Let us assume that the resource $\{n,l\}$ is not allocated to the transmitter k, but it belongs to a higher order in the preference list. According to this assumption, the tuple (k,n,l) will block μ. Since the position of the resource $\{n,l\}$ in the preference profile of k is higher than any resource $\{\hat{n},\hat{l}\}$ that is matched by μ; that is, $\{n,l\} \succeq_k \mu(k)$, transmitter k must select $\{n,l\}$ before the algorithm terminates. Note that the resource $\{n,l\}$ is not assigned to transmitter k in the matching outcome μ. This implies that k is unassigned with the resource $\{n,l\}$ (see line 9 in Algorithm 8.1) and $(k,\hat{n},\hat{l})$ is a better assignment. As a result, the tuple (k,n,l) will not block μ, which contradicts our assumption. The proof concludes since no blocking pair exists and, therefore, the matching outcome μ leads to a stable matching.

It is worth mentioning that the assignment is stable at each iteration of Algorithm 8.1. Since, after evaluation of the utility, the preference profiles are updated and the matching subroutine is repeated, a stable allocation is obtained at each iteration.

8.4.4.2 Optimality

The optimality property of the stable matching approach can be observed using the definition of weak Pareto optimality. Let $\mathcal{R}_\mu$ denote the sum-rate obtained by matching μ. A matching μ is weak Pareto optimal if there is no other matching $\hat{\mu}$ that can achieve a better sum-rate; that is, $\mathcal{R}_{\hat{\mu}} \geq \mathcal{R}_\mu$ [26].

Theorem 8.4.5 ***The stable matching-based resource allocation algorithm is weak Pareto optimal.***

Proof. Let us consider μ to be the stable allocation obtained by Algorithm 8.1. For instance, let $\hat{\mu}$ be an arbitrary stable outcome better than μ; in other words, $\hat{\mu}$ can achieve a better sum-rate. Since the allocation $\hat{\mu}$ is better than μ, there exists at least one resource $\{\hat{n},\hat{l}\}$ allocated to transmitter k in $\hat{\mu}$, and k is allocated to the resource $\{n,l\}$ in μ. According to our assumption, k prefers $\{\hat{n},\hat{l}\}$ to $\{n,l\}$, and let $\{\hat{n},\hat{l}\}$ be allocated to transmitter $\hat{k}$ in μ. It is obvious that resource $\{\hat{n},\hat{l}\}$ is better than $\{n,l\}$ to transmitter k and $\{k,l\}$ is better than $\{\hat{k},\hat{l}\}$ to resource $\hat{n}$; that is, $\{\hat{n},\hat{l}\} \succeq_k \{n,l\}$ and $\{k,l\} \succeq_{\hat{n}} \{\hat{k},\hat{l}\}$. By the definition of blocking pair, μ is blocked by $(k,\hat{n},\hat{l})$ and hence μ is unstable. This contradicts our assumption that μ is a stable allocation. Since there is no stable outcome $\hat{\mu}$ which is better that μ, by definition μ is an optimal allocation.

8.4.4.3 Complexity

It is possible to show that the stable matching algorithm will iterate for finite number of times.

Theorem 8.4.6 ***The RB allocation subroutine terminates after some finite step T'.***

Proof. Let the finite set $\tilde{\mathcal{X}}$ represent the all possible combinations of transmitter-resource matching where each element $\tilde{x}_k^{(n,l)} \in \tilde{\mathcal{X}}$ denotes the resource $\{n,l\}$ is allocated to the transmitter k. Since no transmitter is rejected by the same resource more than once (see line 9 in Algorithm 8.1), the finiteness of the set $\tilde{\mathcal{X}}$ ensures the termination of the matching subroutine in finite number of steps.

For each underlay transmitter, the complexity to build the preference profile using any standard sorting algorithm is $\mathcal{O}(NL\log(NL))$ (line 8, Algorithm 8.2). Similarly, in line 9, the complexity to output the ordered set of preference profile for the RBs is of $\mathcal{O}(NKL\log(KL))$. Let $\xi = \sum_{k=1}^{K}|\mathcal{P}_k(\mathcal{N},\mathcal{L})| + \sum_{n=1}^{N}|\mathcal{P}_n(\mathcal{K}^{\mathrm{T}},\mathcal{L})| = 2KNL$ be the total length of input preferences in Algorithm 8.1, where $|\mathcal{P}_j(\cdot)|$ denotes the length of profile vector $\mathcal{P}_j(\cdot)$. From Theorem 8.4.6 and also work by O'Malley [29, Ch. 1] it can be shown that, if implemented with suitable data structures, the time complexity of the RB allocation subroutine is linear in the size of input preference profiles: $\mathcal{O}(\xi) = \mathcal{O}(KNL)$. Since the update phase of Algorithm 8.2 runs at most fixed $T < T_{\max}$ iterations, the complexity of the stable matching-based solution is linear in K, N, L.

8.5 Message-passing Approach for Resource Allocation

In the following, we reformulate the resource allocation problem **P8.1** in such a way that can be solved with a message-passing (MP) technique. The MP approach involves computation of the marginals: the messages exchanged between the nodes of a specific graphical model. Among different representations of graphical model, we consider a factor graph based MP scheme. A factor graph is made up of two different types of nodes: function and variable nodes. An edge connects a function (factor) node to a variable node if and only if the variable appears in the function. Mathematically, this can be expressed as follows [30]:

Definition 8.5.1 *A factor graph can be represented by a $\mathcal{V}$-$\mathcal{F}$ bipartite graph where $\mathcal{V} = \{v_1, \cdots v_a\}$ is the set of variable nodes and $\mathcal{F} = \{f_1(\cdot), \cdots f_b(\cdot)\}$ is the set of function (e.g., factor) nodes. The connectivity (e.g., edges) of the factor graph can be represented by an $a \times b$*

binary matrix $\mathbf{E} = [E_{i,j}]$ *where* $E_{i,j} = 1$ *if the variable node i is connected with the factor node j and* $E_{i,j} = 0$, *otherwise.*

8.5.1 Overview of the MP Scheme

Before presenting the detailed resource allocation approach for a heterogeneous scenario, we briefly introduce the generic MP scheme. Details of a factor graph-based MP scheme can be found in the literature [30]. Let us consider the maximization of an arbitrary function $f(v_1,\cdots,v_J)$ over all possible values of the argument, $Z = \max_{\mathbf{v}} f(\mathbf{v})$, where $\mathbf{v} = [v_1,\cdots,v_J]^{\mathrm{T}}$. We denote by $\max_{\mathbf{v}}$ that the maximization is computed over all possible combinations of the elements of the the vector $\mathbf{v}$. The *marginal* of Z with respect to variable v_j is given by $\phi_j(v_j) = \max_{\sim(v_j)} f(\mathbf{v})$ where $\max_{\sim(\cdot)}$ denotes the maximization over all variables except $(\cdot)$. Let us now decompose $f(\mathbf{v})$ into a summation of I functions; that is, $\sum_{i=1}^{I} f_i(\hat{v}_i)$ where $\hat{v}_i$ is a subset of the elements of the vector $\mathbf{v}$ and $\mathbf{f} = [f_1(\cdot),\cdots,f_I(\cdot)]^{\mathrm{T}}$ is the vector of I functions. In addition, let $\mathfrak{f}_j$ represent a subset of functions in $\mathbf{f}$ where the variable v_j appears. Hence the marginal can be rewritten as $\phi_j(v_j) = \mathit{max}_{\sim(v_j)} \sum_{i=1}^{I} f_i(\hat{v}_i)$. According to the max-sum MP strategy the message passed by any variable node v_j to any generic function node $f_i(\cdot)$ is given by $\delta_{v_j \to f_i(\cdot)}(v_j) = \sum_{i' \in \mathfrak{f}_j, i' \neq i} \delta_{f_{i'}(\cdot) \to v_j}(v_j)$. Similarly, the message from function node $f_i(\cdot)$ to variable node v_j is given as $\delta_{f_i(\cdot) \to v_j}(v_j) = \max_{\sim(v_j)} \left(f_i(v_1,\cdots,v_J) + \sum_{j' \in \hat{v}_i, j' \neq j} \delta_{v_{j'} \to f_i(\cdot)}(v_{j'}) \right)$. When the factor graph is cycle free – there is a unique path connecting any two nodes – all the variables nodes $j = \{1,\cdots,J\}$ can compute the marginals as $\phi_j(v_j) = \sum_{i=1}^{I} \delta_{f_i(\cdot) \to v_j}(v_j)$. Utilizing the general distributive law ($\max \sum = \sum \max$) [31] the maximization can be computed as $Z = \sum_{j=1}^{J} \max_{v_j} \phi_j(v_j)$.

8.5.2 Reformulation of the Resource Allocation Problem Utilizing the MP Approach

In order to solve the resource allocation problem **P8.1** that was presented in Section 8.3.3 using MP, we reformulate it as a utility maximization problem. Let us define the reward functions $\mathfrak{W}_n(\mathbf{X})$ and $\mathcal{R}_k(\mathbf{X})$, where the transmission alignment vector $\mathbf{X}$ is given by Eq. (8.7). With the constraint in Eq. (8.5a), we can define $\mathfrak{W}_n(\mathbf{X})$ as follows:

$$\mathfrak{W}_n(\mathbf{X}) = \begin{cases} 0, & \text{if} \sum_{k=1}^{K} \sum_{l=1}^{L} x_k^{(n,l)} g_{k,m_k^*}^{(n)} p_k^{(n)} < I_{\max}^{(n)} \\ -\infty, & \text{otherwise.} \end{cases} \tag{8.9}$$

Similarly to how we dealt with the constraint in Eq. (8.5b) we define $\mathfrak{R}_k(\mathbf{X})$ as

$$\mathfrak{R}_k(\mathbf{X}) = \begin{cases} \sum_{n=1}^{N} \sum_{l=1}^{L} x_k^{(n,l)} B_{\mathrm{RB}} \log_2\left(1 + \gamma_{u_k}^{(n)}\right) & \text{if} \sum_{n=1}^{N} \sum_{l=1}^{L} x_k^{(n,l)} \leq 1 \\ -\infty & \text{otherwise} \end{cases} \tag{8.10}$$

The interpretations of the reward functions in Eqs (8.9) and (8.10) are straightforward. Satisfying the interference constraint in Eq. (8.5a) does not cost any penalty (there is zero reward) in the function $\mathfrak{W}_n(\mathbf{X})$, and in the function $\mathfrak{R}_k(\mathbf{X})$ fulfillment of the RB requirement constraint in Eq. (8.5b) gives the desired data rate. However, both in the functions $\mathfrak{W}_n(\mathbf{X})$ and $\mathfrak{R}_k(\mathbf{X})$, the unfulfilled constraints, given in Eqs (8.5a) and (8.5b), respectively, result in infinite cost.

From the Eqs (8.9) and (8.10), the resource allocation problem **P8.1** can be rewritten as $\max_{\mathbf{X}}\left(\sum_{n=1}^{N}\mathfrak{W}_n(\mathbf{X})+\sum_{k=1}^{K}\mathfrak{R}_k(\mathbf{X})\right)$ and the optimal transmission allocation vector is therefore given by

$$\mathbf{X}^* = \underset{\mathbf{X}}{\operatorname{argmax}}\left(\sum_{n=1}^{N}\mathfrak{W}_n(\mathbf{X})+\sum_{k=1}^{K}\mathfrak{R}_k(\mathbf{X})\right) \tag{8.11}$$

Since our goal is to obtain a distributed solution for this resource allocation problem, we focus on a single transmission alignment allocation variable: $x_k^{(n,l)}$. From Eq. (8.11) we obtain $x_k^{(n,l)*} = \underset{x_k^{(n,l)}}{\operatorname{argmax}}\, \phi_k^{(n,l)}(x_k^{(n,l)})$ where the marginal $\phi_k^{(n,l)}(x_k^{(n,l)})$ is given by

$$\phi_k^{(n,l)}\left(x_k^{(n,l)}\right) = \max_{\sim\left(x_k^{(n,l)}\right)}\left(\sum_{n=1}^{N}\mathfrak{W}_n(\mathbf{X})+\sum_{k=1}^{K}\mathfrak{R}_k(\mathbf{X})\right) \tag{8.12}$$

As mentioned in the previous section, $\max_{\sim(x_k^{(n,l)})}$ denotes the maximization over all variables in $\mathbf{X}$ except $x_k^{(n,l)}$. The marginalization in Eq. (8.12) can be computed in a distributed way, where each node conveys the solution of a local problem to one another by passing information messages according to the max sum MP strategy. Note that according to our system model the underlay transmitters and the resources (transmission alignments) can form a bipartite graph; each transmission alignment $\{n,l\}$ can be assigned to any of the K transmitters as long as interference to the MUEs on RB n is below threshold. Without loss of generality, let us consider a generic transmission alignment – an RB-power level pair – $\{n,l\} \in \mathcal{N}\times\mathcal{L}$ and an underlay transmitter $k \in \mathcal{K}^{\mathrm{T}}$. Using the function in Eq. (8.9) and utilizing the max-sum MP strategy presented in Section 8.5.1, it is possible to show that the message delivered by the resource $\{n,l\}$ to the transmitter k can be expressed as [32]:

$$\begin{aligned}\delta_{\{n,l\}\to k}\left(x_k^{(n,l)}\right) &= \max \sum_{k'\in\mathcal{K}^{\mathrm{T}},k'\neq k} \delta_{k'\to\{n,l\}}\left(x_{k'}^{(n,l)}\right)\\ \text{subject to: } &\sum_{k=1}^{K}\sum_{l=1}^{L} x_k^{(n,l)} g_{k,m_k^*}^{(n)} p_k^{(n)} < I_{\max}^{(n)}\end{aligned} \tag{8.13}$$

Note that the term $\delta_{k\to\{n,l\}}(x_k^{(n,l)})$ in the Eq (8.13) denotes the message from transmitter k to the resource $\{n,l\}$, which can be written as [32]:

$$\begin{aligned}\delta_{k\to\{n,l\}}\left(x_k^{(n,l)}\right) &= x_k^{(n,l)} R_{u_k}^{(n,l)} + \max \sum_{\substack{\{n',l'\}\in\mathcal{N}\times\mathcal{L}\\ n'\neq n,l'\neq l}} x_k^{(n',l')} R_{u_k}^{(n',l')} + \delta_{\{n',l'\}\to k}\left(x_k^{(n',l')}\right)\\ \text{subject to: } &\sum_{n=1}^{N}\sum_{l=1}^{L} x_k^{(n,l)} \leq 1\end{aligned} \tag{8.14}$$

where $R_{u_k}^{(n,l)} = B_{\mathrm{RB}}\log_2(1+\Gamma_k^{(n,l)})$ and $\Gamma_k^{(n,l)} \triangleq \gamma_{u_k}^{(n)}\big|_{p_k^{(n)}=l}$.

The interpretations of Eqs (8.13) and (8.14) are as follows: the messages $\delta_{\{n,l\}\to k}(1)$ and $\delta_{k\to\{n,l\}}(1)$ carry the information relative to the use of the resource $\{n,l\}$ by the transmitter k, while the messages $\delta_{\{n,l\}\to k}(0)$ and $\delta_{k\to\{n,l\}}(0)$ carry the information relative to the lack of transmission over the resource $\{n,l\}$ by the transmitter k. In order to obtain both the messages $\delta_{\{n,l\}\to k}(x_k^{(n,l)})$ and $\delta_{k\to\{n,l\}}(x_k^{(n,l)})$, it is required to solve the local optimization problem relative to the allocation variable $x_k^{(n,l)}$.

Based on the discussions in Section 8.5.1, the link-wise marginal in Eq. (8.12) can be written as [32]:

$$\phi_k^{(n,l)}\left(x_k^{(n,l)}\right) = \delta_{\{n,l\}\to k}\left(x_k^{(n,l)}\right) + \delta_{k\to\{n,l\}}\left(x_k^{(n,l)}\right) \tag{8.15}$$

and hence the transmission allocation variable is given by

$$x_k^{(n,l)*} = \underset{x_k^{(n,l)}}{\operatorname{argmax}}\, \phi_k^{(n,l)}\left(x_k^{(n,l)}\right) \tag{8.16}$$

At each iteration of the MP-based resource allocation algorithm, at most one message passes through the edge of any given direction – from transmitters to resources or from resources to transmitters – and at each iteration the messages are updated by replacing the previous message sent on the same edge in the same direction [32]. When both the messages given by Eqs (8.13) and (8.14) are available, the marginal can be computed using Eq. (8.15) and the transmission allocation variable is obtained with Eq. (8.16).

8.5.3 *Effective Implementation of MP Scheme in a Practical Heterogeneous Network*

It is worth noting that sending messages from resources to transmitters (and vice versa) requires actual transmission on the radio channel. In a practical LTE-A-based 5G system, since the exchange of messages actually involves effective transmissions over the channel, the MP scheme described in the preceding section might be limited by the signaling overhead due to transfer of messages between the transmitters and resources. In the following, we observe that the amount of message signaling can be significantly reduced by some algebraic manipulations. Since the messages carry the information regarding whether any resource is used by any underlay transmitter, each transmitter k actually delivers a real-valued vector with two elements, $\delta_{k\to\{n,l\}} = [\delta_{k\to\{n,l\}}(1), \delta_{k\to\{n,l\}}(0)]^{\mathrm{T}}$, and each resource $\{n,l\}$ delivers the vector $\delta_{\{n,l\}\to k} = [\delta_{\{n,l\}\to k}(1), \delta_{\{n,l\}\to k}(0)]^{\mathrm{T}}$. Let us now rewrite the message $\delta_{k\to\{n,l\}}(x_k^{(n,l)})$ using the utility function introduced in Eq. (8.8) as follows:

$$\delta_{k\to\{n,l\}}\left(x_k^{(n,l)}\right) = x_k^{(n,l)}\mathfrak{U}_k^{(n,l)} + max \sum_{\substack{\{n',l'\}\in\mathcal{N}\times\mathcal{L} \\ n'\neq n, l'\neq l}} x_k^{(n',l')}\mathfrak{U}_k^{(n',l')} + \delta_{\{n',l'\}\to k}\left(x_k^{(n',l')}\right) \tag{8.17}$$

By subtracting the constant term $\sum_{\substack{\{n',l'\}\in\mathcal{N}\times\mathcal{L}\\ n'\neq n, l'\neq l}} \delta_{\{n',l'\}\to k}(0)$ from the both sides of Eq. (8.17) we can obtain the following:

$$\delta_{k\to\{n,l\}}\left(x_k^{(n,l)}\right) - \sum_{\substack{\{n',l'\}\in\mathcal{N}\times\mathcal{L}\\ n'\neq n,\, l'\neq l}} \delta_{\{n',l'\}\to k}(0) = x_k^{(n,l)}\mathfrak{U}_k^{(n,l)} + \max \sum_{\substack{\{n',l'\}\in\mathcal{N}\times\mathcal{L}\\ n'\neq n,\, l'\neq l}} x_k^{(n',l')}\mathfrak{U}_k^{(n',l')} + \delta_{\{n',l'\}\to k}\left(x_k^{(n',l')}\right) - \delta_{\{n',l'\}\to k}(0) \tag{8.18}$$

Let us now introduce the parameter $\psi_{\{n,l\}\to k} = \delta_{\{n,l\}\to k}(1) - \delta_{\{n,l\}\to k}(0)$ defined as the normalized message. For instance, consider the vector

$$\Psi_k = \left[\mathfrak{U}_k^{(1,1)} + \psi_{\{1,1\}\to k}, \cdots, \mathfrak{U}_k^{(1,L)} + \psi_{\{1,L\}\to k}, \cdots, \mathfrak{U}_k^{(N,L)} + \psi_{\{N,L\}\to k}\right]^{\mathrm{T}}$$

Let us denote by $\left\langle \upsilon_{\{n',l'\}\to k}\right\rangle_{\sim\{n,l\}}$ the maximal entry of the vector Ψ_k without considering the term $\mathfrak{U}_k^{(n,l)} + \psi_{\{n,l\}\to k}$. It can be noted that the terms within the summation in Eq. (8.18) are either 0 (when $x_k^{(n,l)} = 0$) or $\mathfrak{U}_k^{(n',l')} + \psi_{\{n',l'\}\to k}$ (when $x_k^{(n,l)} = 1$). Since each transmitter requires only a single transmission alignment, when the variable $x_k^{(n,l)} = 0$, only one term in the summation of Eq. (8.18) is non-zero. For the case $x_k^{(n,l)} = 1$, no term within the summation of Eq. (8.18) is non-zero. Consequently, for $x_k^{(n,l)} = 0$, the maximum rate will be achieved if

$$\delta_{k\to\{n,l\}}(0) - \sum_{\substack{\{n',l'\}\in\mathcal{N}\times\mathcal{L}\\ n'\neq n, l'\neq l}} \delta_{\{n',l'\}\to k}(0) = \left\langle \upsilon_{\{n',l'\}\to k}\right\rangle_{\sim\{n,l\}} \tag{8.19}$$

Similarly, when $x_k^{(n,l)} = 1$, the maximum is given by

$$\delta_{k\to\{n,l\}}(1) - \sum_{\substack{\{n',l'\}\in\mathcal{N}\times\mathcal{L}\\ n'\neq n, l'\neq l}} \delta_{\{n',l'\}\to k}(0) = \mathfrak{U}_k^{(n,l)} \tag{8.20}$$

Since by definition $\psi_{k\to\{n,l\}} = \delta_{k\to\{n,l\}}(1) - \delta_{k\to\{n,l\}}(0)$, from Eqs (8.19) and (8.20), the normalized messages from the transmitter k to the resource $\{n, l\}$ can be derived as

$$\begin{aligned}\psi_{k\to\{n,l\}} &= \mathfrak{U}_k^{(n,l)} - \left\langle \upsilon_{\{n',l'\}\to k}\right\rangle_{\sim\{n,l\}} \\ &= \mathfrak{U}_k^{(n,l)} - \left\langle \mathfrak{U}_k^{(n',l')} + \psi_{\{n',l'\}\to k}\right\rangle_{\sim\{n,l\}}\end{aligned} \tag{8.21}$$

Likewise, it can be shown that the normalized message sent from the resource $\{n, l\}$ to transmitter k becomes [32]:

$$\psi_{\{n,l\}\to k} = \delta_{\{n,l\}\to k}(1) - \delta_{\{n,l\}\to k}(0) = -\max_{k'\in\mathcal{K}^{\mathrm{T}}, k'\neq k} \psi_{k'\to\{n,l\}} \tag{8.22}$$

For any arbitrary graph, the allocation variables may keep oscillating and might not converge to any fixed point, and the MP scheme may require some heuristic approach to terminate it. However, in the context of loopy graphical models, by introducing a suitable weight, the messages given by Eqs (8.21) and (8.22) perturb to a fixed point [32, 33]. Accordingly, Eqs (8.21) and (8.22) can be rewritten as [32]:

$$\psi_{k\to\{n,l\}} = \mathfrak{U}_k^{(n,l)} - \omega\left\langle \mathfrak{U}_k^{(n',l')} + \psi_{\{n',l'\}\to k}\right\rangle_{\sim\{n,l\}} - (1-\omega)\left(\mathfrak{U}_k^{(n,l)} + \psi_{\{n,l\}\to k}\right) \tag{8.23}$$

$$\psi_{\{n,l\}\to k} = -\omega \max_{k'\in\mathcal{K}^{\mathrm{T}}, k'\neq k} \psi_{k'\to\{n,l\}} - (1-\omega)\psi_{k\to\{n,l\}} \tag{8.24}$$

where $\omega \in (0,1]$ denotes the weighting factor for each edge. Notice that when $\omega = 1$, the messages given by Eqs (8.23) and (8.24) reduce to the original formulation: Eqs (8.21) and (8.22), respectively. Given the normalized messages $\psi_{k\to\{n,l\}}$ and $\psi_{\{n,l\}\to k}$ for $\forall k, n, l$, the node marginals for the normalized messages can be calculated as $\tau_k^{(n,l)} = \psi_{k\to\{n,l\}} + \psi_{\{n,l\}\to k}$ and hence from Eq. (8.16) the transmission alignment allocation can be obtained as

$$x_k^{(n,l)*} = \begin{cases} 1 & \text{if } \tau_k^{(n,l)} > 0 \text{ and } I^{(n)} < I_{\max}^{(n)} \\ 0 & \text{otherwise.} \end{cases} \tag{8.25}$$

8.5.4 Algorithm Development

In line with our discussions and from the expressions derived in Section 8.5.3, the MP-based resource allocation approach is outlined in Algorithm 8.3. The underlay transmitters and the resources (MBS) exchange the messages in an iterative manner. The MBS assigns the resource to the transmitters considering the node marginals, as well as the interference experienced on the RBs. The algorithm terminates when the sum data rate reaches a steady value; in other words, the allocation vector **X** remains the same in successive iterations.

Algorithm 8.3 Resource allocation using message passing

Initialization:

1: Estimate the CSI parameters from previous time slot.
2: Each underlay transmitter $k \in \mathcal{K}^{\mathrm{T}}$ selects a transmission alignment randomly and reports to MBS.
3: Initialize $t := 1, \psi_{k\to\{n,l\}}(0) := 0, \psi_{\{n,l\}\to k}(0) := 0$ for $\forall k, n, l$.

Update:

4: **while** $\mathbf{X}(t) \neq \mathbf{X}(t-1)$ **and** t less than some predefined threshold $T_{\max}$ **do**
5: Each underlay transmitter $k \in \mathcal{K}^{\mathrm{T}}$ sends the message
$\psi_{k\to\{n,l\}}(t) = \mathfrak{U}_k^{(n,l)}(t-1) - \omega\left\langle \mathfrak{U}_k^{(n',l')}(t-1) + \psi_{\{n',l'\}\to k}(t-1)\right\rangle_{\sim\{n,l\}} - (1-\omega)(\mathfrak{U}_k^{(n,l)}(t-1) + \psi_{\{n,l\}\to k}$
$(t-1))$ for $\forall \{n,l\} \in \mathcal{N} \times \mathcal{L}$ to the MBS.

For all the resource $\forall\{n,l\} \in \mathcal{N} \times \mathcal{L}$, MBS sends messages

$$\psi_{\{n,l\}\to k}(t) = -\omega \max_{k' \in \mathcal{K}^{\mathrm{T}}, k' \neq k} \psi_{k' \to \{n,l\}}(t-1) - (1-\omega)\psi_{k\to\{n,l\}}(t-1)$$

to each underlay transmitter $k \in \mathcal{K}^{\mathrm{T}}$.

Each underlay transmitter $k \in \mathcal{K}^{\mathrm{T}}$ computes the marginals as $\tau_k^{(n,l)}(t) = \psi_{k\to\{n,l\}}(t) + \psi_{\{n,l\}\to k}(t)$ for $\forall\{n,l\} \in \mathcal{N} \times \mathcal{L}$ and reports to the MBS.

/ MBS calculates the allocation vector according to Eq. (8.25) */*

Set $x_k^{(n,l)} := 0$ for $\forall k, n, l$ */* Initialize the variable to obtain final allocation */*

for each $k \in \mathcal{K}^{\mathrm{T}}$ and $\{n,l\} \in \mathcal{N} \times \mathcal{L}$ **do**

if $\tau_k^{(n,l)}(t) > 0$ **then**

Set $x_k^{(n,l)} := 1$ */* Assign the resource to the transmitter */*

$\mathfrak{I}^{(n)} := \sum_{k'=1}^{K} \sum_{l'=1}^{L} x_{k'}^{(n,l')} g_{k',m_{k'}^*}^{(n)} p_{k'}^{(n)}$. */*Calculate interference in RB n */*

if $\mathfrak{I}^{(n)} \geq I_{\max}^{(n)}$ **then**

repeat

$\{\hat{k},\hat{l}\} := \underset{k' \in \mathcal{K}^{\mathrm{T}}, l' \in \mathcal{L}}{\operatorname{argmax}}\, x_{k'}^{(n,l')} g_{k',m_{k'}^*}^{(n)} p_{k'}^{(n)}$ */* Most interfering transmitter k with $p_{\hat{k}}^{(n)} = l$ */*

Set $x_{\hat{k}}^{(n,\hat{l})} := 0$. /* Unassigned due to interference threshold violation */

$\mathfrak{I}^{(n)} := \sum_{k'=1}^{K} \sum_{l'=1}^{L} x_{k'}^{(n,l')} g_{k',m_{k'}^*}^{(n)} p_{k'}^{(n)}$. */* Update interference level */*

until $\mathfrak{I}^{(n)} < I_{\max}^{(n)}$

end if

end if

MBS calculates the transmission alignment allocation vector $\mathbf{X}(t) = [x_k^{(n,l)}]_{\forall k,n,l}$ for the iteration t.

Update $t := t+1$.

end while

Transmission:

Use the allocated transmission alignments (the RB and power levels) for data transmission.

8.5.5 *Convergence, Optimality, and Complexity of the Solution*

The convergence, optimality, and complexity of the message passing approach is analyzed in the following subsections.

8.5.5.1 Convergence and Optimality

As presented in the following theorem, the message-passing algorithm converges to fixed messages within fixed number of iterations.

Theorem 8.5.2 ***The marginals and the allocation in Algorithm 8.3 converge to a fixed point.***
Proof. The proof is constructed by utilizing the concept of contraction mapping [34, Ch. 3]. Let the vector $\boldsymbol{y}(t) = [\psi_{1\to\{1,1\}}(t), \cdots, \psi_{k\to\{n,l\}}(t), \cdots \psi_{K\to\{N,L\}}(t)]^{\mathrm{T}}$ represent all the messages exchanged between the transmitters and the resources (MBS) at iteration t. Let us consider the

messages are translated into the mapping $\psi(t+1)=\mathbb{T}(\psi(t))=[\mathbb{T}_1^{(1,1)}(\psi(t)),\cdots,\mathbb{T}_K^{(N,L)}(\psi(t))]^{\mathrm{T}}$. From Eqs (8.23) and (8.24) we can obtain $\psi_{k\to\{n,l\}}(t+1)=\mathbb{T}_k^{(n,l)}(\psi(t))$ as follows:

$$\begin{aligned}\mathbb{T}_k^{(n,l)}\left(\psi(t)\right) &= \omega\left(\mathfrak{U}_k^{(n,l)}(t)-\mathfrak{U}_k^{(n',l')}(t)\right)\\ &+\omega\left(\omega\max_{k'\in\mathcal{K}^{\mathrm{T}},k'\neq k}\psi_{k'\to\{n',l'\}}(t)+(1-\omega)\psi_{k\to\{n',l'\}}(t)\right)\\ &+(1-\omega)\left(\omega\max_{k'\in\mathcal{K}^{\mathrm{T}},k'\neq k}\psi_{k'\to\{n,l\}}(t)+(1-\omega)\psi_{k\to\{n,l\}}(t)\right)\end{aligned} \tag{8.26}$$

For any vectors **u** and **v**, any generic mapping $\mathbb{T}$ is a contraction if $\|\mathbb{T}(\mathbf{u})-\mathbb{T}(\mathbf{v})\|_\infty\le\varepsilon\|\mathbf{u}-\mathbf{v}\|_\infty$, where $\varepsilon<1$ is the modulus of the mapping [34, Ch. 3]. It can be shown that the mapping $\mathbb{T}:\mathbb{R}^{KNF}\to\mathbb{R}^{KNF}$ is a contraction under the maximum norm [33]: $\|\mathbb{T}(\psi)\|_\infty=\max_{k\in\mathcal{K}^{\mathrm{T}},n\in\mathcal{N},l\in\mathcal{L}}\left|\mathbb{T}_k^{(n,l)}(\psi)\right|$. Since the contraction mappings have a unique fixed-point convergence property for any initial vector, the proof concludes with the fact that the message-passing algorithm converges to a fixed marginal and hence to a fixed allocation vector **X**.

The following theorem presents the fixed convergence point of the message passing algorithm is an optimal solution of the original resource allocation problem.

Theorem 8.5.3 ***The allocation obtained by message passing algorithm converges to the optimal solution of resource allocation problem P8.1.***

Proof. The theorem is proved by contradiction. Let us consider that the solution $\tilde{\mathbf{X}}$ obtained by message passing algorithm is not optimal and let $\mathbf{X}^*$ be the optimal solution obtained by solving **P1.1**. Let us further assume that there are $\chi\le|\mathbf{X}|$ entries (e.g., allocations) that differ between $\tilde{\mathbf{X}}$ and $\mathbf{X}^*$. In addition, let $\tilde{\mathcal{N}}\times\tilde{\mathcal{L}}\subseteq\mathcal{N}\times\mathcal{L}$ denote the subset of resources for which two allocations differ. For each $\{\tilde{n},\tilde{l}\}\in\tilde{\mathcal{N}}\times\tilde{\mathcal{L}}$ there is a transmitter $k_{\{\tilde{n},\tilde{l}\}}$ such that $\tilde{x}_{k_{\{\tilde{n},\tilde{l}\}}}^{(\tilde{n},\tilde{l})}=1$ and $x_{k_{\{\tilde{n},\tilde{l}\}}}^{*(\tilde{n},\tilde{l})}=0$, and a transmitter $\ddot{\kappa}_{\{\tilde{n},\tilde{l}\}}\neq\kappa_{\{\tilde{n},\tilde{l}\}}$ such that $\tilde{x}_{\ddot{\kappa}_{\{\tilde{n},\tilde{l}\}}}^{(\tilde{n},\tilde{l})}=0$ and $x_{\ddot{k}_{\{\tilde{n},\tilde{l}\}}}^{*(\tilde{n},\tilde{l})}=1$. Hence, the assignment of resource $\{\tilde{n},\tilde{l}\}$ to transmitter $k_{\{\tilde{n},\tilde{l}\}}$ implies that the marginal $\tau_{\ddot{k}_{\{\tilde{n},\tilde{l}\}}}^{(\tilde{n},\tilde{l})}<0$ and the following set of inequalities hold for each $\{\tilde{n},\tilde{l}\}\in\tilde{\mathcal{N}}\times\tilde{\mathcal{L}}$:

$$\tau_{\ddot{k}_{\{\tilde{n},\tilde{l}\}}}^{(\tilde{n},\tilde{l})}=\omega\left[\left(\mathfrak{U}_{\ddot{k}_{\{\tilde{n},\tilde{l}\}}}^{(\tilde{n},\tilde{l})}+\psi_{\{\tilde{n},\tilde{l}\}\to\ddot{k}_{\{\tilde{n},\tilde{l}\}}}\right)-\left(\mathfrak{U}_{\ddot{k}_{\{\tilde{n},\tilde{l}\}}}^{(n',l')}+\psi_{\{n',l'\}\to\ddot{k}_{\{\tilde{n},\tilde{l}\}}}\right)\right]<0 \tag{8.27}$$

where $\{n',l'\}$ is the resource as represented in Eq. (8.21). According to our assumption, the resource $\{n',l'\}$ also belongs to $\tilde{\mathcal{N}}\times\tilde{\mathcal{L}}$. Hence, $\sum_{\{\tilde{n},\tilde{l}\}\in\tilde{\mathcal{N}}\times\tilde{\mathcal{L}}}\tau_{\ddot{k}_{\{\tilde{n},\tilde{l}\}}}^{(\tilde{n},\tilde{l})}=\omega(\Delta\mathfrak{U}+\Delta\psi)$ where

$$\begin{aligned}\Delta\mathfrak{U} &= \sum_{\{\tilde{n},\tilde{l}\}\in\tilde{\mathcal{N}}\times\tilde{\mathcal{L}}}\left(\mathfrak{U}_{\ddot{\kappa}_{\{\tilde{n},\tilde{l}\}}}^{(\tilde{n},\tilde{l})}-\mathfrak{U}_{k_{\{\tilde{n},\tilde{l}\}}}^{(\tilde{n},\tilde{l})}\right)\\ &= \sum_{k=1}^{K}\sum_{N=1}^{N}\sum_{l=1}^{L}x_k^{*(n,l)}\mathfrak{U}_k^{(n,l)}-\sum_{k=1}^{K}\sum_{N=1}^{N}\sum_{l=1}^{L}\tilde{x}_k^{(n,l)}\mathfrak{U}_k^{(n,l)}\end{aligned} \tag{8.28}$$

and $\Delta\psi=\sum_{\{\tilde{n},\tilde{l}\}\in\tilde{\mathcal{N}}\times\tilde{\mathcal{L}}}\left(\psi_{\{\tilde{n},\tilde{l}\}\to\ddot{\kappa}_{\{\tilde{n},\tilde{l}\}}}-\psi_{\{\tilde{n},\tilde{l}\}\to\kappa_{\{\tilde{n},\tilde{l}\}}}\right)$. After some algebraic manipulations (for details refer to [33]) we can obtain $\frac{2(1-\omega)}{\omega}\sum_{\{\tilde{n},\tilde{l}\}\in\tilde{\mathcal{N}}\times\tilde{\mathcal{L}}}\tau_{\ddot{\kappa}_{\{\tilde{n},\tilde{l}\}}}^{(\tilde{n},\tilde{l})}\le\Delta\mathfrak{U}$. Since $0<\omega<1$ and both the variables $\sum_{\{\tilde{n},\tilde{l}\}\in\tilde{\mathcal{N}}\times\tilde{\mathcal{L}}}\tau_{\ddot{\kappa}_{\{\tilde{n},\tilde{l}\}}}^{(\tilde{n},\tilde{l})}$ and $\Delta\mathfrak{U}$ are positive, our assumption that $\tilde{\mathbf{X}}$ is not optimal is contradicted and the proof follows.

8.5.5.2 Complexity

If the message passing algorithm requires $T < T_{\max}$ iterations to converge, it is straightforward to verify that the time complexity at each MBS is $\mathcal{O}(T\,K\,N\,L)$. Similarly, considering a standard sorting algorithm that outputs the term $\left\langle \mathfrak{U}_k^{(n',l')} + \psi_{\{n',l'\}\to k} \right\rangle_{\sim\{n,l\}}$ in order to generate the message $\psi_{k\to\{n,l\}}$ with worst-case complexity of $\mathcal{O}(N\,L\log(N\,L))$, the overall time complexity at each underlay transmitter is $\mathcal{O}(T(N\,L)^2\log(N\,L))$.

8.6 Auction-based Resource Allocation

Our final approach for the resource allocation problem is the distributed auction algorithm. The allocation using an auction is based on a bidding procedure, where the agents (underlay transmitters) bid for the resources (RB and power level). The transmitters select from the bids for the resources based on the costs (the interference caused to the MUEs) of using the resource. The desired assignment relies on the appropriate selection of the bids. The unassigned transmitters raise the cost of using resource and bid for the resources simultaneously. Once the bids from all the transmitters are available, the resources are assigned to the highest bidder. An overview of auction approach is presented in the following.

8.6.1 Overview of the Auction Approach

In a generic auction based assignment model, every resource j is associated with a cost c_j and each agent i can get the benefit B_{ij} from the resource j. Given the benefit $B_{ij'}$, every agent i who wishes to be assigned to the resource j, needs to pay the cost c_j. The net value (utility) that an agent i can get from the resource j is given by $B_{ij} - c_j$. The auction procedure involves the assignment of agent i to the resource j' that provides the maximal net value:

$$B_{ij'} - c_{j'} = \max_j \left\{ B_{ij} - c_j \right\} \tag{8.29}$$

If the condition given in Eq. (8.29) is satisfied for all the agents i, the assignment and the set of costs are referred to as *equilibrium* [35]. However, in many practical problems, obtaining an equilibrium assignment is not straightforward due to the possibility of cycles. In particular, there may be cases where the agents contend for a small number of equally desirable resources without increasing the cost, which creates a cycle (an infinite loop) in the auction process. To avoid this difficulty, the notion of *almost equilibrium* is introduced in the literature. The assignment and the set of costs are said to be almost equilibrium when the net value for assigning each agent i to resource j' is within a constant $\varepsilon > 0$ of being maximal. Hence, in order to be an almost equilibrium assignment, the following condition needs to be satisfied for all the agents [35]:

$$B_{ij'} - c_{j'} \geq \max_j \left\{ B_{ij} - c_j \right\} - \epsilon \tag{8.30}$$

The condition in Eq. (8.30) is known as ϵ-*complementary slackness*. When $\epsilon = 0$, Eq. (8.30) reduces to ordinary complementary slackness, as given by Eq. (8.29).

For instance, let the variable $\Theta_i = j$ denote that agent i is assigned with the resource j. In addition, let c_{ij} denote the cost that agent i incurs in order to be assigned with resource j and $\mathfrak{b}_{ij}$ is the bidding information (such as highest bidder) available to the agent i about resource j. The auction procedure evolves in an iterative manner. Given the assignment Θ_i, the set of costs $[c_{ij}]_{\forall ij}$, and the set of largest bidders $[\mathfrak{b}_{ij}]_{\forall ij}$ of the previous iteration, the agents locally update the costs and the highest bidders for the current iteration. In particular, the costs $c_{ij}(t)$ and bidding information $\mathfrak{b}_{ij}(t)$ available to the agent i about resource j for iteration t are updated from the previous iteration as follows [36]:

$$c_{ij}(t) = \max_{i',i'\neq i}\left\{c_{ij}(t-1), c_{i'j}(t-1)\right\} \tag{8.31}$$

$$\mathfrak{b}_{ij}(t) = \max_{i^*\in \operatorname{argmax}_{i',i'\neq i}\left\{c_{ij}(t-1), c_{i'j}(t-1)\right\}}\left\{\mathfrak{b}_{i^*j}(t-1)\right\} \tag{8.32}$$

These update equations ensure that the agents will have the updated maximum cost of the resource j (that is, $c_j \triangleq \max_i\{c_{ij}\}$) and the corresponding highest bidder for that resource. Once the update cost and bidding information are available, agent i checks whether the cost of the resource currently assigned to agent i, $c_{i\Theta_i(t-1)}$, has been increased by any other agents. If so, the current assignment obtained from previous iteration may not be at (almost) equilibrium and the agent needs to select a new assignment: $\Theta_i(t) = \operatorname{argmax}_j\{B_{ij}(t) - c_{ij}(t)\}$. In order to update the cost $\Theta_i(t)$ for the new assignment for any iteration t, the agent will use the following cost update rule [36]:

$$c_{ij}(t) = c_{ij}(t-1) + \Delta_i(t-1) \tag{8.33}$$

where Δ_i is given by

$$\Delta_i(t-1) = \max_j\left\{B_{ij}(t-1) - c_{ij}(t-1)\right\} - \max_{j'\neq\Theta_i(t)}\left\{B_{ij'}(t-1) - c_{ij'}(t-1)\right\} + \epsilon \tag{8.34}$$

The variables $\max_j\{B_{ij}(t-1) - c_{ij}(t-1)\}$ and $\max_{j'\neq\Theta_i(t)}\{B_{ij'}(t-1) - c_{ij'}(t-1)\}$ denote the maximum and second maximum net utility, respectively. Note that Δ_i is always greater than zero as $\epsilon > 0$ and by definition $\max_j\{B_{ij}(t-1) - c_{ij}(t-1)\} > \max_{j'\neq\Theta_i(t)}\{B_{ij'}(t-1) - c_{ij'}(t-1)\}$. Since $c_{i\Theta_i(t)}(t)$ is the highest cost for iteration t, agent i can also update the bidding information: $\mathfrak{b}_{i\Theta_i(t)}(t) = i$. Accordingly, the cost update rule using Δ_i as given in Eq. (8.33) ensures that the assignment and the set of costs are almost at equilibrium [36].

8.6.2 *Auction for Radio Resource Allocation*

Based on the discussion in the preceding section, we now present the auction-based resource allocation scheme. We present the cost model and use the concept of auction to develop the resource allocation algorithm in our heterogeneous network setup.

8.6.2.1 Cost Function

Let us consider the utility function given by Eq. (8.8). Recall that the term $w_2(I^{(n)} - I^{(n)}_{\max})$ in Eq. (8.8) represents the cost – the interference caused by underlay transmitters to the MUE – of using the RB n. In particular, when the transmitter k is transmitting with power level l, the cost of using RB n can be represented by

$$\begin{aligned} c_k^{(n,l)} &= w_2\left(I^{(n)} - I^{(n)}_{\max}\right) = w_2\left(\sum_{k'=1}^{K}\sum_{l'=1}^{L} x_{k'}^{(n,l')} g_{k',m_{k'}^*}^{(n)} p_{k'}^{(n)} - I^{(n)}_{\max}\right) \\ &= w_2\left(g_{k,m_k^*}^{(n)} l + \sum_{k'\in\mathcal{K}^{\mathrm{T}}, k'\neq k}\ \sum_{l'=1}^{L} x_{k'}^{(n,l')} g_{k',m_{k'}^*}^{(n)} p_{k'}^{(n)} - I^{(n)}_{\max}\right) \end{aligned} \tag{8.35}$$

Let the parameter $C_k^{(n,l)} = \max\{0, c_k^{(n,l)}\}$ and accordingly the cost $C_k^{(n,l)} = 0$ only if $I^{(n)} \le I^{(n)}_{\max}$. Notice that using the cost term we can represent Eq. (8.8) as

$$\mathfrak{U}_k^{(n,l)} = w_1\mathcal{R}\left(\Gamma_{u_k}^{(n,l)}\right) - w_2\left(I^{(n)} - I^{(n)}_{\max}\right) = B_k^{(n,l)} - c_k^{(n,l)} = B_k^{(n,l)} - C_k^{(n,l)}$$

where $B_k^{(n,l)} = w_1\mathcal{R}(\Gamma_{u_k}^{(n,l)})$, and $c_k^{(n,l)}$ is given by Eq. (8.35). The variable $B_k^{(n,l)}$ is proportional to the data rate achieved by transmitter k using resource $\{n,l\}$. Analogous to the discussion of the previous section, $\mathfrak{U}_k^{(n,l)}$ represents the net benefit that transmitter k obtains from the resource $\{n,l\}$.

Let $\mathfrak{b}_k^{(n,l)}$ denote the local bidding information available to transmitter k for the resource $\{n,l\}$. For notational convenience, let us assume that $\Theta : [k]_{k=1,\cdots,K} \rightarrow [\{n,l\}]_{n=1,\cdots,N;\, l=1,\cdots,L}$ denotes the mapping between the transmitters and the resources; that is, $\Theta_k = \{n,l\}$ represents the assignment of resource $\{n,l\}$ to transmitter k. Hence we represent by $C_k^{\Theta_k}$ the cost of using the resource $\{n,l\}$ obtained by the assignment $\Theta_k = \{n,l\}$. Similarly, given $\Theta_k = \{n,l\}$ the variable $\mathfrak{b}_k^{\Theta_k} \equiv \mathfrak{b}_k^{(n,l)}$ denotes the local bidding information about the resource $\{n,l\}$ available to the transmitter k. Note that $\Theta_k = \{n,l\}$ also implies $x_k^{(n,l)} = 1$. In other words, $\Theta_k = \{n,l\}$ denote the non-zero entries of the vector $\mathbf{x}_k = [x_k^{(n,l)}]_{\forall n,l}$. Since each underlay transmitter k selects only one resource $\{n,l\}$, only a single entry in the vector $\mathbf{x}_k$ is non-zero.

8.6.2.2 Update of Cost and Bidder Information

In order to obtain the updated cost and bidding information, we utilize a concept similar to that given by Eqs (8.31)–(8.34). At the beginning of the auction procedure, each underlay transmitter updates the cost as $C_k^{(n,l)}(t) = \max_{k'\in\mathcal{K}^{\mathrm{T}}, k'\neq k}\{C_k^{(n,l)}(t-1), C_{k'}^{(n,l)}(t-1)\}$. In addition, as described by Eq. (8.34), the information about the maximum bidder is obtained by $\mathfrak{b}_k^{(n,l)}(t) = \mathfrak{b}_{k^*}^{(n,l)}(t-1)$ where $k^* = \operatorname{argmax}_{k'\in\mathcal{K}^{\mathrm{T}}, k'\neq k}\{C_k^{(n,l)}(t-1), C_{k'}^{(n,l)}(t-1)\}$. When the transmitter k needs to select a new assignment $\Theta_k = \{\hat{n},\hat{l}\}$, the transmitter increases the cost of using the resource, $C_k^{(\hat{n},\hat{l})}(t) = C_k^{(\hat{n},\hat{l})}(t-1) + \Delta_k(t-1)$, and $\Delta_k(t-1)$ is given by

$$\Delta_k(t-1) = \max_{\{n',l'\}\in\mathcal{N}\times\mathcal{L}} \mathfrak{U}_k^{(n',l')}(t-1) - \max_{\substack{\{n',l'\}\in\mathcal{N}\times\mathcal{L} \\ n'\neq\hat{n}, l'\neq\hat{l}}} \mathfrak{U}_k^{(n',l')}(t-1) + \epsilon \tag{8.36}$$

where $\epsilon > 0$ indicates the minimum bid requirement parameter. Similar to Eq. (8.34), the term $\max_{\{n',l'\}\in\mathcal{N}\times\mathcal{L}} \mathfrak{U}_k^{(n',l')}(t-1) - \max_{\substack{\{n',l'\}\in\mathcal{N}\times\mathcal{L}\\ n'\neq\hat{n},l'\neq\hat{l}}} \mathfrak{U}_k^{(n',l')}(t-1)$ denotes the difference between the maximum and the second maximum utility values. In the case when the transmitter k does not prefer to be assigned to a new resource, the allocation from the previous iteration will remain unchanged, $\Theta_k(t) = \Theta_k(t-1)$, and consequently, $\mathbf{x}_k(t) = \mathbf{x}_k(t-1)$.

Algorithm 8.4 Auction method for any underlay transmitter k

Input: Parameters from previous iteration: an assignment $\mathbf{X}(t-1) = [\mathbf{x}_1(t-1), \cdots \mathbf{x}_K(t-1)]^{\mathrm{T}}$, aggregated interference $I^{(n)}(t-1)$ for $\forall n$, cost of using resources $\mathbf{C}(t-1) = [C_k^{(n,l)}(t-1)]_{\forall k,n,l}$ and the highest bidders of the resources $\mathfrak{B}(t-1) = [\mathfrak{B}_k(t-1)]_{\forall k}$ where $\mathfrak{B}_k(\cdot) = [\mathfrak{b}_k^{(n,l)}(\cdot)]_{\forall n,l}$.

Output: The allocation variable $\mathbf{x}_k(t) = [x_k^{(n,l)}]_{\forall n,l}$, updated costs $\mathbf{C}_k(t) = [C_k^{(n,l)}(t)]_{\forall n,l}$, and bidding information $\mathfrak{B}_k(t) = [\mathfrak{b}_k^{(n,l)}(t)]_{\forall n,l}$ at current iteration t for the transmitter k.

1: Initialize $\mathbf{x}_k(t) := \mathbf{0}$.
2: For all the resources $\{n,l\} \in \mathcal{N} \times \mathcal{L}$,
- Obtain the transmitter $k^* := \underset{k'\in\mathcal{K}^{\mathrm{T}},k'\neq k}{\operatorname{argmax}}\{C_k^{(n,l)}(t-1), C_{k'}^{(n,l)}(t-1)\}$ and update the highest bidder as $\mathfrak{b}_k^{(n,l)}(t) := \mathfrak{b}_{k^*}^{(n,l)}(t-1)$.
- Update the cost as $C_k^{(n,l)}(t) := \max_{k'\in\mathcal{K}^{\mathrm{T}},k'\neq k} \{C_k^{(n,l)}(t-1), C_{k'}^{(n,l)}(t-1)\}$.

/* *Let* $\Theta_k(t-1)$ *denote the assignment of transmitter k at previous iteration* $t-1$, *that is,* $\Theta_k(t-1)$ *represents the non-zero entry in the vector* $\mathbf{x}_k(t-1)$. *Since each transmitter uses only one transmission alignment, only a single entry in the vector* $\mathbf{x}_k(t-1)$ *is non-zero. When cost is greater than previous iteration and the transmitter k is not the highest bidder, update the assignment* */

3: **if** $C_k^{\Theta_k(t-1)}(t) \geq C_k^{\Theta_k(t-1)}(t-1)$ **and** $\mathfrak{b}_k^{\Theta_k(t-1)}(t) \neq k$ **then**
4: $\{\hat{n},\hat{l}\} := \underset{\{n',l'\}\in\mathcal{N}\times\mathcal{L}}{\operatorname{argmax}} \mathfrak{U}_k^{(n',l')}(t)$. /* *Obtain the best resource for transmitter k**/
5: $\mathfrak{I}^{(\hat{n})} := g_{k,m_k^*}^{(\hat{n})} \hat{l} + I^{(\hat{n})}$. /* Calculate additional interference caused by transmitter k for using RB n */
6: **if** $\mathfrak{I}^{(\hat{n})} < I_{\max}^{(\hat{n})}$ **then**
7: Set $x_k^{(\hat{n},\hat{l})} := 1$. /* *e.g.*, $\Theta_k(t) = \{n,l\}$ */
8: Update the highest bidder for the resource $\{\hat{n},\hat{l}\}$ as $\mathfrak{b}_k^{(\hat{n},\hat{l})}(t) := k$.
9: Increase the cost for the resource $\{\hat{n},\hat{l}\}$ as $C_k^{(\hat{n},\hat{l})}(t) = C_k^{(\hat{n},\hat{l})}(t-1) + \Delta_k(t-1)$ where $\Delta_k(t-1)$ is given by Eq. (8.36).
10: **else**
11: Keep the assignment unchanged from previous iteration, i.e., $\mathbf{x}_k(t) := \mathbf{x}_k(t-1)$.
12: end if
13: **else**
14: Keep the assignment unchanged from previous iteration, i.e., $\mathbf{x}_k(t) := \mathbf{x}_k(t-1)$.
15: **end if**

8.6.3 *Algorithm Development*

Algorithm 8.5 outlines the auction-based resource allocation approach. Each transmitter locally executes Algorithm 8.4 and obtains a temporary allocation. When the execution of Algorithm 8.4 is finished, each underlay transmitter k reports to the MBS the local information: the choices for the resources, $\mathbf{x}_k = [x_k^{(n,l)}]_{\forall n,l}$. Once the information (the output parameters from Algorithm 8.4) from all the transmitters is available to the MBS, the necessary parameters (the input arguments required by Algorithm 8.4) are calculated and broadcast by the MBS. Algorithm 8.4 repeats iteratively until the allocation variable $\mathbf{X} = [\mathbf{x}_k]_{\forall k} = [x_1^{(1,1)}, \cdots, x_1^{(1,L)}, \cdots, x_1^{(N,L)}, \cdots, x_K^{(N,L)}]^{\mathrm{T}}$ for two successive iterations becomes similar.

Algorithm 8.5 Auction-based resource allocation

Initialization:

1: Estimate the CSI parameters from the previous time slot.
2: Each underlay transmitter $k \in \mathcal{K}^{\mathrm{T}}$ randomly selects a transmission alignment and reports to the MBS.
3: MBS broadcasts the assignment of all transmitters, aggregated interference of each RB, the costs and the highest bidders using pilot signals.
4: Initialize number of iterations $t := 1$.

Update:

5: **while** $\mathbf{X}(t) \neq \mathbf{X}(t-1)$ **and** t is less than some predefined threshold $T_{\max}$ **do**
6: Each underlay transmitter $k \in \mathcal{K}^{\mathrm{T}}$ locally runs the Algorithm 8.4 and reports the assignment $\mathbf{x}_k(t)$, the cost $\mathbf{C}_k(t)$ and the bidding information $\mathfrak{B}_k(t)$ to the MBS.
7: MBS calculates the aggregated interference $I^{(n)}(t)$ for $\forall n$, the allocation variable $\mathbf{X}(t)$, information about highest bidders $\mathfrak{B}(t)$, the cost $\mathbf{C}(t)$, and broadcast to the underlay transmitters.
8: Update $t := t+1$.
9: end while

Transmission:

10: Use the resources (the RB and power levels) allocated in the final stage of update phase for data transmission.

8.6.4 *Convergence, Complexity, and Optimality of the Auction Approach*

In the following subsections we analyze the convergence, complexity, and optimality of the solution obtained by the auction algorithm.

8.6.4.1 Convergence and Complexity

For any arbitrary fixed $\varepsilon > 0$, the auction approach is guaranteed to converge to a fixed assignment. The following theorem shows that the auction process terminates within a fixed number of iterations.

Theorem 8.6.1 ***The auction process terminates in a finite number of iterations.***
Proof. According to our system model, each underlay transmitter selects only one transmission alignment. Hence, once each resource receives at least one bid (which implies that each transmitter is assigned to a resource), the auction process must terminate. Now if any resource $\{n, l\}$ receives a bid in $\hat{t}$ iterations, the cost must be greater than the initial price by

$\hat{t}\varepsilon$. As a result, the resource $\{n, l\}$ becomes costly to be assigned when compared to any resource $\{n', l'\}$ that has not received any bid yet. The argument follows that there are two possibilities:

- the auction process terminates in a finite iterations with each transmitter assigned to a resource, regardless of every resource receives a bid
- the auction process continues for a finite number of iterations, each resource will receive at least one bid, and therefore the algorithm terminates.

At termination, the solution (allocation) obtained is almost at equilibrium: the condition in Eq. (32) is satisfied for all the underlay transmitters. Since the algorithm terminates after a finite number of iterations, we can show that the algorithm converges to a fixed allocation and the complexity at each transmitter is linear to the number of resources.

Theorem 8.6.2 ***The auction algorithm converges to a fixed allocation with the number of iterations of*** $\mathcal{O}\left(T\,K\,N\,L\left\lceil \frac{\max_{k,n,l} B_k^{(n,l)} - \min_{k,n,l} B_k^{(n,l)}}{\varepsilon} \right\rceil\right)$.

Proof. The proof follows from an argument similar to the one presented in Theorem 8.6.1. In the worst case, the total number of iterations in which a resource can receive a bid is no more than $\Upsilon = \left\lceil \frac{\max_{k,n,l} B_k^{(n,l)} - \min_{k,n,l} B_k^{(n,l)}}{\varepsilon} \right\rceil$ [36]. Since each bid requires $\mathcal{O}(N\,L)$ iterations, and each iteration involves a bid by a single transmitter, the total number of iterations in Algorithm 8.5 is of $\mathcal{O}(K\,N\,L\,\Upsilon)$. For convergence, the allocation variable $\mathbf{X}$ needs to be unchanged for at least $T \geq 2$ consecutive iterations. Hence, the overall running time of the algorithm is $\mathcal{O}(T\,K\,N\,L\,\Upsilon)$.

Note that for any transmitter node $k \in \mathcal{K}^{\mathrm{T}}$, the complexity of the auction process given by Algorithm 8.4 is linear with number of resources for each of the iterations.

8.6.4.2 Optimality

In the following we show that the data rate obtained by the auction algorithm is within $K\varepsilon$ of the maximum data rate obtained by solving the original optimization problem P8.1.

Theorem 8.6.3 ***The data rate obtained by the distributed auction algorithm is within*** $K\epsilon$ ***of the optimal solution.***

Proof. We construct the proof using an approach similar to that presented by Zavlanos et al. [36]. The data rate obtained by any assignment will satisfy the following condition:

$$\sum_{k=1}^{K} R_{u_k} \leq \sum_{\{n,l\}\in\mathcal{N}\times\mathcal{L}} \hat{C}^{(n,l)} + \sum_{k=1}^{K} \max_{\{n,l\}\in\mathcal{N}\times\mathcal{L}} \left\{ B_k^{(n,l)} - \hat{C}^{(n,l)} \right\} \tag{8.37}$$

where $\hat{C}^{(n,l)} = \max_{k'\in\mathcal{K}^{\mathrm{T}}} C_{k'}^{(n,l)}$, $B_k^{(n,l)} = w_1 \mathcal{R}(\Gamma_{u_k}^{(n,l)})$ and R_{u_k} is given by Eq. (8.4). The inequality given by Eq. (8.37) is satisfied since the first term in the right-hand side of the inequality, namely $\sum_{\{n,l\}\in\mathcal{N}\times\mathcal{L}} \hat{C}_{(n,l)}$, is equal to $\sum_{k=1}^{K}\sum_{n=1}^{N}\sum_{l=1}^{L} x_k^{(n,l)} C_k^{(n,l)}$ and the second term is not less than $\sum_{k=1}^{K}\sum_{n=1}^{N}\sum_{l=1}^{L} x_k^{(n,l)}(B_k^{(n,l)} - \hat{C}^{(n,l)})$. Let the variable $A^* \triangleq \max_{\mathbf{X}} \sum_{k=1}^{K} R_{u_k} = \sum_{k=1}^{K}\sum_{n=1}^{N}\sum_{l=1}^{L} x_k^{(n,l)} B_{\text{RB}} \log_2(1+\gamma_{u_k}^{(n)})$ denote the optimal achievable data rate. In addition, let the variable D^* be defined as

$$D^* \triangleq \min_{\substack{\hat{C}^{(n,l)} \\ \{n,l\}\in\mathcal{N}\times\mathcal{L}}} \left\{ \sum_{\{n,l\}\in\mathcal{N}\times\mathcal{L}} \hat{C}^{(n,l)} + \sum_{k=1}^{K} \max_{\{n,l\}\in\mathcal{N}\times\mathcal{L}} \left\{ B_k^{(n,l)} - \hat{C}^{(n,l)} \right\} \right\} \tag{8.38}$$

Hence from Eq. (8.37), we can write $A^* \le D^*$. Since the final assignment and the set of costs are almost at equilibrium, for any underlay transmitter k, the condition $\sum_{n=1}^{N}\sum_{l=1}^{L} x_k^{(n,l)}(B_k^{(n,l)} - \hat{C}^{(n,l)}) \ge \max_{\{n,l\}\in\mathcal{N}\times\mathcal{L}} \{B_k^{(n,l)} - \hat{C}^{(n,l)}\} - \varepsilon$ will hold. Consequently, we can obtain the following inequality:

$$\begin{aligned} D^* &\le \sum_{k=1}^{K} \left(\sum_{n=1}^{N}\sum_{l=1}^{L} x_k^{(n,l)} \hat{C}^{(n,l)} + \max_{\{n,l\}\in\mathcal{N}\times\mathcal{L}} \left\{ B_k^{(n,l)} - \hat{C}^{(n,l)} \right\} \right) \\ &\le \sum_{k=1}^{K}\sum_{n=1}^{N}\sum_{l=1}^{L} x_k^{(n,l)} B_k^{(n,l)} + K\epsilon \le \sum_{k=1}^{K} R_{u_k} + K\epsilon \le A^* + K\epsilon \end{aligned} \tag{8.39}$$

Since $A* \le D*$, the data rate achieved by the auction algorithm is within $K\epsilon$ of the optimal data rate A^* and the proof follows.

8.7 Qualitative Comparison of the Resource Allocation Schemes

In this section, we compare the different resource allocation schemes discussed above based on several criteria (e.g., flow of algorithm execution, information requirement and algorithm overhead, complexity and optimality of the solution, convergence behavior, etc.). We term the centralize solution (which can be obtained by solving the optimization problem **P1.1**) as COS (centralized optimal scheme) and compare it with the distributed solutions. A comparison among the resource allocation schemes is presented in Table 8.2.

8.8 Summary and Conclusion

We have presented three comprehensive distributed solution approaches for future 5G cellular mobile communication systems. Considering a heterogeneous multi-tier 5G network, we have developed distributed radio resource allocation algorithms using three different mathematical models: stable matching, message passing, and the auction method. The convergence, complexity, optimality properties of these distributed solutions are also briefly analyzed. To this end, a qualitative comparison of these schemes is illustrated.

The solution tools presented in this chapter are also applicable to resource allocation problems in other enabling technologies for 5G systems. In particular, they open up new opportunities to investigate other network models, such as resource allocation problems for wireless virtualization [37] and cloud-based radio access networks [38]. In such systems, these

Table 8.2 Comparison of the different resource allocation approaches.

Criterion	Schemes			
	COS	Stable matching	Message passing	Auction method
Type of solution	Centralized	Distributed	Distributed	Distributed
Algorithm execution	MBS solves the resource optimization problem (**P8.1**)	MBS and underlay transmitters locally update the preference profiles; MBS runs the matching subroutine	MBS and underlay transmitters iteratively exchange the messages; MBS computes the marginals and selects allocation	Each underlay transmitters locally runs the auction subroutine, MBS collects the parameters from all the transmitters and broadcasts parameters required for the auction subroutine
Optimality	Optimal	Weak Pareto optimal	Optimal subject to the weight ω	Within $K\epsilon$ of the optimal
Complexity	$\mathcal{O}((N\,L)^K)$ at the MBS	$\mathcal{O}(T\,N\,L\log(N\,L))$ at the transmitters, $\mathcal{O}(T\,K\,N\,L)$ at the MBS	$\mathcal{O}(T(N\,L)^2\log(N\,L))$ at the transmitters, $\mathcal{O}(T\,K\,N\,L)$ at the MBS	For each iteration linear with N, L at the transmitters, overall running time $\mathcal{O}(T\,K\,N\,L\,\Upsilon)$
Convergence behavior	N/A	Converges to a stable matching and hence to a fixed allocation	Converges to a fixed marginal and to a fixed allocation	Converges to a fixed allocation within $K\epsilon$ of the optimal
Information required by the MBS	Channel gains (CSI parameters) between all the links of the network	The preference profiles and the channel gains $G_k^{(n)} = [g_{k,m_k^*}^{(n)}]_{\forall k,n}$	The messages $[\psi_{k\to\{n,l\}}]_{\forall k,n,l}$ and the channel gains $G_k^{(n)} = [g_{k,m_k^*}^{(n)}]_{\forall k,n}$	The channel gains $G_k^{(n)} = [g_{k,m_k^*}^{(n)}]_{\forall k,n}$, local assignments $\mathbf{x}_k$, the cost $\mathbf{C}_k$, and the bidding information $\mathfrak{B}_k$ for $\forall k$
Algorithm overhead	High (exponential) computational complexity, requirement of all CSI parameters of the network	Build the preference profiles, exchange information to update preference profiles, execution of matching subroutine	Calculation and exchange of messages, computation of the marginals	Computation and exchange of the parameters, $I^{(n)}$ for $\forall n$, the allocation vector $\mathbf{X}$, information about highest bidders $\mathfrak{B}$, the cost vector $\mathbf{C}$

modeling tools need to be customized based on the objectives and constraints of the resource allocation problem.

In addition to the solutions presented, there are a few game theoretical models that have not been covered in this chapter. However, these models can also be considered as potential distributed solution tools too. In contrast to traditional cooperative and non-cooperative games, the game models, such as mean field games [39, 40] and evolutionary games [41], are scalable by nature, and hence applicable to modeling large heterogeneous 5G networks. Utilizing those advanced game models for resource allocation problems and analyzing the performance of 5G systems in terms of data rates and spectrum and energy efficiency could be an interesting area of research.

References

[1] A. Osseiran, F. Boccardi, V. Braun, et al. (2014) Scenarios for 5G mobile and wireless communications: The vision of the METIS project. *IEEE Communications Magazine*, **52** (5), 26–35.

[2] P. Demestichas, A. Georgakopoulos, D. Karvounas, et al. (2013) 5G on the horizon: Key challenges for the radio-access network. *IEEE Vehicular Technology Magazine*, **8** (3), 47–53.

[3] W. H. Chin, Z. Fan, and R. Haines (2014) Emerging technologies and research challenges for 5G wireless networks. *IEEE Wireless Communications*, **21** (2), 106–112.

[4] M. Tehrani, M. Uysal, and H. Yanikomeroglu (2014) Device-to-device communication in 5G cellular networks: Challenges, solutions, and future directions. *IEEE Communications Magazine*, **52** (5), 86–92.

[5] N. Bhushan, J. Li, D. Malladi, et al. (2014) Network densification: The dominant theme for wireless evolution into 5G. *IEEE Communications Magazine*, **52** (2), 82–89.

[6] N. Himayat, S. Yeh, A. Panah, et al. (2014) Multi-radio heterogeneous networks: Architectures and performance. *International Conference on Computing*, Networking and Communications (ICNC), pp. 252–258.

[7] C.-X. Wang, F. Haider, X. Gao, et al. (2014) Cellular architecture and key technologies for 5G wireless communication networks. *IEEE Communications Magazine*, **52** (2), 122–130.

[8] T. Rappaport, S. Sun, R. Mayzus, et al. (2013) Millimeter wave mobile communications for 5G cellular: It will work! *IEEE Access*, **1**, 335–349,.

[9] S. Talwar, D. Choudhury, K. Dimou, et al. (2014) Enabling technologies and architectures for 5G wireless. *IEEE MTT-S International Microwave Symposium (IMS)*, pp. 1–4.

[10] V. Subramanian, R. Berry, and R. Agrawal (2010) Joint scheduling and resource allocation in CDMA systems. *IEEE Transactions on Information Theory*, **56** (5), 2416–2432.

[11] S.-J. Oh, D. Zhang, and K. Wasserman (2003) Optimal resource allocation in multiservice CDMA networks. *IEEE Transactions on Wireless Communications*, **2** (4), 811–821.

[12] K. Pedersen, Y. Wang, S. Strzyz, and F. Frederiksen (2013) Enhanced inter-cell interference coordination in co-channel multi-layer LTE-advanced networks. *IEEE Wireless Communications*, **20** (3), 120–127.

[13] Qualcomm Inc. (2011) White paper: LTE Advanced: Heterogeneous networks. URL: www.qualcomm.com/documents/lte-heterogeneous-networks.

[14] D. Lee, H. Seo, B. Clerckx, et al. (2012) Coordinated multipoint transmission and reception in LTE-advanced: Deployment scenarios and operational challenges. *IEEE Communications Magazine*, **50** (2), 148–155.

[15] Y. Zhou, L. Liu, H. Du, et al. (2014) An overview on intercell interference management in mobile cellular networks: From 2G to 5G. *IEEE International Conference on Communication Systems (ICCS)*, pp. 217–228.

[16] Huawei (2014) White paper: 5G: A technology vision. URL: www.huawei.com/5gwhitepaper/ (accessed August 5, 2016).

[17] E. Hossain, M. Rasti, H. Tabassum, and A. Abdelnasser (2014) Evolution toward 5G multi-tier cellular wireless networks: An interference management perspective. *IEEE Wireless Communications*, **21** (3), 118–127.

[18] M. Hasan and E. Hossain (2013) Resource allocation for network-integrated device-to-device communications using smart relays. *IEEE Global Communications Conference Workshops (GC Wkshps)*, pp. 597–602.

[19] M. Arslan, J. Yoon, K. Sundaresan, et al. (2013) A resource management system for interference mitigation in enterprise OFDMA femtocells. *IEEE/ACM Transactions on Networking*, **21** (5), 1447–1460.

[20] H. Zhang, X. Chu, W. Ma, et al., Resource allocation with interference mitigation in OFDMA femtocells for co-channel deploymenta. *EURASIP Journal on Wireless Communications and Networking*, **2012** (1). doi: http://dx.doi.org/10.1186/1687-1499-2012-289.

[21] W.-H. Kuo and W. Liao (2008) Utility-based radio resource allocation for QoS traffic in wireless networks. *IEEE Transactions on Wireless Communications*, **7** (7), 2714–2722.
[22] E. Hossain, L. B. Le, and D. Niyato (2013) *Radio Resource Management in Multi-tier Cellular Wireless Networks*. John Wiley & Sons, Ltd, Chichester.
[23] P. Semasinghe, E. Hossain, and K. Zhu (2015) An evolutionary game for distributed resource allocation in self-organizing small cells. *IEEE Transactions on Mobile Computing*, **14** (2), 274–287.
[24] G. Fodor, E. Dahlman, G. Mildh, et al. (2012) Design aspects of network assisted device-to-device communications. *IEEE Communications Magazine*, **50** (3), 170–177.
[25] K. Son, S. Lee, Y. Yi, and S. Chong (2011) REFIM: A practical interference management in heterogeneous wireless access networks. *IEEE Journal on Selected Areas in Communications*, **29** (6), 1260–1272.
[26] E. Jorswieck (2011) Stable matchings for resource allocation in wireless networks. *17th International Conference on Digital Signal Processing (DSP)*, pp. 1–8.
[27] D. Gale and L.S. Shapley (1962) College admissions and the stability of marriage, *American Mathematical Monthly*, **18**, 9–15.
[28] A.E. Roth and M.A.O. Sotomayor (1992) *Two-sided Matching: A Study In Game-Theoretic Modeling And Analysis*. Cambridge University Press,
[29] G. O'Malley (2007) *Algorithmic Aspects of Stable Matching Problems*. PhD dissertation, University of Glasgow.
[30] F.R. Kschischang, B.J. Frey, and H.A. Loeliger (2008) Factor graphs and the sum-product algorithm. *IEEE Transactions on Information Theory*, **47** (2), 498–519.
[31] S. Aji and R. McEliece (2000) The generalized distributive law. *IEEE Transactions on Information Theory*, **46** (2), 325–343.
[32] A. Abrardo, M. Belleschi, P. Detti, and M. Moretti (2011) A min-sum approach for resource allocation in communication systems. *IEEE International Conference on Communications (ICC)*, pp. 1–6.
[33] M. Moretti, A. Abrardo, and M. Belleschi (2014) On the convergence and optimality of reweighted message passing for channel assignment problems. *IEEE Signal Processing Letters*, **21** (11), 1428–1432.
[34] D.P. Bertsekas and J.N. Tsitsiklis (1989) *Parallel and Distributed Computation: Numerical Methods*. Prentice-Hall, Inc..
[35] D. Bertsekas (2001) Auction algorithms. In: C. Floudas and P. Pardalos (eds) *Encyclopedia of Optimization*, Springer, New York, pp. 73–77.
[36] M. Zavlanos, L. Spesivtsev, and G. Pappas (2008) A distributed auction algorithm for the assignment problem. *47th IEEE Conference on Decision and Control*, pp. 1212–1217.
[37] C. Liang and F. Yu (2015) Wireless network wirtualization: A survey, some research issues and challenges. *IEEE Communications Surveys and Tutorials*, **17** (1), 358–380.
[38] M. Hadzialic, B. Dosenovic, M. Dzaferagic, and J. Musovic (2013) Cloud-RAN: Innovative radio access network architecture. *55th IEEE International Symposium (ELMAR)*, pp. 115–120.
[39] M. Manjrekar, V. Ramaswamy, and S. Shakkottai (2014) A mean field game approach to scheduling in cellular systems. *IEEE International Conference on Computer Communications*, pp. 1554–1562.
[40] H. Tembine, R. Tempone, and P. Vilanova (2012) Mean field games for cognitive radio networks. *American Control Conference (ACC)*, pp. 6388–6393.
[41] H. Tembine, E. Altman, R. El-Azouzi, and Y. Hayel (2010) Evolutionary games in wireless networks. *IEEE Transactions on Systems, Man, and Cybernetics, Part B: Cybernetics*, **40** (3), 634–646.

Additional Reading

5G and Heterogeneous Networks

J. Andrews, S. Buzzi, W. Choi, et al. (2014) What will 5G be? *IEEE Journal on Selected Areas in Communications*, **32** (6), 1065–1082.
B. Bangerter, S. Talwar, R. Arefi, and K. Stewart (2014) Networks and devices for the 5G era. *IEEE Communications Magazine*, **52** (2), 90–96.
P. Pirinen (2014) A brief overview of 5G research activities. *1st International Conference on 5G for Ubiquitous Connectivity (5GU)*, pp. 17–22.
Y. Lee, T. Chuah, J. Loo, and A. Vinel (2014) Recent advances in radio resource management for heterogeneous LTE/LTE-A networks. *IEEE Communications Surveys and Tutorials*, **16** (4), 2142–2180.

Stable Matching

K. Iwama and S. Miyazaki (2008) A survey of the stable marriage problem and its variants, in International *Conference on Informatics Education and Research for Knowledge-Circulating Society (ICKS)*, pp. 131–136.

X. Feng, G. Sun, X. Gan, et al. (2014) Cooperative spectrum sharing in cognitive radio networks: a distributed matching approach. *IEEE Transactions on Communications*, **62** (8), 2651–2664.

A. Leshem, E. Zehavi, and Y. Yaffe (2012) Multichannel opportunistic carrier sensing for stable channel access control in cognitive radio systems. *IEEE Journal on Selected Areas in Communications*, **30** (1), 82–95.

Message Passing

M. Hasan and E. Hossain (2014) Distributed resource allocation for relay-aided device-to-device communication: A message passing approach. *IEEE Transactions on Wireless Communications*, **13** (11), 6326–6341.

A. Abrardo, M. Belleschi, P. Detti, and M. Moretti (2012) Message passing resource allocation for the uplink of multi-carrier multi-format systems. *IEEE Transactions on Wireless Communications*, **11** (1), 130–141.

K. Yang, N. Prasad, and X. Wang (2018) A message-passing approach to distributed resource allocation in uplink DFT-Spread-OFDMA systems, *IEEE Transactions on Communications*, **59** (4), 1099–1113.

Auction Algorithm

Y. Zhang, C. Lee, D. Niyato, and P. Wang (2013) Auction approaches for resource allocation in wireless systems: A survey, *IEEE Communications Surveys and Tutorials*, **15** (3), 1020–1041.

I. Koutsopoulos and G. Iosifidis (2010) Auction mechanisms for network resource allocation. *8th International Symposium on Modeling and Optimization in Mobile, Ad Hoc and Wireless Networks (WiOpt)*, pp. 554–563.

K. Yang, N. Prasad, and X. Wang (2009) An auction approach to resource allocation in uplink OFDMA systems. *IEEE Transactions on Signal Processing*, **57** (11), 4482–4496.

M. Bayati, B. Prabhakar, D. Shah, and M. Sharma (2007) Iterative scheduling algorithms. *26th IEEE International Conference on Computer Communications. (INFOCOM)*, pp. 445–453.

9

Device-to-Device Communications

Andreas F. Molisch,[1] Mingyue Ji,[2] Joongheon Kim,[3] Daoud Burghal,[1]
and Arash Saber Tehrani[1]
[1] *University of Southern California, Los Angeles, CA, USA*
[2] *University of Utah, Salt Lake City, UT, USA*
[3] *Chung-Ang University, Korea*

9.1 Introduction and Motivation

For many years, the worlds of infrastructure-based wireless communications on one hand, and ad-hoc networks on the other hand, remained essentially separate. Infrastructure-based communications occurs commonly in licensed bands, and are carefully planned. Ad-hoc networks operating in the same band were, at best, seen as interference sources that might need to be accommodated. However, in the past 5 years, a new paradigm has emerged: device-to-device D2D communications, where devices – non-infrastructure wireless nodes – talk directly with each other (similar to ad-hoc networks), but they do so in coordination with, and possibly under the instruction of, the infrastructure nodes; see also Doppler et al. (2009b) and Wei et al. (2014b).

D2D is motivated largely by the eternal quest for higher spectral efficiencies. The change in usage patterns for phones has greatly increased the demand for communications between nearby devices. Phones are a major repository of photographs, music, movies, and so on. It has become common for friends to electronically exchange such media when they are nearby, often in the course of a conversation. Thus, devices wishing to communicate with other devices that might just be meters away has become a common scenario. It is obviously wasteful for two such devices to talk to each other through a base station (BS) that might be a kilometer away; both energy and spectral resources can be saved by direct communications. Since they need to communicate only over a short distance, the capacity of the direct link tends to be higher, even at low transmit powers. Even more importantly, the same time/frequency resources can be re-used in other parts of the cell.

Towards 5G: Applications, Requirements and Candidate Technologies, First Edition.
Edited by Rath Vannithamby and Shilpa Talwar.

Companion website: www.wiley.com/go/vannithamby/towards5g

Furthermore, D2D can extend the coverage range of an infrastructure node – BS or access point (AP) – through multi-hop communications. In other words, devices might act as relay stations for other devices, extending cell ranges and increasing the efficiency of processes such as multicast. Last, but not least, D2D provides robustness as it can continue to function even in the absence of infrastructure. The development of D2D communications in the LTE standard was originally motivated by the requirement of the US frequency regulator, the Federal Communications Commission, that operators provide such emergency communications capability.

Creating functioning D2D networks requires many different components, which are reviewed in this chapter. Devices first need to discover who their neighbors (potential communication partners) are, and determine how good the communication channel is with them (see Section 9.3). Based on this information, the system then has to decide:

- which devices employ D2D communications for their needs, and which talk directly to the infrastructure, a process called mode selection (see Section 9.4.1)
- which devices get assigned which power and spectral resources for the chosen mode: resource allocation (see Section 9.4.2)
- which devices are scheduled at what times: scheduling (see Section 9.5).

If the distance that can be covered by a direct link is too short for communication with a desired partner, multiple hops might be required (see Section 9.6). A quantitative design of all of these features requires, of course, a good understanding of the underlying wireless propagation channels (see Section 9.2).

D2D communications have drawn attention in international standardization; the most important result is the LTE Direct standard discussed in Section 9.7. Last, but not least, we will discuss various applications and pay particular attention to D2D communications in conjunction with video coding, one of the most promising developments for radical improvement of area spectral efficiency, see Sections 9.8 and 9.9.

This chapter aims to give an overview of the main challenges and solution methods for D2D communications. The literature list aims to provide some key examples of relevant papers, but does not lay a claim to completeness; such a goal would be impossible to reach within the constraints of a chapter when more than 500 papers on this topic have already been published at the time of writing in mid-2015. For further references see also the survey by Asadi et al. (2014). In any case, the literature is quickly outdated as papers continue to be published at a brisk pace in this very vibrant area.

9.2 Propagation Channels

D2D radio channels have fundamentally different properties to those of conventional, device-to-infrastructure (D2I) cellular channels. The main reason for this is that most often both the receive antenna and the transmit antenna are located at low heights, and hence there is more interaction with objects in the close neighborhood of the devices. The difference is especially pronounced for outdoor links, where a BS would be high above ground (typically 10 m for microcell, and up to 100 m for macrocells), while all devices are at street level. Consequently, over-the-rooftop propagation is not a viable mechanism, and street canyon propagation is more strongly affected by shadowing objects such as cars and trucks. One important class of outdoor D2D systems is vehicle-to-vehicle (V2V) communications, which implies strong mobility. A different class is that of people located, more or less stationary, in outdoor cafes, plazas, and so on. In indoor situations, the difference between D2I and D2D propagation mechanisms is less pronounced, and the range of validity for many indoor channel models includes the D2D case.

9.2.1 Pathloss

A first impact of the different propagation conditions is the path-loss model. For outdoor situations, the path-loss exponents are generally in the range $1.6-2$ when the two devices communicating with each other are in the same street (in urban, suburban, or rural environments) or on the same highway; see Karedal et al. (2011). When the two devices are on orthogonal streets, the model of Mangel et al. (2011), based on extensive measurements in and around Munich, Germany, proposes

$$PL(d_r,d_t,w_r,x_t,i_s)=C+i_sL_{SU}$$
$$+\begin{cases} 10\log_{10}\left(\left(\dfrac{d_t^{E_T}}{(x_tw_r)^{E_S}}\dfrac{4\pi d_r}{\lambda}\right)^{E_L}\right), & \text{if } d_r\le d_b \\ 10\log_{10}\left(\left(\dfrac{d_t^{E_T}}{(x_tw_r)^{E_S}}\dfrac{4\pi d_r^2}{\lambda d_b}\right)^{E_L}\right), & \text{if } d_r> d_b \end{cases} \quad (9.1)$$

where d_t and d_r denote the distance of the TX and RX to the intersection center, respectively, w_r is the width of the RX street, and x_t is the distance of the TX to the wall. In the model $C = 3.75\,\text{dB}$ is the so-called curve shift, $L_{SU} = 2.94\,\text{dB}$ is the sub-urban loss, $i_s = 0$ is the urban loss factor, $i_s = 1$ the sub-urban loss factor, $E_L = 2.69$ is the loss exponent, $E_S = 0.81$ is the street exponent, $E_T = 0.957$ is the TX distance exponent, and finally $d_b = 180\,\text{m}$ is the breakpoint distance. The model was later validated by independent measurement data by Abbas et al. (2013) with good agreement, but also with the suggestion to include a random offset parameter to reflect variations between different street intersections; random variations were also suggested in Li et al. (2012b) based on the comparison of ray-tracing results with the model of Schack et al. (2011).

Shadowing also requires some new and improved modeling. Traditionally, the correlation of the shadowing between the different links originating at the BS has been modeled; this can be described in a rather straightforward way through spatial correlation functions. For D2D communications, we have to determine the correlation between link pairs where both the transmitters and receivers are moving. Furthermore, we have to describe the shadowing for the three-link scenario: link from the BS to device A, BS to device B, and device A to device B. Proper implementation of the shadowing in a physically meaningful way can be difficult, and requires careful definition of stationarity regions for shadowing (compare Li et al. (2012b)). For communication between two pedestrians holding devices, the body shadowing plays an important role; it is noteworthy that a "rotational" shadowing – in other words, shadowing depending on the orientation of the person – occurs in addition to a "translational" shadowing as a person is moving on a trajectory (walking); see Cotton (2015) and Karedal et al. (2008).

9.2.2 *Delay Dispersion*

As far as the delay dispersion is concerned, the rms delay spread (second central moment of the power-delay profile) varies as a function of location, and can be modeled as a random variable (log-normally distributed in most cases). While most measurements have been performed for V2V systems, the results are anticipated to be similar for quasi-stationary outdoor environments. Mean rms delay spreads are on the order of 100–200 ns for rural and suburban environments, and up to 400 ns for urban environments. Detailed reviews of the measurement results for both path loss and delay dispersion for various environments can be found in Mecklenbrauker et al. (2011) and Molisch et al. (2009). Delay spreads in safety-related scenarios for V2V communication are reported in Bernado et al. (2014). High rms delay spreads are observed when the line-of-sight component is blocked and there are large reflecting objects close to the road. Mean rms delay spreads of 170 ns are reported for longer periods, but there are also extreme values of the rms delay spread just below 1 μs.

9.2.3 *Temporal Variations*

There are two distinct groups of D2D channels, depending on the dynamics of the nodes; that is, if the devices themselves are moving or not. In the first case, devices at *both* link ends can move, sometimes very fast. In addition scatterers and shadowing objects can also move. This is the case for V2V channels, for example, for which extensive research has shown that the channel statistics typically change over time and hence the conventional assumption of wide-sense-stationary uncorrelated scattering is only fulfilled for rather short time intervals, and moderate frequency intervals. In Renaudin (2013) a stationarity bandwidth of around 50 MHz and a stationarity time corresponding to a movement of 10–40 m were suggested, whereas Bernado et al. (2012) reported somewhat larger values for the stationarity bandwidth, but smaller values for the stationarity time. Delay spreads and Doppler spreads in safety-related scenarios for V2V communication are reported in Bernado et al. (2014). High Doppler spreads typically arise in drive-by situations, with mean rms Doppler spreads of 300 Hz are reported over a longer time, but with extreme values of just below 1,000 Hz.

In order to handle the non-stationarities from a channel-modeling perspective, the most straightforward solution is often to use a geometry-based stochastic channel model (GSCM; see Molisch et al., 2003), where the non-stationarities are automatically taken care of and modeled by the randomly placed scatterers in the environment. Examples of GSCMs for V2V scenarios can be found in Walter et al. (2014), Karedal et al. (2009), and Renaudin et al. (2010). Another approach is based on tapped delay lines, such that the location (delay) of the taps is either adjusted continuously, or a birth/death process of the taps is implemented; the latter method is used, for example, in the IEEE 802.11p channel model; see Acosta-Marum and Ingram (2007).

In static or nomadic scenarios, the two nodes do not change with respect to each other. This occurs, for example, in machine-to-machine communications (static nodes) or peer-to-peer

(WiFi Direct) links between laptops (nomadic scenarios). In that case the Doppler spectrum of the channel is determined by moving objects in the surroundings. Typically the Doppler spread is low, as many of the dominant scatterers are static as well and hence the coherence time of the channel can be quite large; see Karedal et al. (2010). However, one important aspect that should not be forgotten is the shadowing or blocking of the radio signal by other objects, as the coherence time of this process can be very large as well.

A survey of modeling approaches is provided by Cheng et al. (2015). Finally, future D2D communications might be operating in the mm-wave bands. The propagation characteristics of such channels have seen some initial investigations, for example, by Al-Hourani et al. (2014), but more will be needed.

9.3 Neighbor Discovery and Channel Estimation

9.3.1 Neighbor Discovery

Neighbor discovery is the first task that must be performed in a network, since this enables all other the operations mentioned in Section 9.1. It is noteworthy that devices have to make themselves known to the infrastructure (that is, the BS), similar to the normal association process in cellular networks. However, this does not obviate the need for the devices to discover their neighbors, which is a much more complicated process: each device has to find all of its neighbors and this means that if there are N devices in the cell, the BS has to discover N devices, but each device has to discover whether the other $N-1$ devices are its neighbors, leading to a total of $N \times (N-1)$ potential links that must be discovered and decided. Fast discovery of neighbors is thus essential.

The situation is further compounded by mobility of devices, which changes the neighbor structure; due to the short range of D2D links, the neighborhood of a device changes much faster than the association to different BSs. Thus, neighborhood discovery must be repeated sufficiently often such that sufficient knowledge of the network topology is maintained.

Neighbor discovery can be classified into:

- network based
- direct discovery.

In a network-based discovery, the discovering device informs the BS of its interest in a local service – a file, say – and then assuming the BS is aware of the proximity of a suitable device it can inform the discovering device of the possibility of a neighbor device. Examples of such discovery techniques are proposed in Zhang et al. (2015) and Prasad et al. (2014). In direct neighbor discovery, the devices autonomously search for all possible neighbor devices. Note that in both cases the device has to announce itself or query the availability of devices through a packet or a signature; see, for example, Choi et al. (2014) and Zou et al. (2014).

To provide a more precise mathematical description, let us define a neighbor relation as follows: two nodes are neighbors if each can hear the other with a power greater than a certain threshold Γ, which depends on the communication channel and the purpose of the communication:

$$N(v_i) = \left\{ v_j : \gamma_t g_t g_r |h_{ij}|^2 \geq \Gamma \right\}, \tag{9.2}$$

where γ_t, g_t, g_r are the transmission power, receiver, and transmitter antenna gains, respectively (equal for all nodes), and h_{ij} is the channel gain between nodes v_i and v_j.

Neighbor-discovery algorithms can be categorized in different ways. We first distinguish between randomized and deterministic protocols. In the randomized algorithm, the nodes randomly choose to either transmit or listen in each time slot so that each node gets a chance to hear its neighbors, as well as to be heard by its neighbors.[1]

One of the most influential early approaches was the "birthday protocol", in which each node transmits randomly, in k out of n available timeslots; listens in another k slots, and remains idle for the remaining slots; see McGlynn and Borbash (2001). The probability that one node hears another (when two nodes are behaving in a similar manner) is

$$1-\frac{\binom{n-k}{k}}{\binom{n}{k}} \tag{9.3}$$

Other algorithms that fall into the randomized category include directional antenna neighbor discovery, and slotted random transmission and reception, see Vasudevan et al. (2005) and Borbash et al. (2007), respectively. Several approaches have been proposed to modify the popular IEEE802.11 MAC protocol for directional wireless links by modifying the IEEE802.11MAC protocol; see Ramanathan et al. (2005) and Takai et al. (2002). They use the directional virtual carrier sensing concept, which extends the IEEE802.11 distributed coordinated function to directional wireless networks.

In the deterministic algorithms, on the other hand, nodes transmit their identifiers according to a pre-determined transmission pattern. Methods such as:

- compressed sensing (CS); see Luo and Guo (2008); Tehrani et al. (2013) and Zhang et al. (2013a)
- ZigZag; see Tehrani and Caire (2014)

are deterministic algorithms. The main idea of the ZigZag algorithm, which can outperform random access discovery methods both in reliability and speed, is that instead of randomly transmitting or receiving, nodes transmit and receive according to a pre-determined on-off pattern. Thus, the nodes may cancel the known identifiers from received signals to retrieve identifiers of other neighbors. The principle is shown in Figure 9.1, where nodes v_1, v_2, and v_3 are neighbors of v_0 with identifiers (IDs) s_1, s_2, and s_3. Assuming the nodes transmit their IDs according to Figure 9.1, node nu_0 can discover the three neighbors.

The challenge consists of designing the transmission patterns so that the discovery time remains short while all nodes manage to detect all their neighbors with high probability.

In CS neighbor discovery the similarity of neighbor discovery and sparse signal recovery is exploited; see Tehrani et al. (2013). A key observation is that number of neighbors a node has is much smaller than the total number of nodes in the network and thus the zero/one vector $\mathbf{x}_i$

[1] Obviously, nodes that are subject to the half-duplex constraint (almost all currently existing devices) cannot listen and transmit at the same time.

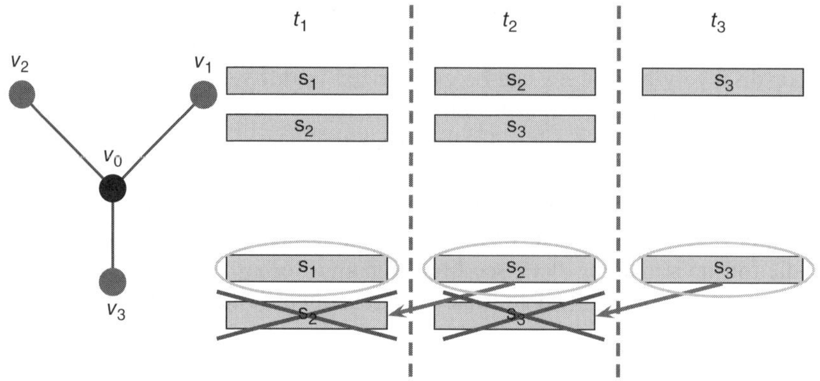

Figure 9.1 Operating principle of Zig-Zag.

with elements set to one for neighbors of node i is sparse. The advantage of applying CS is that the receivers are simple energy detectors. For building the transmission pattern of the nodes in CS, the authors use the columns of the measurement matrices recently built for compressed sensing by Tehrani and Caire (2014). Moreover, they use a linear programming formulation known as basis pursuit, to detect the neighbors; compare Candes and Tao (2005); Donoho (2006). Sparse signal recovery was also used by Tang et al. (2014); the authors chose Zhadoff–Chu sequences to be the signature of the devices, since these have good orthogonality properties. The authors used block sparse Bayesian learning techniques to detect the neighbors and estimate the channels. Zou et al. (2014) used a simpler signature design, based on a Galois field, in which a receiving device needs only do integer comparison to detect the neighbors.

9.3.2 Channel Estimation

In contrast to conventional cellular networks, where only the channel state information (CSI) between devices and the BS is required, in D2D networks the CSI *between* devices is needed. This poses new challenging problems; the BS cannot infer the state of the links between devices from direct observations. For example, the link between two devices could be bad due to shadowing, while their links to the BS are fine. Thus, it is necessary to perform estimation of the channels between devices, and subsequently report the results to the BS. Yet the question remains of how often, and in what form, pilot transmission and the associated CSI training should occur; temporal variability of the channel variation and the broadcast effect make "straightforward" solutions inefficient. For instance, time division multiple access (TDMA), in which each device in the cell gets a separate timeslot for transmission of its pilot sequence, reduces the interference between devices but increases channel outdatedness due to the long times between pilot sequences of each particular device. On the other hand, contention-based techniques, such as CSMA/CS, might reduce the delay of CSI acquisition but the quality of estimated channels might be poor due to interference. For these reasons the acquisition of CSI between *all* devices is usually referred to as being a cumbersome problem or impractical. Consequently, some recent work, restrains the channel knowledge to the average gain, which can be obtained from the path loss as function of distance – see, for example, Fodor and Reider

(2011) – or by outdated sets of previous measurements, for example, acquired during the neighbor discovery process – see Tang et al. (2014) – or estimate only a subset of the available devices; see Wu et al. (2010). Other papers constrain the CSI estimate to the currently active links Pradini et al. (2014). Another important practical aspect is the realization of the estimation process within the constraints of existing standards, especially LTE; see Lee et al. (2014).

To provide a general formulation, the acquisition of CSI can also be viewed as a scheduling problem, where every device has to transmit a pilot signal to its neighbors. The traditional scheduling problem has been studied thoroughly in mobile ad-hoc networks, for the purpose of maximizing the throughput of data communication. However, the quality of the CSI has an impact on the *subsequent* communication, leading to a change in the objective function. Also, many scheduling schemes discussed in the literature assume a knowledge of interference levels between devices, information that is not available in the CSI acquisition problem.

To tackle these problems, Burghal and Molisch (2016) developed the Location Aware Training Scheme (LATS), which exploits geographical information in order to optimize the scheduling of training sequences for CSI acquisition. LATS uses the well-known concept of spatial reuse to jointly trade off the interference and delay. The effect of distance is evident: the interference power received by device m decays as a function of the distance between a transmitter and a receiver m. One possible approach for scheduling the pilot tones would be to determine the optimum sets of devices that can transmit simultaneously without exceeding an interference threshold. However, such a scheduling is known to be extremely complex (usually NP-hard); see Section 9.5. To simplify the problem, LATS groups devices into segments of equal size on which it then performs scheduling.

The detailed setup of the scheme can be seen from Figure 9.2. The cell is divided into equal-sized square segments with side length L_s. The communication disk the area centered

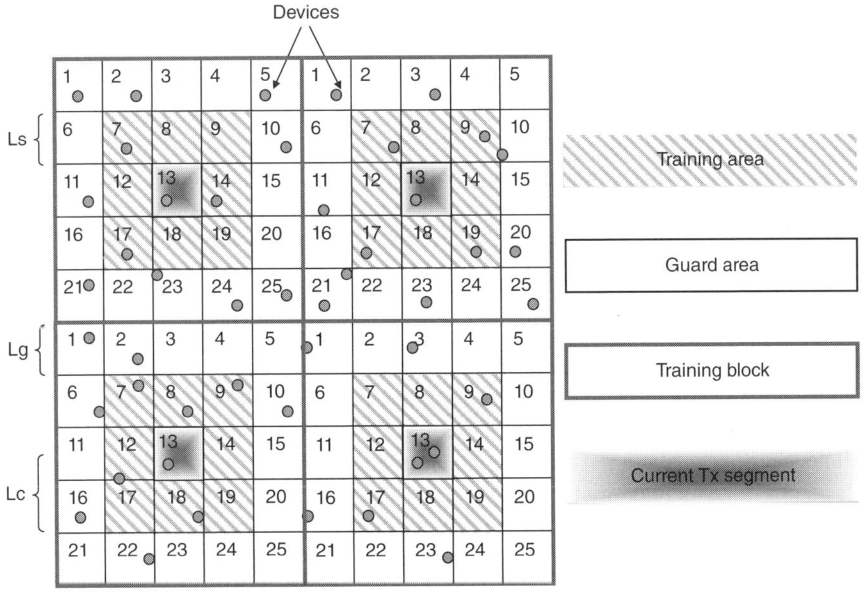

Figure 9.2 Simplified segmented cell and LATS cell components. In this figure $S_n = 1$ and $G_n = 2$; devices in training set 13 are allowed to transmit.

around a device within which neighbors lie – is assumed to consist of an integer number of segments. In order to reduce interference, LATS defines the minimum separation distance between virtual devices that are scheduled simultaneously through use of a guard distance.

By writing the normalized mean square error (NMSE) as a function of the geometry parameters (segment size, communication disk length, guard distance) as well as the mobility parameters, optimization of those geometry parameters is achieved through minimization of the NMSE.

9.4 Mode Selection and Resource Allocation

9.4.1 Mode Selection

A key question in D2D systems is which links should be handled by D2D connections, and which should be handled by cellular connections, a problem often called "mode selection" in the relevant literature. For mode selection, in general there are two D2D communication modes: D2D mode and cellular mode. The D2D mode means that the communication from device to device is direct. Cellular mode is where the D2D communications are relayed by the BS.

The D2D links can be categorized based on the spectrum in which D2D communication occurs: in-band D2D and out-of-band D2D.

In-band D2D. In-band D2D technologies use the cellular spectrum for both D2D and cellular links. The motivation for choosing in-band communication is usually the high control over licensed cellular spectrum. Some researchers, for example Akkarajitsakul et al. (2012) and Doppler et al. (2009a), argue that the interference in the unlicensed spectrum is uncontrollable, which can create constraints for QoS provisioning in multimedia applications. In-band D2D can be further divided into underlay and overlay categories. In underlay D2D, cellular networks and D2D networks share the same spectrum (non-orthogonal) whereas overlay D2D networks allow communication between cellular networks and D2D networks with orthogonal spectrum. In-band D2D can improve the spectrum efficiency of cellular networks by reusing spectrum resources (underlay) or allocating dedicated cellular resources to D2D links that show better performance (overlay). The key disadvantage of in-band D2D is the interference caused by D2D links to cellular networks (and vice versa). This interference can be mitigated by introducing computationally intensive high-complexity resource allocation methods, as outlined next.

Out-of-band D2D. Out-of-band D2D technologies exploit unlicensed spectrum. The motivation behind using out-of-band D2D communication is to eliminate the interference issue between D2D and cellular links. Using unlicensed spectrum requires an extra interface and usually adopts other wireless technologies such as WiFi Direct, or 60-GHz millimeter-wave radio. Some contributions, for example Golrezaei et al. (2014), suggest giving control of the second interface/technology to the cellular network. In contrast others, for example Wang and Rengarajan (2013), propose keeping only cellular communications controlled while leaving the D2D communications to the users, who are termed *autonomous*. Out-of-band D2D uses unlicensed spectrum, which makes the potential interference between D2D and cellular users irrelevant. On the other hand, out-of-band D2D may suffer from the uncontrolled nature of unlicensed spectrum. It should be noted that currently only cellular devices with multiple wireless interfaces (e.g., LTE, WiFi, or 60-GHz millimeter-wave) are able to use out-of-band D2D, and therefore D2D users can have simultaneous D2D and cellular networks. This situation might change with future operation of LTE in the unlicensed bands.

Mode selection is to switch a potential D2D pair between the different modes mentioned previously. For example, the D2D pairs can be switched between underlay in-band mode, out-of-band mode, and cellular mode – see Doppler et al. (2010) – or just between underlay in-band mode and cellular mode; see Lin et al. (2013a).

Commonly, the interference at the receiver is treated as noise, such that the quality of the link can be measured as received SINR. We denote $\mathrm{SINR}_{i,j}$ as the SINR between transmitter i and receiver j in the network; it is defined as

$$\mathrm{SINR}_{i,j} \triangleq \frac{P_{i,j}}{I + N_0}, \tag{9.4}$$

where $P_{i,j}$ is the received power spectral density of node j, N_0 is the power spectral density of of the additive while circularly symmetric Gaussian noise, and I is the interference. Notice that I is the summation of all the interference from both other D2D communications and the communications between users and BS. The instantaneous link rate $R_{i,j}$ between transmitter i and receiver j is given by

$$R_{r,j} = \log\left(1 + \mathrm{SINR}_{i,j}\right), \tag{9.5}$$

and the ergodic rate is given by

$$\bar{R}_{i,j} = \mathbb{E}\left[\log\left(1 + \mathrm{SINR}_{i,j}\right)\right], \tag{9.6}$$

where the expectation is over all the channel realizations. Generally, the goal is to maximize the sum rate or minimize the individual power for D2D communications subject to the constraint of a SINR threshold for the cellular users and/or D2D users, or a constraint of the maximum outage probability of the cellular users.

Depending on the assumptions of I, we can have different optimization problems. When we assume only one D2D link in a single-cell network, in other words we consider each D2D link separately, then the interference term I consists only of one term and the optimization can be done in a comparatively straightforward way. For example, Doppler et al. (2010) consider selection among three modes, which are underlay in-band mode, out-of-band mode, and cellular mode, where the underlay in-band mode considers both uplink reuse and downlink reuse. If, as assumed in that paper, there is only one D2D pair and one cellular user, it is easy to write the system sum rate for each of the three cases. For instance, if the overlay in-band mode is uplink reuse, then we denote the sum rates for the overlay in-band model, out-of-band mode, and cellular mode as $R_{\text{in-band}}, R_{\text{out-band}}, R_{\text{cellular}}$. Then, by fixing the transmitting power for all the users, we pick a D2D communication mode such that $\max\{R_{\text{in-band}},, R_{\text{out-band}},, R_{\text{cellular}}\}$ is achieved. When the transmitting powers for the users are variable, Feng et al. (2014) consider a similar model with a minimum rate constraint for both the D2D pair and the cellular user, and then propose a bisection algorithm that is adopted to solve the quasiconvex optimization problems in the out-of-band and cellular modes through transforming the original problem into a sequence of convex feasibility problems. For the in-band mode, a concave-convex procedure is used to solve the "difference of convex" optimization problem.

When multiple D2D links are active (which is the usual case), computation of the distribution of the interference I becomes the main difficulty in obtaining the analytical results of the achievable average sum rate. To solve this problem, the Poisson point processes (PPP) model for the distribution of BSs and devices can be used. Counterintuitively, by introducing this additional randomness, the independence of the positions of nodes (BS or devices) allows application of tools from stochastic geometry, so that the coverage probability (the probability that SINR is larger than some threshold) and the average achievable rate can be computed analytically. Hence, SINR can be used to perform mode selection of the devices. In practice, simpler methods can also be used. For example, if there are only two modes for the potential D2D pairs – underlay in-band mode and cellular mode –the selection criterion can be based on distance. For instance, let the distance between the D2D pairs be D. Given γ_I, then if $D \geq f(\gamma_I)$, the cellular mode is used; otherwise the underlay in-band mode is used. For example, Lin et al. (2013a) and Lin et al. (2014) use such a model and obtain the analytical formula of the sum rate given γ_I.

Power control also needs to be considered, as it controls the interference emanating from a D2D link, as well as the possible throughput on that link. Lee et al. (2015) consider a random network model using stochastic geometry and develop power-control algorithms such that multiple D2D links may communicate successfully while guaranteeing reliable communication for the existing cellular link. A joint optimization of model selection and power allocation is made in Hakola et al. (2010), which derives the system equations given the information of the network such as link rate, noise levels, SINR, and mode selection for the D2D pairs. By using these system equations, one can obtain both the optimal power allocation and mode selection for the devices such that the SINR criteria for both D2D users and cellular users are satisfied. Lin et al. (2013a) also investigate mode selection as part of a larger analytical investigation of the performance. When the precise information of the network cannot be obtained, Yu et al. (2009a) consider a network within a disk of radius R, where there is one cellular user uniformly distributed in this disk, one D2D user whose position is fixed, and the position of the other D2D user is uniformly distributed within a disk of radius L and centered by the previous D2D user. By using this model, these authors are able to derive the distribution of SINR such that by using only the statistics of the network, one can evaluate the sum throughput by guaranteeing the outage probability of the cellular user. ElSawy et al. (2014) present an analytical framework for D2D-enabled uplink cellular networks with a flexible mode-selection scheme, which is based on a biasing factor that accounts for both the D2D link quality and the cellular link quality and controls the extent to which the D2D communication is enabled in the network, along with truncated channel inversion power control.

9.4.2 Resource Allocation

Mode selection is also closely connected to the "resource allocation" problem; in other words, which spectral resources should be assigned to each link. A considerable number of papers exists for this problem, which differ both in their mathematical approach and in the optimization criterion. One approach (for in-band setups) is to consider the D2D links as "interferers" to a victim BS, and ensure that the interference is minimized or at least kept within certain boundaries, as shown by Zulhasnine et al. (2010), where the D2D links can coexist with both uplink and/or downlink of the cellular communications. Another criterion is to optimize the throughput in the cell (irrespective of which mode is used for the communication), or to impose fairness or minimum performance criteria for each communication link; see Duong et al. (2013) and Le (2012). A survey of the resource-allocation problem can be found in Lei et al. (2014).

The mathematical techniques to tackle the optimization are described in the following.

Mixed-integer Nonlinear Programming. The mixed-integer nonlinear programming (MINLP) approach is described in Zulhasnine et al. (2010). The goal is to maximize the sum throughput of the whole system including the cellular users and the D2D users. To make the problem tractable, instead of maximizing the sum throughput directly, the authors formulate the problem in terms of maximizing the summation of all SINR, as shown in Eq. (9.4), subject to the constraints consisting of the minimum throughput requirements of both the cellular users and the D2D users. MINLP is notoriously hard to solve in general; the authors propose a heuristic greedy algorithm such that the interference is minimized in each step (see also next item). Yu et al. (2014) formulate an optimization problem that can be decomposed into two subproblems: power control and joint mode selection, and channel assignment. Similar to Zulhasnine et al., an heuristic low-complexity algorithm is developed. Bansal et al. (2014) develop a holistic approach to efficient offloading with D2D traffic. The approach consists of a two timescale solution: first, a coarse timescale dynamic fractional frequency reuse scheme that uses D2D traffic for both downlink and uplink jointly among interfering sectors, and a fine timescale scheduling solution that intelligently schedules cellular and D2D traffic jointly across downlink and uplink resources. In this case, polynomial time approximation algorithms are designed with a constant approximation ratio.

Greedy Selection. Greedy selection is a popular heuristic for solving complicated optimization problems, where in a first step the "best" solution (user, mode) is picked without consideration of the impact this has on other users. Subsequent selections are also based on finding the "best", taking into account the interference from the already-decided users, but disregarding the effect the selection has on later-decided users. Lee et al. (2013) formulate the problem as maximizing the spatial reuse of the resources or maximizing the number of user resources, while satisfying the required number of resource blocks allocated to each D2D link. This problem can be converted to a multiple-set-covering problem, which can be approximately solved by a greedy user-selection algorithm. Wen et al. (2013) aim to jointly optimize the mode selection and resource allocation by assuming the same priority for both cellular and D2D users. Due to the hardness of the problem, a greedy selection algorithm based on a criterion that incorporates a priority factor, which can be computed by a proportional fairness function, is proposed to achieve a promising performance, especially when the system load is heavy and the channel condition between the devices is poor.

Exclusion Zones. Another approach is to define an "exclusion zones" around active links. Min et al. (2011) define a δ_D-interference limited area (ILA) where the interference–to-signal ratio at the D2D receiver is greater than the designed threshold, δ_D. This approach prohibits scheduling simultaneously the D2D pair and cellular users that are located in the δ_D-ILA. In addition, Chen et al. (2012a) use the idea of "exclusion zones" around active links for both cellular and D2D users. By giving the cellular users high priority, they design a resource-allocation scheme by maximizing the total throughput of the system subject to picking only the links satisfying the "exclusion zones".

Graph-based Methods. Zhang et al. (2013b) propose a resource allocation algorithm based on a so-called interference-aware graph. Different from the commonly used interference graph only based on interference relationships among different communication links, the interference-aware graph consists of three properties: link attribute (the vertex belongs to a cellular link or a D2D link), resource attribute (ordered resource blocks associated with each vertex by its SNR), and cluster attribute (the cluster of resource blocks to which the vertex is assigned). By using this graph, an iterative resource allocation algorithm can be used to approximately maximize the throughput with polynomial-time complexity. Approaches based on graph coloring and bipartite matching can be found in Xu et al. (2015) and Han et al. (2014), respectively.

Particle Swarm Optimization. Su et al. (2013) formulate the problem as a non-convex maximization problem, where the goal is to maximize the sum throughput instead of maximizing the sum SINR as done by Zulhasnine et al. (2010). The main constraints are the power constraints for both cellular and D2D users, and the minimum sum throughput constraint of the D2D users. Su et al. (2013) also let the variables be both mode selection and resource allocation. It can be seen that the problem is difficult to solve, and Su et al. (2013) provide an algorithm based on particle swarm optimization with the changes dealing with the discrete values of the resource allocation and the sum-rate constraint of the D2D users.

Game-theoretic Methods. Xu et al. (2012b) use the tools from game theory to solve the resource-allocation problem. Specifically, their proposed algorithm is based on a so-called sequential second price auction, where each user makes a bid based on the value of the current resource unit, and the BS allocates this unit to the user with the largest bid but charges the value of the second largest bid. The auction ends when all the resource units are assigned. Xu et al. (2014), formulate the problem of joint subcarrier and power allocation for a D2D underlay system as a reverse iterative combinatorial auctions game. Yin et al. (2015) propose a pricing-based game-theoretic method that is applied to model the problem. To control inter-layer interference caused by D2D transmission, the BS sets a price for interference on each subchannel occupied by the cellular users based on its data-rate requirement. Then the D2D pairs compete for the subchannels with different power-allocation strategies as a non-cooperative Nash game. To guarantee the data-rate requirement of the D2D pairs, sufficient conditions for existence and uniqueness of a Nash equilibrium are derived. Other game theoretic methods whose objective is either data-rate or energy optimization can also be found in Li et al. (2014a), Xiao et al. (2014), Zhu and Hossain (2015), Zhou et al. (2015), Maghsudi and Stanczak (2014), Ye et al. (2015), and Wu et al. (2014).

Subgradient Approach. Peng et al. (2012) formulate the problem as the maximization of the sum throughput subject to rate requirements and a power constraint in an OFDMA-based system. Hence, the problem is a joint optimization of power and resource allocation. Obviously, this is a non-convex optimization problem. Peng et al. (2012) intuitively find the optimal solution of the dual problem by using a subgradient algorithm. It can be shown that as the number of subcarriers in the OFDMA system tends to infinity, the duality gap becomes zero asymptotically.

9.5 Scheduling

This section consists of three subsections, focusing scheduling algorithms in in-band D2D, scheduling algorithms in out-of-band D2D, and the most well-known D2D scheduling algorithms including FlashLinQ and ITLinQ.

9.5.1 In-band D2D

It is helpful to categorize these algorithms with respect to their optimization criteria, which encompass spectral efficiency, power efficiency, and fairness. For each of these groups, a variety of mathematical methods have been applied.

Scheduling for Spectral Efficiency. As discussed in the literature, many of these papers use their own mathematical methods for analyzing their own spectral-efficient scheduling algorithms including:

- the Chen–Stein method; see Peng et al. (2009)
- mixed integer and linear programming; see Janis et al. (2009)
- convex optimization; see Min et al. (2011)
- bipartite matching; see Chen et al. (2012b)
- Kuhn–Munkres algorithm; see Yu and Tirkkonen (2012)
- Han–Kobayashi; see Doppler et al. (2010)
- Newton's method, see Osseiran et al. (2011)
- Lagrangian multipliers; see Pei and Liang (2013a)
- game theory; see Xu et al. (2012a)
- auction algorithm; see Zulhasnine et al. (2010)
- particle swarm optimization; see Liu et al. (2012).

Scheduling for Power Efficiency. Papers in this area use the following formulations for designing power-efficient scheduling algorithms in in-band D2D:

- heuristic search; see Jung et al. (2012) and Xiao et al. (2011)
- exhaustive search; see Belleschi et al. (2011)
- linear programming (see Yu et al. (2009b).

Scheduling for Fairness. Xu et al. (2012b) use auction algorithms for their own objective in terms of fairness. The algorithm proposed by Nguyen and Rao (2015) also discusses fairness-aware scheduling as well as introducing methods for exploiting multiuser diversity in cellular networks. In addition, the algorithm proposed by Park et al. (2014) also aims for fairness with distributed iterative methods.

Scheduling for Quality-of-Service (QoS). As presented by Huang et al. (2016), prioritized QoS-aware scheduling policies are analyzed with stochastic network calculus methods.

There are also algorithms for overlay in-band D2D scheduling.

Scheduling for Energy Efficiency. The energy-efficient scheduling algorithm proposed by Fodor et al. (2012) is for uplink D2D scheduling over LTE platforms. It improves the performance from 0.8 bps/Hz/mW up to 20 bps/Hz/mW.

Scheduling for Spectrum Efficiency. The spectrum efficient scheduling algorithm proposed by Li et al. (2012a) is for uplink D2D. It solves the convex optimization formulation and improves cell throughput over an underlay mode.

9.5.2 Out-of-band D2D

For out-of-band D2D, various scheduling algorithms have been designed and proposed including scheduling based on file availability, scheduling for reducing channel sensing overheads, scheduling with the communicaion via cluster heads, and scheduling for improving fairness.

Scheduling Based on File Availability. Golrezaei et al. (2012a, 2014) point out the similarities among video content requests of cellular users. Because of this, users might be able to get the desired content from other devices. One algorithm divides the cell into smaller regions (clusters), allowing only one link per cluster, and employs FDMA-type frequency reuse for interference avoidance; files requested, but not available for transmission in the same cluster, are transmitted directly from the BS. More details of this setup are given in Section 9.9.

Scheduling for Reducing Channel Sensing Overheads. When the same band (say, 2.45 GHz ISM) is used for D2D networking in LTE, simultaneous channel contention from both D2D and WiFi users reduces the network throughput. Therefore, Zhou et al. (2013b) propose a joint group-wise channel sensing and resource pre-allocation scheme for contending and utilizing the ISM band resources together with WiFi. In the proposed scheme, the D2D links with differentiated QoS requirements are grouped and pre-scheduled to approximately fill an overall flexible resource block. Then, a representative device is appointed for resource scheduling, and contending for resource allocation to avoid intra-group collisions.

Communication via Cluster Head. A cluster-based approach, which can improve throughput and energy efficiency, was proposed in Asadi and Mancuso (2013). Clusters are formed among cellular users who are in range for D2D communications. For each cluster, only the cluster member with the highest cellular channel quality becomes a cluster head, and then can communicate with the cellular BS. The cluster head is also responsible for forwarding the cellular traffic of the cluster members to the cellular BS.

Scheduling for Improving Fairness. Cai et al. (2013) propose a scheduler to consider both time-varying channel and users' random mobility in cellular networks. They consider a scenario where the cellular BS broadcasts deadline-based content to different group of users. Users move randomly within the cell and users of the same group are considered to be able to communicate directly at high rate when they are close to each other for allowing content exchange during a contact period. In each slot, the cellular BS dynamically selects a group of users to broadcast content. If the service rate is too high for some users to successfully receive the content, these users will exploit D2D networking to fetch content from nearby users.

The authors formulate the scheduling problem with the objective of maximizing the group utility function; and then solve the maximization problem under the assumption of statistically homogeneous and heterogeneous mobility scenarios.

9.5.3 *FlashLinQ and ITLinQ*

Currently, the best-known D2D scheduling protocol is FlashLinQ, which was proposed by Wu et al. (2010). It is a distributed algorithm that schedules D2D links according to their priorities, such that the higher-priority links do not suffer from significant interference from possibly scheduled lower-priority links. Theoretically, it can guarantee the maximum number of activated D2D links, as analyzed using the theory of stochastic geometry by Baccelli et al. (2012). The ITLinQ algorithm of Naderializadeh and Avestimehr (2014) builds on these insights for when it is information-theoretically optimal to treat interference as noise, using them to define information-theoretic independent sets in which the interference between users can be optimally treated as noise.

The main objective of these scheduling algorithms is to find a maximal feasible subset of D2D links – a set of D2D transmissions that can simultaneously co-exist with each other while maintaining a sufficiently large signal-to-interference ratio (SIR) – among all the given one-hop D2D links that have data to transmit. In the following, we use a binary interference (protocol) model. In this model, if a given D2D link A interferes "too strongly" with another link B, then link B cannot be scheduled at all; by "too strongly" we define that if the D2D link A transmits, then the received signal power at the receiver of link B from the transmitter of link A is higher than the interference threshold γ_T.

The interference situation is pictorially described by an interference graph, in which links are denoted as vertices $\{A,B,\cdots\}$; and edges $\{(A,B),\cdots\}$ represent "too strong" interference between two links, see Figure 9.3.

In addition, interference from different concurrently active transmitters accumulates and the interference from each transmitter depends on the transmit power as well as the channel state. Thus, obviously, finding the best transmission set is very difficult.

The FlashLinQ scheduler is designed to find a feasible subset (ideally a maximal one) in a distributed manner, under the assumption that all channel states are known and no power control

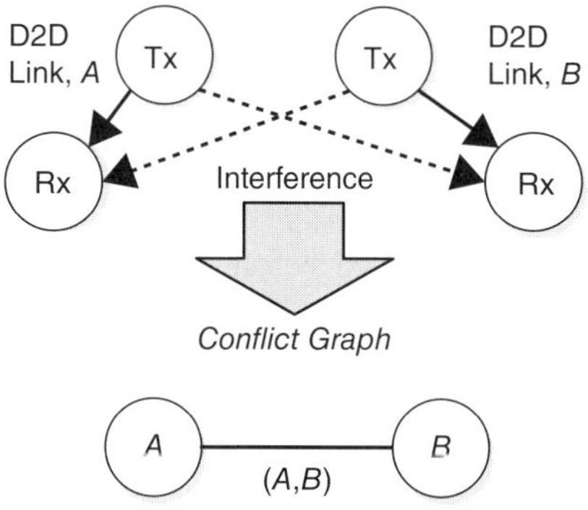

Figure 9.3 Conflict graph construction for FlashLinQ.

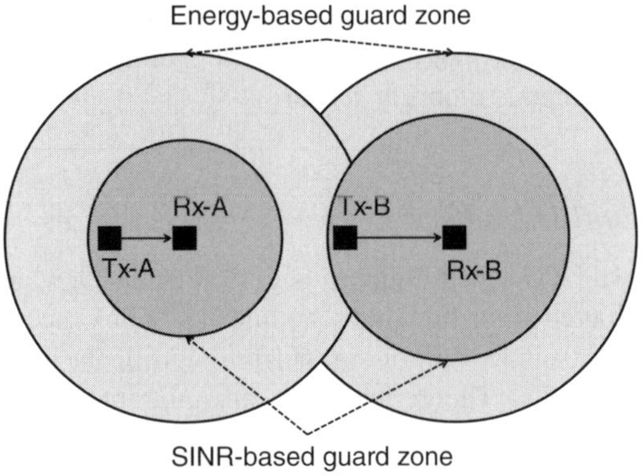

Figure 9.4 SINR-based guard zone versus energy-based guard zone. Source: Reprinted from Ji (2016b) with permission from IEEE.

is used. This is achieved by assigning (possibly randomized) priorities to the different links. The highest-priority link is scheduled for certain. The second-highest priority link is scheduled only if:

- it does not create excessive interference to the receiver of the higher-priority link
- it does not suffer excessive interference from the transmitter of the higher-priority link.

Then the next link (in order of priority) is considered according to a similar criterion, and so on. In other words, for a given D2D transmitter-receiver pair (as denoted by Tx-A and Rx-A in Figure 9.4) we need to determine which other D2D links (Tx-B and Rx-B in Figure 9.4) are allowed to simultaneously transmit without generating too much interference at Rx-A. An energy-based guard zone scheduling in this example can schedule only one of the two Tx-A and Tx-B to their corresponding receivers, whereas both Tx-A and Tx-B can transmit data if SINR-based guard zone scheduling (referred to simply as SINR-based scheduling) is used. According to theoretical evaluations based on stochastic geometry, this scheduling guarantees the maximum number of scheduled D2D links in a slot; see Baccelli et al. (2012).

Suppose that D2D links are deployed on a two-dimensional infinite Euclidean plane R^2. The transmitters are located in the network according to a homogeneous Poisson point process. The receivers are assumed to be randomly located around the transmitters. Let $Z \in R^2$ be the location of a given transmitter. Then, the receiver is located at $Z + D_Z$ where D_Z is a random displacement from the transmitter. For each transmitter, assume that the receiver displacements D_Z are independent and identically distributed (i.i.d.) according to a certain distribution. Let H_{Z_1,Z_2} indicate the channel gain between the transmitter located at Z_1 and the receiver corresponding to the transmitter located at Z_2. Notice that H_{Z_1,Z_2} and H_{Z_2,Z_1} need not be the same. Assume that H_{Z_1,Z_2} can be determined as follows:

$$H_{Z_1,Z_2} = \frac{F_{Z_1,Z_2}}{\left|Z_1 - Z_2 - D_{Z_2}\right|^{\alpha}}, \tag{9.7}$$

where F_{Z_1,Z_2} is the fading effects between the transmitter Z_1 and the corresponding receiver at Z_2. α is the path-loss exponent where $2 \le \alpha \le 4$. To simplify the notation, H_Z and F_Z are denoted as the channel gain and fading component between the transmitter at location Z and its intended receiver, located at $Z + D_Z$. We also assume that the link is interference-limited, so that the effects of noise can be neglected. The FlashLinQ mechanism can be interpreted as each receiver having a guard zone, which is a circular region where no other transmitter can simultaneously transmit. Link A conflicts with (that is, is a neighbor to) link B if and only if link A's transmitter falls into the guard zone of link B's receiver or link B's transmitter falls into the guard zone of link A:

$$\frac{\frac{P_Z F_Z}{|D_Z|^\alpha}}{\frac{P_0 F_{0,Z}}{|Z + D_Z|^\alpha}} > \gamma_T \text{ and } \frac{\frac{P_0 F_0}{|D_0|^\alpha}}{\frac{P_Z F_{Z,0}}{|Z + D_0|^\alpha}} > \gamma_T. \tag{9.8}$$

Clearly, it is critical to define the right radius of the guard zones to maximize the spatial reuse of the system; heuristics have been discussed in the literature.

The ITLinQ algorithm by Naderializadeh and Avestimehr (2014) exploits insights into when it is information-theoretically optimal to treat interference as noise, and uses it to schedule users simultaneously only if it results in interference that is optimally treated as noise. A simple sufficient condition is as follows

$$\frac{\sqrt{\text{SNR}}}{\text{INR}} \geq \gamma \tag{9.9}$$

whereas the condition of distributed scheduling of FlashLinQ is as follows:

$$\frac{\text{SNR}}{\text{INR}} \geq \gamma \tag{9.10}$$

where SNR and INR denote the signal-to-noise ratio and interference-to-noise ratio, respectively. The performance of ITLinQ was analyzed with n transmitter-receiver D2D pairs. The transmitters were located uniformly at random in a circle of radius R. Each receiver was located within a distance $r_n = r_0 n^{-\beta}$ of its corresponding transmitter, where $\beta < 0$. Assume that the channel gain at distance r is equal to $h_0 r^{-\alpha}$ (path loss). The ITLinQ can almost surely achieve a fraction of γ of the capacity region within a gap of k bits where

$$\begin{cases} \gamma = \dfrac{\sqrt{3}\pi R^2}{2\gamma^2} n^{\beta-1}, k \leq \dfrac{\sqrt{3}\pi R^2}{2\gamma^2} \dfrac{\log 3n}{n^{1-\beta}}, & \mathit{if}\ 0 < \beta < 1, \\ \gamma = \dfrac{\ln(\ln n)}{\ln n}, k \leq \log(\ln n), & \mathit{if}\ \beta = 1, \\ \gamma = \dfrac{1}{\left\lfloor \left| \dfrac{1}{\beta - 1} \right| + \dfrac{1}{2} \right\rfloor + 1}, k = \dfrac{\log 3n}{\left\lfloor \left| \dfrac{1}{\beta - 1} \right| + \dfrac{1}{2} \right\rfloor + 1}, & \mathit{if}\ \beta > 1, \end{cases} \tag{9.11}$$

when $n \to \infty$.

9.6 Multi-hop D2D

The literature mentioned previously considers the situation where there is a defined pair of devices that wish to communicate with each other. D2D can also be used for relaying. This is when a (relay) device sends information it gets from the BS to one or more destination devices. An overview of the system model is shown in Figure 9.5. In that case, relay selection becomes a new issue that needs to be optimized. Pei and Liang (2013b) study a network consisting of one BS, two D2D users (D_n and $D_{\bar{n}}$) and one cellular user (CU). In this network, two-way BS–CU and D2D communications are considered, which means that both the uplink and downlink between the BS and CU can be active. Similarly, the D2D pair can also communicate with each other. However, the D2D users have a direct link to each other, to the BS and to the CU, while the BS and CU do not have a direct link. The proposed protocol is a two-step scheme. By fixing D_n as the relay for the cellular transmission between the BS and CU, in the first step (time slot), we do the following transmissions: BS $\rightarrow D_n$, CU$\rightarrow D_n$, and $D_{\bar{n}} \rightarrow D_n$. In the second step, we do the following transmissions: $D_n \rightarrow$BS, $D_n \rightarrow$CU, and $D_n \rightarrow D_{\bar{n}}$. By treating all the interference as noise, we can compute the Pareto boundary of the throughput region for both CU and D2D users by optimizing the transmit power at the BS and CU and the power splitting factor at the relay D2D users. Therefore, given the required throughput for the D2D users, we can pick one of the D2D users as the relay that gives a higher throughput for the CU. If user n is selected as the relay node, by denoting the throughput of CU as R^n_{CU}, the throughput of the D2D users as R^n_{D2D}, and the requirement of the D2D pair as R^n_{D}, the optimization relay selection problem is given by:

$$\max_{\hat{n}} R^{\hat{n}}_{\mathrm{CU}} \tag{9.12}$$

$$\textit{subject to}: R^{\hat{n}}_{\mathrm{D2D}} = R_{\mathrm{D}}, \tag{9.13}$$

$$\hat{n} = \{n, \bar{n}\}. \tag{9.14}$$

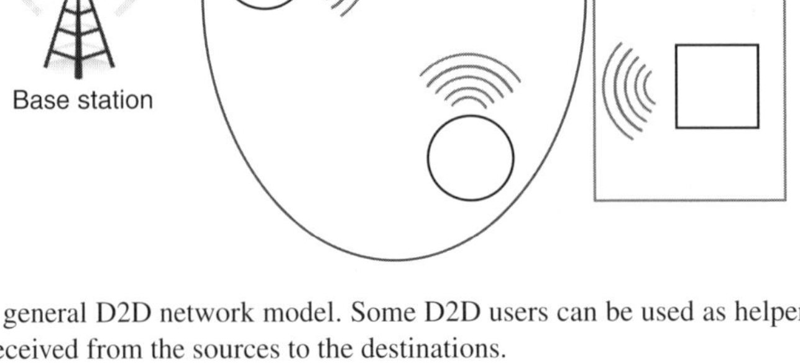

Figure 9.5 The general D2D network model. Some D2D users can be used as helper nodes that relay the information received from the sources to the destinations.

Xia et al. (2013) propose a similar idea for a network where the devices help to relay the information of the D2D communications. The relay selection operates in three steps:

1. The BS selects a set of relay nodes $\mathcal{R}_D^1$ whose channel gain (or distances) to both BS and the destination is greater than a predesigned parameter.
2. The BS selects the D2D relay nodes in $\mathcal{R}_D$ that have the best channel efficiency η defined as,

$$\eta = \frac{\text{total amount of desired data}}{\text{total bandwidth needed}}, \tag{9.15}$$

   and we denote the set of relay nodes selected by the BS in the second step as $\mathcal{R}_D^2$.
3. The BS picks the best relay node in $\mathcal{R}_D^2$ according to the best channel gain (or the nearest distance) to both BS and the destination.

Hasan et al. (2014) generalize the analysis to a multi-user and multi-relay network, and propose a distributed solution for resource allocation by maximizing network sum throughput, which can be shown to be a convex program. Li et al. (2014b) propose a dynamic graph optimization framework, which is formulated as a maximization of a flow function and falls into the category of linear programming. When cooperation between the D2D transmitter and the BS is allowed, ben Slimane (2014) proposes that the D2D transmitter can employ superposition coding, where it transmits a linear combination of its private information and the decoded information transmitted by the BS for the cellular user. The objective is to minimize the assigned power for cooperation while the cellular user can achieve at least its direct link rate.

Chien et al. (2012) investigate how D2D can be used to help broadcast transmission, through proper scheduling of re-broadcasts from devices with good channels. Specifically, both BS and D2D users are assumed to have multiple antennas and the receiver structure is assumed to be zero-forcing (ZF) for simplicity. In the first hop, the transmitter (BS) selects the D2D user with highest effective SNR after ZF reception at the destined receiver. We denote the selected D2D user as $u(1)$. In the second hop, $u(1)$ re-encodes its decoded information and then multicasts to the remaining users. Also in this second (retransmission) phase, the destined user is picked according to highest effective SNR, which can be obtained by using ZF with an interference cancellation scheme. Then the procedure continues until all the users successfully decode the information. A similar approach is applied by Zhou et al. (2013a), where instead of retransmitting to only one user completely based on the channel condition, the D2D users are grouped into clusters, and the communication can only take place within one group.

When a D2D direct link is desired but has low channel quality, the D2D source–destination transmission can also be satisfied through multihop. In other words, the range of D2D connections can also be extended through multi-hop transmission; this then involves the determination of the optimal routes. Kaufman et al. (2013) consider such a scenario where the network information is distributed by so-called route discovery packets in a random access manner in order to establish a multi-hop (including single-hop) route between a D2D source-destination pair. In detail, Kaufman et al. (2013) assume a network consisting of CUs, D2D users, and BSs, where the D2D communications take place during the uplink frame of the CUs. Similarly, the goal is to guarantee that the SINR of both CUs and D2D users is greater than certain required thresholds. The D2D source–destination pairs can be implemented using either single-hop or multihop based on the channel condition. There are two main steps of the protocol. First is the power control for the D2D users. This can be done by simply using the SINR threshold of the links by CUs and channel inversion. Second, the D2D users have to discover a route (single-hop/multihop) to their intended destinations. In this work, the discovery packets were transmitted using a CSMA/CA protocol to ensure that only one D2D pair accesses the channel at a given time. By using the discovery packets, each D2D user can measure the received SINR. If the node satisfying the required SINR condition is the destination of the D2D pair, then the single-hop route exists and the D2D pair will take place by single-hop transmission. If this node is not the destination, then it will continue the discovery process and relay the information if it can find a D2D receiver such that the required SINR threshold is satisfied. This process continues until the entire route is found.

In addition, Wei et al. (2014a) consider a simpler scenario with three users, where instead of exchanging information directly between two devices, one device may help the other two users to exchange information using an analog network coding technique. The results show that this method is more energy efficient when the D2D pair is far away from the BS.

Ren et al. (2013) also jointly consider the geometric information, interference constraint, and D2D rate requirements. The goal of their algorithm is to minimize the number of hops, while both the SINR requirements for CUs and D2D users are satisfied. They show that this problem can be converted to the problem of searching for the shortest path through a graph, which can be solved by Dijkstra's algorithm. Ren et al. (2013) further proposed another low-complexity algorithm, namely the maximum rate towards destination (MR-D) routing algorithm, achieving a similar performance as Dijkstra's algorithm. The MR-D routing algorithm is based on the heuristic that each receiving node for the current hop is chosen such that it satisfies the SINR requirement and, more interestingly, it obtains the maximum link rate to the destination, which is closely related to the geometric information.

A related approach considers a structure in which all communication runs via cluster heads. Hassan et al. (2013) proposed a mobile cloud (MC) architecture, where the devices are classified into primary cluster head (PCH), secondary cluster head (SCH), and standard devices. The coverage area of the MC is divided into multiple logical regions, where there is one region containing the PCH, and each of other regions has one SCH. Communication within a logical region is considered to be one-hop. The PCH can distribute the data locally in its own region and also exchange data with other SCHs. Moreover, the PCH also takes charge of the MAC functions within the MC (error control, retransmissions). Similarly, an SCH is to also to distribute data locally in its own logical region, and exchange data with other SCHs or the PCH. In addition, it also performs error control in its logical region. The requirements of the PCH and SCH are based on SINR. Hassan et al. (2013) propose a PCH/SCH selection algorithm based on residual energy of devices, power consumed for transmission, and processing and SINR. Moreover, the Zhou et al. paper (2013a), discussed previously, can also be considered as a D2D cluster network, where the CH is the node that can decode the multicast message from the BS successfully. Although this network structure is similar to conventional ad-hoc networks, there are several key differences. First, all the users are CUs and require data from the BS. The PCH/SCHs only help the BS to relay the information to all the users, where the D2D communication takes place. Second, the BS is the central controller that selects the PCH/SCH and schedules all the transmissions between the PCK/SCH and users.

It is worth mentioning that the first prototype of multihop D2D networks for use in emergency communications was successfully developed by Nishiyama et al. (2014). In this study, the relays are smartphones and the prototype was tested in a field experiment.

9.7 Standardization

A critical aspect of D2D communications is the integration into future standards. WiFi Direct, a mode of the well-known IEEE 802.11 standard, is well established and has been shown experimentally to enable efficient communications; see Camps-Mur et al. (2013). However, it is intended mainly for traditional peer-to-peer communications. A lot of attention has been drawn by the FlashLinQ system developed by Qualcomm; see Wu et al. (2013). This is a system that can schedule users without the intervention of a BS, based on priorities and transmission qualities for each link; in a greedy manner, the "most important" link is scheduled. Through a two-way information exchange (compare Section 9.5), the next link in the priority list can determine whether it can be scheduled, or would create too much interference (above a certain threshold), and so on. A number of papers has dealt with D2D optimization specifically in the context of FlashLinQ:

- distributed link scheduling based on an on-off interference map; see Kang et al. (2013)
- power control; see Bae et al. (2013)
- settings of thresholds in the link scheduling scheme of FlashLinQ; see Yoon et al. (2012).

FlashLinQ is also a candidate for the D2D mode in the LTE cellular standard (dubbed "LTE Direct"), though this standardization is still ongoing. For an overview of recent standardization developments, see, for example, Fodor et al. (2012); Pyattaev et al. (2013); Raghothaman et al. (2013) and Yang et al. (2013). The complete FlashLinQ operation consists of three separate synchronized steps (see Figure 9.6):

- peer discovery
- connection scheduling
- data transmission with local transmit power control.

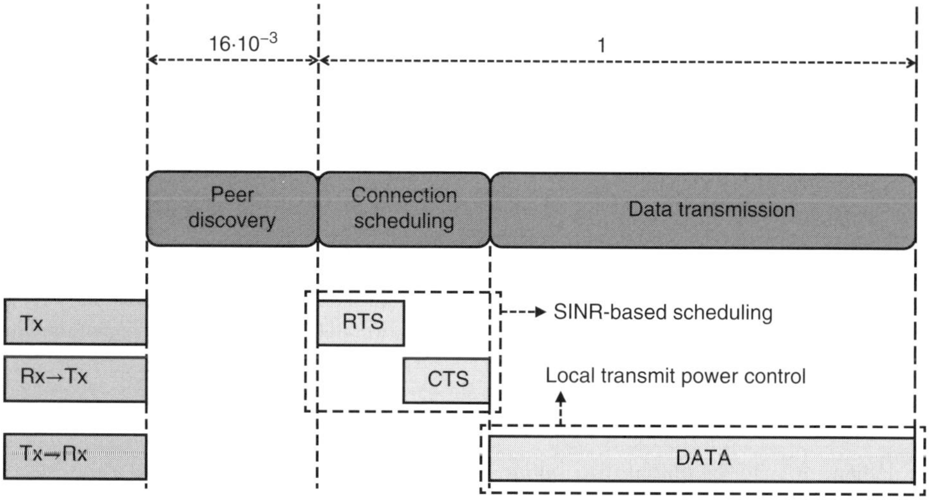

Figure 9.6 FlashLinQ: a synchronous distributed scheduler.

The FlashLinQ operation is synchronized among all given devices. Therefore, all devices are doing peer discovery, connection scheduling, and data transmission (if scheduled) at the same time. Suppose that the transmitter of a D2D pair wants to transmit data to the other device; in other words, its desired receiver. Then, the transmitter should do the peer discovery (the first step) to find out its desired receiver. After finding out its desired receiver, the transmitter sends an RTS-like single tone signal over the OFDM physical layer toward its desired receiver for connection scheduling; in other words, the second step. When the receiver receives the signal, it replies with a CTS-like single tone signal. Based on the received signal strength of this tone, the transmitter determines how much transmit power should be allocated (SIR-based scheduling); in other words, data transmission with local transmit power control, which is the third step of FlashLinQ operation.

The fundamental concept of FlashLinQ was introduced for consideration into the 4G LTE standardization process with the name of LTE-Direct. LTE-Direct was initially proposed as a part of 3GPP Release 12. Besides the topics mentioned previously, LTE-Direct considers a variety of topics required to provide a complete system design, for example assignment of resources in the FDD uplink or TDD resources for peer discovery; a cellular BS (if present) assigns a part of the discovery resources via the system information block to authorized LTE-Direct capable devices. For more details, see the final part of the LTE-Direct standard when it becomes available.

9.8 Applications

D2D communications systems are essential in emerging applications including content distribution in social networks, video distribution, roadside infostations, emergency communications, distributed storage systems, and smart grids.

9.8.1 Content Distribution in Social Networks

Consider an in-band D2D communication system, which consists of two layers: a logical network layer and a physical network layer; see Zhang et al. (2013c). The first logical network layer is named OnSN, which is a logical connection among social network service participants for content distribution. For example, if two users are connected over a social network service, then they are connected in OnSN. Once the logical network link is accessed or established, the packets of content for sharing should be transmitted to target mobile users over the actual physical network layer, named OffSN. For example, two users can be connected over OffSN if both can establish a wireless D2D link. Taking advantage of the social ties, OffSN represents the physical network layer in which to transmit the requested content. The mobile users access content links in an increasing order of their labels. In OnSN, each content link is spread out according to its popularity from frequent users to regular users. Frequent users have a high online activity, and thus they are the main source of influence and information dissemination in the social network.

9.8.2 Video Distribution

The demand for video content distribution has increased as video sharing over the Internet has become enormously popular. This demand has considerable impacts on the load of the network infrastructure due to its high bandwidth requirements. As more users access videos from their mobile devices – smartphones and tablets – the load on the current capacity-limited wireless infrastructure will be more significant. Based on observations from local video = sharing scenarios, Xing et al. (2009) studied the tradeoffs between the usage of WiFi ad-hoc mode and infrastructure mode for video streaming in D2D links. Their experimental results show the improvements in latency, jitter, and loss rate in D2D video distribution. More importantly, the results also show that direct D2D video delivery is more stable than non-D2D approaches, where different factors affect the performance, leading to unpredictable quality. Militano et al. (2014) provide a comprehensive discussion of different solutions for multicast content distribution services involving groups of users in close proximity in a modern LTE-A cellular system. The study focuses on two system parameters: the energy spent and the content-delivery latency time. The results show that the proposed D2D-based solutions can enhance the multicast scheme, introducing 40% time savings and 80% energy savings compared with the state-of-the-art schemes. In addition, a fountain–code-based approach is presented by Cao and Maaref (2014), who aimed to satisfy the heterogeneous QoE requirements. A graph-based flow-maximization approach was introduced by Li et al. (2014d) to enhance content-downloading performance. Cloud-central-storage video-distribution algorithms have been studied, for example, by Li et al. (2014c) and Danihelka et al. (2014) A more detailed description of D2D using caching for video streaming is given in Section 9.9.

9.8.3 Roadside Infostations

Roadside infostations aim to deliver an efficient and scalable delivery system for location-aware data for mobile users, typically those in cars. Kone et al. (2013) define location-aware content as local information of interest to travelers on the go. For example, a roadside infostation could be deployed a few miles ahead of a highway exit. As the vehicle passes by, it can download brochures describing local entertainment options and information via D2D links, where the two communicating devices are infostation and vehicle. Once the car downloads all this information, passengers could peruse it and decide on their best option. These types of content are typically organized as a collection or hierarchical list of files, each providing details of an item or location of interest. Given the high rates of mobility, this system aims for the maximization of the number of complete files delivered to each vehicle to optimize the offline browsing experience.

9.8.4 Emergency Communications

Disasters, such as earthquakes in Sichuan, China (2008), in Haiti (2010), and in Fukushima, Japan (2011) have shown that the need for information and communication without infrastructure for emergency situations. In such large-scale disasters, communication infrastructure is generally useless due to the damage to BSs and backbone networks. Therefore, user-centric D2D communications and networking technologies are obviously one possible and essential solution for this purpose; see Macuha et al. (2013). Note that in such a case "D2D" really becomes "peer-to-peer" ad-hoc networking, as no interaction with infrastructure is used. It is noteworthy that LTE-Direct was motivated by a mandate of government regulators for infra-structure-less communications, but that it foresees also more efficient infrastructure-facilitated communications for non-disaster communications. This emergency communication application of D2D can also be integrated within 4G LTE wireless networks, as illustrated by Gomez et al. (2014), and next generation 5G millimeter-wave cellular networks, as presented by Qiao et al. (2015).

9.8.5 Distributed Storage Systems

Mobile data traffic is growing tremendously; total global mobile traffic was about 885 petabytes per month at the end of 2012, and is expected to keep increasing. In addition, data files can be stored and retrieved from the mobile users themselves in order to offload traffic from the infrastructure network as the storage or memory capacity of mobile devices increases.

As stated by Pääkkönen et al. (2013), consider a cellular network where mobile users roam freely in and out of a geographically limited area. Assume that the mobile users themselves can store (cache) data and they can, upon request, transmit data to one another. A set of mobile users that are within a specified distance of each other forms a storage community, or a local network. The local mobile users can communicate with each other in a D2D mode, without the help of the BS. Also, the BS can be used to transmit data to the mobile users, but there is no need to relay data from a mobile user to another mobile user via the BS. This principle is somewhat similar to the storage/D2D communications for video proposed by Golrezaei et al. (2011), see also Section 9.9.

9.8.6 Smart Grids

The application of D2D communications in smart grids is a new paradigm for designing and managing the electrical power system in order to improve efficiency, enhance service quality, and save costs in power generation and distribution. D2D communications can be applied to solve a number of challenges concerning interoperability, scalable inter-networking, self-organization, and security, as mentioned by Niyato et al. (2011).

9.9 D2D for Video

As outlined in Section 9.8, multimedia content delivery is a very important application for D2D. Due to its importance, we describe here in more detail recent work that has shown that network throughput for wireless delivery of video files can be greatly enhanced by D2D communications in conjunction with caching on the devices; see Golrezaei et al. (2012a). In particular, these studies propose systems where each device has an installed storage capacity for caching videos, and caches independently, according to a certain optimal distribution, a subset of popular video files. When a user needs a file not already present in its own cache, it obtains it from one of its neighbors through a spectrally efficient, short-range D2D link. As user density increases, the aggregate storage capacity of the D2D network increases linearly with the number of users, while the average communication distance decreases and the spatial reuse increases. For these reasons, D2D networks for video delivery are scalable, such that both demand and throughput increase linearly with user density. An overview of D2D for video is given by Molisch et al. (2014). Papers about this topic include Golrezaei et al. (2012a,b, 2011, 2012c, 2014, 2012d); Ji et al. (2013a,b, 2015, 2016a,b) and Kim et al. (2013, 2014).

We now describe a mathematical model for such a system.[2] Unlike the D2D communication problems discussed previously, in this case, we need to design a "caching scheme" consisting of the caching placement and the delivery scheme. In the following, we will describe two types of scheme. The first one involves independent and random caching of entire files and unicasting delivery. The second uses dedicated vector (caching packets of files) caching placement and coded multicasting (the source device multicasts coded packets) for delivery. In the following, the theoretical results are obtained under the protocol channel assumption: if a node i transmits a packet to node j, then the transmission is successful if and only if:

- The distance between i and j is less than r.
- Any other node k transmitting simultaneously is at distance $d(k,j) \geq (1+\Delta)r$ from the receiver j, where $r, \Delta > 0$ are protocol parameters.

In practice, nodes send data at some constant rate C_r bit/s/Hz, where C_r is a non-increasing function of the transmission range r. In the following discussions, we focus on the simple protocol channel model. Nevertheless, it is worth mentioning that subsequent work by Altieri et al. (2014) and Malak and Al-Shalash (2014) has generalized these result to the random channel model using stochastic geometry.

9.9.1 *Random Caching and Unicasting*

We consider a network consisting of n devices, each of which can cache a fixed number M of video files, which will be used to serve other devices within a transmission range upon request. The devices that cannot be served by D2D communication can obtain the requested files via conventional cellular transmission from the macro BS. We let the number of users in the network be n and the library size be m. The demand distribution is assumed to follow a Zipf distribution, which has been shown to be a good model for the popularity of video files. Under this model, the probability $P_r(f)$ that the fth file is requested is inversely proportional to its rank:

$$P_r(f) = \frac{\frac{1}{f^{\gamma_r}}}{\sum_{j=1}^{m} \frac{1}{j^{\gamma_r}}}, \quad 1 \leq f \leq m. \tag{9.16}$$

When a random and independent caching placement scheme is considered, we divide the network into clusters and the D2D communications can only take place inside the corresponding cluster. Given the cardinality of the number of users in each cluster g_c, Ji et al. (2015) find that the optimal caching distribution P_c^* that maximizes the probability that any user finds their requested file inside its own cluster is given (for a node arrangement on a rectangular grid as described earlier) by

[2] For simplicity, we neglect the issue of video rate adaptation.

$$P_c^*(f) = \left[1 - \frac{\nu}{z_f}\right]^+, \quad f = 1, \ldots, m, \tag{9.17}$$

where $\nu = \frac{m^* - 1}{\sum_{f=1}^{m^*} \frac{1}{z_f}}$, $z_f = P_r(f)^{\frac{1}{M(g_c-1)-1}}$, $m^* = \min\left\{\frac{M}{\gamma_r} g_c, m\right\}$ and $[\Lambda]^+ = \max[\Lambda, 0]$.

The delivery scheme is as follows. Inside a cluster, all the potential links (the links established by one source–destination pair) are served in a round-robin manner. Different clusters will be scheduled in a TDMA manner. It can be seen that it is possible that users may not find the requested video files in their own cluster, so they cannot be served by the delivery scheme. We call this event an outage. When $Mn \geq m$ (the whole library can be cached in the network), for arbitrarily small outage probability, Ji et al. (2015) show that the per-user throughput scales as $\Theta(\frac{M}{m})$. This means that the per-user throughput is independent of the number of users (or in other words, the network throughput increases linearly with the number of users). Furthermore, the throughput grows linearly with M. This can be very attractive since, for example, in order to double the throughput, instead of increasing the bandwidth or power, we can just double the (relatively cheap) storage capacity per user.

Interestingly, this result for throughput coincides with the achievable throughput using subpacketized caching and coded multicasting algorithms obtained by Ji et al. (2016b) and Maddah-Ali and Niesen (2014). However, the practical performance in realistic channels can be quite different; see Section 9.9.3.

9.9.2 Coded Caching and Multicasting

One important property of the scheme described previously is that both the caching placement and the delivery scheme apply an uncoded approach. The gain of the throughput is mainly obtained by spatial reuse (TDMA). At this point, a natural question to ask is whether coded multicasting for D2D transmissions can provide an additional gain, or whether the coding gain and the spatial reuse gain can accumulate. Ji et al. (2016b) designed a scheme with (deterministic or random) subpacketized caching and network-coded delivery for D2D caching networks. The schemes are best explained by the example shown in Figure 9.7, where we assume no spatial reuse, or only one transmission per time-frequency slot is allowed, but that the transmission range can cover the whole network. In this example there are three users, three files and $M = 2$, achieving 1/2 transmissions in term of file. We divide each file into six packets (e.g. A is divided into $A_1, \cdots, A_6$.). User 1 requests A; user 2 requests B, and user 3 requests C. The cached packets are shown in the rectangles under each user. For the delivery phase, user 1 transmits $B_3 \oplus C_1$; user 2 transmits $A_5 \oplus C_2$, and user 3 transmits $A_6 \oplus B_4$. The normalized number of transmissions is $3 \cdot \frac{1}{6} = \frac{1}{2}$, which is also information theoretically optimal for this network.

This scheme can be generalized to any n, m, M. Without spatial reuse, for zero outage, the achievable normalized number of transmissions such that every user can successfully decode is $\frac{m}{M}\left(1 - \frac{M}{m}\right)$,[3] which is surprisingly almost the same as the result shown by Maddah-Ali and

[3]We normalize the number of transmissions by the file size, which is assumed to be the same for all the files.

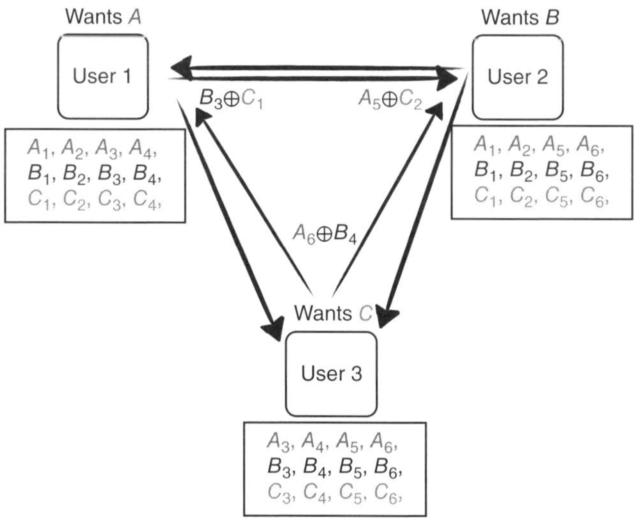

Figure 9.7 Subpacketized caching and network-coded delivery of Ji et al. (2016b).

Niesen (2014), where instead of D2D communications, one central server (BS) with access to all the video files multicasts coded packets. In addition, it also has the same scaling law as the throughput of the system with decentralized random caching and an uncoded delivery scheme described in the previous subsection.[4] Moreover, it can be shown that there is no further gain when spatial reuse is also exploited. In other words, the gains of spatial reuse and coding cannot accumulate. Intuitively, if spatial reuse is not allowed, a complicated caching scheme can be designed such that one transmission can be useful for as many users as possible. If the transmission range is reduced and coding is done in one cluster, then the number of users benefitted from one transmission is reduced but the D2D transmissions can operate simultaneously at a higher rate. Moreover, the complexity of caching subpacketization and coding can also be reduced. Hence the benefit of coding depends on the actual physical layer throughput (bits/s/Hz) and the caching/coding complexity, rather than throughput scaling laws.[5]

9.9.3 *Simulation Results*

It is interesting to compare, using more realistic channel models, D2D caching with conventional unicasting, harmonic broadcasting. and coded multicasting. Consider a network of size 600×600 m, and let $n = 10{,}000$ users be distributed uniformly. The file library has size $m = 300$; that is, there are 300 popular movies and TV shows that are refreshed on a daily basis at off-peak times by the conventional cellular network. The storage capacity per user is $M = 20$ and the parameter for the Zipf distribution is $\gamma_r = 0.4$. The channel model is a mixture of models from WINNER-II (2007); for more details see Ji et al. (2016a). The simulation results of the throughput-outage tradeoff for different schemes are given in Figure 9.8. We observe that in

[4] Notice that the reciprocal of the number of transmissions is proportional to the throughput under our protocol model assumption.

[5] An extensive analysis of the performance for D2D multicast is given by Lin et al. (2013b).

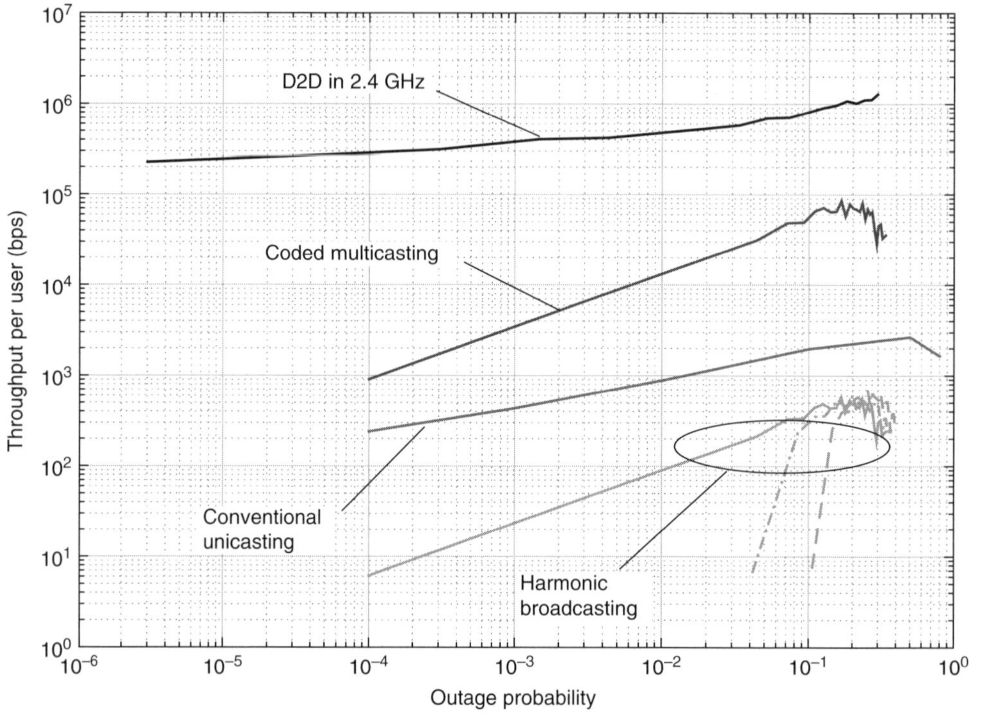

Figure 9.8 Simulation results for the throughput-outage tradeoff for different schemes under the realistic indoor/outdoor propagation environment (for details, see Ji et al. (2016a)). For harmonic broadcasting with only the m' most popular files, solid line: $m' = 300$; dash-dot line: $m' = 280$; dash line: $m' = 250$. We have $n = 10{,}000$, $m = 300$, $M = 20$ and $\gamma_r = 0.4$.

this realistic propagation scenario (which is not a protocol model) the D2D single-hop caching network with simple transmission scheme can provide both large throughput – sufficient for streaming video at standard definition quality – and low outage probability. Also, the D2D caching scheme significantly outperforms the other schemes for all the regimes of outage probability. This performance gain is particularly notable with respect to conventional unicasting and harmonic broadcasting from the BS, which are representative of current technology. We also note the distinct performance advantages compared to coded multicasting, despite the fact that the two schemes have the same scaling laws. The main reason for this is that the capacity of multicasting is limited by the "weakest link" between the BS and the various mobile devices, while for the D2D transmission schemes, short-distance transmission, which usually has high SNR, shallow fading, and thus high capacity, determines the overall performance.

It is also worthwhile to notice that the scheduling scheme used in the simulations is based on the clustering structure and the interference avoidance (TDMA). Advanced interference management schemes such as:

- FlashLinQ; see Wu et al. (2010)
- ITLinQ; see Naderializadeh and Avestimehr (2014)

may provide additional gains for the D2D caching networks. In particular, Naderializadeh et al. (2014) show that by using a slightly modified version of Eq (1.17), ITLinQ delivery can provide a noticeable gain compared with the TDMA-based scheduling proposed in Ji et al. (2016a).

9.10 Conclusions

The field of D2D communications is relatively young, but has strong connections to the well-established area of ad-hoc (peer-to-peer) networking, and it is important to find what distinguishes D2D from ad-hoc networks. A key point here is the interaction with (and possible control by) a cellular network. In other words, D2D might:

- enable central control of D2D transmission through control by the BS
- compete for resources with D2I links (for in-band communications)
- allow off-loading of D2I traffic.

All these aspects make D2D different.

The chapter has provided an overview of the key problems in D2D research: neighbor discovery, channel estimation, mode selection, resource allocation, and scheduling. None of these can be considered to be definitely solved, despite the large variety of solutions that have been proposed for each of them; there are too many different operating conditions, and many of the problems are too complex, to lend themselves to a single, simple, and optimum solution. The array of mathematical techniques that have been brought to bear on each of them is an indication of the challenges researchers are facing.

D2D will also be influential in standardization. Both the dominant WiFi and the cellular standards have defined D2D modes, and 5G will certainly contain some form of D2D as well; see Mumtaz et al. (2014) and Tehrani et al. (2014). Security for such communications will also be a major issue, as both threats to secure digital communications, and the public awareness of the necessity for suitable security, will increase; see, for example, Alam et al. (2014). Finally, D2D will also continue to find new applications: the chapter has mentioned the wide variety that has been defined up to now, but there is reason to think that other usages will find D2D an attractive mode of communication.

Acknowledgments

Parts of this work were financially supported by the National Science Foundation under grants CNS-1457340 and CIF 1423140, and by Intel under the “Higher Denser Wilder” 5G research program.

References

Abbas T, Thiel A, Zemen T, Mecklenbräuker CF and Tufvesson F 2013 Validation of a non-line-of-sight path-loss model for V2V communications at street intersections *13th International Conference on ITS Telecommunications, Tampere, Finland, 2013-11-05*.

Acosta-Marum G and Ingram M 2007 Six time- and frequency- selective empirical channel models for vehicular wireless LANs. *IEEE Vehicular Technology Magazine* **2**(4), 4–11.

Akkarajitsakul K, Phunchongharn P, Hossain E and Bhargava V 2012 Mode selection for energy-efficient D2D communications in LTE-Advanced networks: A coalitional game approach *Communication Systems (ICCS), 2012 IEEE International Conference on*, pp. 488–492.

Al-Hourani A, Chandrasekharan S and Kandeepan S 2014 Path loss study for millimeter wave device-to-device communications in urban environment *Communications Workshops (ICC), 2014 IEEE International Conference on*, pp. 102–107.

Alam M, Yang D, Rodriguez J and Abd-Alhameed R 2014 Secure device-to-device communication in LTE-A. *Communications Magazine, IEEE* **52**(4), 66–73.

Altieri A, Piantanida P, Vega LR and Galarza CG 2014 A stochastic geometry approach to distributed caching in large wireless networks *Wireless Communications Systems (ISWCS), 2014 11th International Symposium on*, pp. 863–867.

Asadi A and Mancuso V 2013 Energy efficient opportunistic uplink packet forwarding in hybrid wireless networks *Proceedings of the Fourth International Conference on Future Energy Systems*, pp. 261–262.

Asadi A, Wang Q and Mancuso V 2014 A survey on device-to-device communication in cellular networks. *Communications Surveys Tutorials, IEEE* **16**(4), 1801–1819.

Baccelli F, Li J, Richardson T, Shakkottai S, Subramanian S and Wu X 2012 On optimizing CSMA for wide area ad hoc networks. *Queueing Systems: Theory and Applications* **72**(1–2), 31–68.

Bae SJ, Kim DH, Choi BG and Chung MY 2013 Transmission power control for flashlinq device-to-device communication system *Vehicular Technology Conference (VTC Spring), 2013 IEEE 77th*, pp. 1–5.

Bansal T, Sundaresan K, Rangarajan S and Sinha P 2014 R2D2: Embracing device-to-device communication in next generation cellular networks *INFOCOM, 2014 Proceedings IEEE*, pp. 1563–1571.

Belleschi M, Fodor G and Abrardo A 2011 Performance analysis of a distributed resource allocation scheme for D2D communications *GLOBECOM Workshops (GC Wkshps), 2011 IEEE*, pp. 358–362.

Bernado L, Zemen T, Tufvesson F, Molisch A and Mecklenbrauker C 2012 The (in-) validity of the WSSUS assumption in vehicular radio channels *Personal Indoor and Mobile Radio Communications (PIMRC), 2012 IEEE 23rd International Symposium on*, pp. 1757–1762.

Bernado L, Zemen T, Tufvesson F, Molisch A and Mecklenbrauker C 2014 Delay and Doppler spreads of nonstationary vehicular channels for safety-relevant scenarios. *Vehicular Technology, IEEE Transactions on* **63**(1), 82–93.

Borbash SA, Ephremides A and McGlynn MJ 2007 An asynchronous neighbor discovery algorithm for wireless sensor networks. *Ad Hoc Networks* **5**(7), 998–1016.

Burghal, D and Molisch AF (2016) Efficient channel state information acquisition for device-to-device networks. *Wireless Communications, IEEE Transactions on*, **15**(2), 965–979.

Cai H, Koprulu I and Shroff N 2013 Exploiting double opportunities for deadline based content propagation in wireless networks *INFOCOM, 2013 Proceedings IEEE*, pp. 764–772.

Camps-Mur D, Garcia-Saavedra A and Serrano P 2013 Device-to-device communications with Wi-Fi direct: overview and experimentation. *Wireless Communications, IEEE* **20**(3), 96–104.

Candes EJ and Tao T 2005 Decoding by linear programming. *Information Theory, IEEE Transactions on* **51**(12), 4203–4215.

Cao Y and Maaref A 2014 Device cooperation-assisted scalable video multicast with heterogeneous qoe guarantees *Wireless Communications Systems (ISWCS), 2014 11th International Symposium on*, pp. 733–738.

Chen X, Chen L, Zeng M, Zhang X and Yang D 2012a Downlink resource allocation for device-to-device communication underlaying cellular networks *Personal Indoor and Mobile Radio Communications (PIMRC), 2012 IEEE 23rd International Symposium on*, pp. 232–237.

Chen X, Chen L, Zeng M, Zhang X and Yang D 2012b Downlink resource allocation for device-to-device communication underlaying cellular networks *Personal Indoor and Mobile Radio Communications (PIMRC), 2012 IEEE 23rd International Symposium on*, pp. 232–237.

Cheng X, Li Y, Ai B, Yin X and Wang Q 2015 Device-to-device channel measurements and models: a survey. *Communications, IET* **9**(3), 312–325.

Chien CC, Su HJ and Li HJ 2012 Device-to-device assisted downlink broadcast channel in cellular networks *Wireless Personal Multimedia Communications (WPMC), 2012 15th International Symposium on*, pp. 85–89.

Choi KW, Lee H and Chang SC 2014 Discovering mobile applications in device-to-device communications: Hash function-based approach *Vehicular Technology Conference (VTC Spring), 2014 IEEE 79th*, pp. 1–5.

Cotton SL 2015 Human body shadowing in cellular device-to-device communications: Channel modeling using the shadowed fading model. *Selected Areas in Communications, IEEE Journal on* **33**(1), 111–119.

Danihelka J, Giustiniano D and Hossmann T 2014 Hycloud: A system for device-to-device content distribution controlled by the cloud *A World of Wireless, Mobile and Multimedia Networks (WoWMoM), 2014 IEEE 15th International Symposium on*, pp. 1–5.

Donoho DL 2006 Compressed sensing. *Information Theory, IEEE Transactions on* **52**(4), 1289–1306.

Doppler K, Rinne M, Janis P, Ribeiro C and Hugl K 2009a Device-to-device communications; functional prospects for LTE-Advanced networks *Communications Workshops, 2009. ICC Workshops 2009. IEEE International Conference on*, pp. 1–6.

Doppler K, Rinne M, Wijting C, Ribeiro C and Hugl K 2009b Device-to-device communication as an underlay to LTE-Advanced networks. *Communications Magazine, IEEE* **47**(12), 42–49.

Doppler K, Yu CH, Ribeiro C and Janis P 2010 Mode selection for device-to-device communication underlaying an LTE-Advanced network *Wireless Communications and Networking Conference (WCNC), 2010 IEEE*, pp. 1–6.

Duong Q, Shin Y and Shin OS 2013 Resource allocation scheme for device-to-device communications underlaying cellular networks *Computing, Management and Telecommunications (ComManTel), 2013 International Conference on*, pp. 66–69.

ElSawy H, Hossain E and Alouini MS 2014 Analytical modeling of mode selection and power control for underlay D2D communication in cellular networks. *Communications, IEEE Transactions on* **62**(11), 4147–4161.

Feng D, Yu G, Yuan-Wu Y, Li G, Feng G and Li S 2014 Mode switching for device-to-device communications in cellular networks *Signal and Information Processing (GlobalSIP), 2014 IEEE Global Conference on*, pp. 1291–1295.

Fodor G and Reider N 2011 A distributed power control scheme for cellular network assisted D2D communications *Global Telecommunications Conference (GLOBECOM 2011), 2011 IEEE*, pp. 1–6.

Fodor G, Dahlman E, Mildh G, Parkvall S, Reider N, Miklos G and Turanyi Z 2012 Design aspects of network assisted device-to-device communications. *Communications Magazine, IEEE* **50**(3), 170–177.

Golrezaei N, Dimakis A and Molisch A 2012a Device-to-device collaboration through distributed storage *Global Communications Conference (GLOBECOM), 2012 IEEE*, pp. 2397–2402.

Golrezaei N, Dimakis AG and Molisch AF 2012b Wireless device-to-device communications with distributed caching *Information Theory Proceedings (ISIT), 2012 IEEE International Symposium on*, pp. 2781–2785.

Golrezaei N, Dimakis AG, Molisch AF and Caire G 2011 Wireless video content delivery through distributed caching and peer-to-peer gossiping *Signals, Systems and Computers (ASILOMAR), 2011 Conference Record of the Forty Fifth Asilomar Conference on*, pp. 1177–1180.

Golrezaei N, Ji M, Molisch AF, Dimakis AG and Caire G 2012c Device-to-device communications for wireless video delivery *Forty Sixth Asilomar Conference on Signals, Systems and Computers (ASILOMAR)*, pp. 930–933.

Golrezaei N, Mansourifard P, Molisch A and Dimakis A 2014 Base-station assisted device-to-device communications for high-throughput wireless video networks. *Wireless Communications, IEEE Transactions on* **13**(7), 3665–3676.

Golrezaei N, Molisch AF and Dimakis AG 2012d Base-station assisted device-to-device communications for high-throughput wireless video networks *Communications (ICC), 2012 IEEE International Conference on*, pp. 7077–7081.

Gomez K, Goratti L, Rasheed T and Reynaud L 2014 Enabling disaster-resilient 4G mobile communication networks. *Communications Magazine, IEEE* **52**(12), 66–73.

Hakola S, Chen T, Lehtomaki J and Koskela T 2010 Device-to-device (D2D) communication in cellular network-performance analysis of optimum and practical communication mode selection *Wireless Communications and Networking Conference (WCNC), 2010 IEEE*, pp. 1–6.

Han J, Cui Q, Yang C and Tao X 2014 Bipartite matching approach to optimal resource allocation in device to device underlaying cellular network. *Electronics Letters* **50**(3), 212–214.

Hasan M, Hossain E and Kim DI 2014 Resource allocation under channel uncertainties for relay-aided device-to-device communication underlaying LTE-A cellular networks. *Wireless Communications, IEEE Transactions on* **13**(4), 2322–2338.

Huang J, Sun Y, Xiong Z, Duan Q, Zhao Y, Cao X and Wang W 2016 Modeling and analysis on access control for device-to-device communications in cellular network: A network calculus based approach. *Vehicular Technology, IEEE Transactions on*, **65**(3), 1615–1626.

Janis P, Koivunen V, Ribeiro C, Korhonen J, Doppler K and Hugl K 2009 Interference-aware resource allocation for device-to-device radio underlaying cellular networks *Vehicular Technology Conference, 2009. VTC Spring 2009. IEEE 69th*, pp. 1–5.

Ji M, Caire G and Molisch A 2013a Optimal throughput-outage trade-off in wireless one-hop caching networks. *Information Theory Proceedings (ISIT), 2013 IEEE International Symposium on*, pp. 1461–1465.

Ji M, Caire G and Molisch AF 2013b Fundamental limits of distributed caching in D2D wireless networks. *Information Theory Workshop (ITW), 2013 IEEE*, pp. 1–5.

Ji M, Caire G and Molisch AF 2015 The throughput-outage tradeoff of wireless one-hop caching networks. *Information Theory, IEEE Transactions on*, **61**(12), 6833–6859.

Ji M, Caire G and Molisch AF 2016a Fundamental limits of caching in wireless D2D networks. *Information Theory, IEEE Transactions on*, **62**(2), 849–869.

Ji M, Caire G and Molisch AF 2016b Wireless device-to-device caching networks: basic principles and system performance. *Selected Areas in Communications, IEEE Journal on*, **34**(1), 176–189.

Jung M, Hwang K and Choi S 2012 Joint mode selection and power allocation scheme for power-efficient device-to-device (D2D) communication *Vehicular Technology Conference (VTC Spring), 2012 IEEE 75th*, pp. 1–5.

Kang JW, Lim C and Kim SH 2013 A distributed link scheduling with on-off interference map for device-to-device communications *ICT Convergence (ICTC), 2013 International Conference on*, pp. 1065–1066.

Karedal J, Almers P, Johansson A, Tufvesson F and Molisch A 2010 A MIMO channel model for wireless personal area networks. *IEEE Transactions on Wireless Communications* **9**(1), 245–255.

Karedal J, Czink N, Paier A, Tufvesson F and Molisch A 2011 Path loss modeling for vehicle-to-vehicle communications. *Vehicular Technology, IEEE Transactions on* **60**(1), 323–328.

Karedal J, Johansson AJ, Tufvesson F and Molisch AF 2008 A measurement-based fading model for wireless personal area networks. *Wireless Communications, IEEE Transactions on* **7**(11), 4575–4585.

Karedal J, Tufvesson F, Czink N, Paier A, Dumard C, Zemen T, Mecklenbrauker C and Molisch A 2009 A geometry-based stochastic MIMO model for vehicle-to-vehicle communications. *Wireless Communications, IEEE Transactions on* **8**(7), 3646–3657.

Kaufman B, Lilleberg J and Aazhang B 2013 Spectrum sharing scheme between cellular users and ad-hoc device-to-device users. *Wireless Communications, IEEE Transactions on* **12**(3), 1038–1049.

Kim J, Meng F, Chen P, Egilmez HE, Bethanabhotla D, Molisch AF, Neely MJ, Caire G and Ortega A 2013 Adaptive video streaming for device-to-device mobile platforms *Proceedings of the 19th Annual International Conference on Mobile Computing & Networking*, pp. 127–130.

Kim J, Caire G and Molisch AF 2014 Quality-aware streaming and scheduling for device-to-device video delivery. *Networking, IEEE/ACM Transactions on*, **PP**(99), 1-1.

Kone V, Zheng H, Rowstron A, O'Shea G and Zhao B 2013 Measurement-based design of roadside content delivery systems. *Mobile Computing, IEEE Transactions on* **12**(6), 1160–1173.

Le LB 2012 Fair resource allocation for device-to-device communications in wireless cellular networks *Global Communications Conference (GLOBECOM), 2012 IEEE*, pp. 5451–5456.

Lee DH, Choi KW, Jeon WS and Jeong DG 2013 Resource allocation scheme for device-to-device communication for maximizing spatial reuse *Wireless Communications and Networking Conference (WCNC), 2013 IEEE*, pp. 112–117.

Lee K, Kang W and Choi HJ 2014 A practical channel estimation and feedback method for device-to-device communication in 3GPP LTE system *Proceedings of the 8th International Conference on Ubiquitous Information Management and Communication* ICUIMC '14.

Lee N, Lin X, Andrews J and Heath R 2015 Power control for D2D underlaid cellular networks: Modeling, algorithms, and analysis. *Selected Areas in Communications, IEEE Journal on* **33**(1), 1–13.

Lei L, Kuang Y, Shen X, Lin C and Zhong Z 2014 Resource control in network assisted device-to-device communications: solutions and challenges. *Communications Magazine, IEEE* **52**(6), 108–117.

Li J, Lei M and Gao F 2012a Device-to-device (D2D) communication in MU-MIMO cellular networks *Global Communications Conference (GLOBECOM), 2012 IEEE*, pp. 3583–3587.

Li Y, Jin D, Yuan J and Han Z 2014a Coalitional games for resource allocation in the device-to-device uplink underlaying cellular networks. *Wireless Communications, IEEE Transactions on* **13**(7), 3965–3977.

Li Y, Song C, Jin D and Chen S 2014b A dynamic graph optimization framework for multihop device-to-device communication underlaying cellular networks. *Wireless Communications, IEEE* **21**(5), 52–61.

Li Y, Sun L and Wang W 2014c Exploring device-to-device communication for mobile cloud computing *Communications (ICC), 2014 IEEE International Conference on*, pp. 2239–2244.

Li Y, Wang Z, Jin D and Chen S 2014d Optimal mobile content downloading in device-to-device communication underlaying cellular networks. *Wireless Communications, IEEE Transactions on* **13**(7), 3596–3608.

Li Z, Wang R and Molisch A 2012b Shadowing in urban environments with microcellular or peer-to-peer links *Antennas and Propagation (EUCAP), 2012 6th European Conference on*, pp. 44–48.

Lin X, Andrews J and Ghosh A 2014 Spectrum sharing for device-to-device communication in cellular networks. *Wireless Communications, IEEE Transactions on* **13**(12), 6727–6740.

Lin X, Andrews JG and Ghosh A 2013a A comprehensive framework for device-to-device communications in cellular networks. *arXiv preprint arXiv:1305.4219*.

Lin X, Ratasuk R, Ghosh A and Andrews JG 2013b Modeling, analysis and optimization of multicast device-to-device transmissions. *Wireless Communications, IEEE Transactions on*, **13**(8), 4346–4359.

Liu Z, Peng T, Chen H and Wang W 2012 Optimal D2D user allocation over multi-bands under heterogeneous networks *Global Communications Conference (GLOBECOM), 2012 IEEE*, pp. 1339–1344.

Luo J and Guo D 2008 Neighbor discovery in wireless ad hoc networks based on group testing *Communication, Control, and Computing, 2008 46th Annual Allerton Conference on*, pp. 791–797.

Macuha M, Amgalan D and Derham T 2013 Device-centric approach to improve resiliency of emergency communication *Humanitarian Technology Conference (R10-HTC), 2013 IEEE Region 10*, pp. 124–129.

Maddah-Ali M and Niesen U 2014 Fundamental limits of caching. *Information Theory, IEEE Transactions on*, **60**(5), 2856–2867.

Maghsudi S and Stanczak S 2014 Transmission mode selection for network-assisted device to device communication: A levy-bandit approach *Acoustics, Speech and Signal Processing (ICASSP), 2014 IEEE International Conference on*, pp. 7009–7013.

Malak D and Al-Shalash M 2014 Optimal caching for device-to-device content distribution in 5G networks *Globecom Workshops (GC Wkshps), 2014*, pp. 863–868.

Mangel T, Klemp O and Hartenstein H 2011 5.9 GHz inter-vehicle communication at intersections: a validated non-line-of-sight path-loss and fading model. *EURASIP Journal on Wireless Communications and Networking* **2011**(1), 182.

McGlynn MJ and Borbash SA 2001 Birthday protocols for low energy deployment and flexible neighbor discovery in ad hoc wireless networks *Proceedings of the 2nd ACM International Symposium on Mobile Ad Hoc Networking & Computing*, pp. 137–145 ACM.

Mecklenbrauker C, Molisch A, Karedal J, Tufvesson F, Paier A, Bernado L, Zemen T, Klemp O and Czink N 2011 Vehicular channel characterization and its implications for wireless system design and performance. *Proceedings of the IEEE* **99**(7), 1189–1212.

Militano L, Condoluci M, Araniti G, Molinaro A, Iera A and Fitzek F 2014 Wi-fi cooperation or D2D-based multicast content distribution in LTE-A: A comparative analysis *Communications Workshops (ICC), 2014 IEEE International Conference on*, pp. 296–301.

Min H, Lee J, Park S and Hong D 2011 Capacity enhancement using an interference limited area for device-to-device uplink underlaying cellular networks. *Wireless Communications, IEEE Transactions on* **10**(12), 3995–4000.

Molisch A, Tufvesson F, Karedal J and Mecklenbrauker C 2009 A survey on vehicle-to-vehicle propagation channels. *Wireless Communications, IEEE* **16**(6), 12–22.

Molisch AF, Caire G, Ott D, Foerster JR, Bethanabhotla D and Ji M 2014 Caching eliminates the wireless bottleneck in video-aware wireless networks. *arXiv preprint arXiv:1405.5864*.

Molisch AF, Kuchar A, Laurila J, Hugl K and Schmalenberger R 2003 Geometry-based directional model for mobile radio channels – principles and implementation. *European Transactions on Telecommunications* **14**(4), 351–359.

Mumtaz S, Saidul Huq K and Rodriguez J 2014 Direct mobile-to-mobile communication: Paradigm for 5G. *Wireless Communications, IEEE* **21**(5), 14–23.

Naderializadeh N and Avestimehr AS 2014 ITLinQ: A new approach for spectrum sharing in device-to-device communication systems. *Selected Areas in Communications, IEEE Journal on*, **32**(6), 1139–1151.

Naderializadeh N, Kao D and Avestimehr A 2014 How to utilize caching to improve spectral efficiency in device-to-device wireless networks *Communication, Control, and Computing (Allerton), 2014 52nd Annual Allerton Conference on*, pp. 415–422.

Nguyen P and Rao B 2015 Fair scheduling policies exploiting multiuser diversity in cellular systems with device-to-device communications. *Wireless Communications, IEEE Transactions on*, **14**(9), 4757–4771.

Nishiyama H, Ito M and Kato N 2014 Relay-by-smartphone: realizing multihop device-to-device communications. *Communications Magazine, IEEE* **52**(4), 56–65.

Niyato D, Xiao L and Wang P 2011 Machine-to-machine communications for home energy management system in smart grid. *Communications Magazine, IEEE* **49**(4), 53–59.

Osseiran A, Xiao M, Ben Slimane S, Skoglund M and Manssour J 2011 Advances in wireless network coding for IMT-Advanced and beyond *Wireless Communication, Vehicular Technology, Information Theory and Aerospace Electronic Systems Technology (Wireless VITAE), 2011 2nd International Conference on*, pp. 1–6.

Pääkkönen J, Hollanti C and Tirkkonen O 2013 Device-to-device data storage for mobile cellular systems. *arXiv preprint arXiv:1309.6123*.

Park SH, Ryu H and Lim C 2014 Distributed iterative scheduling for D2D communications *Wireless Communications and Networking Conference Workshops (WCNCW), 2014 IEEE*, pp. 254–259.

Pei Y and Liang YC 2013a Resource allocation for device-to-device communication overlaying two-way cellular networks *Wireless Communications and Networking Conference (WCNC), 2013 IEEE*, pp. 3346–3351.

Pei Y and Liang YC 2013b Resource allocation for device-to-device communications overlaying two-way cellular networks. *Wireless Communications, IEEE Transactions on* **12**(7), 3611–3621.

Peng B, Hu C, Peng T and Wang W 2012 Optimal resource allocation for multi-D2D links underlying OFDMA-based communications *Wireless Communications, Networking and Mobile Computing (WiCOM), 2012 8th International Conference on*, pp. 1–4.

Peng T, Lu Q, Wang H, Xu S and Wang W 2009 Interference avoidance mechanisms in the hybrid cellular and device-to-device systems *Personal, Indoor and Mobile Radio Communications, 2009 IEEE 20th International Symposium on*, pp. 617–621.

Pradini A, Fodor G, Miao G and Belleschi M 2014 Near-optimal practical power control schemes for D2D communications in cellular networks *Networks and Communications (EuCNC), 2014 European Conference on*, pp. 1–5.

Prasad A, Kunz A, Velev G, Samdanis K and Song J 2014 Energy-efficient D2D discovery for proximity services in 3GPP LTE-Advanced networks: Prose discovery mechanisms. *Vehicular Technology Magazine, IEEE* **9**(4), 40–50.

Pyattaev A, Johnsson K, Andreev S and Koucheryavy Y 2013 Proximity-based data offloading via network assisted device-to-device communications *Vehicular Technology Conference (VTC Spring), 2013 IEEE 77th*, pp. 1–5.

Qiao J, Shen X, Mark J, Shen Q, He Y and Lei L 2015 Enabling device-to-device communications in millimeter-wave 5G cellular networks. *Communications Magazine, IEEE* **53**(1), 209–215.

Raghothaman B, Deng E, Pragada R, Sternberg G, Deng T and Vanganuru K 2013 Architecture and protocols for LTE-based device to device communication *Computing, Networking and Communications (ICNC), 2013 International Conference on*, pp. 895–899.

Ramanathan R, Redi J, Santivanez C, Wiggins D and Polit S 2005 Ad hoc networking with directional antennas: a complete system solution. *Selected Areas in Communications, IEEE Journal on* **23**(3), 496–506.

Ren P, Du Q and Sun L 2013 Interference-aware routing for hop-count minimization in wireless D2D networks *Communications in China – Workshops (CIC/ICCC), 2013 IEEE/CIC International Conference on*, pp. 65–70.

Renaudin O 2013 *Experimental Channel Characterization for Vehicle-to-Vehicle Communication Systems* PhD thesis, Université Catholique de Louvain, Belgium.

Renaudin O, Kolmonen V, Vainikainen P and Oestges C 2010 Non-stationary narrowband MIMO inter-vehicle channel characterization in the 5-GHz band. *Vehicular Technology, IEEE Transactions on* **59**(4), 2007–2015.

Schack M, Nuckelt J, Geise R, Thiele L and Kurner T 2011 Comparison of path loss measurements and predictions at urban crossroads for C2C communications *Antennas and Propagation (EUCAP), Proceedings of the 5th European Conference on*, pp. 2896–2900.

Shalmashi S and Ben Slimane S 2014 Cooperative device-to-device communications in the downlink of cellular networks *Wireless Communications and Networking Conference (WCNC), 2014 IEEE*, pp. 2265–2270.

Su L, Ji Y, Wang P and Liu F 2013 Resource allocation using particle swarm optimization for D2D communication underlay of cellular networks *Wireless Communications and Networking Conference (WCNC), 2013 IEEE*, pp. 129–133.

Takai M, Martin J, Bagrodia R and Ren A 2002 Directional virtual carrier sensing for directional antennas in mobile ad hoc networks *Proceedings of the 3rd ACM International Symposium on Mobile Ad Hoc Networking and Computing*, pp. 183–193.

Tamoor-ul Hassan S, Ashraf MI and Katz MD 2013 Mobile cloud based architecture for device-to-device (D2D) communication underlying cellular network *Wireless Days (WD), 2013 IFIP*, pp. 1–3.

Tang H, Ding Z and Levy B 2014 Enabling D2D communications through neighbor discovery in LTE cellular networks. *Signal Processing, IEEE Transactions on* **62**(19), 5157–5170.

Tehrani AS and Caire G 2014 Zigzag neighbor discovery in wireless networks *Information Theory (ISIT), 2014 IEEE International Symposium on*, pp. 1386–1390.

Tehrani AS, Dimakis AG and Caire G 2013 Optimal deterministic compressed sensing matrices *Acoustics, Speech and Signal Processing (ICASSP), 2013 IEEE International Conference on*, pp. 5895–5899.

Tehrani M, Uysal M and Yanikomeroglu H 2014 Device-to-device communication in 5G cellular networks: challenges, solutions, and future directions. *Communications Magazine, IEEE* **52**(5), 86–92.

Vasudevan S, Kurose J and Towsley D 2005 On neighbor discovery in wireless networks with directional antennas *INFOCOM 2005. 24th Annual Joint Conference of the IEEE Computer and Communications Societies. Proceedings IEEE*, vol. **4**, pp. 2502–2512.

Walter M, Shutin D and Fiebig UC 2014 Delay-dependent Doppler probability density functions for vehicle-to-vehicle scatter channels. *Antennas and Propagation, IEEE Transactions on*, **62**(4), 2238–2249.

Wang Q and Rengarajan B 2013 Recouping opportunistic gain in dense base station layouts through energy-aware user cooperation *World of Wireless, Mobile and Multimedia Networks (WoWMoM), 2013 IEEE 14th International Symposium and Workshops on*, pp. 1–9.

Wei L, Hu R, Li Q and Wu G 2014a Energy-efficiency of multi-hop device-to-device communications underlaying cellular networks *Communications (ICC), 2014 IEEE International Conference on*, pp. 5486–5491.

Wei L, Hu R, Qian Y and Wu G 2014b Enable device-to-device communications underlaying cellular networks: challenges and research aspects. *Communications Magazine, IEEE* **52**(6), 90–96.

Wen S, Zhu X, Zhang X and Yang D 2013 Qos-aware mode selection and resource allocation scheme for device-to-device (D2D) communication in cellular networks *Communications Workshops (ICC), 2013 IEEE International Conference on*, pp. 101–105.

WINNER-II 2007 D1. 1.2, WINNER II channel models.

Wu D, Wang J, Hu R, Cai Y and Zhou L 2014 Energy-efficient resource sharing for mobile device-to-device multimedia communications. *Vehicular Technology, IEEE Transactions on* **63**(5), 2093–2103.

Wu X, Tavildar S, Shakkottai S, Richardson T, Li J, Laroia R and Jovicic A 2010 Flashlinq: A synchronous distributed scheduler for peer-to-peer ad hoc networks *Communication, Control, and Computing (Allerton), 2010 48th Annual Allerton Conference on*, pp. 514–521.

Wu X, Tavildar S, Shakkottai S, Richardson T, Li J, Laroia R and Jovicic A 2013 Flashlinq: A synchronous distributed scheduler for peer-to-peer ad hoc networks. *Networking, IEEE/ACM Transactions on* **21**(4), 1215–1228.

Xia W, Shao S and Sun J 2013 Relay selection strategy for device to device communication *Information and Communications Technologies (IETICT 2013), IET International Conference on*, pp. 318–323.

Xiao X, Tao X and Lu J 2011 A QoS-aware power optimization scheme in OFDMA systems with integrated device-to-device (D2D) communications *Vehicular Technology Conference (VTC Fall), 2011 IEEE*, pp. 1–5.

Xiao Y, Chen KC, Yuen C and DaSilva L 2014 Spectrum sharing for device-to-device communications in cellular networks: A game theoretic approach *Dynamic Spectrum Access Networks (DYSPAN), 2014 IEEE International Symposium on*, pp. 60–71.

Xing B, Seada K and Venkatasubramanian N 2009 An experimental study on Wi-Fi ad-hoc mode for mobile device-to-device video delivery *INFOCOM Workshops 2009, IEEE*, pp. 1–6.

Xu C, Song L, Han Z, Li D and Jiao B 2012a Resource allocation using a reverse iterative combinatorial auction for device-to-device underlay cellular networks *Global Communications Conference (GLOBECOM), 2012 IEEE*, pp. 4542–4547.

Xu C, Song L, Han Z, Zhao Q, Wang X and Jiao B 2012b Interference-aware resource allocation for device-to-device communications as an underlay using sequential second price auction *Communications (ICC), 2012 IEEE International Conference on*, pp. 445–449.

Xu C, Song L, Zhu D and Lei M 2014 Subcarrier and power optimization for device-to-device underlay communication using auction games *Communications (ICC), 2014 IEEE International Conference on*, pp. 5526–5531.

Xu Y, Liu Y and Li D 2015 Resource management for interference mitigation in device-to-device communication. *Communications, IET* **9**(9), 1199–1207.

Yang MJ, Lim SY, Park HJ and Park NH 2013 Solving the data overload: Device-to-device bearer control architecture for cellular data offloading. *Vehicular Technology Magazine, IEEE* **8**(1), 31–39.

Ye Q, Al-Shalash M, Caramanis C and Andrews J 2015 Distributed resource allocation in device-to-device enhanced cellular networks. *Communications, IEEE Transactions on* **63**(2), 441–454.

Yin R, Yu G, Zhang H, Zhang Z and Li G 2015 Pricing-based interference coordination for D2D communications in cellular networks. *Wireless Communications, IEEE Transactions on* **14**(3), 1519–1532.

Yoon HW, Kim JS, Bae SJ, Choi BG and Chung MY 2012 Performance analysis of Flashlinq with various yielding threshold values *ICT Convergence (ICTC), 2012 International Conference on*, pp. 477–478.

Yu CH and Tirkkonen O 2012 Device-to-device underlay cellular network based on rate splitting *Wireless Communications and Networking Conference (WCNC), 2012 IEEE*, pp. 262–266.

Yu CH, Tirkkonen O, Doppler K and Ribeiro C 2009a On the performance of device-to-device underlay communication with simple power control *Vehicular Technology Conference, 2009. VTC Spring 2009. IEEE 69th*, pp. 1–5.

Yu CH, Tirkkonen O, Doppler K and Ribeiro C 2009b Power optimization of device-to-device communication underlaying cellular communication *Communications, 2009. ICC '09. IEEE International Conference on*, pp. 1–5.

Yu G, Xu L, Feng D, Yin R, Li G and Jiang Y 2014 Joint mode selection and resource allocation for device-to-device communications. *Communications, IEEE Transactions on* **62**(11), 3814–3824.

Zhang B, Li Y, Jin D, Hui P and Han Z 2015 Social-aware peer discovery for D2D communications underlaying cellular networks. *Wireless Communications, IEEE Transactions on* **14**(5), 2426–2439.

Zhang L, Luo J and Guo D 2013a Neighbor discovery for wireless networks via compressed sensing. *Performance Evaluation, Selected papers from WiOpt 2011* **70**(7), 457–471.

Zhang R, Cheng X, Yang L and Jiao B 2013b Interference-aware graph based resource sharing for device-to-device communications underlaying cellular networks *Wireless Communications and Networking Conference (WCNC), 2013 IEEE*, pp. 140–145.

Zhang Y, Pan E, Song L, Saad W, Dawy Z and Han Z 2013c Social network enhanced device-to-device communication underlaying cellular networks *Communications in China Workshops (CIC/ICCC), 2013 IEEE/CIC International Conference on*, pp. 182–186.

Zhou B, Hu H, Huang SQ and Chen HH 2013a Intracluster device-to-device relay algorithm with optimal resource utilization. *Vehicular Technology, IEEE Transactions on* **62**(5), 2315–2326.

Zhou B, Ma S, Xu J and Li Z 2013b Group-wise channel sensing and resource pre-allocation for LTE D2D on ISM band *Wireless Communications and Networking Conference (WCNC), 2013 IEEE*, pp. 118–122.

Zhou Z, Dong M, Ota K, Shi R, Liu Z and Sato T 2015 Game-theoretic approach to energy-efficient resource allocation in device-to-device underlay communications. *Communications, IET* **9**(3), 375–385.

Zhu K and Hossain E 2015 Joint mode selection and spectrum partitioning for device-to-device communication: A dynamic Stackelberg game. *Wireless Communications, IEEE Transactions on* **14**(3), 1406–1420.

Zou K, Wang M, Yang K, Zhang J, Sheng W, Chen Q and You X 2014 Proximity discovery for device-to-device communications over a cellular network. *Communications Magazine, IEEE* **52**(6), 98–107.

Zulhasnine M, Huang C and Srinivasan A 2010 Efficient resource allocation for device-to-device communication underlaying LTE network *Wireless and Mobile Computing, Networking and Communications (WiMob), 2010 IEEE 6th International Conference on*, pp. 368–375.

10

Energy-efficient Wireless OFDMA Networks

Cong Xiong and Geoffrey Ye Li
Georgia Institute of Technology, Atlanta, GA, USA

10.1 Overview

Information and communication technology is playing an important role in global greenhouse gas emissions. It is reported that the total energy consumed by the infrastructure of cellular wireless networks, wired communication networks, and the Internet takes up more than 3% of the world electric energy consumption nowadays and the portion is expected to increase rapidly in the future. As an important part of information and communication technology, wireless communications are responsible for greenhouse gas emissions and resultant environmental issues. Meanwhile, widespread application of wireless services requires ubiquitous access and has triggered rapidly booming energy consumption in wireless networks. Such an escalation of energy consumption also causes high operational expenditure from electricity bills for operators and unsatisfactory user experience due to the limited battery capacity of wireless devices, besides a large amount of greenhouse gas emissions. Energy-efficient wireless networks, which emphasize energy efficiency (EE) as well as spectral efficiency (SE), have been proposed as an effective solution and are becoming the mainstream for future wireless network design. Unfortunately, EE and SE do not always coincide and may conflict, even in new-generation wireless networks such those based on orthogonal frequency division multiple access (OFDMA).

OFDMA has been extensively studied from the SE perspective and has been adopted for next-generation wireless communication networks, such as WiMAX and the 3GPP LTE. In OFDMA wireless networks, system resources, such as subcarriers and transmit power, can be properly allocated to different users to achieve high performance. While OFDMA can provide high throughput and SE, its EE is has previously not been of much concern. To keep pace with

Towards 5G: Applications, Requirements and Candidate Technologies, First Edition.
Edited by Rath Vannithamby and Shilpa Talwar.

Companion website: www.wiley.com/go/vannithamby/towards5g

the trend for energy-efficient network design, it is necessary for OFDMA to guarantee a high level of EE at the same time. Considering the wide deployment of OFDMA wireless networks in the world, it would save a lot of energy if OFDMA wireless networks are also energy-efficient.

This chapter focuses on energy-efficient transmission and resource allocation techniques for OFDMA networks. In Section 10.2, the origin of EE and energy-efficient wireless network is reviewed. Some interesting open problems in EE are also discussed. Then, the fundamental principles of energy-efficient design for OFDMA networks are discussed in the following sections. As the first step, the fundamental interrelationship between EE and SE in downlink OFDMA networks is investigated, and the impacts of channel gain and circuit power on the EE–SE relationship are analyzed in Section 10.3. A general EE–SE optimization framework is proposed, where the overall EE, SE, and per-user quality-of-service (QoS) are all considered. Under this framework, EE proves to be quasiconcave in SE and decreases with SE when SE is large enough. These findings form very helpful guidelines for designing energy- and spectrum-efficient OFDMA. In Section 10.4, the tradeoff between EE and delay for delay-sensitive traffic in downlink OFDMA networks is analyzed by integrating information theory with the concept of effective capacity. The relationship between spectrum-efficient and energy-efficient designs, and the impact of system parameters, including circuit power and delay exponents, on the overall performance, are also discussed.

To facilitate the application of energy-efficient resource allocation, energy-efficient resource allocation strategies for both downlink and uplink OFDMA networks are investigated in Sections 10.5 and 10.6, respectively. For the downlink transmission, the generalized EE is maximized, while for the uplink case the minimum individual EE is maximized, both under prescribed per-user minimum data rate requirements. For both scenarios, an optimal solution is derived, and then a computationally efficient suboptimal approach is developed by exploring the inherent structure and property of the energy-efficient design.

Finally, concluding remarks and further reading materials are given in Section 10.7.

10.2 Energy Efficiency and Energy-efficient Wireless Networks

Energy efficiency is commonly defined as information bits per unit transmit energy and has been studied from the information-theoretic perspective for various scenarios (Verdú 2002). For an additive white Gaussian noise (AWGN) channel, it is well known that for a given transmit power, P, and system bandwidth, B, the channel capacity is $R = \frac{1}{2}\log_2\left(1+\frac{P}{N_0 B}\right)$ bits per real dimension or degrees of freedom (DOF), where N_0 is the noise power spectral density. According to the Nyquist sampling theory, DOF per second is $2B$. Therefore, the channel capacity is $C = 2BR$ bits per second. Consequently, EE is (Chen et al. 2011; Polyanskiy et al. 2010):

$$\eta_{EE} = \frac{C}{P} = \frac{2R}{N_0\left(2^{2R}-1\right)}. \tag{10.1}$$

From Eq. (10.1), it is obvious that η_{EE} decreases monotonically with R, with $(\eta_{EE})_{\max} = 1/(N_0 \ln 2)$ as $R \to 0$, and $(\eta_{EE})_{\min} = 0$ as $R \to \infty$. And EE and SE has a simple tradeoff relationship as shown in Figure 10.1 (Chen et al. 2011).

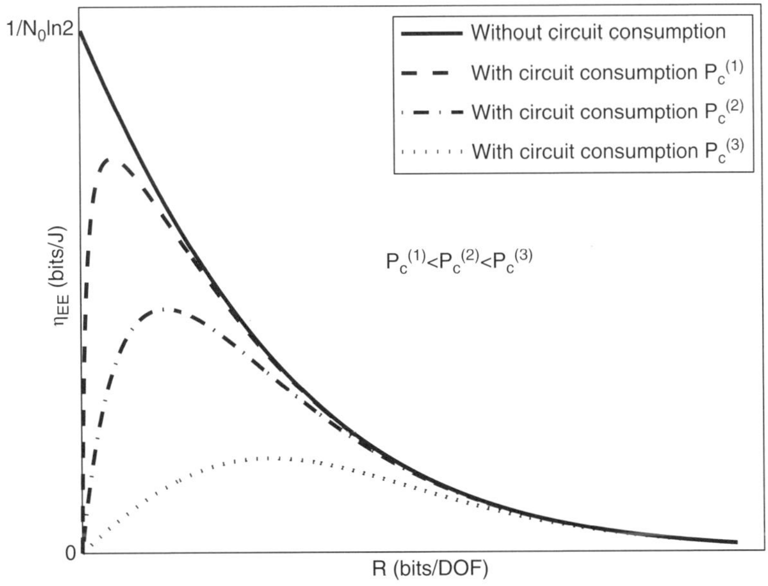

Figure 10.1 Tradeoff between EE and rate in an AWGN channel.

The EE bounds derived from the information-theoretic analysis might not be achieved in practical systems as a result of performance loss of capacity-approaching channel codes, imperfect knowledge of channel state information (CSI) (Gursoy 2009), cost of synchronization (Chandar et al. 2010), and transmission associated electronic circuit energy consumption (Cui et al. 2004, 2005; Isheden and Fettweis 2010; Miao et al. 2010, 2008). Among these factors, electronic circuit energy consumption changes the fundamental tradeoff between EE and data rate. When circuit energy consumption is considered, EE is redefined as information bits per unit energy (not only transmit energy), where an additional circuit power factor, P_c, needs to be added in the denominator of Eq. (10.1). Accordingly, the η_{EE}-versus-R curve will turn from a cup shape to a bell shape as shown in Figure 10.1 from (Chen et al. 2011).

However, EE–SE relation for general multi-user networks, including downlink OFDMA network, is much more complicated and not clear yet. And how to allocate system resource to tradeoff EE and SE efficiently for these networks is a nontrivial question.

10.3 Energy Efficiency and Spectral Efficiency Tradeoff in OFDMA

In this section, the fundamental tradeoff relationship in OFDMA between EE and SE will be investigated. We address the EE–SE tradeoff issue in downlink OFDMA networks. We build a general EE–SE tradeoff framework, prove that EE is quasiconcave in SE, and further discuss the impact of channel power gain and circuit power on the EE–SE relation.

Consider the downlink of a single cell OFDMA network consisting of K active users. The total bandwidth B is equally divided into N subcarriers, each with a bandwidth of $W = B/N$. Let $\mathcal{K} = \{1,2,\cdots,K\}$ and $\mathcal{N} = \{1,2,\cdots,N\}$ denote the sets of active users and all subcarriers, respectively.

Assume that one subcarrier is exclusively assigned to at most one user to avoid interference among different users. Denote the transmit power of the kth user on the nth subcarrier as $p_{k,n}$. It is obvious that there is at most one $k \in \mathcal{K}$ for each $n \in \mathcal{N}$ such that $p_{k,n} > 0$. Then, the maximum achievable data rate of the kth user on the nth subcarrier is

$$r_{k,n} = W \log_2 \left(1 + \frac{p_{k,n} g_{k,n}}{N_0 W} \right), \tag{10.2}$$

where $g_{k,n} = |h_{k,n}|^2$ is the channel power gain of the kth user on the nth subcarrier, $h_{k,n}$ is the corresponding frequency response, which is assumed to be accurately known at the transmitter, and N_0 is the single-sided noise spectral density. Consequently, the overall system throughput and the total transmit power are

$$R = \sum_{k=1}^{K} \sum_{n=1}^{N} \rho_{k,n} r_{k,n} \quad \text{and} \quad P = \sum_{k=1}^{K} \sum_{n=1}^{N} \rho_{k,n} p_{k,n}, \tag{10.3}$$

respectively, where $\rho_{k,n} \in \{1,0\}$ indicates whether the nth subcarrier is allocated to the kth user (by 1) or not (by 0). Obviously, any feasible subcarrier assignment indictor matrix, $\rho = [\rho_{k,n}]_{K \times N}$, and power allocation matrix, $\boldsymbol{P} = [p_{k,n}]_{K \times N}$, should satisfy

$$\rho \in \rho \triangleq \left\{ \left[\rho_{k,n} \right]_{K \times N} \middle| \rho_{k,n} \in \{0,1\}, \sum_{k=1}^{K} \rho_{k,n} \le 1 \right\}, \tag{10.4}$$

$$\boldsymbol{P} \in \mathcal{P} \triangleq \left\{ \left[p_{k,n} \right]_{K \times N} \middle| p_{k,n} \ge 0, \sum_{k=1}^{K} \sum_{n=1}^{N} p_{k,n} \le P_{\max} \right\}, \tag{10.5}$$

respectively, where $P_{\max}$ represents the peak total transmit power at the transmitter.

Besides transmit power, the energy consumption at a transmitter also includes circuit energy consumption that is incurred by signal processing and active circuit blocks, such as analog-to-digital converter, digital-to-analog converter, synthesizer, and mixer (Arnold et al. 2010; Cui et al. 2004; Isheden and Fettweis 2010). Circuit energy consumption can be divided into a static (fixed) part and a dynamic part that depends on parameters of active links (Arnold et al. 2010; Isheden and Fettweis 2010). Here the transmission-associated circuit consumption can be in general modeled as a linear function of throughput (Isheden and Fettweis 2010):

$$P_c = P_s + \xi R, \tag{10.6}$$

where P_s is the static circuit power and ξ is a constant denoting dynamic power consumption per unit data rate. Note that the power loss by the power amplifier with efficiency $\frac{1}{\zeta}$ to generate the transmit power, P, is not counted in the circuit power, P_c, but will be considered together with transmit power, P, later.

For a downlink OFDMA network, EE and SE are

$$\eta_{EE} = \frac{R}{\zeta P + P_c} \tag{10.7}$$

and

$$\eta_{SE} = \frac{R}{B}, \tag{10.8}$$

respectively, where R, P, and P_c are determined by Eqs (10.3), (10.4), (10.5), and (10.6), respectively.

To obtain a high EE as well as a desirable SE, and to guarantee QoS for each user with the limited bandwidth and transmit power resources, it is reasonable to maximize EE under a satisfactory minimum overall throughput requirement, $\breve{R}$, and a series of (minimum) rate requirements, $\breve{R}_k$'s, which depend on the types of the traffic of the corresponding users. Since the capability of providing differentiated services is an important feature for future wireless networks, a heterogeneous traffic model including both real-time and non-real-time traffic is considered in our work. For users with real-time services (Wang et al. 2007), such as video conferencing and online gaming, fixed rates $\breve{R}_i$s are required. For users with non-real-time services (Wang et al. 2007), such as file transfers and online video, only minimum rate requirements $\breve{R}_j$s are demanded. Accordingly, the optimization problem can be formulated as:

$$\max_{\rho\in\varrho,\mathbf{P}\in\mathcal{P}} \eta_{EE} \left(= \frac{\sum_{k=1}^{K}\sum_{n=1}^{N}\rho_{k,n} r_{k,n}}{\sum_{k=1}^{K}\sum_{n=1}^{N}\rho_{k,n}\left(\zeta p_{k,n} + \xi r_{k,n}\right) + P_s} \right) \tag{10.9}$$

subject to

$$\sum_{k=1}^{K}\sum_{n=1}^{N}\rho_{k,n} r_{k,n} \geq \breve{R}, \tag{10.10}$$

$$\sum_{n=1}^{N}\rho_{k,n} r_{k,n} = \breve{R}_k, \forall k \in \mathcal{K}_1, \tag{10.11}$$

$$\sum_{n=1}^{N}\rho_{k,n} r_{k,n} \geq \breve{R}_k, \forall k \in \mathcal{K}_2, \tag{10.12}$$

where $\mathcal{K}_1 = \{1, 2, \cdots, K_0 - 1\}$ and $\mathcal{K}_2 = \{K_0, K_0 + 1, \cdots, K\}$ represent the sets of $K_0 - 1 (K_0 \geq 1)$ real-time users and $K - K_0 + 1$ non-real-time users among all K active users, respectively. Without loss of generality, assume that $\breve{R} \geq \sum_{k=1}^{K} \breve{R}_k$; otherwise, Eq. (10.10) can be removed.

10.3.1 Fundamentals of the EE–SE Relationship

The following theorem reveals the relationship between EE and SE for downlink OFDMA (Xiong et al. 2011).

Theorem 10.1 *For any given SE,* $\eta_{SE} \geq \frac{\breve{R}}{B}$*, achieved with subcarrier allocation matrix,* $\rho \in \varrho$*, and power allocation matrix,* $\mathbf{P}$*, that satisfy all constraints but not necessarily including the* *peak transmit power one in Eqs (10.9)–(10.12), the maximum EE,* $\eta_{EE}^{*}(\eta_{SE}) = \max_{\rho \in \rho, p_{k,n} \geq 0} \eta_{EE}(\eta_{SE})$, *is strictly quasiconcave in* η_{SE} *if there is a sufficiently large number of subcarriers. Moreover, in the SE region* $\left[\frac{\breve{R}}{B}, \frac{\hat{R}}{B}\right]$, *the EE,* $\eta_{EE}^{*}(\eta_{SE})$

(i) strictly decreases with η_{SE} *and is maximized at* $\eta_{SE} = \frac{\breve{R}}{B}$ *if*

$$\left.\frac{\mathrm{d}\eta_{EE}^{*}(\eta_{SE})}{\mathrm{d}\eta_{SE}}\right|_{\eta_{SE}=\frac{\breve{R}}{B}} \leq 0,$$

(ii) strictly increases with η_{SE} *and is maximized at* $\eta_{SE} = \frac{\hat{R}}{B}$ *if*

$$\left.\frac{\mathrm{d}\eta_{EE}^{*}(\eta_{SE})}{\mathrm{d}\eta_{SE}}\right|_{\eta_{SE}=\frac{\breve{R}}{B}} > 0 \quad \text{and} \quad \left.\frac{\mathrm{d}\eta_{EE}^{*}(\eta_{SE})}{\mathrm{d}\eta_{SE}}\right|_{\eta_{SE}=\frac{\hat{R}}{B}} \geq 0,$$

(iii) first strictly increases and then strictly decreases with η_{SE} *and is maximized at* $\eta_{SE} = \frac{R_{EE,\max}}{B}$ *if*

$$\left.\frac{\mathrm{d}\eta_{EE}^{*}(\eta_{SE})}{\mathrm{d}\eta_{SE}}\right|_{\eta_{SE}=\frac{\breve{R}}{B}} > 0 \quad \text{and} \quad \left.\frac{\mathrm{d}\eta_{EE}^{*}(\eta_{SE})}{\mathrm{d}\eta_{SE}}\right|_{\eta_{SE}=\frac{\hat{R}}{B}} < 0,$$

where $\hat{R}$ *is the maximum throughput under all constraints in Eqs (10.9)–(10.12), and* $R_{EE,max}$ *is the throughput that corresponds to the maximum EE,* $\eta_{EE}^{\max}$, *under all constraints but not necessarily including the peak transmit power one in Eqs (10.9)–(10.12).*

By changing the minimum overall throughput, $\breve{R}$, and solving the problem in Eqs (10.9)–(10.12), the EE–SE curve and their tradeoff relationship can be obtained. It has been proved that EE is a quasiconcave function of SE (Xiong et al. 2011). The quasiconcavity of EE on SE indicates the existence and the uniqueness of the globally optimal EE, η_{EE}^{opt}, of the problem in Eqs (10.9)–(10.12).

Figure 10.2 illustrates three possible EE–SE curves. In Case 1, the optimal EE, η_{EE}^{opt}, is reached when R_{opt} equals $\breve{R}$; in Case 2, the optimal EE, η_{EE}^{opt}, is reached when R_{opt} equals $\hat{R}$; in Case 3, the optimal EE, η_{EE}^{opt}, is reached when R_{opt} equals $R_{EE,\max}$, where R_{opt} is the optimal throughput of the problem in Eqs (10.9)–(10.12), $\hat{R}$ is the maximum throughput under all constraints in Eq. (10.9)–(10.12), and $R_{EE,\max}$ is the throughput that corresponds to the maximum EE, $R_{EE,\max}$, under all constraints but not including the peak transmit power one in Eqs (10.9)–(10.12). Note that $R_{EE,\max}$ equals η_{EE}^{opt} in Cases 1 and 3 while it is greater than the latter in Case 2.

More importantly, as a result of this quasiconcavity, the problem in Eqs (10.9)–(10.12) can be decomposed into two layers and solved iteratively:

- Inner layer: for a given SE, η_{SE}, find the maximum EE, $\eta_{EE}^{*}(\eta_{SE})$, and its derivative, $\frac{\mathrm{d}\eta_{EE}^{*}(\eta_{SE})}{\mathrm{d}\eta_{SE}}$.
- Outer layer: find the optimal EE, η_{EE}^{opt}, by bisection search.

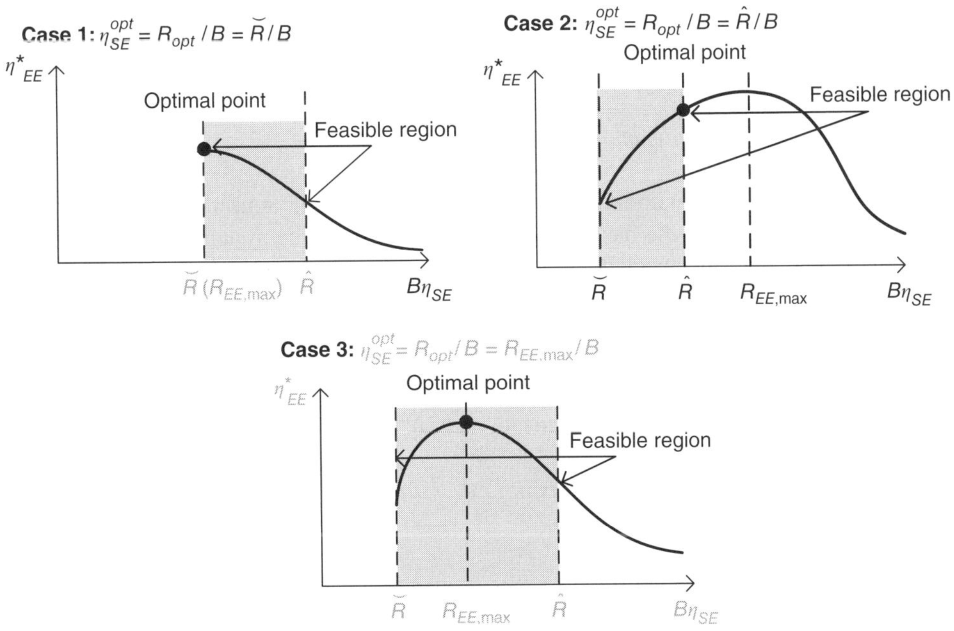

Figure 10.2 EE–SE relation in downlink OFDMA.

The bisection search in the outer-layer is clear and easy. The key lies in the inner-layer algorithm that finds $\eta^*_{EE}(\eta_{SE})$ and (the sign of) $\frac{d\eta^*_{EE}(\eta_{SE})}{d\eta_{SE}}$. For a given total throughput, the inner-layer subproblem to find $\eta^*_{EE}(\eta_{SE})$ is equivalent to minimizing the total transmit power. Approaches like convex relaxation (Boyd and Vandenberghe 2004) and Lagrange dual decomposition (Seong et al. 2006; Zhou et al. 2010) can be applied to reach the solution. Hence the problem in Eqs (10.9)–(10.12) can be successfully solved by the aforementioned joint inner- and outer-layer optimization.

10.3.2 Impacts of System Parameters on the EE–SE Tradeoff

The impact of channel power gain and circuit power on the EE–SE relationship is summarized in the following statements and proved by Xiong et al. (2011). Note that the $\eta^*_{EE}(\eta_{SE})$-versus-η_{SE} curve is exclusively referred to as the EE–SE curve throughout this chapter.

Theorem 10.2 *For any fixed SE, the EE is non-decreasing with channel power gain. Consequently, both the optimal EE, η_{EE}^{opt}, and the global maximum EE, η_{EE}^{max}, are non-decreasing with channel power gain $g_{k,n}$.*

Theorem 10.3 *For any fixed SE, the EE strictly decreases with circuit power. Consequently, both the optimal EE, η_{EE}^{opt}, and the global maximum EE, η_{EE}^{max}, strictly decrease with the circuit* *power, P_c. Besides, the optimal SE, $\eta_{SE}^{opt} \triangleq \frac{R_{opt}}{B}$, is non-decreasing with the static circuit power, P_s, while $\eta_{SE}^{max} \triangleq \frac{R_{EE,max}}{B}$ strictly increases with the static circuit power, P_s. However, both η_{SE}^{opt} and η_{SE}^{max} are independent of the dynamic circuit power factor ξ.*

Theorem 10.2 coincides with intuition. From it, the scheduling of users with better channel quality, such as users near the base station, helps to improve the system EE as well as SE. Theorem 10.3 is rather unexpected and not that straightforward for the independence of η_{SE}^{max} and η_{SE}^{opt} on the dynamic circuit power factor ξ. From it, high SE may be utilized without much loss in EE in the case that the static circuit power and/or dynamic circuit power factor are large. This is because large static circuit power indicates large η_{SE}^{max} and η_{SE}^{opt}, while a large dynamic power factor makes the EE–SE curve low and flat – EE is insensitive to the change of SE – although it does not impact the locations of η_{SE}^{max} and η_{SE}^{opt}.

Theorem 10.4 *The optimal EE, η_{EE}^{opt}, is not necessarily achieved at $\eta_{SE} = \frac{\breve{R}}{B}$ even if circuit power, P_c, is zero.*

Theorem 10.4 indicates that, in the case that transmit power (absolutely) dominates power consumption, it is not necessarily the most energy-efficient scheme to transmit with minimum power to merely realize the minimum throughput requirement and guarantee the QoS of each user. In fact, it is easy to verify that Theorem 10.1 still holds without considering the circuit power in Eqs (10.9)–(10.12). And it is well worth noting that in the single-user case ($K = 1$), η_{EE}^{opt} is always achieved at the minimum throughput, which is consistent with the conclusion of Miao et al. (2010) and can be proved similarly.

Note that all conclusions (theorems and properties) in this section, except for the independence of η_{SE}^{opt} and η_{SE}^{max} on dynamic circuit power factor ξ, also hold in the more general case that circuit power is a convex function of throughput, which can proved in a similar way to the linear ones.

Figure 10.3 demonstrates the EE–SE relation in the case that two real-time and four non-real-time users with an identical average channel-gain-to-noise ratio (CNR) of 20 dB compete for 72 data subcarriers. Different static circuit powers (15 W, 30 W) and dynamic circuit power factors (0 W/Mbps, 2 W/Mbps) are considered. No maximum transmit power constraint or specific minimum overall throughput requirement is imposed to investigate the performance limit. From Figure 10.3a, the EE–SE relation has a bell-shape curve and is also quasiconcave, since the upper bound and the lower bound almost perfectly match and they are in a bell shape. Indeed, this quasiconcavity is the foundation of the proposed JIOO method. Figure 10.3b also demonstrates the influence of the circuit power on the EE–SE relation. From there, η_{EE}^{opt} decreases with the circuit power while the η_{SE}^{opt} increases with the static circuit power but is independent of the dynamic circuit power factor. Figure 10.3b plots the η_{EE}^{opt}-versus-$\breve{\eta}_{SE}$ curves corresponding to the curves in Figure 10.3a. It indicates that for different minimum throughput requirements, $B\breve{\eta}_{SE}$, which are smaller than or equal to a certain threshold, optimal energy-efficient designs lead to the same throughput. However, if the minimum throughput requirement is larger than the threshold, the most energy-efficient transmission strategy is to operate exactly at the required minimum throughput.

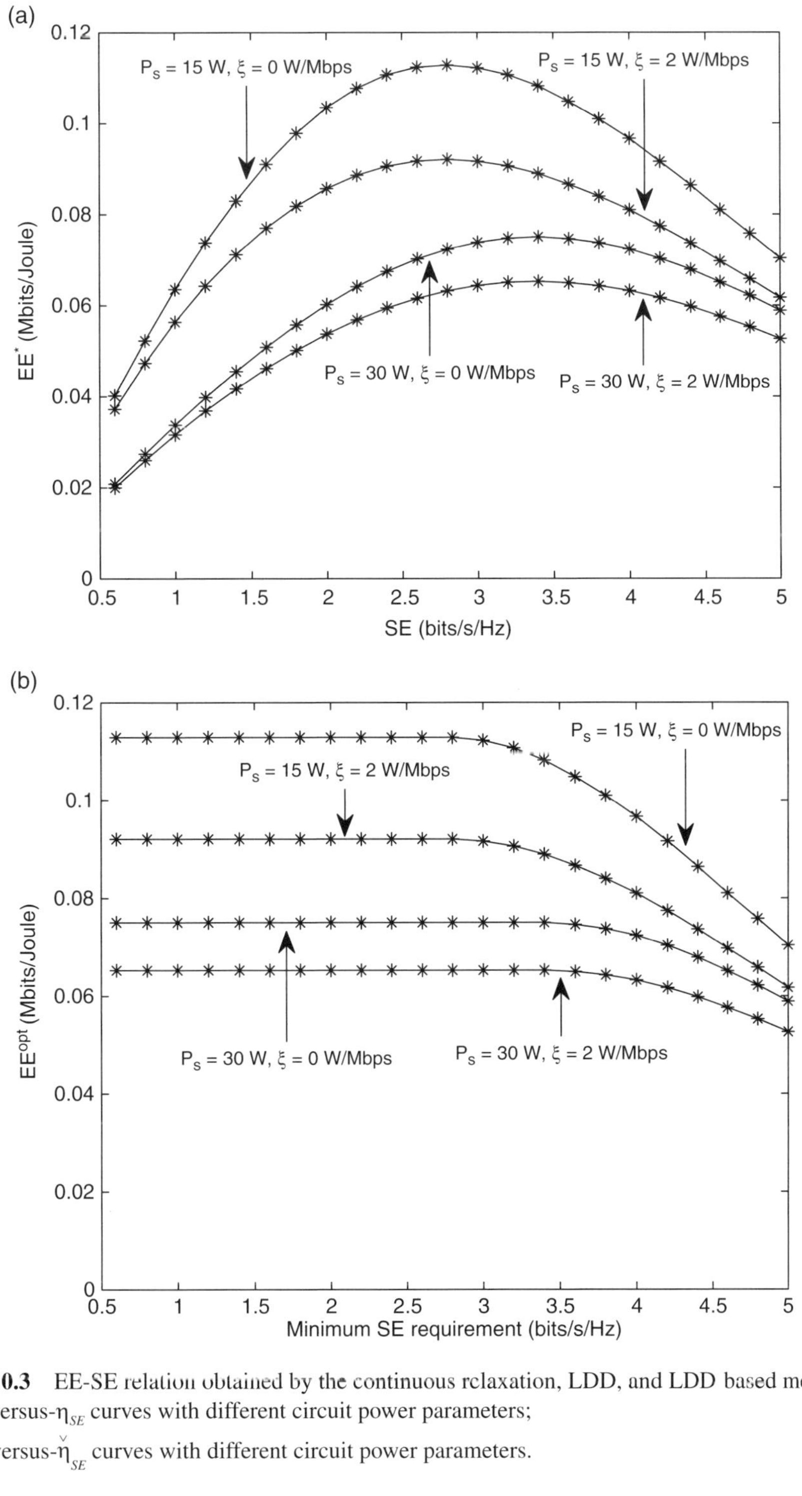

Figure 10.3 EE-SE relation obtained by the continuous relaxation, LDD, and LDD based methods; (a) η_{EE}^{*}-versus-η_{SE} curves with different circuit power parameters;
(b) η_{EE}^{opt}-versus-$\check{\eta}_{SE}$ curves with different circuit power parameters.

10.4 Energy Efficiency, Power, and Delay Tradeoff in OFDMA

In the previous section, we discussed the EE and SE tradeoff. In this section, the emphasis is put on the delay requirements instead of the rate requirements, to ensure the QoS of users. We will address optimal energy-efficient resource allocation in a downlink network, with the consideration of statistical delay guarantees for delay-sensitive traffic and also include its spectrum-efficient design counterpart for comparison. We model the problems as maximizing effective capacity-based throughput and EE under statistical delay constraints. The intrinsic quasiconcavity of EE on transmit power is proved and exploited for the solution, which is further demonstrated to be relatively tight when the number of subcarriers is larger than that of users. We also study the fundamental tradeoff relation between EE and delay, the connection between spectrum-efficient and energy-efficient designs, and the impact of system parameters including circuit power and delay exponents.

Consider a single-cell downlink OFDMA system consisting of K active users with heterogeneous services. Assume that the channel power gain of each user on each subcarrier is subject to a stationary, ergodic, and uncorrelated discrete-time process and perfect CSI is available at the receiver, while only the statistics of the channel power gain are known at the transmitter for the sake of reducing the CSI feedback overhead of a group of subcarriers.

Then, the maximum instantaneous service/transmission rate of the kth user on the nth subcarrier in one frame with duration T is:

$$r_{k,n} = TW \log_2\left(1 + \frac{g_{k,n} p_{k,n}}{N_0 W}\right)(\text{bits / frame}), \tag{10.13}$$

Although the exact value of $g_{k,n}$ is not known at the transmitter (only the statistics are known), rateless codes, such as Luby transform (LT) and Raptor codes, enable the transmitter to adapt its rate to the channel realization and achieve $r_{k,n}$ without requiring CSI at the transmitter (Gursory et al. 2009). For systems without such codes, service rates are smaller than $r_{k,n}$, and the results in this paper can serve as upper bounds on SE and EE, respectively. Then, the aggregate service rate for the kth user is $R_k = \sum_{n=1}^{N} \rho_{k,n} r_{k,n}$. Obviously, a feasible subcarrier assignment indictor matrix, $\rho = [\rho_{k,n}]_{K\times N}$, should satisfy the same constraint as in Eq. (10.4).

Denote $\theta_k > 0$ as the statistical QoS exponent of the kth user. The QoS exponent, θ_k, characterizes the steady-state delay violation probability of the kth user by

$$\Pr\left(D_k \geq D_k^{\max}\right) \approx e^{-\theta_k c D_k^{\max}}, \tag{10.14}$$

where D_k represents the delay and is a random variable, $D_k^{\max}$ is the delay bound, and c is a constant determined by the arrival and service processes (Qiao et al. 2011; Tang and Zhang 2008). Clearly, a larger QoS exponent implies a more stringent QoS guarantee and vice versa. Then, the effective capacity is the maximum constant arrival rate that the service can support under the statistical QoS requirement, specified by θ_k, of the kth user and can be expressed as

$$\begin{aligned}
\mathrm{E}_{\mathrm{C}}^{k}(\theta_k) &\triangleq -\frac{1}{\theta_k}\ln\mathbb{E}\left(e^{-\theta_k R_k}\right) \\
&= -\frac{1}{\theta_k}\ln\mathbb{E}\left(e^{-\theta_k\sum_{n=1}^{N}\rho_{k,n}r_{k,n}}\right), \\
&= -\frac{1}{\theta_k}\sum_{n=1}^{N}\ln\mathbb{E}\left(e^{-\theta_k\rho_{k,n}r_{k,n}}\right)
\end{aligned} \tag{10.15}$$

where $\mathbb{E}(\cdot)$ is the expectation operator. Let $\gamma_{k,n} \triangleq \frac{g_{k,n}}{N_0 B}$ be the CNR of the kth user on the nth subcarrier and $f_{k,n}(\gamma)$ be the probability density distribution of $\gamma_{k,n}$. Since the distribution of $g_{k,n}$ is known at the transmitter, $f_{k,n}(\gamma)$ is also known. Assume that $f_{k,n}(\gamma)$ is continuously differentiable in γ, which is true for almost all practical situations. Therefore, the expectation in Eq. (10.15) with respect to the random variables $\gamma_{k,n}$ can be easily evaluated with the following expression:

$$\mathrm{E}_{\mathrm{C}}^{k}(\theta_k) = -\frac{1}{\theta_k}\sum_{n=1}^{N}\ln\left[\int_0^{\infty} e^{-\theta_k\rho_{k,n}r_{k,n}} f_{k,n}\left(\gamma_{k,n}\right)\mathrm{d}\gamma_{k,n}\right]. \tag{10.16}$$

In practical networks, the total transmit power of a base station is both nonnegative and bounded. Thus, any possible power allocation matrix, $\boldsymbol{P} = [p_{k,n}]_{K\times N}$, should be subject to the same constraint as in Eq. (10.5). Then, the overall transmit power for the kth user and the total transmit power are

$$P_k = \sum_{n=1}^{N} p_{k,n} \text{ and } P = \sum_{k=1}^{K} P_k,$$

respectively.

The overall power consumption at the base station is given by (Cui et al. 2005):

$$P_{tot} = \zeta P + P_c. \tag{10.17}$$

Here we consider a simpler linear circuit power consumption model, where P_c is assumed to be a constant, and ignore the data-rate-related circuit power consumption in Eq. (10.6).

For the downlink OFDMA network, the overall effective capacity for a given QoS exponent vector $\theta \triangleq [\theta_k]_{K\times 1}$ is $\mathrm{E}_{\mathrm{C}}^{(\theta)}(\rho, \boldsymbol{P}) \triangleq \sum_{k=1}^{K}\mathrm{E}_{\mathrm{C}}^{k}(\theta_k)$. Then the EE for the downlink OFDMA network is defined as the ratio of the overall effective capacity to the total consumed energy as follows:

$$\begin{aligned}
\eta_{EE}^{(\theta)}(\rho, \boldsymbol{P}) &\triangleq \frac{\mathrm{E}_{\mathrm{C}}^{(\theta)}(\rho, \boldsymbol{P})}{TP_{tot}} \ (\text{bits / Joule}) \\
&= \frac{\sum_{k=1}^{K} -\frac{1}{\theta_k}\sum_{n=1}^{N}\ln\mathbb{E}\left(e^{-\theta_k\rho_{k,n}r_{k,n}}\right)}{T(\zeta P + P_c)}.
\end{aligned} \tag{10.18}$$

Compared to the conventional definition of EE as the ratio of the overall throughput/transmission rate to the total consumed energy (Chen et al. 2011; Miao et al. 2009), this EE definition considers the QoS requirements specified by $\boldsymbol{\theta}$ and serves as a QoS-guaranteed EE metric. Then, the energy-efficient design can be formulated as maximizing the EE under the statistical QoS guarantees as follows:

$$\hat{\eta}_{EE}^{(\theta)} \triangleq \max_{\rho\in\varrho, \boldsymbol{P}\in\mathcal{P}} \eta_{EE}^{(\theta)}(\rho, \boldsymbol{P}), \tag{10.19}$$

subject to

$$-\frac{1}{\theta_k}\sum_{n=1}^{N} \ln \mathbb{E}\left(e^{-\theta_k \rho_{k,n} r_{k,n}}\right) \geq \mathrm{E}_{\mathrm{C}}^{k,\min}, \forall k. \tag{10.20}$$

The problem is a combinatorial integer programming problem and is, in general, NP-hard. To make it more tractable, continuous relaxation of $\rho_{k,n}$ and concave relaxation of $\mathrm{E}_{\mathrm{C}}^{(\theta)}(\boldsymbol{\rho},\boldsymbol{P})$ to be jointly concave in $\boldsymbol{\rho}$ and $\boldsymbol{P}$ are helpful. Eqs (10.4) and (10.13), and $\mathrm{E}_{\mathrm{C}}^{(\theta)}(\boldsymbol{\rho},\boldsymbol{P})$ can be modified into (Rhee and Cioffi 2000; Yu and Coffi 2002):

$$\tilde{\rho} \in \tilde{\varrho} \triangleq \left\{ \left[\tilde{\rho}_{k,n}\right]_{K\times N} \middle| \tilde{\rho}_{k,n} \in [0,1], \sum_{k=1}^{K} \tilde{\rho}_{k,n} \leq 1 \right\}, \tag{10.21}$$

$$\tilde{r}_{k,n} = TW \log_2\left(1 + \frac{\gamma_{k,n} P_{k,n}}{\tilde{\rho}_{k,n} N_0 W}\right), \tag{10.22}$$

and

$$\tilde{\mathrm{E}}_{\mathrm{C}}^{(\theta)}(\tilde{\rho}, \boldsymbol{P}) \triangleq \sum_{k=1}^{K} -\frac{1}{\theta_k}\sum_{n=1}^{N} \ln \mathbb{E}\left(e^{-\theta_k \tilde{\rho}_{k,n} \tilde{r}_{k,n}}\right), \tag{10.23}$$

respectively. After these relaxations, the optimization problem turns into

$$\max_{\tilde{\rho}\in\tilde{\varrho}, \boldsymbol{P}\in\mathcal{P}} \frac{\tilde{\mathrm{E}}_{\mathrm{C}}^{(\theta)}(\tilde{\rho}, \boldsymbol{P})}{T(\zeta P + P_c)}, \tag{10.24}$$

subject to

$$-\frac{1}{\theta_k}\sum_{n=1}^{N} \ln \mathbb{E}\left(e^{-\theta_k \tilde{\rho}_{k,n} \tilde{r}_{k,n}}\right) \geq \mathrm{E}_{\mathrm{C}}^{k,\min}, \forall k. \tag{10.25}$$

It is obvious that the problem in Eqs (10.24) and (10.25) yields an upper bound on the maximum EE of the original problem in Eqs (10.19) and (10.20).

10.4.1 Relationship between EE and Transmit Power

As preparation for solving the problem in Eqs (10.24) and (10.25), define $\hat{\tilde{\mathrm{E}}}_{\mathrm{C}}^{(\theta)}(P)$ as follows:

$$\hat{\tilde{\mathrm{E}}}_{\mathrm{C}}^{(\theta)}(P) \triangleq \max_{\tilde{\rho}\in\tilde{\varrho},p_{k,n}\geq 0} \tilde{\mathrm{E}}_{\mathrm{C}}^{(\theta)}(\tilde{\rho},\boldsymbol{P}), \tag{10.26}$$

under the constraints of Eq. (10.25) and $\sum_{k=1}^{K}\sum_{n=1}^{N} p_{k,n} \leq P$. Then, the following lemma can be proved (Xiong et al. 2013).

Lemma 10.1 *The maximum modified effective capacity,* $\hat{\tilde{\mathrm{E}}}_{\mathrm{C}}^{(\theta)}(P)$*, for a given total transmit power, P, is strictly concave in P.*

Based on Lemma 10.1, the following theorem can be easily derived (Xiong et al. 2013). It reveals the intrinsic property of delay-guaranteed energy-efficient transmission and is the foundation of the proposed optimal solution framework for the problem in Eqs (10.19)–(10.20).

Theorem 10.5 *The upper bound on the maximum EE,* $\hat{\eta}_{EE}^{(\theta)}(P) \triangleq \hat{\mathrm{E}}_{\mathrm{C}}^{(\theta)}(P)/(T(\zeta P+P_c))$*, for a given total transmit power, P, namely,*

$$\hat{\tilde{\eta}}_{EE}^{(\theta)}(P) \triangleq \frac{\hat{\tilde{\mathrm{E}}}_{\mathrm{C}}^{(\theta)}(P)}{T(\zeta P+P_c)}, \tag{10.27}$$

has the following properties

(i) $\hat{\tilde{\eta}}_{EE}^{(\theta)}(P)$ *is continuously differentiable and quasiconcave in P.*

(ii) $\hat{\tilde{\eta}}_{EE}^{(\theta)}(P)$ *either decreases or first increases and then decreases with P.*

$$(iii)\quad \frac{\mathrm{d}\hat{\tilde{\eta}}_{EE}^{(\theta)}(P)}{\mathrm{d}P}\begin{cases} >0 & \text{if } \hat{\tilde{\eta}}_{EE}^{(\theta)}(P) < \dfrac{1}{\zeta T}\dfrac{\mathrm{d}\hat{\tilde{\mathrm{E}}}_{\mathrm{C}}^{(\theta)}(P)}{\mathrm{d}P} \\ =0 & \text{if } \hat{\tilde{\eta}}_{EE}^{(\theta)}(P) = \dfrac{1}{\zeta T}\dfrac{\mathrm{d}\hat{\tilde{\mathrm{E}}}_{\mathrm{C}}^{(\theta)}(P)}{\mathrm{d}P}, \\ <0 & \text{if } \hat{\tilde{\eta}}_{EE}^{(\theta)}(P) > \dfrac{1}{\zeta T}\dfrac{\mathrm{d}\hat{\tilde{\mathrm{E}}}_{\mathrm{C}}^{(\theta)}(P)}{\mathrm{d}P} \end{cases}$$

and the derivative satisfies Eq. (10.28),

$$\frac{\mathrm{d}\hat{\tilde{\mathrm{E}}}_{\mathrm{C}}^{(\theta)}(P)}{\mathrm{d}P} = \max_{k,n}\left\{ \tilde{\rho}_{k,n}^{*} TW \log_2 e \frac{\displaystyle\int_0^\infty e^{-\theta_k \tilde{\rho}_{k,n}^{*} TW \log_2\left(1+\frac{p_{k,n}^{*}\gamma_k}{\tilde{\rho}_{k,n}^{*}}\right)} \frac{\gamma_{k,n}}{\tilde{\rho}_{k,n}^{*}+p_{k,n}^{*}\gamma_{k,n}} f_{k,n}(\gamma_{k,n})\mathrm{d}\gamma_{k,n}}{\displaystyle\int_0^\infty e^{-\theta_k \tilde{\rho}_{k,n}^{*} TW \log_2\left(1+\frac{p_{k,n}^{*}\gamma_k}{\tilde{\rho}_{k,n}^{*}}\right)} f_{k,n}(\gamma_{k,n})\mathrm{d}\gamma_{k,n}} \right\}. \tag{10.28}$$

where $\tilde{p}^*_{k,n}$ and $p^*_{k,n}$ are the optimal power and subcarrier allocation of Eq. (10.27), respectively.

Theorem 10.5 demonstrates the quasiconcavity of $\hat{\tilde{\eta}}^{(\theta)}_{EE}(P)$, which is an upper bound on the EE $\hat{\eta}^{(\theta)}_{EE}(P)$, in the transmit power, P. It implies the existence and the uniqueness of the global maximum; that is, $\hat{\tilde{\eta}}^{(\theta)}_{EE}$, for Eqs (10.24) and (10.25). More importantly, as a result of the quasiconcavity, the problem in Eqs (10.24) and (10.25) can be decomposed into two layers and solved iteratively by the JIOO algorithm, in a similar way to the one discussed in Section 10.3, as follows:

(i) Inner layer: for a given transmit power, $P \le P_{\max}$, find the maximum EE, $\hat{\tilde{\eta}}^{(\theta)}_{EE}(P)$, and (the sign of) its derivative, $\frac{\mathrm{d}\hat{\tilde{\eta}}^{(\theta)}_{EE}(P)}{\mathrm{d}P}$.
(ii) Outer layer: search for the transmit power, P_{opt}, that results in the maximum, $\hat{\tilde{\eta}}^{(\theta)}_{EE}$, by derivative-aided bisection power search.

10.4.2 EE and Delay Tradeoff

The EE-delay tradeoff can be stated in the following theorem (Xiong et al. 2013).

Theorem 10.6 *The maximum EE, $\hat{\eta}^{(\theta)}_{EE}$, in Eq. (10.20) is a strictly and monotonically decreasing function of all delay exponents, θ_k.*

From Theorem 10.6, EE and delay performance cannot be improved simultaneously. Moreover, the EE can be maximized when no delay guarantee is required.

Figures 10.4a and b compare the EE and the overall effective capacity of the energy-efficient and the spectrum-efficient designs with different CNRs. It can also be concluded that the energy-efficient design greatly outperforms the spectrum-efficient design in EE and the large QoS exponents reduce the EE and the overall effective capacity.

10.5 Energy-efficient Resource Allocation for Downlink OFDMA

This section will focus on the optimal and low-complexity suboptimal resource allocation algorithms for high EE in downlink OFDMA networks. We model the problem as the maximization of generalized EE under QoS requirements. We first give the optimal solution then develop a low-complexity suboptimal solution by exploring the inherent structure and property of the energy-efficient design.

We still consider a single-cell downlink OFDMA network with K active users. Define the generalized EE for the downlink transmission as the weighted total delivered bits per unit energy:

$$\eta^{DL}_{EE} \triangleq \frac{\sum_{k\in\mathcal{K}} \omega_k R_k}{\zeta P + P_c}, \tag{10.29}$$

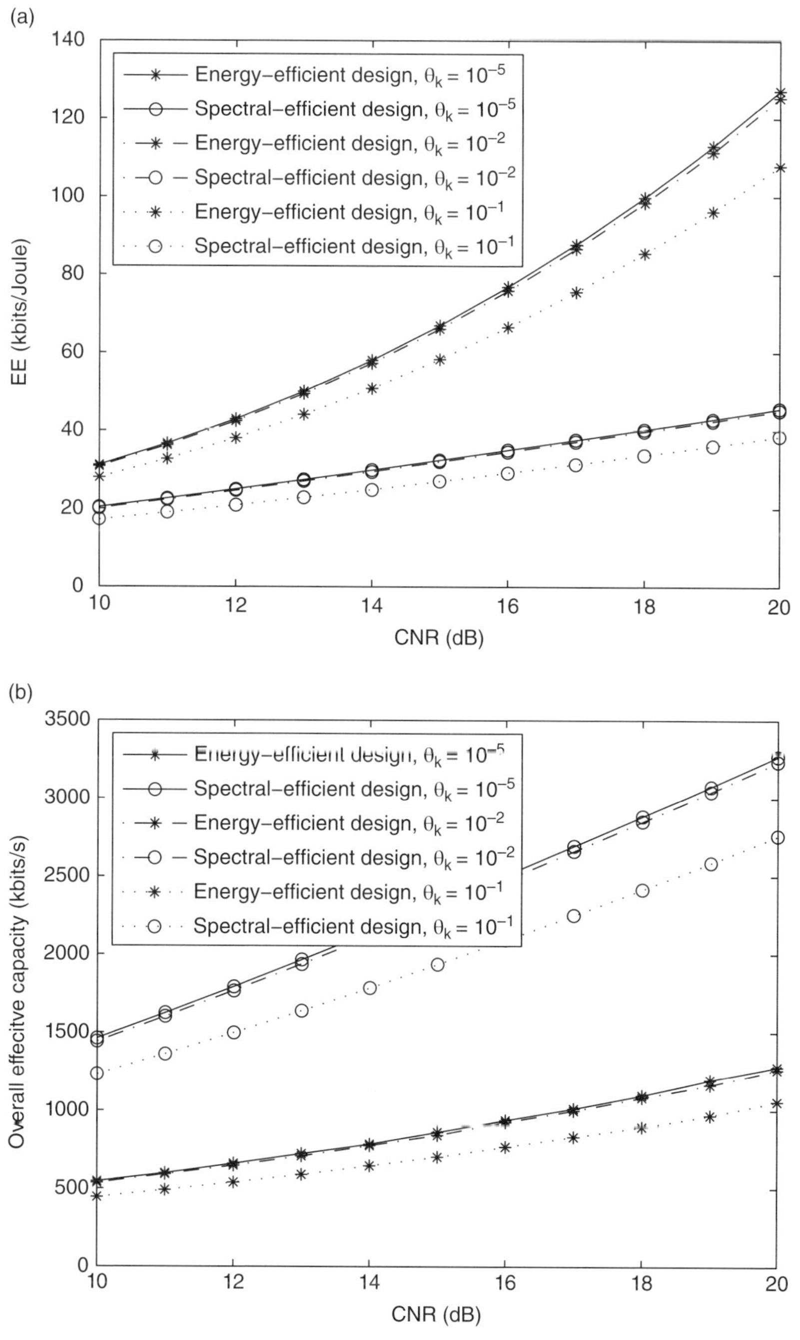

Figure 10.4 EE and overall effective capacity versus the average CNR: (a) EE versus the the average CNR; (b) overall effective capacity versus the average CNR.

where the predetermined weights, ω_k, can potentially provide a certain level of priority and/or fairness among the users and the generalized EE still reflects the notion of the conventional EE that is defined as $\frac{\sum_{k\in\mathcal{K}} R_k}{\zeta P + P_c}$ where $\omega_1 = \omega_2 = \cdots = \omega_K = 1$ (Isheden and Fettweis 2010; Miao et al. 2010, 2009; Prabhu and Daneshrad 2010). To provide different service priorities and guarantee QoS for each user, the generalized EE under a series of traffic-related minimum rate requirements, $\breve{R}_k$, and the peak transmit power, $P_{\max}$, is considered. The generalized EE optimization problem for the downlink transmission can be mathematically formulated as

$$\hat{\eta}_{EE}^{DL} \triangleq \max_{\rho\in\varrho, \boldsymbol{P}\in\mathcal{P}} \frac{\sum\limits_{k\in\mathcal{K}} \omega_k \sum\limits_{n\in\mathcal{N}} \rho_{k,n} r_{k,n}}{\sum\limits_{k\in\mathcal{K}} \sum\limits_{k\in\mathcal{K}} \zeta p_{k,n} + P}, \tag{10.30}$$

subject to

$$\sum_{n\in\mathcal{N}} \rho_{k,n} r_{k,n} \geq \breve{R}_k, \forall\, k \in \mathcal{K}, \tag{10.31}$$

where $\hat{\eta}_{EE}^{DL}$ represents the optimal downlink EE.

10.5.1 Optimal Energy-efficient Resource Allocation

Xiong et al. (2012) proved that a tight upper bound on EE is quasiconcave in the total transmit power. $\hat{\eta}_{EE}^{UB}(P)$ is an upper bound on the EE, defined as $\hat{\eta}_{EE}^{DL}(P) \triangleq \max_{\rho\subset\varrho} \hat{\eta}_{EE}^{(\rho)}(P)$ under the constraints of Eq. (10.31), in the total transmit power, P. Its quasiconcavity implies the existence and the uniqueness of the global maximum, $\hat{\eta}_{EE}^{UB}$. Similar to the previous sections, as a result of the quasiconcavity, the problem in Eqs (10.30) and (10.31) can be decomposed into two layers and solved iteratively by the JIOO optimization, in a similar way to the one discussed in Section 10.3, as follows:

(i) Inner layer: for a given transmit power, $P \leq P_{\max}$, find the maximum EE, $\hat{\eta}_{EE}^{UB}(P)$, and (the sign of) its derivative, $\frac{\mathrm{d}\hat{\eta}_{EE}^{UB}(P)}{\mathrm{d}P}$.
(ii) Outer layer: search for the transmit power that results in the maximum, $\hat{\eta}_{EE}^{UB}$, by bisection power search like the bisection power search in the BPA algorithm.

10.5.2 Low-complexity Suboptimal Energy-efficient Resource Allocation

Although the convex programming is numerically stable, its computational complexity depends on the number of optimizing variables, which can be large if the number of subcarriers and/or the number of users are/is large. Each inner-layer optimization is a weighted sum-rate (WSR) maximization problem. It can be solved using the classic water-filling procedure (Goldsmith 2005), which is equivalent to the Lagrange multiplier method. Each inner-layer WSR maximization needs water-filling at least $\mathcal{O}(NK(1/\delta^2))$ times for δ-optimality (Mokari et al. 2010). The complexity of water-filling procedure is fixed; let its complexity be unit here for simplicity. The overall complexity also depends on the number of iterations in the outer-layer, N_{OL}, and is $\mathcal{O}(N_{OL}NK(1/\delta^2))$. To reduce the complexity, the underlying property of the problem in Eqs (10.30) and (10.31) is first exploited and then, based on this property, a novel low-complexity suboptimal algorithm is proposed to solve problems like the one in Eqs (10.30) and (10.31). The following statement is given without proof due to space restrictions. Interested readers can refer to Xiong et al. (2012) for more details.

The optimal EE, $\hat{\eta}_{EE}^{DL}$, in Eq. (10.30) is always equal to

$$\hat{\eta}_{EE}^{DL} \equiv \max_{\rho\in\varrho,\boldsymbol{P}\in\mathcal{P},\boldsymbol{\alpha}\in\boldsymbol{\alpha}} \left\{ \min_{k\in\mathcal{K}} \frac{\omega_k \sum_{n\in\mathcal{N}} \rho_{k,n} r_{k,n}}{\sum_{n\in\mathcal{N}} \zeta p_{k,n} + \alpha_k P_c} \right\}, \tag{10.32}$$

subject to

$$\sum_{n\in\mathcal{N}} \rho_{k,n} r_{k,n} \geq \breve{R}_k, \forall k \in \mathcal{K}, \tag{10.33}$$

where $\boldsymbol{\alpha} \triangleq \left\{ [\alpha_k]_{K\times 1} \,\middle|\, \sum_{k\in\mathcal{K}} \alpha_k = 1; \alpha_k \in \mathbb{R} \right\}$.

The property in Eq. (10.32), which is proved in Xiong et al. (2012), illustrates the structure of the optimal solution in a split form. From it, the following statement is readily obtained.

For any fixed $\alpha \in \boldsymbol{\alpha}$, the optimal EE, $\hat{\eta}_{EE}^{DL}$, in Eq. (10.30), is lower bounded by

$$\hat{\eta}_{EE}^{DL} \geq \max_{\rho\in\varrho,\boldsymbol{P}\in\mathcal{P}} \left\{ \min_{k\in\mathcal{K}} \frac{\omega_k \sum_{n\in\mathcal{N}} \rho_{k,n} r_{k,n}}{\sum_{n\in\mathcal{N}} \zeta p_{k,n} + \alpha_k P_c} \right\}, \tag{10.34}$$

subject to

$$\sum_{n\in\mathcal{N}} \rho_{k,n} r_{k,n} \geq \breve{R}_k, \forall k \in \mathcal{K}. \tag{10.35}$$

Assume $\alpha^{\text{opt}} = [\alpha_k^{\text{opt}}]_{K\times 1}$ corresponds to the optimal EE, $\hat{\eta}_{EE}^{DL}$, in Eqs (10.30) and (10.32). Then, α_k^{opt} can be intuitively regarded as a portion of the static circuit power incurred individually by the kth user when the maximum EE is achieved. Based on Eqs (10.32) and (10.34), instead of directly optimizing the EE, the energy-efficient design can maximize the minimum individual EE; that is, the objective function $\min_{k\in\mathcal{K}} \frac{\omega_k \sum_{n\in\mathcal{N}} \rho_{k,n} r_{k,n}}{\sum_{n\in\mathcal{N}} \zeta p_{k,n} + \alpha_k P_c}$, under a certain properly chosen $\alpha \in \boldsymbol{\alpha}$, and get a satisfactory EE.

This idea enables one to split joint and complex optimization objectives in Eq. (10.30) into a series of relatively isolated and simple objectives in Eq. (10.32). It makes subcarrier assignment easier, using heuristic algorithms, because the maximum of its individual EE, $\eta_{EE,k}^{DL} \triangleq \frac{\omega_k \sum_{n\in\mathcal{N}} \rho_{k,n} r_{k,n}}{\sum_{n\in\mathcal{N}} \zeta p_{k,n} + \alpha_k P_c}$ for the kth user, depends only on its own parameters and the subcarriers it will occupy, but not on the power-adaptation strategies of other users. Here, a greedy subcarrier assignment approach – the maximizing-EE-lower-bound-based downlink subcarrier assignment (MDSA) algorithm – is proposed. This is outlined in Xiong et al. (2012; Table II). The key idea of the MDSA algorithm is to iteratively assign the subcarriers, aiming at maximizing the minimum individual EE, $\min_{k\in\mathcal{K}} \eta_{EE,k}^{DL}$, under the QoS requirement. With the promising subcarrier assignment obtained by the MDSA algorithm, further improvement can be achieved by using the bisection power search (BPA) algorithm to find the corresponding EE-optimal power-adaptation strategy. The complexity of the MDSA algorithm for a given α is roughly $\mathcal{O}(N_{OL}N)$ times that of water-filling.

Next, an effective way to determine the initial α for the MDSA algorithm is suggested. Let $\overline{g}_k \triangleq \mathbb{E}_n(g_{k,n})$ be the average CNR of the kth user and N_k be the number of subcarriers assigned to the kth user. Assume that each user undergoes flat fading with a CNR of $\overline{g}_k$ and solve the following energy-efficient resource-allocation problem:

$$\overline{\eta}_{EE}^{DL} \triangleq \max_{N_k\in\mathbb{N}, P_k>0} \frac{\sum_{k\in\mathcal{K}} \omega_k N_k W \log_2\left(1+\frac{\overline{g}_k P_k}{N_k}\right)}{\sum_{k\in\mathcal{K}} \zeta P_k + P_c}, \tag{10.36}$$

subject to

$$N_k W \log_2\left(1+\frac{\overline{g}_k P_k}{N_k}\right) \geq \breve{R}_k, \forall k\in\mathcal{K}, \tag{10.37}$$

$$\sum_{k\in\mathcal{K}} N_k = N, \tag{10.38}$$

$$\sum_{k\in\mathcal{K}} P_k \leq P_{\max}. \tag{10.39}$$

By relaxing the N_ks from positive integers to positive real numbers, a similar quasiconcave relation can be proved and thus the problem in Eqs (10.36)–(10.39) can be precisely solved by the JIOO framework, as in Eq. (10.30). The complexity of obtaining $\boldsymbol{\alpha}^{ini}$ is not more than $\mathcal{O}(N_{OL}K(1/\delta^2))$ (Boyd and Vandenberghe 2004; Mokari et al. 2010).

In Table 10.1, the complexity of this near-optimal, low-complexity approach is given for comparison with the suboptimal alternative.

Figure 10.5 evaluates the EE of the energy-efficient design, which optimizes the generalized EE, and the spectrum-efficient design, which maximizes the WSR with the same constraints expect for the objectives in the downlink transmission. The energy-efficient design

Table 10.1 Complexity comparison for downlink transmission.

Algorithm	Complexity
Near-optimal: JIOO based on convex programming	$\mathcal{O}(\frac{1}{\delta^2}N_{OL}NK)$
Suboptimal: MDSA and BPA	$\mathcal{O}(\frac{1}{\delta^2}N_{OL}K + N_{OL}N + N_{OL})$

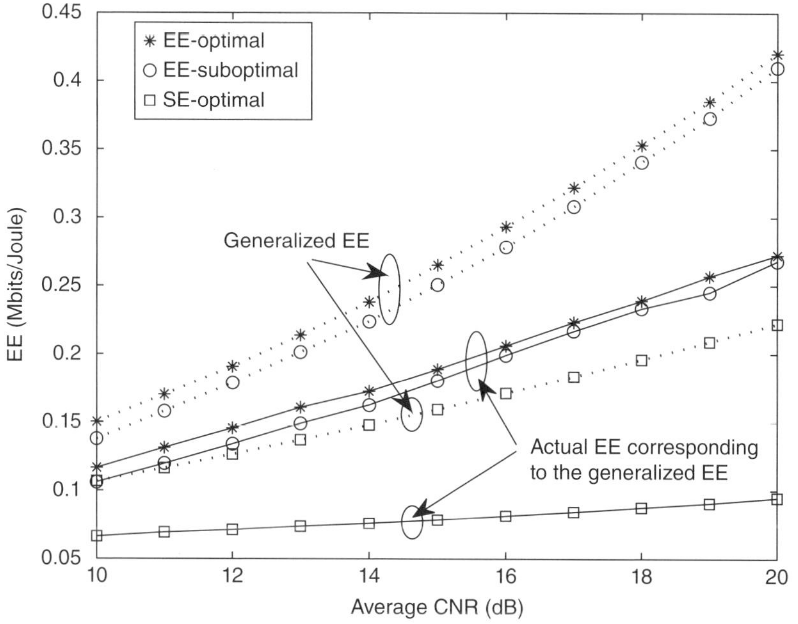

Figure 10.5 Comparison of the EE for the downlink transmission schemes.

significantly improves EE compared to the spectrum-efficient design. And the suboptimal energy efficient scheme based on the MDSA and the BPA algorithms results in an EE that is at least 90% of the optimal EE. The resultant actual/conventional EE corresponding to the generalized EE of each scheme is also plotted. From the figure, the actual EE of the EE-suboptimal scheme is also close to that of the EE-optimal one.

10.6 Energy-efficient Resource Allocation for Uplink OFDMA

This section will focus on the optimal and low-complexity suboptimal resource allocation algorithms for high EE in uplink OFDMA networks. We model the problem as the maximization of the minimum individual EE under QoS requirements. Similar to downlink OFDMA, we also first give the optimal solution then develop a low-complexity suboptimal solution by exploring the inherent structure and properties of the energy-efficient design.

Consider a single-cell uplink OFDMA network with K active users. For uplink transmission, the overall power consumption at the kth uplink user is modeled as

$$P_k^{tot} = \zeta_k P_k + P_k^c, \qquad (10.40)$$

where ζ_k and P_k^c are the reciprocal of drain efficiency of power amplifier and circuit power, respectively.

Any possible power allocation matrix, $\boldsymbol{P} = [p_{k,n}]_{K\times N}$, should be subject to:

$$\boldsymbol{P} \in \mathcal{P} \triangleq \left\{ \left[p_{k,n} \right]_{K\times N} \middle| p_{k,n} \geq 0, \sum_{n\in\mathcal{N}} p_{k,n} \leq P_k^{\max} \right\}, \qquad (10.41)$$

where $P_k^{\max}$ represents the maximum total transmit power at the kth user for uplink transmission.

For the uplink scenario, the minimum individual EE is maximized, which guarantees satisfying EE as much as possible, even for the worst user. The individual EE of the kth user for the uplink transmission is conventionally defined as

$$\eta_{EE,k}^{UL} \triangleq \frac{R_k}{\zeta_k P_k + P_k^c}, \tag{10.42}$$

as in Miao et al. (2008). Mathematically, the EE optimization problem for the uplink transmission can be expressed as

$$\hat{\eta}_{EE}^{UL} \triangleq \max_{\rho \in \varrho, \boldsymbol{P} \in \mathcal{P}} \left\{ \min_{k \in \mathcal{K}} \frac{\sum_{n \in \mathcal{N}} \rho_{k,n} r_{k,n}}{\sum_{n \in \mathcal{N}} \zeta_k p_{k,n} + P_k^c} \right\}, \tag{10.43}$$

subject to

$$\sum_{n \in \mathcal{N}} \rho_{k,n} r_{k,n} \geq \breve{R}_k, \forall k \in \mathcal{K}. \tag{10.44}$$

10.6.1 Optimal Energy-efficient Resource Allocation

Like the downlink case, the uplink problem in Eqs (10.43) and (10.44) is, in general, a complicated integer programming problem. However, once the subcarrier assignment is fixed, the maximization of EE for each user can be regarded as a single-user case of Eq. (10.30) and can be readily solved by the BPA algorithm. By exhaustively searching all feasible subcarrier assignments, the optimal solution to the problem in Eqs (10.43) and (10.44) can be found by choosing the one with the maximum $\min_{k \in \mathcal{K}} \frac{\sum_{n \in \mathcal{N}} \rho_{k,n} r_{k,n}}{\sum_{n \in \mathcal{N}} \zeta_k \rho_{k,n} p_{k,n} + P_{c,k}}$. The total complexity is about $\mathcal{O}(KN_{OL}K^N) = \mathcal{O}(N_{OL}K^{N+1})$ times that of water-filling. Nevertheless, this kind of brute-force method is usually too computationally expensive to be practical.

10.6.2 Low-complexity Suboptimal Energy-efficient Resource Allocation

The formulation of the problem in Eqs (10.43) and (10.44) appears very similar to that of the problem in Eqs (10.32)–(10.34) except that the circuit power is now naturally consumed by each user individually ($\boldsymbol{\alpha}$ is not introduced) and the transmit power limit for each user is now separately imposed. Therefore, joint power allocation across different users is not needed any more. Hence, the problem in Eqs (10.43) and (10.44) can be similarly solved by the MDSA algorithm with some modification. The modified subcarrier assignment approach is known as the maximizing-minimum-EE-based uplink subcarrier assignment (MUSA) algorithm (Xiong et al. 2012). The basic idea of the MUSA algorithm is to iteratively assign the subcarriers, aiming at maximizing the minimum individual EE, $\min_{k \in \mathcal{K}} \eta_{EE,k}^{UL}$, under both

Table 10.2 Complexity comparison for uplink transmission.

Algorithm	Complexity
Optimal: brute-force search based on BPA	$\mathcal{O}(N_{OL}K^{N+1})$
Suboptimal: MUSA	$\mathcal{O}(N_{OL}N)$

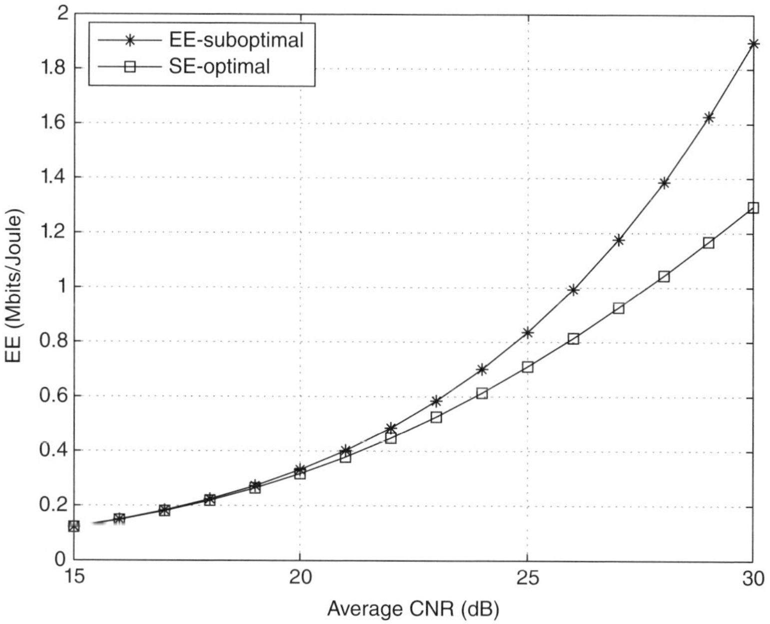

Figure 10.6 Comparison of the EE for the uplink transmission schemes.

the QoS requirement and the transmit-power constraint. In Table 10.2, the complexity of the optimal and low-complexity suboptimal solutions are listed for comparison.

Figure 10.6 illustrates the EE of the spectrum-efficient design in Rhee and Cioffi (2000) and the energy-efficient design based on the suboptimal MUSA algorithm, respectively. From it, the performance differences in EE between the energy-efficient design and the spectrum-efficient design increase with the CNR.

10.7 Concluding Remarks

In this section, the fundamental interrelationship between EE and SE in downlink OFDMA networks as well as the impacts of channel gain and circuit power on the EE–SE relationship are analyzed. It is found that EE is quasiconcave in SE and decreases with SE when SE is large enough. For delay-sensitive traffic, it is found that EE is quasiconcave in transmit power. The EE-delay tradeoff and impacts of system parameters, such as circuit power and delay exponents, are also analyzed. For energy-efficient resource allocation in both downlink and uplink single-cell OFDMA networks, it is found that the most energy-efficient transmission usually does not necessarily transmit at the maximum transmit power while spectrum-efficient transmission usually transmits at the maximum transmit power.

For interested readers, more relevant references are worth studying to get a more complete view of energy-efficient wireless network. For example, in Bae and Stark (2009a), the EE–SE tradeoff with consideration of circuit power has been studied for energy-constrained wireless multihop networks with a single source–destination pair. It reveals that the common power (on each hop) strategy yields better EE–SE tradeoffs than the common rate (on each hop) one. Furthermore, at higher rates, fewer hops are good for EE, while at lower rates more hops are preferred. For two-hop AWGN relay channels, the EE–SE tradeoff for different relaying strategies with different forwarding methods have been investigated and compared (Bae and Stark 2009b). It has been demonstrated that noncooperative relaying outperforms cooperative in EE at lower rate regions while the latter is better at higher rate regions.

Héiot et al. (2012) characterized EE for the single-user multi-input multi-output (MIMO) Rayleigh fading channel by a generic closed-form approximation, which exhibits a greater accuracy for a wider range of SE values and antenna configurations. Another unified metric for EE and SE tradeoff design in point to point wireless networks was proposed by Deng et al. (2013). Both EE and SE are simultaneously optimized through a multi-object optimization problem, where the Pareto-optimal set is further characterized. He et al. (2012) revealed the tradeoff between EE and SE in downlink multiuser distributed antenna systems (DAS). In a later paper (He et al. 2013), the relationship between EE and SE in DAS with proportional fairness among users was systematically investigated using a multi-criteria optimization method.

Joung et al. (2014) analyzed the impact of the power amplifier on the SE–EE tradeoff of orthogonal frequency division multiplex (OFDM) systems. They showed that a practical PA has an SE–EE tradeoff that has a turning point, and decreases sharply after its maximum EE point.

References

Arnold O, Richter F, Fettweis GP and Blume O 2010 Power consumption modeling of different base station types in heterogeneous cellular networks *Proc. of 19th Future Network & MobileSummit 2010*, Florence, Italy.

Bae C and Stark WE 2009a End-to-end energy-bandwidth tradeoff in multihop wireless networks. *IEEE Trans. Inf. Theory* **55**(9), 4051–4066.

Bae C and Stark WE 2009b On the energy-bandwidth tradeoff for AWGN relay channels *Proc. IEEE Military Commun. Conf. (MILCOM'09)*, Boston, MA, US.

Boyd S and Vandenberghe L 2004 *Convex Optimization*. Cambridge University Press.

Chandar V, Tchamkerten A and Tse D 2010 Asynchronous capacity per unit cost *Proc. IEEE Int. Symp. Inf. Theory (ISIT'10)*, Austin, USA, pp. 280–284.

Chen Y, Zhang S, Xu S and Li GY 2011 Fundamental tradeoffs on green wireless networks. *IEEE Commun. Mag.* **49**(6), 30–37.

Cui S, Goldsmith A and Bahai A 2004 Energy-efficiency of MIMO and cooperative MIMO techniques in sensor networks. *IEEE J. Sel. Areas Commun.* **22**(6), 1089–1098.

Cui S, Goldsmith A and Bahai A 2005 Energy-constrained modulation optimization. *IEEE Trans. Wireless Commun.* **4**(5), 2349–2360.

Deng L, Rui Y, Cheng P, Zhang J, Zhang Q and Li M 2013 A unified energy efficiency and spectral efficiency tradeoff metric in wireless networks. *IEEE Commun. Lett.* **17**(1), 55–58.

Goldsmith, A. 2005 *Wireless Communications*. Cambridge University Press 2005

Gursory MC, Qiao D and Velipasalar S 2009 Analysis of energy efficiency in fading channels under QoS constraints. *IEEE Trans. Wireless Commun.* **8**(8), 4252–4263.

Gursoy MC 2009 On the capacity and energy efficiency of training-based transmissions over fading channels. *IEEE Trans. Inf. Theory* **55**(10), 4543–4567.

Héiot F, Imran MA and Tafazolli R 2012 On the energy efficiency-spectral efficiency trade-off over the MIMO rayleigh fading channel. *IEEE Trans. Commun.* **60**(5), 1345–1356.

He C, Sheng B, Zhu P and You X 2012 Energy efficiency and spectral efficiency tradeoff in downlink distributed antenna systems. *IEEE Wireless Commun. Lett.* **1**(3), 153–156.

He C, Sheng B, Zhu P, You X and Li GY 2013 Energy- and spectral-efficiency tradeoff for distributed antenna systems with proportional fairness. *IEEE J. Sel. Areas Commun.* **31**(5), 894–902.

Isheden C and Fettweis GP 2010 Energy-efficient multi-carrier link adaptation with sum rate-dependent circuit power *Proc. IEEE Global Telecommun. Conf. (Globecom'10)*, Miami, FL, US.

Joung J, Ho CK and Sun S 2014 Spectral efficiency and energy efficiency of OFDM systems: Impact of power amplifiers and countermeasures. *IEEE J. Sel. Areas Commun.* **32**(2), 208–220.

Miao G, Himayat N and Li GY 2010 Energy-efficient link adaptation in frequency-selective channels. *IEEE Trans. Commun.* **58**(2), 545–554.

Miao G, Himayat N, Li GY and Bormann D 2008 Energy efficient design in wireless OFDMA *Proc. IEEE Int. Commun. Conf. (ICC'08)*, Beijing, China.

Miao G, Himayat N, Li GY and Swami A 2009 Cross-layer optimization for energy-efficient wireless communications: a survey. *Wiley J. Wireless Commun. Mobile Comput.* **9**(4), 529–542.

Mokari N, Javan MR and Navaie K 2010 Cross-layer resource allocation in OFDMA systems for heterogeneous traffic with imperfect CSI. *IEEE Trans. Veh. Technol.* **59**(2), 1011–1017.

Polyanskiy Y, Poor HV and Verdú S 2010 Mimimum energy to send *k* bits with and without feedback *Proc. IEEE Int. Symp. Inf. Theory (ISIT'10)*, Austin, USA, pp. 221–225.

Prabhu RS and Daneshrad B 2010 An energy-efficient water-filling alglorithm for OFDM systems *Proc. IEEE Int. Conf. Commun. (ICC'10)*, Cape Town, South Africa.

Qiao D, Gursoy MC and Veipasalar S 2011 Energy efficiency in the low-SNR regime under queueing constraints and channel uncertainty. *IEEE Trans. Commun.* **59**(7), 2006–2017.

Rhee W and Cioffi JM 2000 Increase in capacity of multiuser OFDM system using dynamic subcarrier allocation *Proc. IEEE Vehicular Technology Conf. (VTC 2000)*, Tokyo, Japan, pp. 1085–1089.

Seong K, Mohseni M and Cioffi JM 2006 Optimal resource allocation for OFDMA downlink systems *Proc. IEEE Int. Symp. Inf. Theory (ISIT'06)*, Seattle, USA.

Tang J and Zhang X 2008 Cross-layer-model based adaptive resource allocation for statistical QoS guarantees in mobile wireless networks. *IEEE Trans. Wireless Commun.* **7**(6), 2318–2328.

Verdú S 2002 Spectral efficiency in the wideband regime. *IEEE Trans. Inf. Theory* **48**(6), 1319–1343.

Wang X, Giannakis GB and Marques AG 2007 A unified approach to QoS-guaranteed scheduling for channel-adaptive wireless networks. *Proc. IEEE* **95**(12), 2410–2431.

Xiong C, Li GY, Liu YL, Chen Y and Xu SG 2013 Energy-efficient design for downlink OFDMA with delay-sensitive traffic. *IEEE Trans. Wireless Commun.* **12**(6), 3085–3095.

Xiong C, Li GY, Zhang SQ, Chen Y and Xu SG 2012 Energy-efficient resource allocation in OFDMA networks. *IEEE Trans. Commun.* **60**(12), 3767–3778.

Xiong C, Li GY, Zhang S, Chen Y and Xu S 2011 Energy- and spectral-efficiency tradeoff in downlink OFDMA networks. *IEEE Trans. Wireless Commun.* **10**(1), 3874–3886.

Yu W and Coffi JM 2002 FMDA capacity of Gaussian multiple-access channel with ISI. *IEEE Trans. Commun.* **50**(1), 102–111.

Zhou X, Li GY, Li D, Wang D and Soong ACK 2010 Probabilistic resource allocation for opportunistic spectrum access. *IEEE Trans. Wireless Commun.* **9**(9), 2870–2879.

11

Advanced Multiple-access and MIMO Techniques

NOMA sections: Anass Benjebbour,[1] Anxin Li,[2] Kazuaki Takeda,[1] Yoshihisa Kishiyama,[1] and Takehiro Nakamura[1]
SV-MIMO sections: Yuki Inoue,[1] Yoshihisa Kishiyama,[1] and Takehiro Nakamura[1]
[1] *NTT DoCoMo Inc., Tokyo, Japan*
[2] *DoCoMo Beijing Communications Laboratories Co., Ltd, Beijing, China*

11.1 Introduction

In cellular mobile communications systems, the design of the radio access technology (RAT) is one important aspect of improving system performance – spectrum utilization efficiency, system capacity, user fairness – in a cost-effective manner. Radio access technologies are typically characterized by the multiple-access scheme used: frequency division multiple access (FDMA), time division multiple access, code division multiple access, orthogonal frequency division multiple access (OFDMA), and spatial division multiple access (SDMA). These schemes provide the means for multiple users to access and share the system resources simultaneously. In the 3.9th and 4th generation (4G) mobile communications systems such as Long-Term Evolution (LTE) and LTE-Advanced [1–3], standardized by the 3rd Generation Partnership Project (3GPP), orthogonal multiple access (OMA) based on OFDMA or single carrier (SC)-FDMA was adopted. On the other hand, SDMA and multiple-input multiple-output (MIMO) are also enabled as key technologies in order to improve the spectrum efficiency of LTE/LTE-Advanced. For example, new transmission modes for single-user MIMO (SU-MIMO) and multi-user MIMO (MU-MIMO) were specified. Nevertheless, for 5G, including future LTE enhancements, more advanced designs are required in order to further control/mitigate intra-cell and/or inter-cell interference and thus boost the system

Towards 5G: Applications, Requirements and Candidate Technologies, First Edition.
Edited by Rath Vannithamby and Shilpa Talwar.

Companion website: www.wiley.com/go/vannithamby/towards5g

performance. In particular, further enhancements of cellular systems using existing macro-cell base stations over lower frequency bands are essential in order to continue to maintain basic coverage, while using newly deployed base stations over new, higher frequency bands to boost system capacity. However, one major problem that limits further cellular enhancements of macro cells is the limitations on the installation space for additional antennas at the macro base station site. Antennas for macro base stations are installed on buildings or antenna towers and are generally not designed with extra space to accommodate more antennas. Therefore, for macro-cell base stations over lower frequency bands, new RAT designs with improved system performance but no need for additional space for installing new antenna equipment are very important.

Several inter-cell interference coordination/mitigation techniques were studied and specified in LTE/LTE-Advanced, with the goal of improving system performance. Basically, there are two main approaches to handle inter-cell interference. One approach is inter-cell interference coordination and avoidance techniques at the network side. Examples include inter-cell interference coordination [3] and coordinated multi-point (CoMP) transmission/reception [4]. Another approach is interference suppression or mitigation techniques at the receiver side, such as an interference rejection combining (IRC) receiver based on the minimum mean-squared error (MMSE) criterion, for which performance requirements were established in LTE Release 11 [5]. For the MMSE-IRC receiver specified in Release 11, the statistics (covariance matrix) of inter-cell interference signals from other base stations were estimated and used for interference rejection without any network signaling of the transmission parameters of other base stations. In LTE Release 12, further enhancements for interference suppression were considered and network-assisted interference cancellation and suppression (NAICS) is specified [6]. In NAICS, network coordination among different base stations is assumed and the parameters of other interfering base stations can be signaled to the user from the serving base station. As a result, more advanced non-linear receivers, incorporating maximum-likelihood detection (MLD) [7] and successive interference cancellation (SIC) [8], can be applied for inter-cell interference cancellation.

On the other hand, MIMO technology plays an important role in improving spectrum efficiency, system capacity, and peak data rates of cellular systems. Conventional MIMO technology enables spatial multiplexing of multiple data streams over low-correlation channels with orthogonally polarized or widely separated antennas. For example, to achieve Gbps-order data rates using spatial multiplexing over LTE-Advanced, with 64QAM modulation and 100-MHz bandwidth, a minimum of four antenna branches is generally required. However, the widely used dual polarized antennas for macro cells can only provide two antenna branches because of the limitations on installation space. Thus antenna size needs to be reduced in order to accommodate more than two branches. However, with a smaller antenna size, the coverage area is reduced because the antenna gain and pattern is related to its size.

To solve such problems, three-dimensional (3D) MIMO techniques have received a great deal of attention recently with the introduction of active antenna systems [9]. 3D MIMO techniques control antennas not only in azimuth but also in elevation by using two-dimensional (2D) planar antenna array structures [10, 11]. Related to 3D MIMO, elevation beamforming and full dimension (FD)-MIMO have been proposed as topics for study in 3GPP [12, 13]. Both elevation beamforming and FD-MIMO would provide beam gains and interference reductions by exploiting the additional vertical direction of beam control for capacity enhancements and/or coverage extensions. In addition, the 3D channel model is being studied in order to define and emulate the propagation characteristics in both horizontal and vertical directions, including 3D path-loss, line of sight (LOS) and Non-line of sight (NLOS) ratio, dependency of path-loss and LOS probability on height, and short-term fading [14, 15].

In this chapter we explain two promising technologies, related to advanced interference management and 3D MIMO, which are being investigated as ways to improve system performance of macro cells over existing frequency bands by exploring new dimensions:

- non-orthogonal multiple access (NOMA), which is the intentional non-orthogonalization by exploiting power-domain user multiplexing
- smart vertical MIMO (SV-MIMO), which is the advanced control and grouping of antennas in the horizontal and vertical dimensions.

NOMA is a multiplexing scheme that utilizes an additional new domain, the power domain, which is not sufficiently utilized in previous systems as shown in Figure 11.1. Power-domain utilization is enabled by the evolution in the processing power of devices. For downlink NOMA, non-orthogonality is intentionally introduced via power-domain user multiplexing, while user de-multiplexing is obtained via the allocation of large power differences between paired users at the transmitter side and the application of successive interference cancellation (SIC) at the receiver side. The channel gain – path-loss, received signal to interference plus noise ratio (SINR) – difference among multiple users is translated into multiplexing gains by superposing in the power domain the transmit signals of multiple users with large channel-gain differences. As a result, both the users of high and low channel gains benefit. Indeed, although the multiplexed users are allocated less power because of power sharing, they both benefit from being scheduled more often or allocated more bandwidth. NOMA has been intensively investigated in recent years for both downlink and uplink [16–24].

On the other hand, as shown in Figure 11.2, SV-MIMO uses adaptive groupings of vertical antenna components according to the reception quality for mobile devices in the transmission area in order to improve performance while maintaining coverage. Therefore, it can enable a single-size antenna – two-antenna system – to achieve a system throughput equivalent to that of a four-antenna system. Using SV-MIMO, it is possible to achieve 1.2 Gbps transmission with a single-size antenna using LTE-Advanced [25].

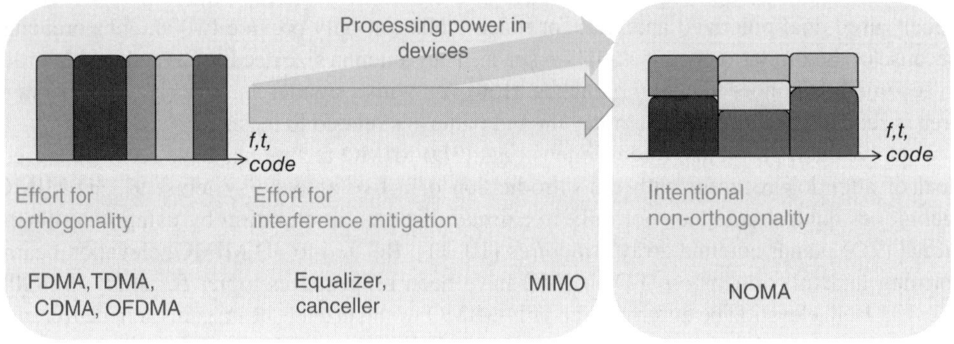

Figure 11.1 NOMA for intentional intra-cell non-orthogonality.

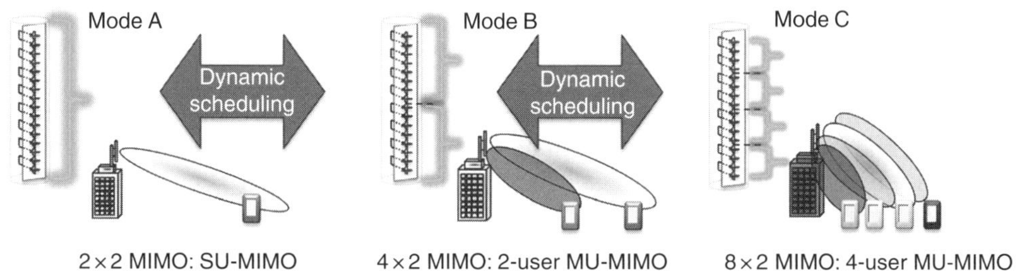

Figure 11.2 SV-MIMO with adaptive grouping of vertical antennas.

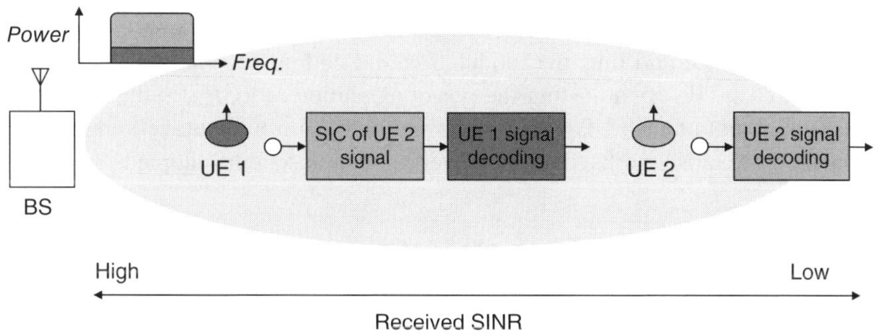

Figure 11.3 Downlink NOMA with SIC applied at UE receiver.

11.2 Non-orthogonal Multiple Access

11.2.1 Concept

Figure 11.3 illustrates downlink NOMA for the case of one base station (BS) and two user equipments (UE).

For simplicity, we assume first the case of single transmit and receive antennas. The overall system transmission bandwidth is assumed to be 1 Hz. The BS transmits a signal for UEi ($i=1, 2$), d_i, where $E[|d_i|^2]=1$, with transmit power P_i and the sum of P_i is equal to P. In NOMA, d_1 and d_2 are superposed as follows:

$$d = \sqrt{P_1}d_1 + \sqrt{P_2}d_2 \tag{11.1}$$

Thus, the received signal at UEi is represented as

$$y_i = h_i d + w_i, \tag{11.2}$$

where h_i is the complex channel coefficient between UEi and the BS. Term w_i denotes additive white Gaussian noise (AWGN) including inter-cell interference. The power spectral density of w_i is $N_{0,i}$.

In downlink NOMA, the SIC process is implemented at the UE receiver. The optimal order for decoding is the order of increasing channel gain normalized by noise and inter-cell interference power, $|h_i|^2/N_{0,i}$ (called simply *channel gain*) [9]. Based on this order, we assume that any user can correctly decode the signals of other users whose decoding order comes before the corresponding user. Thus, UEi can remove the inter-user interference from the jth user whose $|h_j|^2/N_{0,j}$ is lower than $|h_i|^2/N_{0,i}$. In a two-UE case, assuming that $|h_1|^2/N_{0,1} > |h_2|^2/N_{0,2}$, UE2 does not perform interference cancellation since it comes first in the decoding order. UE1 first decodes d_2 and subtracts its component from received signal y_1; then it decodes d_1 without interference from d_2. Assuming successful decoding and no error propagation, the throughput of UEi, R_i, is represented as

$$R_1 = \log_2\left(1 + \frac{P_1|h_1|^2}{N_{0,1}}\right), R_2 = \log_2\left(1 + \frac{P_2|h_2|^2}{P_1|h_2|^2 + N_{0,2}}\right). \tag{11.3}$$

From Eq. (11.3), it can be seen that power allocation for each UE greatly affects the user throughput performance and thus the modulation and coding scheme (MCS) used for data transmission of each UE. By adjusting the power allocation ratio, P_1/P_2, the BS can flexibly control the throughput of each UE. Clearly, the overall cell throughput, cell-edge throughput, and user fairness are closely related to the power-allocation scheme adopted.

11.2.1.1 Comparison with Orthogonal User Multiplexing

For OMA, the bandwidth of α $(0<\alpha<1)$ Hz is assigned to UE1 and the remaining bandwidth, $1-\alpha$ Hz, is assigned to UE2. The throughput of UEi, R_i, is represented as

$$R_1 = \alpha\log_2\left(1 + \frac{P_1|h_1|^2}{\alpha N_{0,1}}\right), R_2 = (1-\alpha)\log_2\left(1 + \frac{P_2|h_2|^2}{(1-\alpha)N_{0,2}}\right). \tag{11.4}$$

In NOMA, the performance gain compared to OMA increases when the difference in channel gains – the path loss between UEs – is large. For example, in Figure 11.4 we assume a two-UE case with a cell-interior UE and a cell-edge UE, where $|h_1|^2/N_{0,1}$ and $|h_2|^2/N_{0,2}$ are set to 20 and 0 dB, respectively. For OMA with equal bandwidth and equal transmission power allocated to each UE ($\alpha=0.5$, $P_1=P_2=1/2P$), the user rates are calculated according to Eq. (11.4) as $R_1=3.33$ and $R_2=0.50$ bps, respectively. On the other hand, in NOMA, when the power allocation is conducted as $P_1=1/5P$ and $P_2=4/5P$, the user rates are calculated according to Eq. (11.3) as

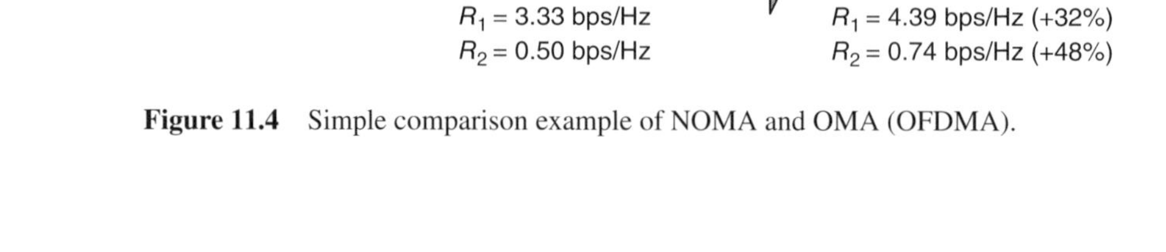

Figure 11.4 Simple comparison example of NOMA and OMA (OFDMA).

$R_1 = 4.39$ and $R_2 = 0.74$ bps, respectively. The corresponding gains of NOMA over OMA are 32% and 48% for UE1 and UE2, respectively. According to the simple two-UE example before, NOMA provides a higher sum rate than OMA. As later shown by system-level simulations, this can indeed be generalized to the case of multiple users, with a sophisticated multi-user proportional fairness scheduling being used. Note also that NOMA can be applied to more than two-UE multiplexing, but only with marginal gains compared to the two-UE case [21].

11.2.1.2 Motivations and Benefits of NOMA

We envisage NOMA as a promising candidate multiple access scheme in the future for the following motivations and benefits.

Exploitation of Channel-gain Differences Among Users

Unlike OMA (OFDMA), where channel gain difference is translated into multi-user diversity gains via frequency-domain scheduling, in NOMA the channel-gain difference is translated into multiplexing gains by superposing in the power-domain the transmit signals of multiple users of different channel gains. By exploiting the channel gain difference in NOMA, both UEs of high and low channel gains are in a win-win situation. Indeed, UEs with high channel gain (bandwidth-limited UEs) lose a little by being allocated less power, but gain much more by being allocated more bandwidth, while UEs with low channel gain (power-limited UEs) also lose only a little by being allocated a little less power and "effective" bandwidth (because of being interfered by the signal designated to the other UEs with high channel gain) but gain much more by being allocated more bandwidth. This win-win situation is also the main reason why NOMA gains over OMA increase when the difference in channel gains between NOMA paired UEs becomes larger.

Intentional Non-orthogonality via Power-domain User Multiplexing and Advanced Receiver Processing

NOMA is a multiplexing scheme that utilizes an additional new domain – the power domain – which has not been sufficiently utilized in previous systems. Non-orthogonality is intentionally introduced via power-domain user multiplexing, but interestingly, quasi-orthogonality can still be achieved. In fact, user demultiplexing is ensured via the allocation of large power differences between paired UEs and the application of SIC in the power domain. The UE with high channel gain (such as UE1 in Figures 11.3 and 11.4) is allocated less power and the UE with low channel gain (such as UE2 in Figures 11.3 and 11.4) is allocated more power. Such large power differences facilitate the successful decoding (with high probability) and thus the successful cancellation of the signal designated to UE2 (being allocated high power) at the UE1 receiver. In addition, at the UE2 receiver, the signal designated to UE2 is decoded directly by treating the interference from the signal designated to UE1 (being allocated low power) as noise.

On the other hand, NOMA captures well the evolution of device processing capabilities, generally following Moore's law, by relying on more advanced receiver processing such as SIC. In this same spirit, but for the purpose of inter-cell interference mitigation, NAICS based on non-linear receivers is specified in LTE Release 12 [6]. NOMA is a good direction to extend the work on NAICS to the intra-cell scenario in LTE Release 13 and beyond, as it should be much easier to apply a non-linear receiver to deal with intra-cell interference than inter-cell interference. Here, note that the network coordination among different BSs does not need to be considered in NOMA, unlike NAICS. In addition, a non-linear receiver structure similar to a Release 12 NAICS receiver can be reused; In NOMA, however, joint optimization to transmitter and receiver would be needed. For example, the strategy for transmission power allocation is greatly dependent on the capability of interference cancellation, and thus the receiver structure for NOMA should be investigated jointly with transmission power allocation so that the system performance can be maximized.

Robust Performance Gains in Practical Wide-area Deployments with High Mobility
NOMA user multiplexing does not rely much on the knowledge of the transmitter of the instantaneous frequency-selective fading channels, such as the frequency-selective channel quality indicator (CQI) or channel-state information (CSI), which requires fine feedback signaling from the UE side. In NOMA, CSI is used at the receiver for user de-multiplexing and at the transmitter mainly to decide on user pairing and multi-user power allocation. Thus robust performance gains in practical wide-area deployments can be expected irrespective of UE mobility or CSI feedback latency.

11.2.2 Link-level Considerations

In downlink NOMA, different data designated to multiple UEs are multiplexed in the same time and frequency resources, by allocating different transmit powers to different UEs. More specifically, a lower transmission power is allocated to a UE experiencing smaller path loss – the cell-center UE – while larger transmission power is allocated to a UE experiencing greater path loss: the cell-edge UE. In the following, we assume that m_s UEs are multiplexed using the same time and frequency resources and the number of receive antennas is N_r. Without loss of generality, the $N_r \times 1$ received signal vector $\mathbf{y}_i$ of the ith ($i = 1 \sim m_s$) UE is represented by

$$\mathbf{y}_i = \mathbf{h}_i \left(\sum_{m=1}^{m_s} \sqrt{P_m} d_m \right) + \mathbf{w}_i \tag{11.5}$$

where $\mathbf{h}_i$ is the N_r x 1 channel vector of the ith ($i = 1 \sim m_s$) UE, P_m is the mth UE's transmit power with $\sum_{m=1}^{m_s} P_m = 1$, d_m is the data symbol of the mth UE, and $\boldsymbol{w}_i$ are the $N_r \times 1$ AWGN vectors.

For the cell-edge UE, advanced receiver technologies may not necessarily be applied since the received signal power for this UE is greater than that for the cell center UE; that is, the interfering UE. On the other hand, in order to decode the received signal for the cell-center UE, an application of interference cancellation is inevitable since the signal for the cell-center UE is significantly contaminated by that for the cell-edge UE in the same time and frequency resources. In the following, the interference cancellation techniques for recovering data of the cell-center UEs are explained. There have been many studies of interference cancellation techniques such as the maximum likelihood detection (MLD) [7] and successive interference cancellation (SIC) [8]. In LTE-Advanced, advanced receivers have been also studied in the multiple-cell case in order to mitigate inter-cell interference [5, 6].

Here, two types of interference cancellation are considered: symbol-level interference cancellation (SLIC) and codeword level IC (CWIC). Figure 11.5 shows the receiver structure employing SLIC- and CWIC-based interference cancellation for NOMA. The received data

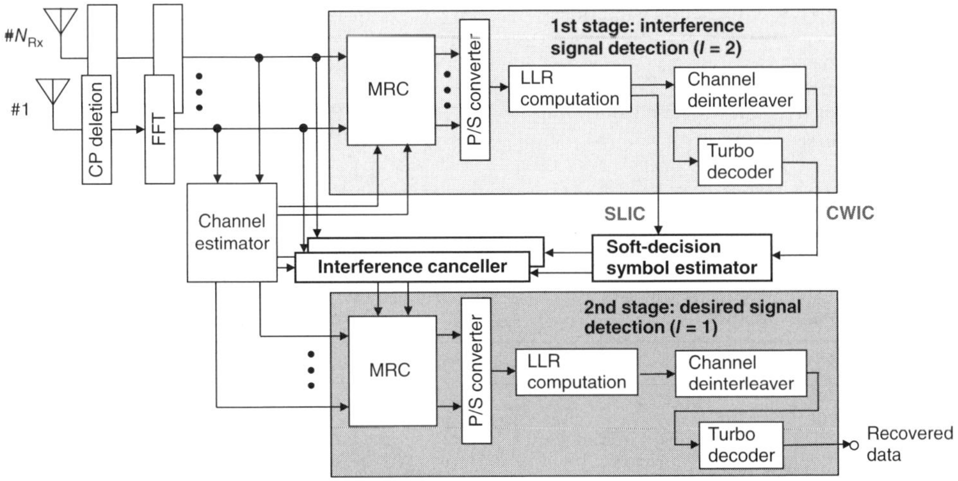

Figure 11.5 Receiver structure employing SLIC and CWIC for NOMA.

symbols for the cell-edge UE are first demodulated by multiplying the received signal with the maximal ratio combining (MRC) weight. The log-likelihood ratios (LLRs) corresponding to those demodulated symbols are calculated. For the SLIC, those LLRs are directly used to generate a symbol replica $\tilde{d}_l$ for the cell-edge UE l. On the other hand, for the CWIC, a sequence of LLRs called a codeword is input to a turbo decoder and a sequence of posteriori-LLRs is generated. After interleaving a sequence of posteriori-LLRs, the interleaved LLRs are used to calculate a soft symbol replica $\tilde{d}_l$ for the cell-edge UEs. The soft symbol replica for the cell-edge UE is subtracted from the received data symbol as follows.

$$\tilde{\mathbf{y}}_i = \mathbf{y}_i - \mathbf{h}_i \left(\sum_{l=1, l \neq i}^{m_s} \sqrt{P_l} \tilde{d}_l \right) \tag{11.6}$$

After multiplying $\tilde{y}_i$ with the MRC weight, turbo decoding is applied to recover the data for the cell-center UE.

In order to verify the effectiveness of the SLIC and CWIC for NOMA, the block error rate (BLER) performances are evaluated by link-level simulation when the number of multiplexed UEs is two. Table 11.1 shows the evaluation assumptions. The number of transmitter and receiver antennas are one and two, respectively. For the CWIC, interference cancellation is performed in each iteration of the turbo decoding. The number of iterations of the turbo decoding is set to six. For practical channel estimation for demodulation, we assume two types of reference signal (RS) based on demodulation RS from LTE-Advanced Release 10: common-RS (C-RS) and UE-specific RS (UE-RS), as shown in Figure 11.6. For C-RS, the same RS is shared among the different UEs multiplexed in the same time and frequency resources. On the other hand, for UE-RS, different transmission powers are allocated to the RSs for different UEs in the same way as power allocation for data, and the RSs for different UEs are multiplexed using different orthogonal cover codes in the same time and frequency resources. Thus, we may observe performance degradation in UE-RS due to loss in the transmission power split among UEs, compared to C-RS. However, for the C-RS, additional control signaling may

Table 11.1 Simulation conditions.

Parameter	Value
Carrier frequency/system bandwidth	2 GHz/10 MHz
Resource allocation	50 RB (= 9 MHz)
Antenna configuration	1-by-2 (uncorrelated)
Channel model	7-path Extended Pedestrian-A
Maximum Doppler frequency	5.55 Hz
Modulation and coding scheme	Desired UE: 16QAM (R=0.59) Interfering UE: QPSK (R=0.59)
Channel estimation	Coherent averaging

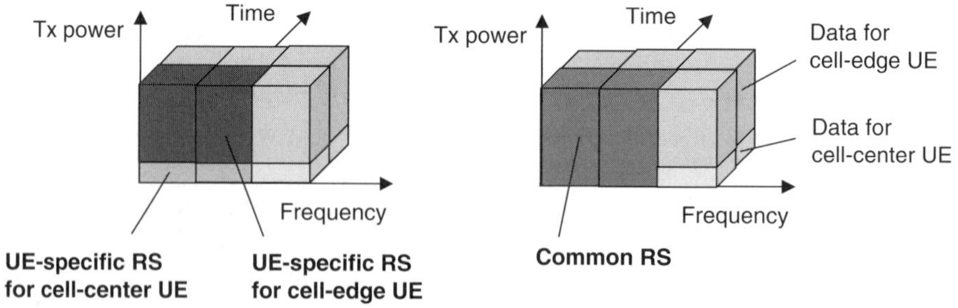

Figure 11.6 Candidate RS structures for NOMA.

need to be explicitly sent to the UE in order to inform it of the transmission power for each UE, which is needed for interference cancellation. Furthermore, UE-RS has good affinity with advanced MIMO techniques such as multi-user (MU)-MIMO since UE-specific precoding weights for MU-MIMO can be also multiplied to the UE-RS.

Figure 11.7 shows the BLER performances of the cell-center UE for ideal SIC, SLIC, and CWIC as a function of signal-to-noise power ratio (SNR) where the signal power after transmission power allocation is considered. In Figure 11.7(a), $P_1 = 1/9$ for the cell-center UE and $P_2 = 8/9$ for the cell-edge UE. We assume 64QAM modulation and coding rate of 0.51 for the cell-center UE and QPSK modulation and coding rate of 0.59 for the cell-edge UE. We find from the figure that almost the same BLER performance is obtained for SLIC and CWIC since the difference in transmission powers is large enough for both receivers to work properly. Figure 11.7(b) shows the BLER performance for a smaller difference in allocated transmission powers between two UEs: $P_1 = 3/10$ for the cell-center UE and $P_2 = 7/10$ for the cell-edge UE. We assume 16QAM modulation and a coding rate of 0.59 for the cell-center UE and QPSK modulation and a coding rate of 0.59 for the cell-edge UE. We observe that the BLER performance for the SLIC is largely degraded – by 6.2 dB – compared to ideal SIC for BLER = 0.1 since the soft replica symbol for the interfering UE is not accurate when the transmission power allocation ratio for the cell-center and cell-edge UEs is 3 : 7. However, an application of CWIC significantly improves the performance of NOMA, by 5.7 dB for BLER = 0.1 since the soft replica symbol is greatly improved by using the output of turbo decoding. We observe from both Figure 11.7(a) and (b) that the performance gap between CWIC and ideal SIC is within 0.7 dB, which means that error propagation is very limited.

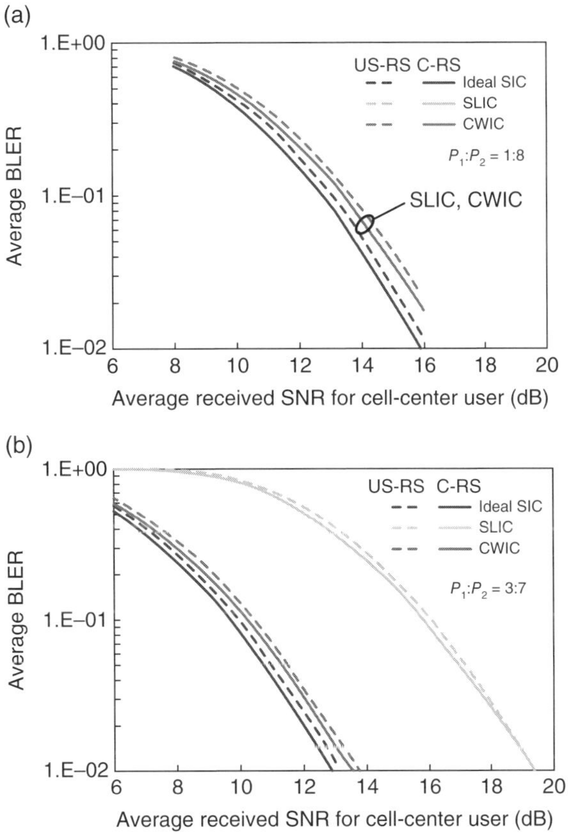

Figure 11.7 BLER performance for SLIC and CWIC: (a) $P_1 : P_2 = 1 : 8$ and (b) $P_1 : P_2 = 3 : 7$.

When comparing C-RS and UE-RS, C-RS exhibits slightly better BLER performance than UE-RS, as expected. However, a gain in the required SNR for achieving BLER = 0.5 from C-RS compared to UE-RS is about 0.2 dB. Thus, the degradation from using UE-RS is very limited under these parameter assumptions. Note, however, that UE-RS does not require additional signaling of the transmission power information and would have a good affinity when NOMA is combined with MIMO.

11.2.3 System-level Considerations

In this section, we discuss practical aspects of system-level performance of NOMA, taking into account multi-user power allocation, signaling overhead, performance in high mobility scenarios, and combination with MIMO. The evaluation results of NOMA performance are presented in a multi-cell system-level simulations [26]. The major simulation parameters assumed are compliant with existing LTE specifications [2]. We employed a 19-hexagonal macrocell model with three sectors per cell. The system bandwidth is 10 MHz, with 48 resource blocks (RBs), and the cell radius is set to 289 m, with an inter-site distance of 500 m.

The locations of the UEs are assigned randomly with a uniform distribution. In the propagation model, we take into account distance-dependent path loss with the decay factor of 3.76, lognormal shadowing with a standard deviation of 8 dB and instantaneous multipath fading. The shadowing correlation between the cells (sectors) is set to 0.5 (1.0). The 3GPP spatial channel model (SCM) "urban macro" is assumed. The baseline maximum Doppler frequency, f_D, is set to 5.55 Hz, which corresponds to 3 km/h at the carrier frequency of 2 GHz. The transmission power of the macrocells is 46 dBm. The antenna gains at the macrocell and UE are 14 dBi and 0 dBi, respectively. One-antenna transmission and two-antenna reception (1×2 SIMO) and maximal ratio combining (MRC) at the UE side are assumed as baseline antenna configuration and receiver. For the case of NOMA with MIMO, two transmit antennas are assumed. The full-buffer traffic model is used and the feedback delay is modeled such that the CQI is not available for scheduling until 4 subframes after the periodic report with a 2-ms interval. Hybrid automatic repeat request (HARQ) is applied. In NOMA, the multi-user scheduler maximizes the multi-user proportional fairness metric [27, 28] and selects the best UE set among all possible UE sets. Full search power allocation and exhaustive user pairing [22] are assumed as baseline. Also, for NOMA, dynamic switching to OMA is assumed with the maximum number of simultaneously paired UEs, m_s, set to 2 ($m_s = 2$). For SIC receiver, codeword-level SIC with no error propagation (ideal SIC) is assumed. The major simulation parameters are summarized in Table 11.2.

Table 11.2 Major simulation parameters.

Cell layout	Hexagonal grid, 19 sites, 3 cells per site, wrap around
Inter-site distance	500 m
Minimum distance between UE and cell site	35 m
Distance dependent path loss	$128.1 + 37.6\log_{10}(r)$ dB
Shadowing correlation	0.5(inter site)/1.0(intra site)
Channel model	3GPP Spatial Channel Model, urban macro
Channel estimation	Ideal
UE speed	3 km/h 20 km/h, 40 km/h, 60 km/h, 80 km/h, 100 km/h
BS total transmission power	46 dBm
Transmit antenna gain and cable loss	14 dBi
UE antenna gain	0 dBi
UE noise figure	9 dB
Thermal noise density	−174 dBm/Hz
Carrier frequency	2 GHz
System bandwidth	10 MHz (48 RBs)
Number of transmitter antenna	1 or 2
Number of receiver antennas	2
Number of UEs per cell	10, 20
Maximum number of multiplexed UEs	1 (OMA) and 2 (NOMA)
Number of subbands	1 (48 RBs, wideband scheduling), or 8 (6 RBs, subband scheduling)
Scheduling algorithm	Proportional fairness (PF)
Control delay in scheduling and AMC	4.0 ms
Channel gain reporting interval	2.0 ms
Traffic model	Full-buffer model

11.2.3.1 NOMA Signaling Overhead

Wideband Versus Subband Scheduling

We explore NOMA performance gains with subband scheduling and subband MCS and compare it to NOMA with wideband scheduling and wideband MCS selection. For the case of subband (wideband) scheduling, the system bandwidth is divided into 8 (1) subbands with 6 RBs (48 RBs) per subband. In Figure 11.8, the cell throughput and cell-edge throughput gains for NOMA over OMA are approximately 40 and 39% for wideband scheduling, and 37 and 32% for subband scheduling, respectively. Thus, similar gains can be maintained for NOMA even with larger numbers of subbands and thus larger frequency-domain scheduling gains. Also note that the performance of all cases is increased according to the number of UEs per cell (10 UEs, 20 UEs) because of the multi-user diversity gain.

In the case of NOMA with subband scheduling, signaling overhead increases linearly with the number of subbands. To reduce signaling overhead, joint encoding of modulation, coding and power set (MCPS) would be beneficial or some signaling could be wideband or long-term while others could remain subband or short-term. In LTE, for example, even when subband scheduling is applied, the same channel coding rate (including rate matching) and data modulation scheme are assumed over all the subbands allocated to each single user, as the average SINR over all the subbands is used for MCS selection. However, for NOMA, such a mismatch between MCS adaptation granularity (wideband) and power allocation granularity (subband) might not allow the full exploitation of NOMA gains with subband scheduling [21]. Thus, such considerations need also to be taken into account.

Multi-user Power Allocation

Because of the power-domain user multiplexing of NOMA, the transmit power allocation (TPA) to one user affects the achievable throughput of not only that user but also the throughput of other users. The best performance of NOMA can obviously be achieved by an exhaustive full search of user pairs and dynamic transmit-power allocations. In case of full-search power

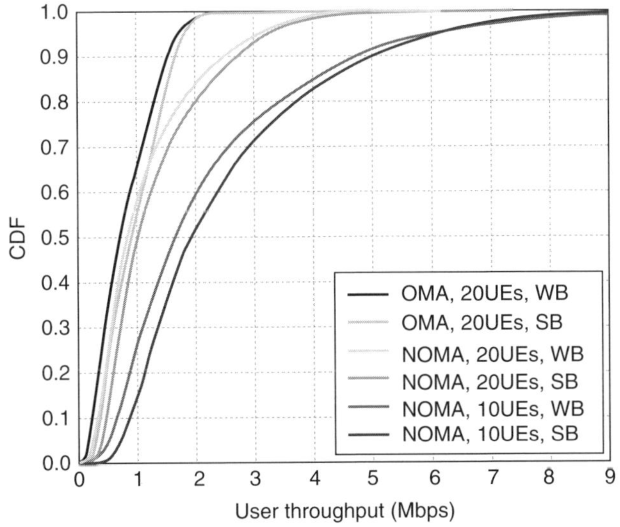

Figure 11.8 CDF of user throughput for OMA ($m_s = 1$) and NOMA ($m_s = 2$) with subband and wideband scheduling.

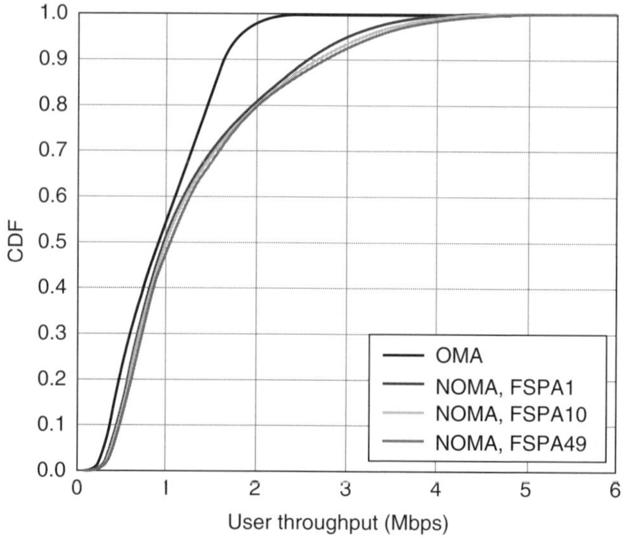

Figure 11.9 CDF of user throughput for OMA ($m=1$) and NOMA ($m=2$) with various numbers of power sets (subband scheduling, 20 UEs).

allocation (FSPA), all possible combinations of power allocations are considered for each candidate user set. FSPA remains computationally complex, however. Also, with such dynamic TPA, the signaling overhead associated with SIC decoding order and power-assignment ratios increases significantly.

In order to reduce the signaling overhead associated with the multi-user transmit-power allocation of NOMA and clarify the degree of impact of user pairing on the performance of NOMA, exhaustive power allocation with different numbers of power sets is explored.

Figure 11.9 shows the performance comparison between OMA and NOMA with different power allocation schemes, with and without user grouping. Here, there are three different numbers of power sets: FSPA with 49, 10 and 1 power set.

The performance gains in the overall cell throughput for NOMA are:

- FSPA49: 37%
- FSPA10: 35%
- FSPA1: 30%.

Thus, even with the 10 and 1 power sets a large portion of the NOMA gains can be maintained.

In NOMA, users with large channel-gain differences (large path-loss differences) are paired with high probability and so, considering practical implementations, user pairing and transmit-power allocations could be simplified by using pre-defined user groupings and fixed per-group power allocations, where users are divided into multiple user groups according to the magnitude of their channel gains using predefined thresholds [22]. Two user groups can be assumed where the threshold for user grouping is 8 dB and power allocations are fixed to a single power set ($0.2P$, $0.8P$). Taking into account the potential saving in signaling overhead, a limited number of power sets and/or pre-defined user groupings can be promising in practical usage.

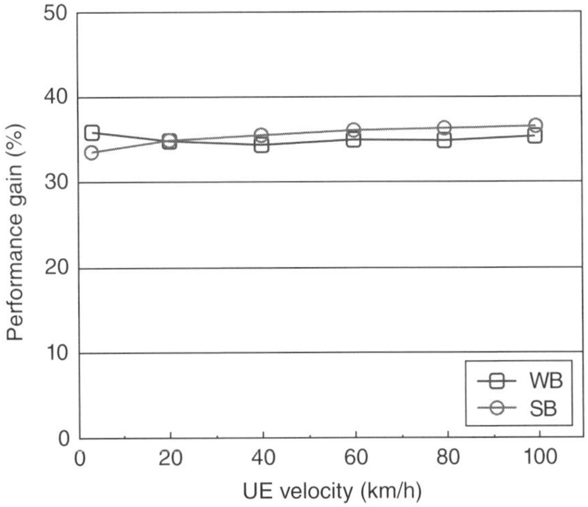

Figure 11.10 NOMA cell throughput gains with various UE speeds (wideband and subband scheduling, 20UEs).

This indeed becomes interesting as the order of successive interference cancellation and information on power allocation do not need to be transmitted in every subframe but rather over a longer timescale.

11.2.3.2 Performance in Low- and High-Mobility Scenarios

Next, we investigate NOMA gains with various UE speeds. Figure 11.10 shows the cell throughput gains and cell-edge throughput gains of NOMA over OMA for wideband and subband scheduling with different user velocities. The number of users per cell is 20 and FSPA with 49 power sets is used for NOMA power allocation. It can be seen that the cell throughput gains of NOMA over OMA are observed over a wide range of UE speeds for both wideband and subband scheduling. Specifically, NOMA is shown to maintain good gains compared to OMA with wideband scheduling in particular. Thus, NOMA can be a promising multiple-access method for providing good robustness to mobility, as it relies mainly on receiver-side CSI and signal processing.

11.2.3.3 Combination of NOMA and MIMO

MIMO is one of the key technologies for improving spectrum efficiency in LTE/LTE-Advanced. In general, MIMO techniques can be categorized into:

- SU-MIMO, where only one UE is served in data transmission
- MU-MIMO, where more than one UE is served in data transmission.

Because MIMO technology exploits the spatial domain and NOMA exploits the power domain, the two technologies can be combined to further boost the system spectrum efficiency. As illustrated in Figure 11.11, there are two major approaches to combine NOMA and MIMO technologies.

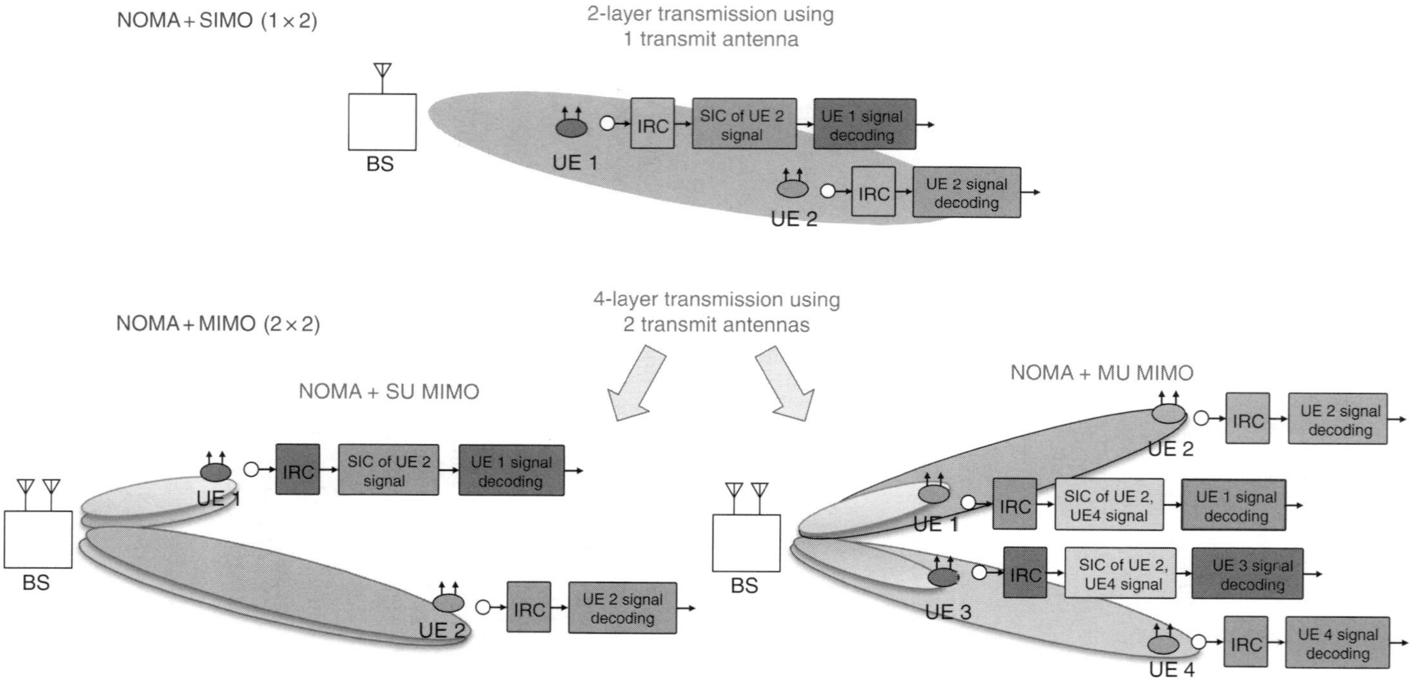

Figure 11.11 NOMA extension from SIMO to MIMO (SU-MIMO and MU-MIMO).

One approach is to use the MU-MIMO technique to create multiple spatial beams and apply the NOMA technique inside each beam. The other approach is to use the NOMA technique to create multiple power levels and apply the SU-MIMO and/or MU-MIMO technique inside each power level. For example, for NOMA with SU-MIMO (2×2), with up to two-user multiplexing in the power domain, non-orthogonal beam multiplexing enables up to four-beam multiplexing using only two transmit antennas. In addition, the combination of NOMA with SU-MIMO can involve both:

- open-loop MIMO, for example space frequency block coding (SFBC) or large delay cyclic delay diversity (CDD)
- closed-loop MIMO based on CSI, including precoder indicator, channel quality indicator, rank indicator, and feedback by users.

When combined with NOMA, open-loop MIMO schemes are expected to provide robust performance in high-mobility scenarios.

Figure 11.12 shows one form of combining downlink NOMA with MIMO using random (opportunistic) beamforming [18, 19]. In this form, the BS transmitter generates multiple beams similarly to MU-MIMO, and superposes multiple UEs within each beam [5]. At the UE-receiver side, two interference cancellation approaches – non-linear SIC and linear interference suppression by interference rejection combining (IRC) [5] – are jointly used. SIC is used for intra-beam user demultiplexing: interference cancellation is applied among the UEs belonging to a group with the same precoding weights. The multiple access scheme within each beam (group) is the same as 1×2 NOMA. IRC is used for inter-beam interference suppression: interference suppression is applied among UE groups with different precoding weights. Interference from other beams is simply suppressed by combining the signals received at the receive antennas of the UE. A key benefit of IRC is that it does not require the decoding of other UE groups in other beams.

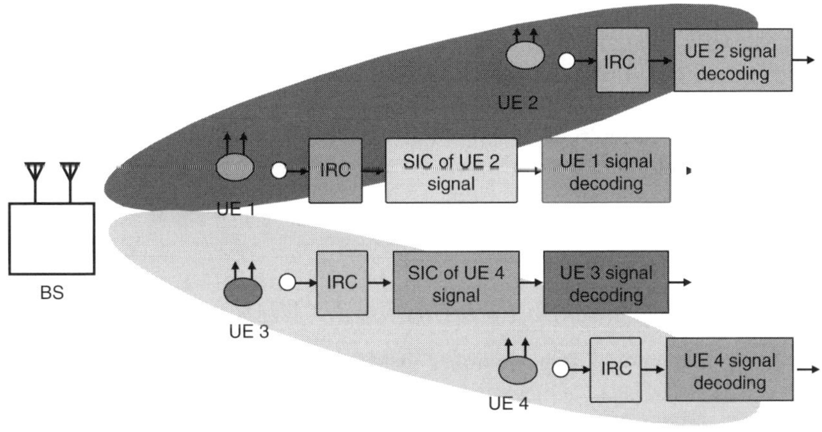

Figure 11.12 NOMA combined with 2×2 MIMO using random beamforming and applying IRC-SIC receivers.

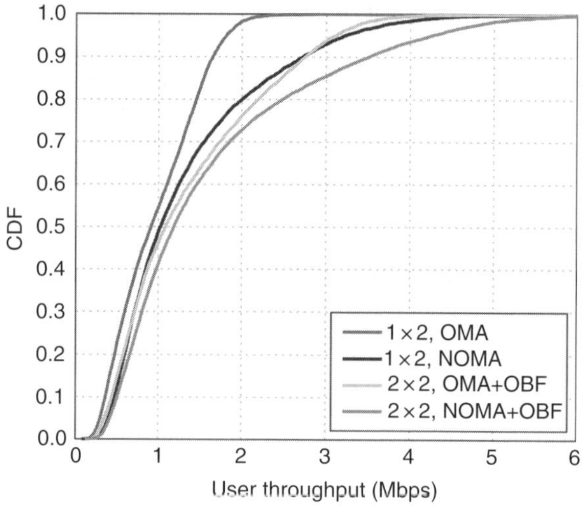

Figure 11.13 System-level evaluation of downlink NOMA combined with MU-MIMO using random beamforming (20 UEs).

Figure 11.13 shows the CDF of the user throughput for NOMA and OMA for 1×2 SIMO (as in previous evaluations), and for 2×2 MIMO using random beamforming at the BS transmitter side and IRC-SIC receiver at the UE side. For the case of OMA with 2×2 MIMO, a single-stream transmission is applied per transmit beam. Figure 11.13 shows that NOMA gains can be maintained almost at the same level irrespective of the antenna configuration; that is, 1×2 SIMO or 2×2 MIMO with random beamforming. One interesting thing observed here is that the performance of NOMA with 1×2 SIMO is very similar to that of OMA with 2×2 MIMO using opportunistic beamforming (OBF). This implies that NOMA with SIC has a similar effect to spatial multiplexing using random beamforming, and NOMA can achieve a competitive level of performance to random beamforming with a smaller number of transmit antennas at the BS. However, the proposed NOMA/MIMO scheme requires a relatively large number of UEs to obtain a sufficient throughput gain for random (opportunistic) beamforming [23].

When the number of UEs per cell is small, it may be better to apply a closed-loop precoding or SU-MIMO approach (Figure 11.11) rather than the random beamforming approach. Therefore, the support of multiple MIMO modes – closed-loop and open-loop SU-MIMO – would be necessary [24].

11.3 Smart Vertical MIMO

11.3.1 Grouping of Antenna Elements for 3D MIMO

MIMO evolution to support a larger number of MIMO antenna branches – more than eight antenna ports – is essential to achieve the target system capacity and peak data rates of 5G. Nevertheless, the widely used dual-polarized antenna in existing macro cells has typically only two antenna ports for each single set of antenna equipment. If a larger number of MIMO branches is to be used, many sets of antenna equipment need to be installed. However, it is difficult to install more antennas because the installation space is limited and wiring becomes complicated. Therefore, to address the problem of antenna installation, grouping of antenna elements becomes an essential technique for future MIMO enhancements. Smart vertical MIMO (SV-MIMO) is a new 3D MIMO transmission technique, which is based on adaptive (smart) grouping of antenna elements in the vertical direction [25]. SV-MIMO enables flexible virtualization of the antenna ports with the transmitted data being mapped to each antenna port adaptively for each frequency and time-resource block using the baseband (BB) signal controller. Hereafter, a configuration of antenna elements grouping is referred to as a "mode". Over the payload resource of the transmitted data, the modes are adapted based on the feedback-channel-quality information from the UE. In addition, these modes can be controlled independently and simultaneously per data or reference-signal resource (e.g., cell specific reference signal (CRS)) in order to maintain coverage while improving system capacity.

In Figure 11.14, we illustrate various configurations for grouping antenna elements for 3D MIMO. *M* is the number of antenna elements in the BS, *N* is the number of antenna elements

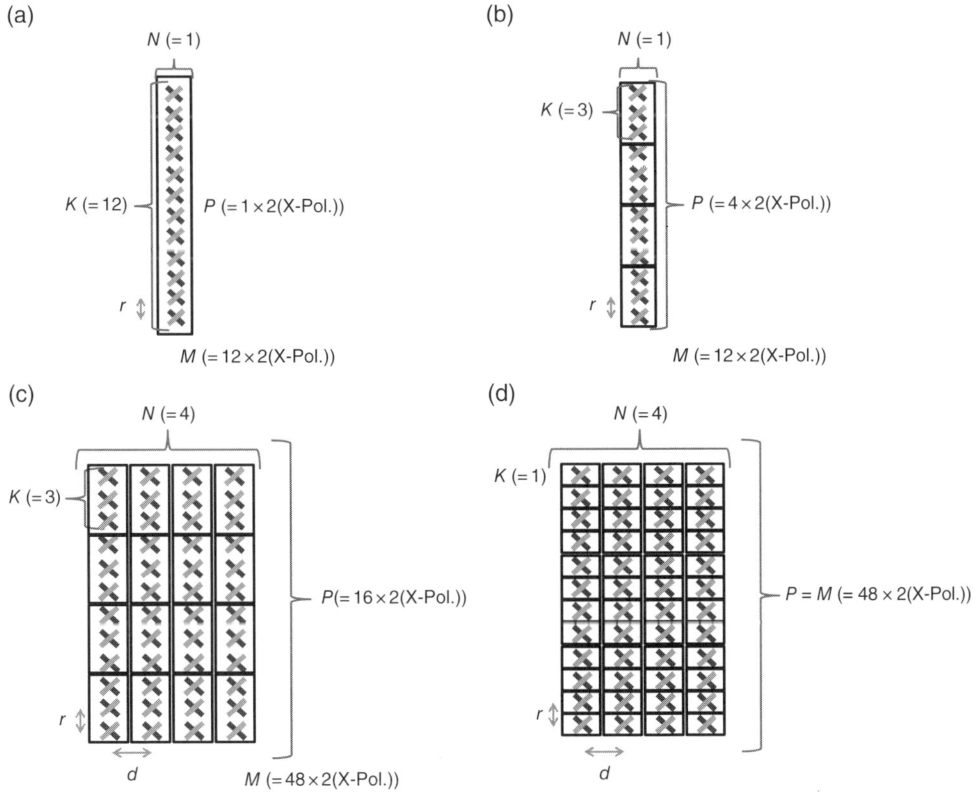

Figure 11.14 Groupings of antenna elements for 3D MIMO: (a) Example of typical MIMO antenna configuration for existing LTE BS ($M=12$, $P=1$); (b) Example of 3D MIMO antenna configuration with vertical antenna grouping ($M=12$, $P=4$); (c) Example of 3D MIMO antenna configuration with horizontal and vertical antenna grouping ($M=48$, $P=16$); and (d) Example of 3D MIMO antenna configuration without antenna grouping ($M=48$, $P=48$).

in a row, and P is the number of antenna ports. Term K is the number of antenna elements in each group for one antenna port. Terms M and P are doubled when the antenna elements are orthogonally polarized. Terms d and r represent the horizontal and vertical antenna-element spacings, respectively. Figure 11.14(a) shows an example of a typical antenna configuration for a current LTE BS. The antenna elements are cross-polarized and all of the 12 antenna elements in the antenna equipment are grouped for each polarization. Figure 11.14(b) shows an example of antenna configuration with a vertical antenna grouping (elevation BF). In this case, three antenna elements are bundled into one antenna-element group. The total number of antenna ports, P, is eight: four sets of two antenna ports with cross polarization (X-pol). Figure 11.14(c) shows an antenna configuration with both vertical and horizontal antenna grouping, such as for FD-MIMO. In this case, four sets of the antenna configuration shown in Figure 11.14(b) are arrayed in a row. Such a configuration can provide beamforming and multiplexing gains while reducing the interference. Figure 11.14(d) shows an example of antenna configuration with full control of all the antenna elements, such as for massive MIMO. The total number of antenna ports, P, is equal to the number of antenna elements ($P=M$), and there are 96 antenna ports. Such an antenna configuration can improve the frequency-usage efficiency through adaptive beamforming for multiple users with highly correlated channels. However, revolutionary advancements in hardware for the BS equipment are needed in order to support the increased number of transceivers required.

11.3.2 Adaptive Grouping of Antenna Elements using SV-MIMO

For future system evolution, it is important to maintain the same coverage area while increasing spectrum efficiency. However, compared to the current antenna configuration of Figure 11.14(a), the antenna configurations of Figures 11.14(b)–(d) will have a different antenna beam pattern for each single antenna port. In particular, the antenna gains of each single antenna port of Figure 11.14(b)–(d) become lower and vertical beamwidth becomes wider than in Figure 11.14(a).

Figure 11.15 shows an example of a virtual antenna port and element configuration in SV-MIMO. In this example, the antenna configuration of Figure 11.14(b) is assumed. Mode A is a mode in which all vertical antenna elements are grouped for each polarization and the number of virtual antenna ports is two as shown in Figure 11.15(a). The antenna beam pattern specifications for Mode A correspond to those for LTE BS antennas, shown in Figure 11.14(a). In Mode B a single antenna is divided into upper and lower groups and the number of virtual antenna ports is four as shown in Figure 11.15(b). In Mode B, the antenna gain of each antenna port becomes lower than that in Mode A. This is because the aperture size of an antenna group in Mode B is smaller than that of Mode A. In Mode C, a single antenna is divided into four groups and the number of antenna ports is eight, with the polarization as shown in Figure 11.15(c). The specification of the antenna beam pattern for Mode C corresponds to the antenna configuration of Figure 11.1(b).

Figure 11.16 shows example usage cases for SV-MIMO for LTE-Advanced using SV-MIMO Modes A, B, and C as shown in Figure 11.15. These modes are virtual since they can be achieved using BB control signaling, and can coexist over the same BS antenna. The antenna hardware in the BS is assumed to use the antenna configuration of Figure 11.14(b). The number of antennas on a user terminal is assumed to be two. The modes are allocated to each resource in the following steps.

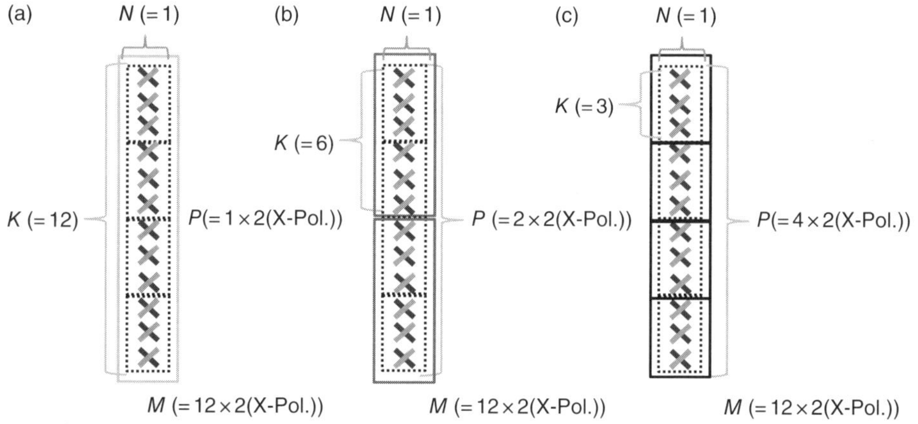

Figure 11.15 Virtual antenna port and element configuration in SV-MIMO: (a) SV-MIMO (Mode A); (b) SV-MIMO (Mode B); and (c) SV-MIMO (Mode C).

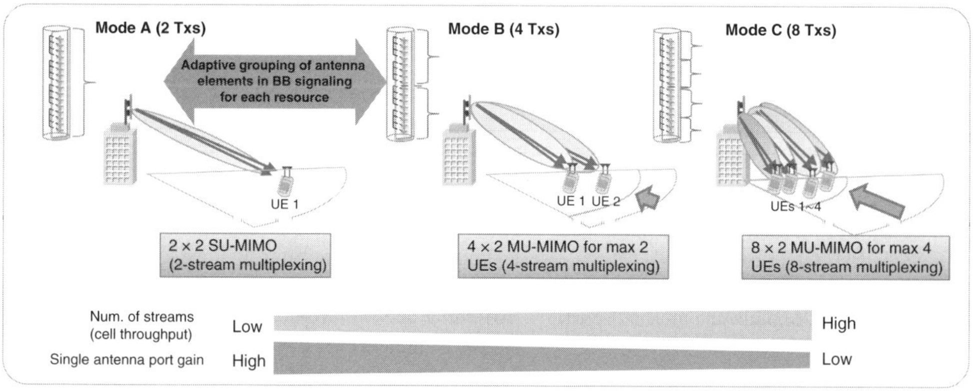

Figure 11.16 Example usage of SV-MIMO.

In the first step, the mode of each resource of the signal is determined according to the type of signal. The mode for:

- a broadcast control signal
- pilot signal, such as CRS, PSS (primary synchronization signal)/SSS (secondary synchronization signal), PDCCH (physical downlink control channel), and PBCH (physical broadcast channel) in LTE-Advanced [30]
- transmission signal to a mobile terminal that supports only conventional terminals for backward compatibility

is fixed to Mode A to maintain coverage. The CSI-RS reference signal used for channel estimation is fixed to Mode B or C.

In the next step, the mode for each resource of UE-specific signals or channels – PDSCH (physical downlink control channel), DM-RS (demodulation reference signal), and EPDCCH (enhanced PDCCH) – in LTE-Advanced [30] is determined according to the channel-quality information fed back from the set of UEs. When Mode B is selected for a resource, the resource can use four-stream multiplexing such as 4×2 MU-MIMO for a maximum of two UEs. When Mode A is selected for a resource, the resource can use a maximum of two-stream multiplexing in 2×2 SU-MIMO. When Mode B is selected for a resource, the resource can use the maximum of four-stream multiplexing in 4×2 MU-MIMO for two UEs. When mode C is selected for a resource, the resource can use the maximum of eight-stream multiplexing in 8×2 MU-MIMO for four UEs.

In summary, with SV-MIMO different signals are transmitted using different modes as shown in Table 11.3:

11.3.3 Performance Evaluation and Field Experiments

To assess the performance of adaptive mode selection in SV-MIMO, we describe some system-level simulation results and experimental results. Figure 11.17 is a reference antenna configuration of four Txs used in the simulation and the experiment. The antenna size of the reference antenna and total number of antenna elements of the antenna becomes larger compared to that of SV-MIMO. Table 11.4 gives the assumptions for the simulation. These are based on 3GPP simulation assumptions using a 3D channel model [13]. The maximum number

Table 11.3 Signals and their modes.

Signal	Mode
Common signals/channels (CRS, PSS/SSS, PDCCH, PBCH)	Fixed to Mode A (two Txs) to maintain coverage and backward compatibility
CSI-RS	Fixed to Mode B or C (four or eight Txs)
UE-specific signals/channels (PDSCH, DM-RS, EPDCCH)	Switching between Modes A, B, and C to obtain adaptive antenna port gains ; Mode A for cell-edge UEs, Modes B and C for cell-center UEs.

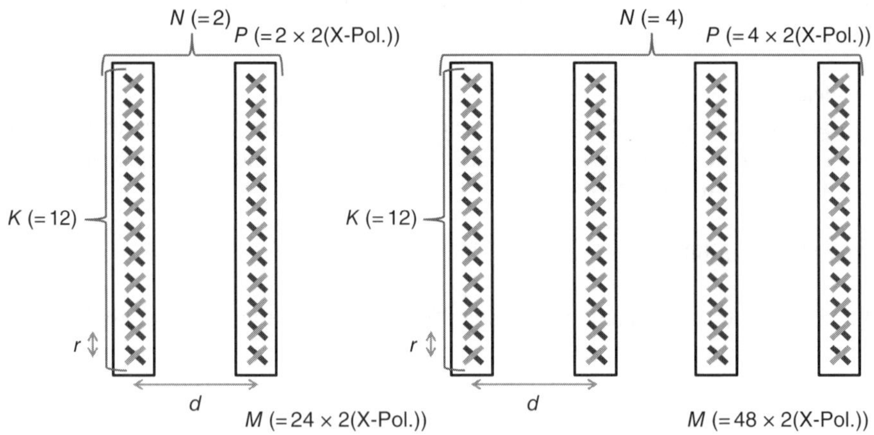

Figure 11.17 Reference antenna configuration (four Txs).

Table 11.4 Simulation assumptions.

Parameter	Value
Network layout	Homogeneous network (19 cell-sites/3 cells)
Carrier frequency	2 GHz
Bandwidth	10 MHz
Transmission power	46 dBm (3D-UMa)
Number of UEs and distribution	Average 10 UEs (3D distributed, 3 km/h)
Number of BS antennas	$M=2\times2$ (V×P) ($K=6$, $r=0.8\lambda$) $M=4\times2$ (V×P) ($K=3$, $r=0.8\lambda$)
Number of UE antennas	2 (P)
Channel model	3D-UMa
Beamforming scheme	Block diagonalization
Receiver	MMSE Wishart IRC
MIMO scheme	SU/MU-MIMO dynamic switching (SU-MIMO up to 2 layers, MU-MIMO up to 4 layers)
Channel estimation	Non-ideal
CSI–RS transmission interval/CSI feedback interval	5 ms (feedback mode 3-2)
Traffic mode	Full buffer
Scheduler	PF
HARQ	Chase combining (8-ms delay)
Control delay	5 ms

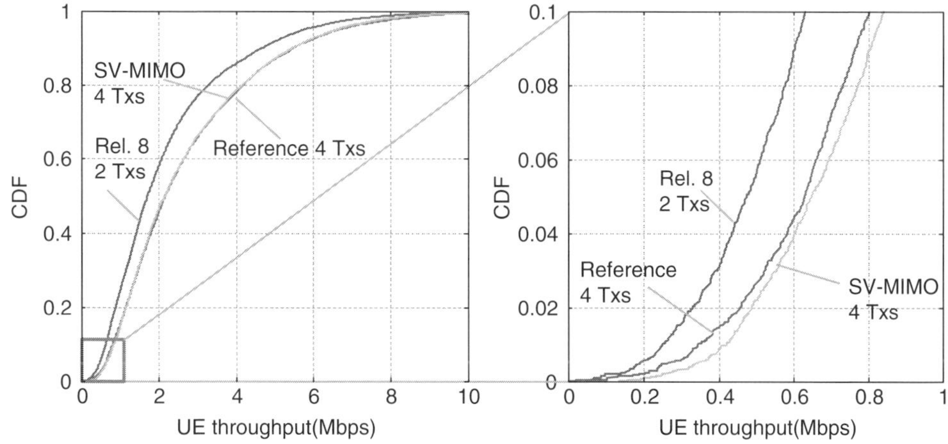

Figure 11.18 System performance of SV-MIMO (urban macrocell: 3D-UMa).

of antenna ports is set to four. Figure 11.18 shows the UE throughput performance for the cumulative distribution function (CDF) under UMa (urban macro-cell) conditions. When comparing the performance of SV-MIMO to that of the reference antenna with the four Txs, the throughput performance is almost the same. The cell-edge throughput is 3.3% greater than that for the reference antenna configuration. When comparing the performance of the two Txs of LTE Release 8 to that of four Txs with SV-MIMO and the reference antenna configuration, there are 22.1% and 37.2% gains of the cell-average and cell-edge throughputs, respectively.

Field experiments were conducted in suburban areas in the city of Yokosuka, Japan. In this experiment, MIMO transmission at a total data rate in excess of 1.2 Gbps for two UEs using only one BS antenna was successfully performed. Moreover, the real-time elevation beam patterns of the BS antenna could be calculated from the measurement precoding vector based on real-time feedback of the downlink channels for SV-MIMO Mode B.

Table 11.5 gives the major radio-link parameters for the experimental trial. The maximum number of antenna ports was set to four. The equipment was based on the LTE-Advanced Release 10 specifications, with an upgrade for explicit feedback of CSI [31]. Figure 11.19 shows measurement *Course #A* for UE #1 and the locations of UE #2 and the BS antenna. The vehicle equipped with UE #2 stopped at points P_1 to P_{18}. In *Course #A*, the majority of the course is under line-of-sight (LOS) conditions and the distance from the BS antenna was approximately 150 m at maximum.

Figure 11.20 shows the fading correlation of vertically polarized antenna ports at the CDF of 50% measured in *Course #A* according to the port spacing in a horizontal array assuming the reference antenna configuration, and in a vertical array antenna assuming SV-MIMO. The antenna spacing defined here is the distance between the centers of the grouped antenna elements for each antenna port. The fading correlation value between the *i*th and *j*th transmitter antennas is calculated as

$$\rho_{ij} = \frac{\left\langle \sum_{l=1}^{L} h_{il}^{\rightarrow} \cdot h_{jl} \right\rangle}{\sqrt{\left\langle \sum_{l=1}^{L} \left| h_{il} \right|^2 \right\rangle} \sqrt{\left\langle \sum_{l=1}^{L} \left| h_{jl} \right|^2 \right\rangle}} \tag{11.7}$$

Table 11.5 Major radio link parameters for measurement.

Parameter	Value
Radio access	OFDMA
Carrier frequency	3.92625 GHz
System bandwidth	100 MHz (20 MHz × 5 CCs)
Number of subcarriers	6,000 (1200 × 5 CCs)
Subcarrier separation	15 kHz
Subframe length	1 ms (14 OFDM symbols)
OFDM symbol duration:	
Effective data	66.67 ms
Cyclic prefix	4.69 ms
BS antennas	$M = 16 \times 2$, $K = 8$
UE antennas	2
Number of UEs	2
Channel coding/decoding	Turbo coding/Max-Log-MAP decoding (6 iterations)
SB size (F_{CSI})	5 RBs (900 kHz)
CSI–RS transmission period (T_{CSI})	5 ms
Signal detection	MLD-based signal detection with interference whitening filter
Control delay for precoding and AMC	10 ms
HARQ	
Packet combining	Incremental redundancy
Max. number of transmissions	3
Round trip delay	8 ms

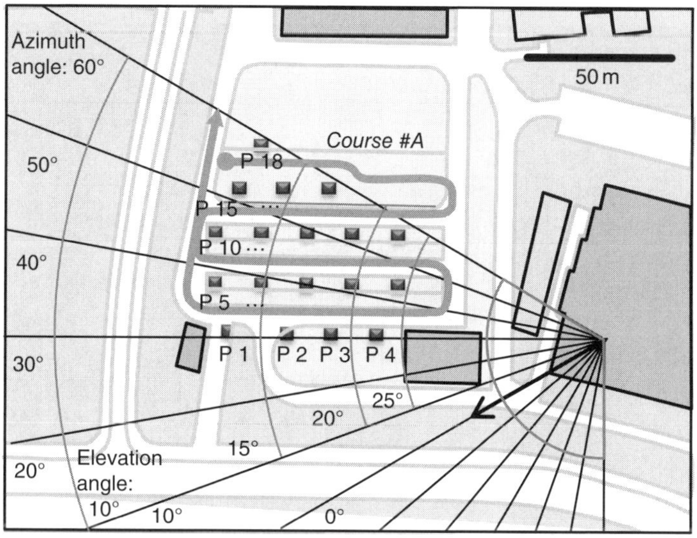

Figure 11.19 Measurement course.

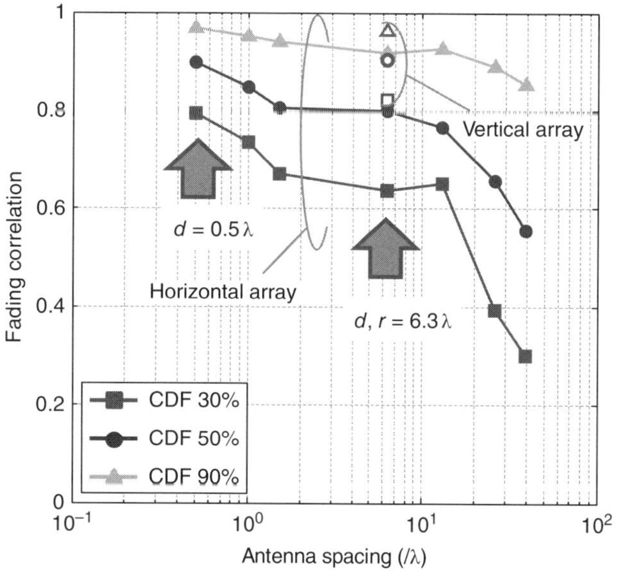

Figure 11.20 Fading correlation of the antenna.

where h_{kl} represents the channel response estimated by the received CSI-RS from the kth transmitter antenna to the lth receiver antenna. In the horizontal array, the value is 0.90 for antenna spacing d of 0.5 λ. It decreases to 0.80 for a spacing of 1.5 λ, and gently decreases to 0.56 for a spacing of 39.3 λ. For the vertical array, the correlation is 0.90 even for antenna spacing r of 6.3 λ.

Figure 11.21 shows the measurement results of the field experiment. A total data rate in excess of 1.2 Gbps for two UEs (mobile stations (MSs)) using only one BS antenna, as

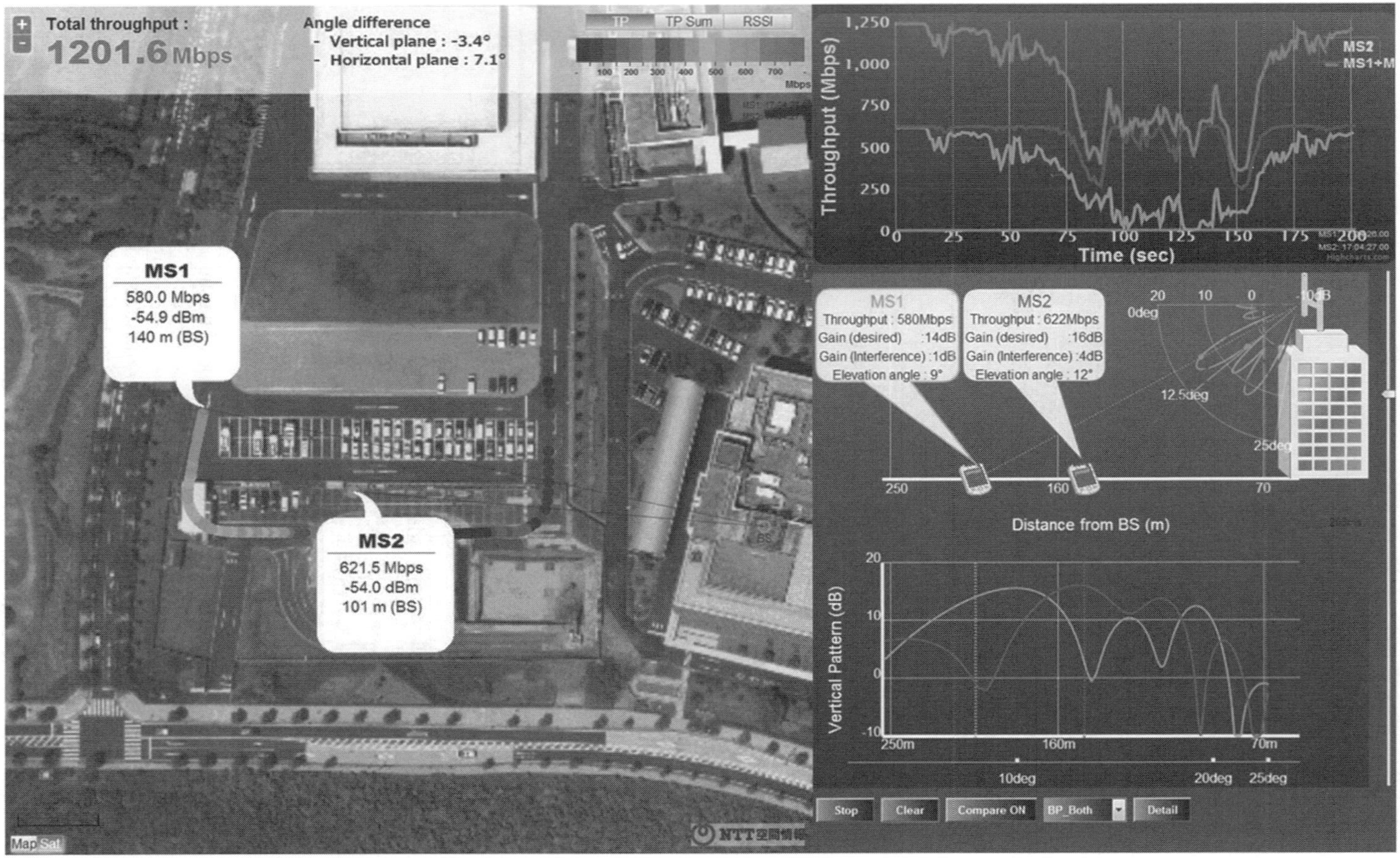

Figure 11.21 Measurement results for suburban environment in Yokosuka, Japan: MS, mobile station.

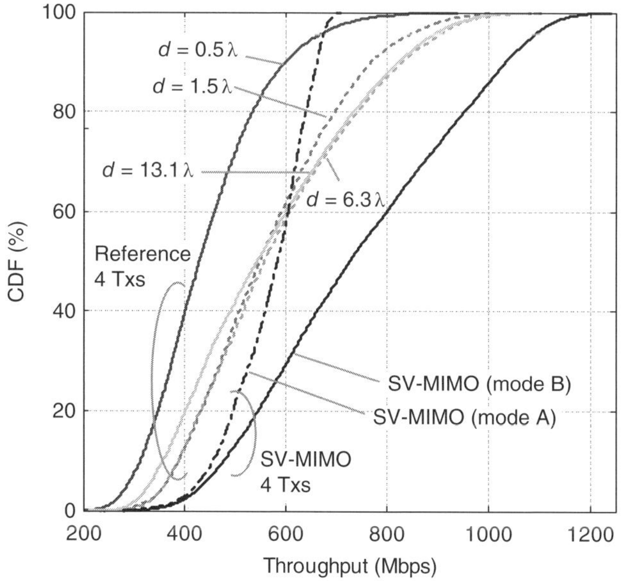

Figure 11.22 Total throughput comparison for BS antennas.

represented in the upper right-hand graph. In the graph, the throughput of UE #1 (MS1), UE #2 (MS2) and the total throughput of UE #1 and UE #2 are shown. The elevation beam patterns of UE #1 (MS1) and UE #2 (MS2) are shown in the lower right-hand graph. Beams are successfully formed toward UE #1 and UE #2. Nulls that cancel the mutual interfering signals are also successfully formed toward UE #2 and UE #1. Furthermore, the interference in the direction of the neighbor cell is reduced by the elevation beam patterns because the beam patterns are reduced by more than 15 dB from the peak in the range of the elevation angle of 0–5°.

Figure 11.22 shows a throughput comparison of CDF between SV-MIMO and the reference antenna. The throughput is the total of all UE #2 points. When comparing the performance of SV-MIMO Mode B to that of *r* with an antenna spacing of 6.3 λ, a 30% throughput gain at the CDF of 50% is achieved using SV-MIMO Mode B. This result agrees with those in Figure 11.20 and in terms of the fading correlation and granularity in the beam and null width. The best performance is achieved using SV-MIMO Mode B in a 4×2 MU-MIMO configuration. A total throughput exceeding 1 Gbps is achieved at a probability of 14%. There are some points that exceed 1.2 Gbps.

11.4 Conclusion

In this chapter, we presented NOMA and SV-MIMO as new advanced interference management and MIMO techniques for future enhancements of spectrum efficiency in lower frequency bands. Regarding NOMA, we presented our design concept and the performance gains over OFDMA. In contrast to the current LTE radio access scheme, NOMA superposes multiple users in the power domain and exploits the channel gain difference between multiplexed users. We discussed the link-level and system-level aspects and the gains of NOMA under practical considerations, such as performance of the SIC receiver, multi-user power allocation, signaling overhead, and performance in high-mobility scenarios, and the combination of NOMA with MIMO. Link-level performance shows that good performance can be achieved for SIC when the power difference allocated to cell-center and cell-edge users is large. Under multiple configurations and setups, the achievable system-level gains are shown to be promising, of the order of 30%, even when practical considerations were taken into account. In addition, we described the design of SV-MIMO, which performs 3D MIMO transmission using antenna elements that are adaptively grouped vertically, according to the type of signal or channel quality at the receiving mobile terminal. SV-MIMO improves the spectrum utilization efficiency without the need for additional cost, or space for installing antenna equipment. Owing to the higher spatial correlation in the vertical domain, a more stable beam nulling toward each UE is feasible, even for higher-mobility scenarios with MU-MIMO. We described the characteristics of adaptive mode selection, some system-level simulation results, and experimental results. These results show that SV-MIMO, by exploiting adaptive model selection, can achieve higher peak data rates and similar effective areas to conventional systems without adaptive antenna grouping.

References

[1] 3GPP (2009) TS36.300, Evolved Universal Terrestrial Radio Access (E-UTRA) and Evolved Universal Terrestrial Radio Access Network (E-UTRAN); Overall description.
[2] 3GPP (2006) TR 25.814 V7.1.0: Physical layer aspects for Evolved UTRA.
[3] 3GPP (2010) TR36.814 V9.0.0: Further advancements for E-UTRA physical layer aspects.
[4] 3GPP (2013) TR36.819 V11.2.0: Coordinated multi-point operation for LTE physical layer aspects (Release 11).
[5] 3GPP (2012) TR36.829 V11.1.0: Enhanced performance requirement for LTE user equipment (UE).
[6] 3GPP (2014) TR 36.866 V.12.0.1: Study on network-assisted interference cancellation and suppression (NAICS) for LTE.
[7] A.V. Zelst, R.V. Nee, and G.A. Awater (2000) Space division multiplexing (SDM) for OFDM systems., *Proc. IEEE VTC2000-Spring*, pp. 1070–1074.
[8] P.W. Wolniansky, G.J. Foschini, G.D. Golden, and R.A. Valenzuela (1998) V-BLAST: an architecture for realizing very high data rates over the rich-scattering wireless channel. *Proceedings of ISSUE*, pp. 295–300.
[9] 3GPP (2013) TR37.840 V.12.1.0: Study of radio frequency (RF) and electromagnetic compatibility (EMC) requirements for active antenna array system (AAS) base station.
[10] O.N.C. Yilmaz, S. Hamalainen, and J. Hamalainen, et al. (2009) System level analysis of vertical sectorization for 3GPP LTE, *Proc. ISWCS*.
[11] Y. Inoue and K. Cho (2011) MIMO base station antenna employing mode selection in vertically split array. *Proc. EUCAP*, pp. 2751–2755.
[12] 3GPP (2012) RP-121413: Study on downlink enhancements for elevation beamforming for LTE, 3GPP TSG RAN meeting #57, Chicago USA.
[13] 3GPP (2012) RP-121410: New SID proposal: study on full dimension MIMO for LTE, 3GPP TSG RAN meeting #57, Chicago USA.
[14] 3GPP (2012) RP-130811: Study on 3D-channel model for elevation beamforming and FD-MIMO studies for LTE.
[15] 3GPP (2014) TR36.873: Study on 3D channel model for LTE, V.1.3.0.
[16] D. Tse and P. Viswanath (2005) *Fundamentals of Wireless Communication*, Cambridge University Press.
[17] T. Takeda and K. Higuchi (2011) Enhanced user fairness using non-orthogonal access with SIC in cellular uplink. *Proc. IEEE VTC2011-Fall*, pp. 1–5.
[18] K. Higuchi and A. Benjebbour (2015) Non-orthogonal multiple access (NOMA) with successive interference cancellation for future radio access. *IEICE Transactions on Communications*, **E98–B** (3), 403–414.

[19] K. Higuchi and Y. Kishiyama (2013) Non-orthogonal access with random beamforming and intra-beam SIC for cellular MIMO downlink. *Proc. IEEE VTC2013-Fall*,.
[20] Y. Saito, Y. Kishiyama, A. Benjebbour, et al. (2013) Non-orthogonal multiple access (NOMA) for cellular future radio access. *Proc. IEEE VTC2013-Spring*, pp. 1–5.
[21] Y. Saito, A. Benjebbour, Y. Kishiyama, and T. Nakamura (2013) System-level performance evaluation of downlink non-orthogonal multiple access (NOMA). *Proc. IEEE PIMRC*, pp. 611–615.
[22] A. Benjebbour, A. Li, Y. Saito, et al. (2013) System-level performance of downlink NOMA for future LTE enhancements. *Proc. IEEE Globecom*, pp. 706–710.
[23] A. Li, A. Benjebbour, and A. Harada (2014) Performance evaluation of non-orthogonal multiple access combined with opportunistic beamforming. *Proc. IEEE VTC2014-Spring*, pp. 1–5.
[24] A. Benjebbour, A. Li, Y. Kishiyama, et al. (2014) System-level performance of downlink NOMA combined with SU-MIMO for future LTE enhancements *Proc. IEEE Globecom*, pp. 706–710.
[25] Y. Inoue, D. Takeda, K. Saito, et al. (2014) Field experimental evaluation of null control performance of MU-MIMO considering smart vertical MIMO in LTE-Advanced downlink under LOS dominant conditions. *IEICE Transactions on Communications*, **E97-B** (10), 2136–2144.
[26] K. Brueninghaus, D. Astely, T. Salzer, S. Visuri, A. Alexiou, S. Karger, and G.A. Seraji (2005) Link performance models for system level simulations of broadband radio access systems. *Proc. IEEE PIMRC 2005*, pp. 2306–2311.
[27] F.P. Kelly, A.K. Maulloo and D.K.H. Tan (1998) Rate control for communication networks: shadow prices, proportional fairness and stability. *Journal of the Operational Research Society*, **49**, 237–252.
[28] M. Kountouris and D. Gesbert (2005) Memory-based opportunistic multi-user beamforming. *Proc. IEEE Int. Symp. Information Theory (ISIT)*, pp. 1426–1430.
[29] P. Viswanath, D.N.C. Tse, and R. Laroia, (2002) Opportunistic beamforming using dumb antennas. *IEEE Transactions on Information Theory*, **48** (6), 1277–1294.
[30] E. Dahlman, S. Parkvall, and J. Sköld (2011) *4G – LTE/LTE-Advanced for Mobile Broadband*. Academic Press.
[31] Y. Kakishima, T. Kawamura, Y. Kishiyama, et al. (2012) Indoor experiments on 4-by-2 multi-user MIMO employing various transmitter antenna arrangements in LTE-Advanced downlink. *Proc. IEEE VTC2012-Fall*, pp. 1–5.

12

M2M Communications

Rapeepat Ratasuk,[1] Amitava Ghosh,[2] and Benny Vejlgaard[3]
[1] *Nokia Bell Labs, New Jersey, USA*
[2] *Nokia Networks, Illinois, USA*
[3] *Nokia Networks, Aalborg, Denmark*

12.1 Chapter Overview

The Internet of Things (IoT) refers to interconnection and exchange of data among devices. This capability will bring about tremendous improvements in user experience and system efficiency. An estimated 50 billion connected devices will be deployed by 2020 [1] and the total machine-to-machine (M2M) revenue is expected to grow from $200 billion in 2011 to $1.2 trillion in 2022 [2]. The ratio of connected things to people is estimated to be 7 : 1 by 2020. Although not all connections will be through a cellular network, it is expected that M2M will account for 22% of cellular connections by 2022 [2].

In this chapter, we will provide an overview of cellular M2M communications and discuss potential solutions for serving M2M traffic. In the 2020 timeframe, 4G Long-Term Evolution (LTE) will remain a strong technology for M2M due to its wide area network, mature ecosystem, high reliability, high performance, and robust features. We will describe the LTE evolution path to serve low data-rate and delay-tolerant M2M services, including the introduction of low-cost devices and features to enhance coverage and battery life. Illustrative analysis of coverage, capacity, and battery life will be provided. In addition, we will discuss how a 5G system using spectrum regime above 6 GHz can be used to serve M2M services that require ultra-low latency and extremely high throughput.

12.2 M2M Communications

The IoT will bring about massive numbers of connected things in diverse areas such as smart cities, consumer electronics, transportation, automotive, healthcare, remote monitoring and management, and public safety. Several broad categories and examples of M2M services are:

Towards 5G: Applications, Requirements and Candidate Technologies, First Edition.
Edited by Rath Vannithamby and Shilpa Talwar.

Companion website: www.wiley.com/go/vannithamby/towards5g

- consumer electronics, including personal multimedia devices, household appliances, wearable devices, audio visual devices, and network equipment
- building automation, including access, monitoring and control, and security
- utilities, including smart metering, sensors, remote monitoring, grid control, and industrial metering
- automotive, including tracking, fleet management, entertainment, navigation, traffic information, emergency assistance, road tolling, and vehicle diagnostics
- healthcare, including monitoring vital signs, remote diagnostics, telemedicine, wearables, emergency response connectivity, and assisted living
- smart cities and transportation, including public safety, public transport and traffic management, parking management, lighting, monitoring, and waste management
- retail, including asset tracking and management, inventory and access control, payment, advertisement, monitoring, and customer support
- manufacturing, including supply chain and order management, distribution, industrial sensor networks, robotics, and monitoring and control
- agriculture, including management and monitoring, sensors, security, and diagnostics.

Table 12.1 lists key M2M sectors and the expected M2M connections in 2022 [2]. The table shows that consumer electronics and building automation are expected to be the two sectors with the most number of M2M connections. Consumer electronics typically require short-range connectivity and are likely to be served using personal or local area networks. Building automation and healthcare can be served using local or wide area networks. Utilities, automotives, and smart cities, however, require wide area networks due to the need for ubiquitous coverage.

Figure 12.1 illustrates M2M requirements along broad criteria of throughput, latency, reliability, cost, and power consumption. M2M requirements vary broadly due to the large number of use cases. For instance, in areas such as smart cities, remote monitoring, or fleet management, a network of sensors is required. These sensors do not require ultra-low latency or high throughput, but must be low cost and consume low power for ultra-long battery life. On the other hand, car-to-car communication will require ultra-low latency as well as high reliability.

Table 12.1 M2M services and connections.

M2M sector	Examples	Connections in 2020	Network
Consumer electronics	Household devices, personal multimedia, wearable devices	35%	Personal, local
Building automation	Building automation and access control, security	34%	Local, wide area
Utilities	Smart meters, sensors, remote monitoring	9%	Wide area
Automotive	Navigation, fleet management, diagnostics	8%	Wide area
Healthcare	Monitoring vital signs, remote diagnostics, wearable	5%	Local, wide area
Smart cities	Environment and public safety, public transport and traffic	4%	Wide area

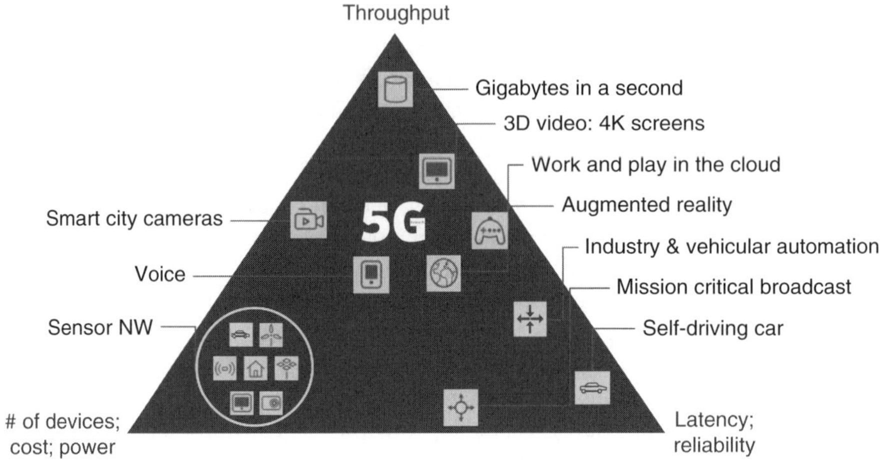

Figure 12.1 M2M services and requirements.

Several types of network can be used to connect things together. These networks may be divided into three broad categories: personal area network, local area network, and wide area network. Personal area networks provide short-range connectivity of up to tens of meters. They can be supported using technologies such as Bluetooth (802.15.1), ZigBee (802.15.4), Body Area Network (802.15.6), and RFID. Local area networks provide short-range connectivity of up to several hundreds of meters. They can be supported using technologies such as WiFi (802.11) as well as cellular 3G/4G/5G using small cell deployment. Wide area M2M connections, however, will be dominated by cellular technologies such as 2G/3G/4G. It is estimated that M2M's share of cellular connections will rise from 3% in 2012 to 22% in 2022 [2]. Other wide area networks include WiMAX (802.16) and satellite. However, they are not expected to play a significant role in providing wide area M2M services.

Figure 12.2 illustrates a cellular deployment with a mixture of a 4G wide area network and 4G/5G local area networks. In this typical scenario, the macro-cell is used to provide wide area coverage while small cells are used to meet capacity demands, for example in hot-spots, and also to address coverage holes. Small cells may also be deployed in clusters as the coverage area of each cell is small. Although the number of M2M devices will outnumber human devices by a factor of up to 7, M2M services are typically low cost and low effort. Therefore, in general the wide area network infrastructure may not be significantly extended for M2M. However, M2M using local area networks is an attractive proposition, especially if the cost of providing network connections is affordable. In LTE, support for home eNB (i.e., small cell) has been introduced. This concept may be used to extend 4G/5G small-cell coverage for M2M as well.

Based on the deployment scenario shown in Figure 12.2, the following network support strategy can be envisioned for M2M:

- LTE evolution to serve most M2M services. LTE can support most M2M sectors such as consumer electronics, building automation, automotive, smart cities, healthcare, and agriculture. Additional features for M2M evolution are described in Section 12.3. These

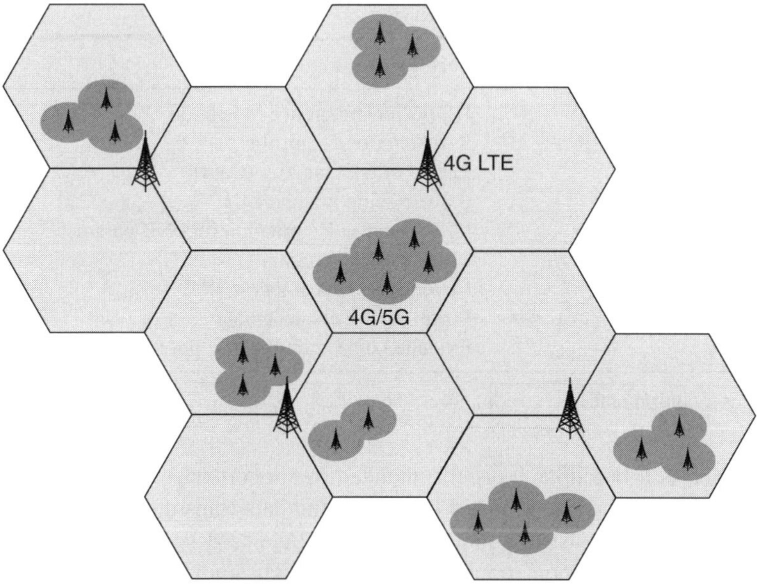

Figure 12.2 Cellular deployment.

features enable low-cost devices, coverage enhancement, and ultra-long battery lives. They are intended to improve LTE effectiveness in serving delay-tolerant and low data rate M2M services.

- 5G system for ultra-low latency, ultra-high reliability, and high-throughput M2M services such as tactile communication, augmented reality, high-definition video, ultra-high-reliability, and other applications requiring near-real-time responses. This new 5G system is described in Section 12.4.

These two systems complement one another and serve to provide underlying infrastructure for the IoT.

12.3 LTE Evolution for M2M

LTE is very attractive for M2M applications for several reasons. First, LTE is a global standard with a mature ecosystem and a large number of service operators and vendors. This leads to reduced costs and robust interoperability. Second, LTE provides wide area network coverage that is comparable to existing 2G/3G systems. There is a large existing installed base for LTE. In addition, service operators are likely to transition their 2G/3G networks in the near future, making LTE a future-proof technology well into the 2020 timeframe. Third, LTE provides high network reliability and high performance. Compared to 2G technology such as GSM/EDGE, LTE provides up to five times higher spectral efficiency in the DL and six times higher spectral efficiency in the UL. Finally, LTE already supports many features that are important to M2M service providers such as security, policy and charging control, QoS management, and so on.

Table 12.2 LTE features for M2M services.

LTE release	Feature
Rel-11	UE power-preference indication
(2012)	RAN overload control
Rel-12	Low-cost UE category (Cat-0)
(2014)	Power saving mode for UE
	UE assistance information for eNB parameter tuning
Rel-13	Low-cost UE category
(expected 2016)	Coverage enhancement
	Power saving enhancement
	Extended discontinuous reception

UE, user equipment; RAN, radio access network.

LTE, however, is a broadband system intended to support data and voice services using large bandwidth. Thus it was optimized for broadband data transmission and not necessarily for M2M services. In contrast to human network traffic, M2M traffic is seen as lower effort and lower cost, with infrequent transmission with small amounts of data, low bandwidth requirements, low power consumption requirements, and supporting a very large number of devices [3]. To ensure that LTE is a competitive system for M2M, several studies have been undertaken to improve it for M2M [4–7]. Based on these studies, enhancements have been introduced, as shown in Table 12.2.

In addition, service differentiation of human and machine traffic can be handled by various techniques in the network. They include:

- time-controlled access of M2M devices, for example allowing some M2M devices to access the network at non-busy times
- barring access to M2M devices with high network utilization.

Similarly, services to M2M devices can be limited if their behavior is not aligned with M2M features [4]. Appropriate QoS classes of identifier can also be configured for machine bearer traffic. Scheduling prioritization, for example through modifying the scheduling metric based on the user equipment (UE) category, and using semi-persistent scheduling to lower overhead.

12.3.1 LTE Features for M2M

In Rel-11, UE power-preference indication and radio access network (RAN) overload control were introduced. UE power-preference indication allows the UE to indicate to the network the preferred power consumption mode of operation [8]. Two modes can be indicated: normal and low-power. This indication is part of the UE assistance information message that is sent as part of radio resource control (RRC) control messages. The details regarding when the UE will sent this indication is left to UE implementation. For example, the UE may indicate a preference for low-power consumption mode when it detects only intermittent background traffic or when it is configured to send best-effort traffic. On the network side, response to this UE power-preference indication is left to network implementation. For example, the network may configure a longer discontinuous reception (DRX) cycle or a shorter network connection release timer to optimize for power saving. It may also release the connection immediately without waiting for the RRC inactivity timer to expire.

Overload control through extended access barring is another feature that was standardized in Rel-11 to prevent M2M devices from overloading the network. Overload situations can happen when a large number of M2M devices try to access the network at the same time. This may be due to large number of simultaneous exception reports, for example by smart meters during power failure, or due to requests for data transmission from many M2M devices at the same time by an application. Without M2M overload control, human traffic will be negatively impacted in these situations. In 3GPP, three different control mechanisms are available: rejecting connection requests in the RAN, rejecting connection requests in the core network, and access barring for M2M devices. As part of the connection set-up request procedure, the network can reject the request. However, this process takes up lots of signaling overhead and is not efficient in situations with significant network overload. In Rel-8, access-class barring is already supported. The mechanism allows for barring of certain UE access classes while allowing high priority access. Extended access barring (EAB) was introduced in Release 11 [8, 9]. In LTE, a UE is randomly assigned to one of 10 access classes. In prior LTE releases, access-class barring applies to all UEs of that access class. EAB, however, allows only M2M devices with mobile-originated traffic that have been configured for EAB to be barred from accessing the cell when it is congested. Other UEs of the same class may still access the system. EAB is implemented through cell-wide broadcasting of extended access-barring information. This allows the network to react quickly to prevent overload by M2M while still continuing to serve human traffic. On the UE side, the UE will be configured for EAB by the home Public Land Mobile Network (PLMN). Similar to RAN overload control, the core network can indicate to eNBs under its service area that the Mobility Management Entity (MME) is overloaded. In this case, the MME can indicate to the eNBs which type of RRC connections should be allowed or rejected.

In Rel-12, a power-saving mode was introduced. In this mode, the UE remains registered to, but not reachable by, the network for mobile terminating traffic, as shown in Figure 12.3 [9]. The UE can be viewed to be in the power-off or sleep mode, and will wake up only when there is data to send or after expiration of a predefined time. This saves a lot of power because the UE does not need to wake up to listen to the paging channel. Currently, the maximum paging cycle is 2.56 s. Without this mode, the UE would have to wake up every 2.56 s to check

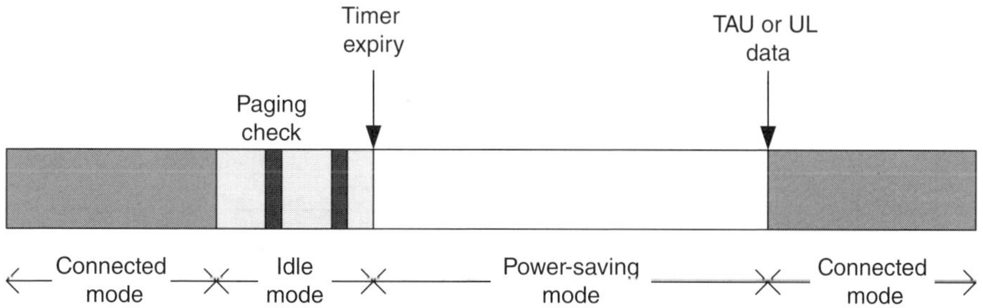

Figure 12.3 Power saving mode.

for paging messages. This can waste a lot of power for M2M devices, which are unlikely to receive pages. This mode is intended for UEs with infrequent mobile-originated data transmissions. It is well suited to M2M services with timed transmissions, such as remote sensing and smart grid applications.

When a UE has data to transmit, the amount of signaling overhead required for sending one IP packet pair (DL and UL) can be quite high, especially for small payload. For instance, a typical payload for a traffic sensor is 1 byte, but the total overhead will be 195 bytes [7]. Several solutions were considered in Rel-12 to reduce this overhead. They included solutions that optimize RRC connection management, control-plane solutions, connectionless data transmission, S1/Iu optimizations, and eNB parameter tuning for small data transfer. Extensive analysis of the different solutions was presented by 3GPP [7]. In Rel-13, the only solution that will be standardized is the introduction of assistance information about the UE and its traffic type/pattern [8]. Assistance information includes expected UE activity behavior, expected activity period, expected idle period, source of UE activity behavior, and expected handover interval. They will help the eNB configure RRC parameters appropriately. Although, the signaling overhead is not reduced, this assistance information can be used to intelligently manage the data connection. For example, for UE with infrequent data transmission, the eNB can configure the inactive timer to enable fast connection release. Similarly, the eNB can keep UEs with frequent transmission of small data in connected mode longer, to minimize the signaling overhead from connection establishment.

In Rel-12, low-cost M2M devices with reduced capabilities have been introduced. Figure 12.4 shows a generic UE architecture. The focus of the cost reduction in Rel-12 is on the modem, which consists of the radio-frequency (RF) and baseband components. Other components such as host CPU, power management unit (PMU), external memory, and input/output interface were not considered in the cost reduction.

The RF block consists of the power amplifier (PA), filters, transceiver, local oscillator (LO), duplexer, and switch. It is estimated that the RF block contributes to 40% of the bill-of-material (BOM) cost of the modem [5]. Within the RF block, the cost breakdown of different components were considered. For example, the PA makes up 25–30% of the RF cost, while the transceiver makes up 40–50%. The baseband block consists of the ADC/DAC, FFT/IFFT,

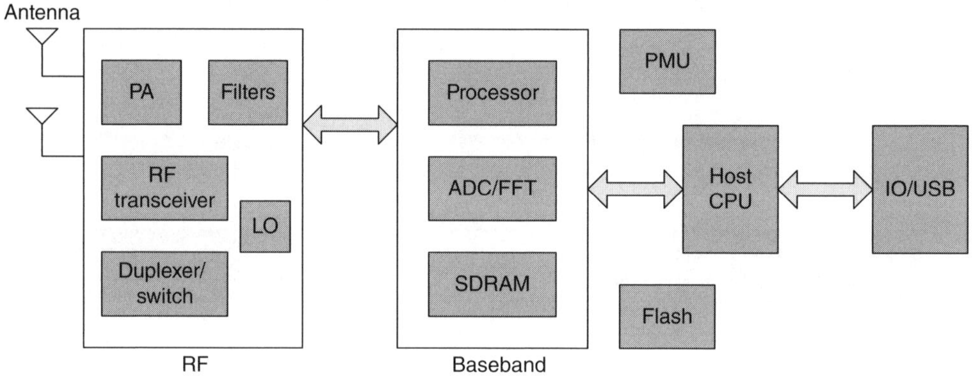

Figure 12.4 Generic UE architecture.

baseband processor, and memory. This block makes up 60% of the BOM cost of the modem [5]. Individual function within the baseband processor – turbo decoding, receiver processing, synchronization, control processing – was considered in order to identify potential cost savings. For example, turbo decoding takes up 5–15% of the baseband cost while receiver processing takes up 20–35%.

Based on this cost-reduction analysis, a new UE category (called Category 0 or Cat-0 UE) intended to support M2M services was introduced in Rel-12. This low-cost UE has the following reduced capabilities:

One Receive (Rx) Antenna: By having one Rx rather than the minimum of two for other UE categories, cost savings can be achieved in both baseband and RF components.

Maximum Transport Block Size: Low-cost UE has a maximum transport block size (TBS) of 1,000 bits in both DL and UL; that is, a low-cost UE can send or receive at most 1,000 bits of unicast traffic per subframe, corresponding to reduced peak data rates of 1 Mbps. In comparison, a Cat-1 UE has peak data rates of 10 Mbps DL and 5 Mbps UL, while a Cat-4 UE has peak data rates of 150 Mbps DL and 75 Mbps UL. Restricting the maximum TBS leads to saving in the baseband components including memory and processor.

Optional Half-duplex FDD Operation: Half-duplex operation allows the duplexer to be removed. In addition, half-duplex only requires one local oscillator to be used for FDD instead of two.

The bill of material (BoM) cost of the modem for this Rel-12 device is approximately 50% of the baseline LTE modem and comparable to that of an EGPRS modem. Peak rate reductions and a single Rx antenna contribute to 42% of this saving, while half-duplex operation contributes 7–10% to the cost saving.

In Rel-13, two features for M2M will be standardized in 3GPP. Both features aim to provide support for massive numbers of devices, very low-cost devices (further cost reductions compared to the Cat-0 UE), improved indoor coverage, and ultra-long battery life. The first feature is enhanced machine-type communications (eMTC) and is described in Section 12.3.1.2. The second feature is the narrowband Internet of Things (NB-IoT) and is described in Section 12.3.1.2. Table 12.3 summarizes key targets of the two features.

Table 12.3 Key targets for eMTC and NB-IoT.

Target	eMTC	NB-IoT
Bandwidth	1.4 MHz	200 kHz
Peak data rates	1 Mbps DL/UL	200 kbps DL, 144 kbps UL
Coverage	156 dB MCL	164 dB MCL
Capacity	>50,000 devices/cell	>50,000 devices/cell
Battery life	>10 years	>10 years
Exception report latency	<10 s	<10 s
Device cost	Low	Low

12.3.1.1 eMTC

In Rel-13, further UE cost reductions will be introduced [10]. Using the Release 12 low-cost UE as the baseline, the Release 13 low-cost UE will have further reduced capabilities as follows:

Reduced UE RF Bandwidth: The RF bandwidth will be 1.4 MHz in DL and UL compared to 20 MHz for other UEs. This will allow the use of narrowband RF components, which are generally cheaper and also consume less power.

Reduced Maximum Transmit Power: This will allow for an integrated PA implementation. Having an integrated PA will provide cost savings since it will allow the design to fit on a single chip. However, the current integrated PA architecture can only deliver a maximum output power in the range of 17–23 dBm. In contrast to the current maximum UE power of 23 dBm, reducing the maximum power will result in lower coverage and may reduce capacity as well.

Reduced Support for DL Transmission Modes: In LTE, 10 different transmission modes are defined corresponding to DL transmissions using:

- a single transmit antenna at eNB (mode 1)
- transmit diversity (mode 2)
- large delay cyclic delay diversity or transmit diversity (mode 3)
- closed-loop spatial multiplexing (mode 4)
- multi-user multiple-input multiple-output (mode 5)
- closed-loop spatial multiplexing using a single transmission layer (mode 6)
- single-antenna port using dedicated reference signals (mode 7)
- dual-layer transmission using dedicated reference signals (mode 8)
- up to eight-layer transmission using dedicated reference signals (modes 9 and 10).

Obviously, low-cost UE with a single receive antenna would not be capable of supporting all these modes. Thus, eliminating some transmission modes will provide some complexity reduction in the baseband processor.

Reducing the UE RF bandwidth to 1.4 MHz provides a cost reduction of approximately 39%. Together with Rel-12 peak rate reduction to 1 Mbps and a single RF antenna, the cost saving is approximately 61%. Half-duplex operations provides another 7–10% cost reduction, while maximum power reduction will provide 2–7% cost reduction (assuming an integrated PA). Transmission-mode reduction can provide 2–10% cost reductions. All together, the techniques can provide a 72–88% reduction in the BoM cost of the modem compared to a Cat-1 UE [11].

The LTE system can be deployed using the following bandwidths: 1.4, 3, 5, 10, 15, and 20 MHz. Currently, all UE categories are capable of accessing up to 20 MHz bandwidth. With the introduction of the Release 13 low-cost UE, however, the device will have RF bandwidth of only 1.4 MHz. This is equivalent to being able receive only six LTE Physical Resource Blocks (PRBs). Although the UE RF bandwidth is 1.4 MHz, it must be able to operate within any system bandwidth (say, 10 or 20 MHz). The following multiplexing structure could be used to support narrowband UE (see also Figure 12.5). In the UL, narrowband UE can be easily

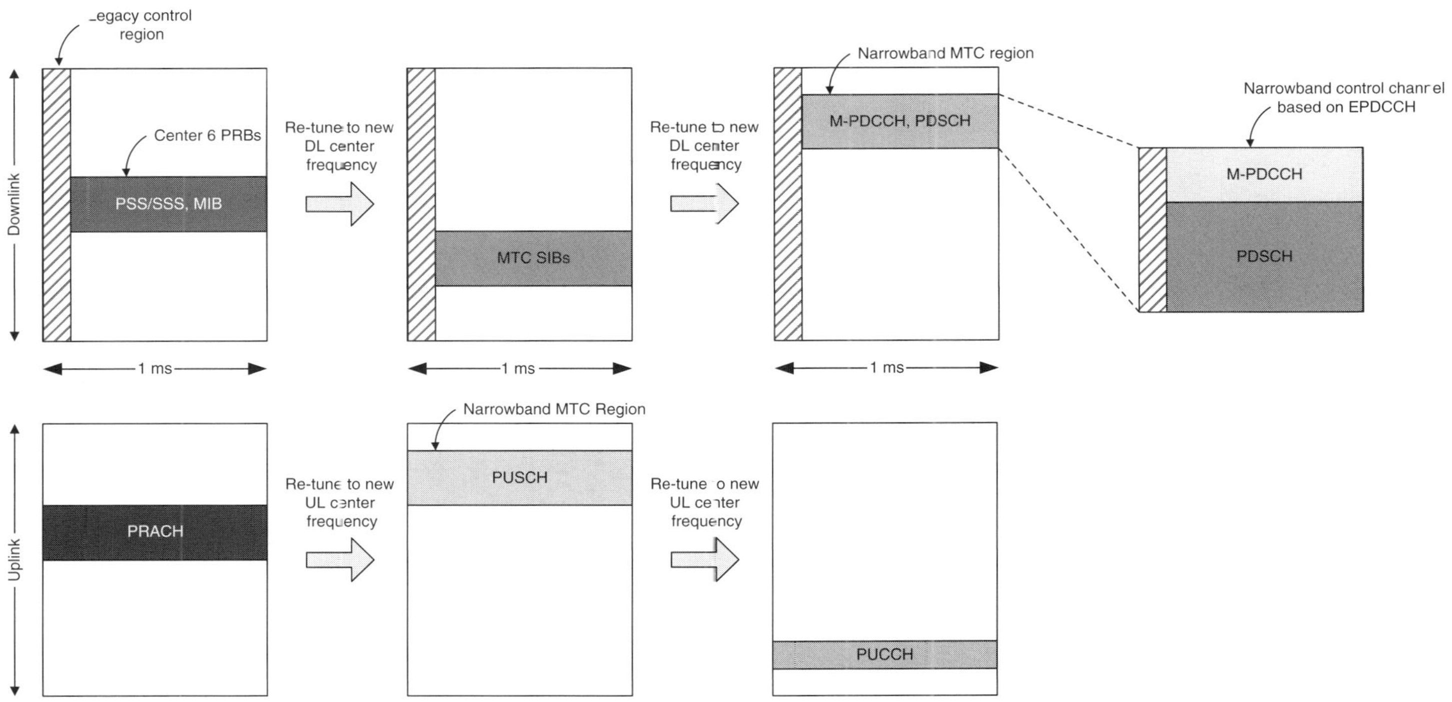

Figure 12.5 Narrowband Rel-13 M2M UE in wider system bandwidth.

accommodated since all the UL channels are multiplexed only in frequency and a half-carrier shift is used instead of DC subcarrier. The following channels are supported in the UL:

- Physical Uplink Shared Channel (PUSCH) used to transmit data
- Physical Uplink Control Channel (PUCCH) used to transmit control information such as acknowledgement and channel state information
- Physical Random Access Channel (PRACH) used for random access.

In the DL, there is a legacy DL control region that spans the entire bandwidth. This region must be avoided by a narrowband Rel-13 low-cost UE so as not to interfere with legacy control. Thus, the DL subframe for this UE contains fewer symbols than regular subframe (say, 11 OFDM symbols compared to 14 for legacy UEs). The following channels are supported in the DL:

- Physical Downlink Shared Channel (PDSCH) used to transmit data
- Physical Broadcast Channel (PBCH) used to transmit master system information block
- Enhanced Physical Downlink Control Channel (EPDCCH) used to transmit DL control information such as scheduling assignment and power control information.

It also contains two synchronization signals:

- Primary Synchronization Signal (PSS)
- Secondary Synchronization Signal (SSS).

When a narrowband M2M UE accesses a wideband LTE system, it will synchronize to the system using the PSS/SSS contained in the center 6 PRBs (corresponding to 1.08 MHz, which is within the RF bandwidth of the UE). It will then decode the Master Information Block (MIB) transmitted on the PBCH. The MIB will tell the UE the DL bandwidth of the wideband system. Subsequent to reading the MIB, it will read the System Information Blocks (SIBs). The SIBs provide the common system information needed for the UE to access the system. This includes information including:

- the PLMN
- whether this cell supports Rel-13 narrowband M2M UEs
- UL carrier frequency and bandwidth
- cell reselection parameters
- broadcast configuration
- common radio configuration parameters.

The EAB is also transmitted on the SIB, which lets the UE know whether the cell is currently barred or not. Subsequent to acquiring the SIBs, the UE will perform random access to let the network know that it is trying to access the system.

After a successful random access procedure, the UE will undergo RRC connection setup and security authentication. As part of the RRC connection setup, the network may re-direct the UE to a different narrowband M2M region within the system bandwidth (each region supporting six PRBs). Multiple regions may be supported based on how many M2M UEs are within the

Table 12.4 Potential coverage enhancement techniques.

Physical channel name	PUCCH	PRACH	PUSCH	PDSCH	SCH	PBCH	EPDCCH
Repetition	×	×	×	×		×	×
PSD boosting				×	×	×	×
Frequency hopping	×	×	×	×			×
Relaxed requirement		×			×		
Overhead reduction							×
HARQ retransmission			×	×			
Multi-subframe channel estimation	×		×	×		×	×
Multiple decoding attempts						×	

network. The UE will then re-tune its center frequency to the appropriate region. The re-tune time is approximately 500 μs. This prevents the UE from dynamically re-tuning from one region to another every subframe – fast hopping – but the UE can perform slow hopping.

In LTE Rel-13, coverage enhancement will also be specified [10]. The goal is to target a relative LTE cell coverage improvement corresponding to 15 dB for FDD for delay-tolerant M2M services. This will provide increased cell coverage areas as well as the ability to support devices in locations with high penetration loss, such as smart meters installed in basements. Table 12.4 lists potential techniques that can be used for coverage extension and applicable channels.

Another key objective in Rel-13 is to reduce power consumption, both in normal coverage and enhanced coverage, to promote ultra-long battery lives [10]. The following power reduction techniques may be considered:

Higher-layer Methods to Reduce Active Time: This includes, for instance, longer paging cycles or UE-initiated access. In UE-initiated access, the UE initiates access to the network periodically and goes to am M2M state (no paging) after the connection release. Any network-originated data is held until the next UE-initiated access occasion. Periodic keep-alive messages are transmitted by the devices. Periodicity configured by the network and UE could provide the preference. Generally, these methods try to reduce the active time of the device as much as possible to minimize power usage. Dynamic schemes that adapt to battery power or life can also be considered here.

Overhead Reduction: Both higher-layer and physical layer overhead reductions may be considered. At the higher layer, signaling overhead reductions using optimized RRC procedures can be used to save battery life. Some of these methods have been studied in Rel-12 but have not been standardized, for example:

- RRC message combining
- RRC connection without U-plan radio bearer establishment
- small data fast path
- connectionless data transmission.

Physical layer overhead reduction (e.g., compact DL control information) can also be used to minimize the time it takes for the UE to receive data. The gain from PHY overhead reduction, however, is expected to be modest compared to the gain from higher-layer overhead reduction.

Reducing System Acquisition Time: This includes synchronization acquisition time and MIB/SIB acquisition time. The system acquisition time can be significant, especially with coverage enhancement. For example, it was determined that the acquisition time for synchronization may be as high as 2 s for each center carrier frequency. Similarly, system acquisition time can be extremely long due to infrequent SIB transmission and extended repetition. New designs for the PSS/SSS and MIB/SIB transmission may substantially reduce the system acquisition time.

Reducing Measurements and Measurement Reports: For devices with limited mobility, only infrequent RRM measurements are needed. UEs in coverage enhancement mode may also be configured to perform measurements and report results infrequently or to skip measurements related to handover procedures and idle-mode mobility. This can limit the active time of the UE and reduce power consumption.

Configurable Coverage Enhancement: Coverage enhancement will be scalable and configurable, with the aim of minimizing the amount of reception and transmission time for each level. Access mechanisms, such as PRACH, may be designed such that the UE does not unnecessarily transmit more than needed. A mechanism to identify the coverage range may be studied such that UEs that are out of coverage range will not waste power trying to connect to the system.

Coverage

Nominal LTE coverage has been well studied [5, 12, 15]. The maximum coupling loss (MCL) for a FDD system with 2Tx–2Rx antennas was shown to be 140.7 dB. Therefore, the target MCL with 15-dB coverage extension is 155.7 dB. This is with respect to nominal LTE coverage, which has been studied by Morrish et al. [2]. Compared to nominal UE, low-complexity UE will have smaller coverage due to reduced capacities:

- the single Rx antenna will lead to approximately 4-dB degradation in performance of the DL channel, due to a lack of receiver combining and diversity gain
- reduced maximum transmit power will lead to a corresponding degradation in coverage of the UL channels; for example, a reduction of the maximum transmit power by 3 dB will require the corresponding gain from coverage enhancement
- reduced UE bandwidth of 1.4 MHz in DL and UL can lead to degradation in performance due to lack of diversity gain; it has been estimated that this results in approximately 1–3 dB loss in performance.

Table 12.5 shows a link budget analysis for an FDD system and the amount of coverage enhancement required for a Rel-13 low-cost M2M UE with 20 dBm maximum power. For this analysis, system bandwidth of 10 MHz is also assumed, so as to provide a fair power comparison with the baseline link budget [5]. From the table, the required coverage enhancements for different channels are shown. They range from 10.4 to 18 dB.

Table 12.6 shows the coverage enhancement targets of different channels for different UE categories. The UL coverage enhancement targets are obviously dependent on the reduction in

Table 12.5 Link budget analysis for LTE M2M (FDD).

Physical channel name	Units	PUCCH	PRACH	PUSCH	PDSCH	SCH	PBCH	EPDCCH
Transmitter								
Max Tx power	dBm	20	20	20	46	46	46	46
(1) Actual Tx power	dBm	20.0	20.0	20.0	32.0	36.8	36.8	32.0
Receiver								
(2) Thermal noise density	dBm/Hz	−174	−174	−174	−174	−174	−174	−174
(3) Receiver noise figure	dB	5	5	5	9	9	9	9
(4) Interference margin	dB	0	0	0	0	0	0	0
(5) Occupied channel bandwidth	kHz	180	1,080	360	360	1,080	1,080	360
(6) Effective noise power[(a)]	dBm	−116.4	−108.7	−113.4	−109.4	−104.7	−104.7	−109.4
(7) Required SINR	dB	−7.8	−10	−4.3	2.6	−3.8	−3.5	2.5
(8) Required coverage enhancement		**11.5**	**17.0**	**18.0**	**16.9**	**10.4**	**10.7**	**16.8**
(9) Receiver sensitivity[(b)]	dBm	−135.7	−135.7	−135.7	−123.7	−118.9	−118.9	−123.7
(10) MCL[(c)]	dB	**155.7**	**155.7**	**155.7**	**155.7**	**155.7**	**155.7**	**155.7**

Notes: [(a)] = (2) + (3) + (4) + 10 log((5)); [(b)] = (6) + (7) − (8); [(c)] = (1) − (9).

Table 12.6 Coverage enhancement targets for different UE categories/types (FDD).

UE category/type	PUCCH	PRACH	PUSCH	PDSCH	PBCH	SCH	(E)PDCCH
Cat-1 UE (2-Rx)	8.5	14.0	15.0	10.3	6.7	6.4	9.6
Cat-0 UE (1-Rx)	8.5	14.0	15.0	14.3	10.7	10.4	13.6
LC-MTC UE (1-Rx, 1.4 MHz BW, 23 dBm)	8.5	14.0	15.0	16.9	10.7	10.4	16.8
LC-MTC UE (1-Rx, 1.4 MHz BW, 20 dBm)	11.5	17.0	18.0	16.9	10.7	10.4	16.8

maximum UE power. On the DL, however, it is seen that the PDSCH and (E)PDCCH performances degrade substantially due to the single Rx antenna and narrow bandwidth. Thus, the required enhancement targets for these two channels are quite large.

Capacity

M2M capacities based on system-level simulations are provided for two simulated 3GPP system deployment scenarios: urban and suburban macro-cell. The macro-cell system simulation scenario is of a traditional 57-cell system setup with wrap-around [14]. The inter-site distances are 500 m and 1,732 m for urban and suburban scenarios, respectively. In our analysis, the devices are assumed to be indoor and therefore experience an indoor penetration loss of 20 dB. The path-loss profiles of the two simulated scenarios are shown in Figure 12.6. To verify the usefulness of our scenarios, the path-loss profile from a real-world system deployment is also shown. It is seen that the real-world profile matches well with the simulated suburban scenario.

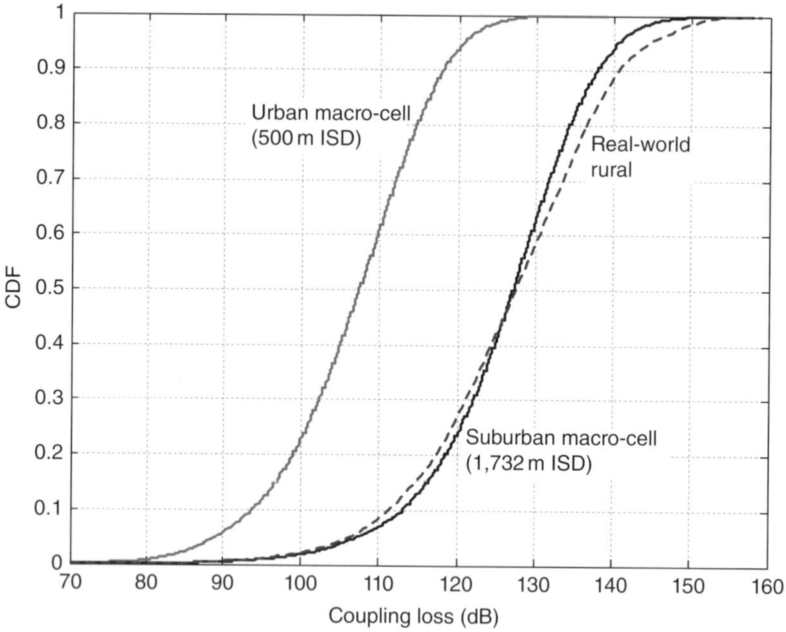

Figure 12.6 Path loss CDF for the deployment scenarios.

Table 12.7 M2M traffic characteristics.

Application	Average message reporting time (s)	Average message size (bytes)
Smart meter	9,000	2,017
Health sensor	60	128
Home security	600	20
Home appliance	24 × 60 × 60	8
Traffic sensor	60	1

Source: IEEE [17].

Some representative M2M traffic characteristics are shown in Table 12.7 [16, 17]. They include services that are likely to be supported using wide-area cellular networks such as smart meters, home security, health sensors, and traffic sensors. Note that only UL traffic characteristics are shown here since typically the UL is the limiting capacity link.

The number of supportable M2M devices per PRB is shown in Table 12.8. Note that one PRB is comprised of 12 subcarriers and occupies 180 kHz. From the table, it is seen that LTE offers very high M2M capacity. In an urban environment, the number of homes per cell has been estimated to be 3,021–12,077 while in a suburban environment the estimate is 4,647–10,456 [17]. If each home has three smart meters (water, gas, electricity) and a home security system, then less than one PRB would be required to support smart meter deployment. In a 10-MHz system, there are 50 available PRBs, so smart meters will take up less than 2% of the

Table 12.8 M2M capacity per PRB.

Application	Urban	Suburban
Smart meter	7.5e4	5.6e4
Health sensor	5.3e3	4.0e3
Home security	1.2e5	9.2e4

system capacity. From the table, it can be seen that when LTE is used to also support machine traffic, there is only a small impact in cell throughput of human traffic. For a 10-MHz LTE deployment, less than 10% of the existing system capacity is expected to be used up by M2M services.

If M2M devices are installed in locations with poor signal quality (say in basements or meter closets), however, then capacity can degrade significantly. For example, if meters are put in the basement where they suffer an 10–15 dB penetration loss in addition to the 20-dB indoor penetration loss, then capacity will reduce by up to 95% [15]. In this case, up to 20% of the system capacity will be required to support smart meters. This would require careful management of system resources when extensive coverage enhancement is to be supported. In this example, several approaches can be taken to limit the impact on system capacity. One approach may be to configure the meters to report less frequently. For example, Table 12.7 shows that a meter will send a report every 2.5 h. Meters that are installed in challenging locations can be configured to report once a day instead. This will reduce their overhead significantly. Another approach may be to allow meters to only access the system during non-busy periods, where there is plenty of spare system capacity.

Battery Life

The UE energy consumption per period is given by the following formula:

$$\text{Energy}_{\text{period}} = P_{tx} \times T_{tx} + P_{rx} \times T_{rx} + P_{\text{leakage}} \times T_{\text{sleep}} + \text{Battery}_{\text{self\_discharge}}$$

where Ptx, Prx, Pleakage are the power consumption during Tx, Rx, and sleep, respectively. The transmission time, Ttx, includes transmitting the random access preamble, RRC connection setup and release messages, and data packet. The reception time, Trx, includes synchronization acquisition, system parameter acquisition time, and RRC connection setup and release messages. Battery life is given by the battery capacity divided by the energy consumption per period. For our analysis, we assume two AA alkaline long-life batteries with energy capacity of 2×2.52 Wh (1.2V, 2.1A)=2×9,072 J. The current drawn and corresponding power consumption values are provided in Table 12.9. During deep sleep mode, there is a gradual loss of energy from the charged capacitors due to leakage current. There is also a reference clock required to keep the UE alive. The minimum leakage current including the slow reference clock is assumed to be 15 μA. The battery also has a self discharge, which is assumed to be 10% in 5 years or 0.16% monthly.

Table 12.10 shows energy consumption for UL data transmission. The data payload is 1,000 bits [5]. For our analysis, we assume the UE has valid system information and therefore there is no need to acquire the SIBs. Since system information is valid for up to 3 h while the data model in [5] assumes regular reporting every 5 min, 30 min, or 1 h, this is a reasonable assumption.

Table 12.9 UE current and power consumption.

Description	Current consumption	Power consumption at 2.4 V
Transmitting power	100 mA	240 mW
Receiving power	20 mA	48 mW
Leakage power	15 μA	36 μW

Table 12.10 UE energy consumption for UL data transmission.

Description	Coupling loss = 140.7 dB				Coupling loss = 155.7 dB			
	Tx/Rx time (ms)		Energy consumption (μJ)		Tx/Rx time (ms)		Energy consumption (μJ)	
	Tx	Rx	Tx	Rx	Tx	Rx	Tx	Rx
Synchronization acquisition		20		960		1,000		48,000
MIB acquisition		40		1,920		1,600		76,800
Preamble + RAR	1	1	240	48	20	70	4,800	3,360
RRC connection setup + security mode command	9	3	2,160	144	1,094	520	262,560	24,960
RRC connection reconfiguration	3	3	720	144	320	610	76,800	29,280
Payload transmission (1,000 bits)	26		6,240		3,414		819,360	
RRC connection release	1	1	240	48	80	130	19,200	6,240
Total	40	68	9,600	3,264	4,928	3,930	1182,720	188,640

Table 12.11 UE energy consumption for paging.

Description	Coupling loss = 140.7 dB				Coupling loss = 155.7 dB			
	Tx/Rx Time (ms)		Energy consumption (μJ)		Tx/Rx time (ms)		Energy consumption (μJ)	
	Tx	Rx	Tx	Rx	Tx	Rx	Tx	Rx
Synchronization		20		960		1,000		48,000
MIB acquisition		40		192		1,600		76,800
Paging		1		48		70		3,360
Total		67		2,928		3,930		128,160

Table 12.10 shows that energy consumption is dominated by the Tx time, especially in a coverage-limited scenario. Furthermore, this table also illustrates the high amount of signaling overhead that is required to send the payload. Table 12.11 shows energy consumption for paging. For paging, it is clear that PSS/SSS and PBCH acquisition times have major impact on energy consumption.

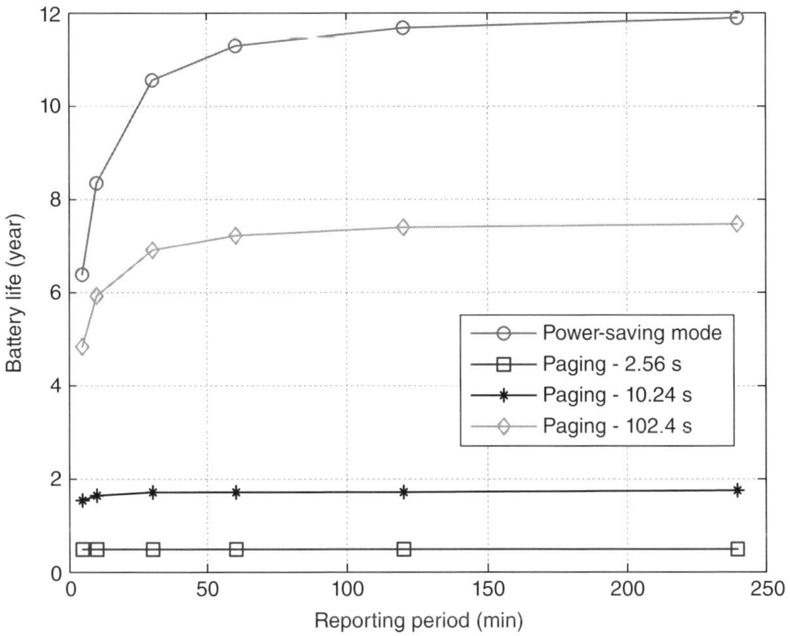

Figure 12.7 UE battery life.

Figure 12.7 illustrates the battery life of a UE at MCL of 140.7 dB. If power-saving mode is used (no paging when the UE is in idle mode), then battery life of 10 years is possible. If paging must be supported to allow the network to reach the UE when there is network originated traffic, the battery life is approximately 0.5 years using the current maximum paging cycle of 2.56 s. If the maximum paging cycle is extended, then significantly better life can be achieved.

12.3.1.2 Narrowband Internet of Things

In Rel-13, the narrowband IoT (NB-IoT) will be introduced for cellular IoT with the following design targets:

- *Improved indoor coverage:* the target is to achieve an extended coverage of 20 dB compared to legacy GPRS devices. This corresponds to achieving target MCL of 164 dB. At this MCL, data rate of at least 160 bps should be supported at the application layer for both the UL and DL.
- *Support of massive number of low throughput devices:* the target is to support at least 52,547 devices within a cell site sector.
- *Reduced complexity:* the goal is to provide ultra-low complexity devices to support IoT applications.
- *Improved power efficiency:* the target is to provide battery life of 10 years with battery capacity of 5 Wh at 164 dB MCL. It is desirable that battery life of 10 years is achieved also for in-band operation.
- *Latency:* exception report latency of 10 s or less is the target for 99% of the devices.

NB-IoT will use 200 kHz of spectrum and support three deployment operations:

1. stand-alone operation using spectrum currently being used by GSM
2. in-band operation using resource blocks within a LTE carrier
3. guard-band operation using spectrum within an LTE carrier's guard band.

For example, NB-IoT can be deployed by re-farming only one GSM channel, or it can be deployed in-band using one PRB from an existing LTE deployment. Figure 12.8 illustrates NB-IoT in-band operation. Here, one PRB outside of the center six PRBs is reserved for NB-IoT. Within this NB-IoT PRB, the following time-frequency resources are reserved for legacy LTE transmission: the first three OFDM symbols are reserved for legacy LTE control region and subcarriers assigned to LTE reference signals are also reserved. NB-IOT signals must not be transmitted in those reserved time-frequency resources. However, NB-IOT can reuse LTE reference signals for channel estimation and demodulation if appropriate.

By making NB-IoT compatible with LTE, it is possible to reuse the same hardware and also to share spectrum without coexistence issues. In addition, NB-IoT can simply plug into the LTE core network. This allows all network services, such as authentication, security, policy, tracking, and charging, to be fully supported. On the device side, in addition to the narrow bandwidth, some simplifications are being considered to reduce device complexity. Examples include using convolution coding in the DL instead of turbo coding and reducing the maximum number of HARQ processes to one or two. To improve coverage, single-tone transmission is being considered in the UL as well as peak-to-average power reduction techniques.

12.3.2 Further Enhancements

Additional enhancements to improve M2M may also be considered, especially on the network and architecture sides. There have been significant numbers of studies done in these areas that have not become part of the LTE standard yet. Some of these enhancements include:

Signaling Overhead Reductions: These include connectionless data transmission, RACH-based small data transmission, RRC connections without user-plane radio bearer establishment, optimized service-request procedures, and lean service-request procedures. These techniques are intended to reduce the signaling overhead associated with data transmission.

Higher-layer Device Power Consumption Optimization: Enhancements include extended DRX in idle mode, long DRX cycles in connected mode, and transmission delays until radio conditions improve. These techniques reduce the transmit/receive time of the device in order to increase battery life.

Group-based Communications: These include group-based messaging using cell broadcast, group-based charging and policy control, and group-based addressing and identifiers. These techniques reduce the overhead and time needed to address large numbers of M2M devices.

Monitoring Enhancements: These include monitoring of specific events, and detection and network responses. They include monitoring via policy and charging control and home

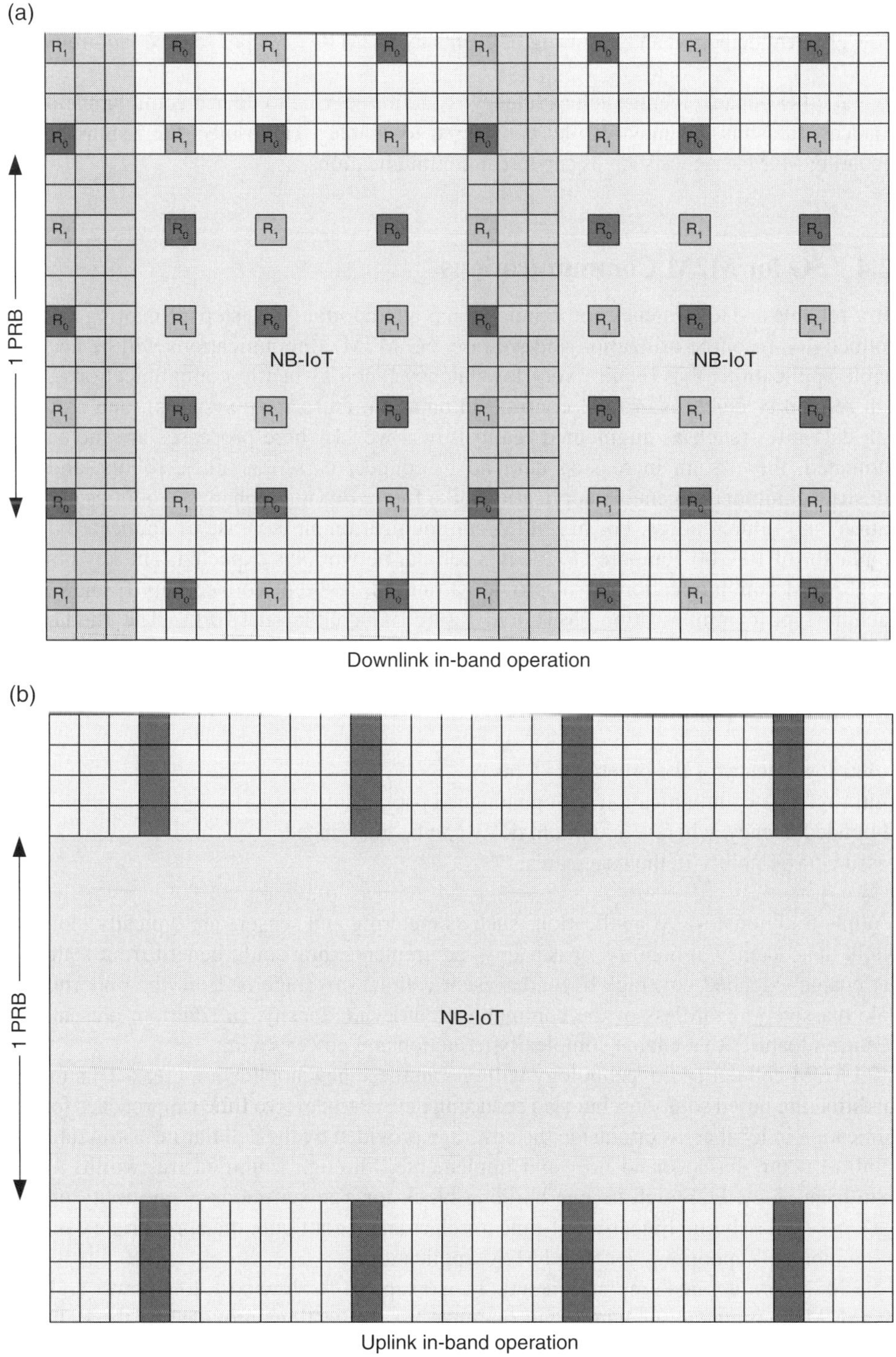

Figure 12.8 NB-IoT in-band operation.

subscriber server, as well as monitoring via MTC interworking function. These techniques allow efficient detection and reporting of events by M2M devices per service requirements.

Several other enhancements such as latency reduction, device to device communication, and broadcast transmission may also be considered to enable LTE to meet the requirements of upcoming M2M services such as car-to-car communications.

12.4 5G for M2M Communications

Ultra-reliable and low-latency communication is an important next step in enabling improved productivity, resource utilization and welfare. 5G M2M communications will be needed to enable applications that require very low latency (such as tactile communications), ultra-high reliability (such as critical control, automation, and safety systems), and extremely high data rates (such as augmented reality). In a world where processes are increasingly automated, there is an increasing demand to connect cars, machines, robots, and other industrial equipment to enable applications like future driving, enhanced robotics and automation, and future energy. For 5G M2M communication an increase of connected devices by a factor of 10–100 compared to today's cellular networks is expected. The key objective of 5G M2M communications is to provide a unified, resilient connectivity framework for machine type communication (MTC) in future 5G cellular networks. The fundamental challenge on this journey is to enable MTC traffic of mixed criticality: varied latency, reliability and capacity demand. The emerging industrial applications are identified as requiring one or more of:

- ultra-low latency of the order of 1–5 ms
- ultra-reliable communication with guaranteed packet delivery
- bounded latency, where a maximum delay can be guaranteed
- assured availability in the usage area.

While traditional M2M applications such as metering and sensors are typically more flexible in their latency, reliability, or capacity requirements they could benefit from techniques that enable extended coverage beyond present cellular coverage or from the opportunity to scale massively regardless of the communication/device density. In addition, present M2M UEs need features that enable complexity reduction and power saving.

5G M2M cellular radio technology will enable these new application areas. This includes infrastructure based solutions, but also emerging device-to-device (D2D) approaches for communication in local areas or outside the coverage provided by the cellular infrastructure. D2D is infrastructure-based or ad hoc, and implemented through a unified trustworthy cellular radio technology. D2D will be one building block for a mixed-mode connectivity platform that can seamlessly/intelligently self-tune to select and reconfigure the appropriate communication mode to support any localized M2M applications.

5G M2M communications will include native support for low-cost, low-power and wide-area M2M communication, which will optimize the applications provided by the LTE M2M evolution. Furthermore, 5G M2M communications will give local and wide area support for M2M applications. The local area M2M communications will enable new applications not addressed by cellular technology today, such as in-house communications, automated factories, and others.

For 5G, the access architecture using spectrum above 6 GHz is a definite possibility. As an example, the availability of huge bandwidth (for example, 13 GHz in the 70–80 GHz band) coupled with the use of large antenna arrays with as many as 128 elements (giving 9–21 dB in link budget gain) at both transmitter and receiver can make this band attractive for deploying high-capacity 5G networks. Generally, the range of 5G systems above 6 GHz is limited. To increase the range, very directional and narrow-beam transmissions will be used. Furthermore, high shadowing losses and blockages may occur at certain locations and/or antenna angles due to the radio wave propagation characteristics at mmWave frequencies. Thus, a cluster deployment, as shown in Figure 12.8, is envisioned. Cluster deployment allows multiple 5G eNBs to cover the same user and geographical area in case of blockage or high shadowing loss.

Figure 12.9 also illustrates 5G deployment with an LTE overlay to provide dual connectivity to both systems. With dual connectivity, the user can be simultaneously connected to both the LTE overlay and the 5G underlay cluster. This approach allows access to the high data rates of the 5G system with the reliability of the LTE overlay.

The bandwidth of the 5G systems can be very large: 1 GHz or 2 GHz in the 30–100 GHz band and 200–400 MHz at bands from 6–20 GHz. As a result, very high data throughputs can be achieved. The frame structure of the 5G system will be designed to meet a 1 ms latency requirement (corresponding to a 10-fold reduction compared to LTE). As an example, consider the frame structure for a 5G mmWave system as shown in Figure 12.10 [18]. The TDM slot size is 100 μs and the slot consists of an UL control portion followed by a DL control portion, then the data portion. In between is the guard period used to allow switching between transmit and receive. The data portion can be used to either transmit or receive data, as well as for backhaul. This allows an integrated wireless backhaul design to be included in the frame structure, increasing efficiency and flexibility.

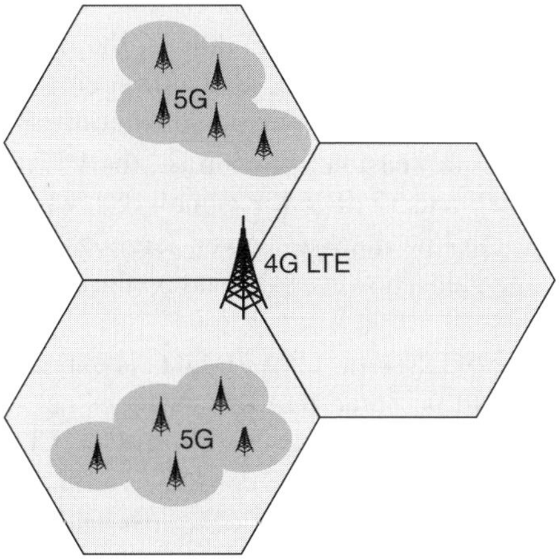

Figure 12.9 5G mmWave cluster deployment.

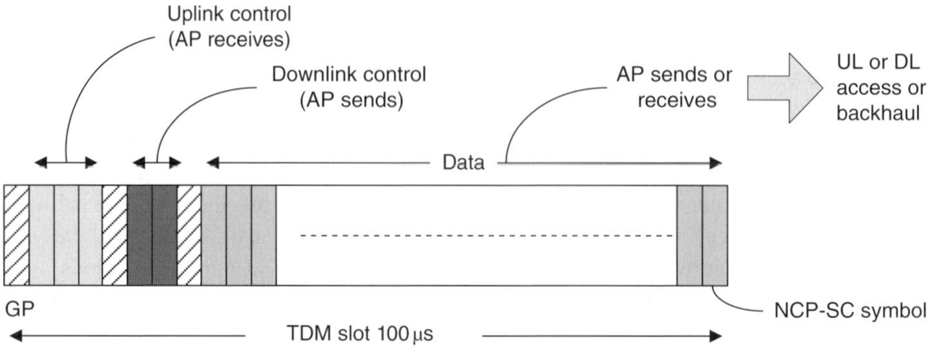

Figure 12.10 Example 5G frame structure.

12.4.1 Coverage

Deploying a local area network at frequencies above 6 GHz can be challenging from a coverage perspective. First, the propagation conditions at very high frequencies are quite challenging as the path loss can be quite high at these bands. In addition to the high path loss, oxygen and water absorption, and rain and foliage can introduce additional losses. Oxygen plus water absorption give a combined 19 dB/km loss at 60 GHz, but only a combined loss of 0.9 dB/km at 70 GHz and 0.52 dB/km at 80 GHz. For local area networks with range of approximately 100 m, however, this loss is only about 2 dB. Losses from foliage and other obstacles, however, may be very severe and may introduce blockages. Ghosh et al. performed measurements at 73.5 GHz to quantify the path loss more accurately [19]. For the scenarios considered, it was determined that the non-line of sight path-loss equation can be given by

$$PL[dB](d) = \alpha + 10\beta \log_{10}(d) + X$$

where $\alpha=82.70$, $\beta=2.69$, and $X=7.55$. The parameter d is the distance given in meters. At a distance of 100 m, the path loss would be approximately 144 dB, which is a significant amount.

In addition to the challenging propagation conditions, the transmit power will be limited due to the use of RFIC semiconductor to reduce size, power consumption, and cost. From the work of Cudak et al. [18], it is seen that approximate maximum transmit powers of 30 dBm in the DL and 19 dBm in the UL are feasible. In contrast, the DL transmit powers in LTE are around 46 dBm for a macro-cell and 37 dBm for a small cell, while the UE transmit power is 23 dBm. To compensate for the low transmit power, large directional antenna arrays at the eNB and UE will be used. The gains, however, depend on the array size as well as the antenna coupling efficiency.

As a result, the range of the 5G systems above 6 GHz is expected to be up to several hundred meters. However, it should be noted that this assumes cell-edge data rates of around 100 Mbps while maintaining ultra-low latency. Similar to LTE, coverage enhancement features may also be standardized. For example, consider a path loss of 144 dB corresponding to a 100-m cell range. If a 15-dB coverage extension is used, then the cell range would increase to 360 m. This, however, would reduce the cell-edge data rate and increase the latency significantly.

12.4.2 *Latency*

For data reporting including system access, LTE requires approximately 100 ms to complete the entire process as shown in Table 12.10. This includes synchronization acquisition, system information acquisition, random access, RRC connection setup and security mode command, payload transmission, then RRC connection release. For a 5G system, there is already a significant reduction due to the 10-fold reduction in subframe length. In addition, a more streamlined access procedure can be considered. For example, in LTE, up to 40 ms may be required to acquire MIBs. A new design will allow this information to be acquired in 100 μs. Some overhead signaling reduction techniques, such as data transmission without user-plane radio bearer establishment, can also be considered. This approach reduces the overhead by approximately 50%, which would help with latency as well. Using all of these techniques, it is expected that the time to connect to the system and transmit data may be reduced from 100 ms in LTE to less than 2 ms for 5G. This represents a 50-fold reduction.

12.4.3 *Capacity*

The capacity of a 5G system at 73.5 GHz is illustrated in Table 12.12 [19]. The simulation scenario is that of an 800×800 m urban rectangular street grid. The number of eNBs covering this area is shown in Table 12.12, with a uniform distribution of access-point (AP) placement. Each AP consists of four sectors, with each sector having two 4×4 antenna arrays. UEs are dropped randomly into the grid. Outage probability is defined as the percentage of UEs with throughput below 100 Mbps.

Results from Table 12.12 show that very high UE throughput can be achieved from a 5G system. It also illustrates potential coverage issues, as large numbers of eNBs are needed to provide ubiquitous coverage. This, however, may not be a problem since the device may rely on an LTE overlay when out of coverage.

12.5 Conclusion

The IoT will bring about massive number of connected things, in diverse areas such as smart cities, consumer electronics, transportation, automotive, healthcare, remote monitoring and management, and public safety. Different M2M services have different requirements with respect to throughput, latency, coverage, reliability, cost, and power consumption. As a result, several types of networks – personal area, local area, and wide area – will be used to connect things together. In the 2020 timeframe, the following wide- and local-area network support strategy can be envisioned. LTE evolution will serve most M2M services: consumer

Table 12.12 5G performance results at 73.5 GHz.

Parameters	72 eNBs	132 eNBs
Average UE throughput	5.6 Gbps	6.5 Gbps
Cell-edge (5%-tile) UE throughput	26.7 Mbps	461 Mbps
Outage probability	5.98%	3.15%

electronics, building automation, automotive, smart cities, health care, and agriculture. The features introduced will enable low-cost devices, coverage enhancement, and ultra-long battery life. They are intended to improve LTE effectiveness in serving delay-tolerant and low-data-rate M2M services. Meanwhile, the 5G system will be used for ultra-low latency, ultra-high reliability, and high-throughput M2M services, such as tactile communication, augmented reality, high-definition video, and other applications requiring a near real-time response.

These two systems complement one another and will serve to provide underlying infrastructure for the IoT.

References

[1] J. Bradley, J. Barbier, and D. Handler (2013) Cisco white paper: Embracing the Internet of everything to capture your share of $14.4 trillion.

[2] J. Morrish, E. Buckland, and M. Hatton (2012) Machina Research Annual Report: M2M global forecast and analysis 2011–22.

[3] 3GPP (2014) TR 23.368: Service requirements for machine-type communications (MTC), v.13.0.0.

[4] 3GPP (2013) TR 23.887: Study on machine-type communications (MTC) and other mobile data applications communications enhancements, v.12.0.0.

[5] 3GPP (2013) TR 36.888: Study on provision of low-cost machine-type communications (MTC) user equipments (UEs) based on LTE, v.12.0.0.

[6] 3GPP (2013) TR 37.868: RAN improvements for machine-type communications, v.11.0.0.

[7] 3GPP (2013) TR 37.869: study on enhancements to machine-type communications (MTC) and other mobile data applications, v.12.0.0.

[8] 3GPP (2014) TS 36.331: Radio resource control (RRC) protocol specification, v.12.3.0.

[9] 3GPP (2013) TS 36.321: Medium access control (MAC) protocol specification, v.11.3.0.

[10] 3GPP (2014) RP-141660, New WI proposal: Further LTE physical layer enhancements for MTC, Ericsson, Nokia Networks, RAN#65, Edinburgh, Scotland.

[11] R. Ratasuk, N. Mangalvedhe, and A. Ghosh (2014) M2M communications in cellular network deployments using LTE. *International Conference on Telecommunications.*

[12] R. Ratasuk, J. Tan, and A. Ghosh (2012) Coverage and capacity analysis for machine type communications in LTE. *Vehicular Technology Conference (VTC Spring), 2012 IEEE 75th*, pp. 1–5.

[13] R. Ratasuk, S. Iraji, K. Hugl, et al. (2013) Performance of low-cost LTE devices for advanced metering infrastructure. *Vehicular Technology Conference (VTC Spring), 2013 IEEE 77th*, pp. 1–5.

[14] 3GPP (2006) TR 25.814: Physical layer aspects for evolved UTRA, v.2.0.0.

[15] R. Ratasuk, Y. Zhang, D. Lei, et al. (2014) Coverage enhancement for M2M communications using LTE. *International Conference on Telecommunications.*

[16] IEEE 802.16 Broadband Wireless Access Working Group (2010) 802.16p-10/0005: Machine to machine (M2M) communications technical report.

[17] IEEE 802.16 Broadband Wireless Access Working Group (2011) 802.16p-10/0014: IEEE 802.16p machine to machine (M2M) evaluation methodology document (EMD).

[18] M. Cudak, A. Ghosh, T. Kovarik, et al. (2013) Moving towards mmWave-based beyond-4G (B-4G) technology. *Vehicular Technology Conference (VTC Spring), 2013 IEEE 77th*, pp. 1–5.

[19] A. Ghosh, T.A. Thomas, M.C. Cudak, et al. (2014) Millimeter-wave enhanced local area systems: a high-data-rate approach for future wireless networks. *IEEE Journal on Selected Areas in Communications*, **32** (6), 1152–1163.

13

Low-latency Radio-interface Perspectives for Small-cell 5G Networks

Toni Levanen,[1] Juho Pirskanen,[2] and Mikko Valkama[1]
[1] *Tampere University of Technology, Tampere, Finland*
[2] *Nokia Networks, Espoo, Finland*

13.1 Introduction to Low-latency Radio-interface Design

Reducing radio access latency is one of the key targets of the future 5G systems [1, 2]. There are two main reasons for needing an ultra-low-latency radio interface in 5G. First of all, if we want to keep the amount of memory required for hybrid automatic repeat request (HARQ) and other physical-layer processes approximately the same as for current solutions, or at least to increase at a slower rate than the actual throughput, we have to decrease the time duration of the transmitted frames as the frequency allocation increases [2]. This unavoidably leads to reduced physical layer latency. Secondly, because new applications and use cases of the future wireless access – cloud services, augmented reality, self driving cars, and so on – require very low latency radio interfaces, we have to provide a physical layer numerology capable of achieving a physical layer (PHY) round-trip-time (RTT) of 1 ms or even less [3]. The 1-ms target is generally considered as a good standard when compared to the LTE-A PHY latencies of the order of 10 ms [4].

The PHY design is dictated by the channel environment and the expected propagation conditions related to the assumed carrier frequency and maximum cell size. Therefore, we start by introducing the expected channel environments for cm-wave channels and mm-wave channels in Section 13.2. In Section 13.2, we will also shortly discuss the expected traffic distributions in 5G small-cell networks (SCN). Then, as the physical limitations of the channel environment are clear, we can optimize the PHY design for this environment with the main emphasis on

Towards 5G: Applications, Requirements and Candidate Technologies, First Edition.
Edited by Rath Vannithamby and Shilpa Talwar.

Companion website: www.wiley.com/go/vannithamby/towards5g

minimizing latency without compromising the system capacity. The good thing about minimizing latency is that it automatically leads to minimized duty cycles, especially with time division duplex (TDD) access and also maximizes the sleeping times of the devices. By minimized duty cycles the active on-times of the devices are minimized as the durations of symbols and frames are reduced. Agile TDD radio access is assumed to be the best solution for small cells, where the number of simultaneously active users may be assumed to be relatively small, especially when compared to macro cells. If the number of active users per cell is relatively high, say in the hotspot use case, by scheduling only a few users per frame we can still gain from the agile TDD because we can push per-user packets through the radio interface faster. And due to the reduced frame duration, we can serve larger groups of users by distributing them in the time domain. The open research question is whether we want to drive the system towards very dynamic interference conditions to alleviate the operation of advanced interference cancellation (AIC) receivers or do we try the stabilize the downlink (DL)–uplink (UL) allocation as the network load increases. By an AIC receiver, we mean a receiver capable of detecting and subtracting one or multiple interfering data streams from the received signal and by interference dynamics, we emphasize here the power dynamics. Time dynamics is also a critical aspect and requires further study. By agile TDD radio access, we emphasize that the DL–UL allocation can be changed on a per-frame basis to meet the instantaneous traffic conditions. LTE-A Release 12 introduces the first steps in this direction by providing for the possibility of utilizing UL subframes for DL transmission based on eNB scheduling decisions [5]. However, this adaptation is very restricted due to stringent backwards compatibility requirements.

Current LTE networks, especially in Europe and the USA, utilize frequency division duplex (FDD), which has specific frequency allocations for uplink and downlink transmissions. However, when new higher frequencies bands are to be used with wider system bandwidths FDD becomes problematic, as it is difficult to find two wideband frequency portions with a suitable duplex distance to be combined into a frequency pair. Additionally, when traffic is highly dynamic and asymmetric between UL and DL, at least at the frame timescale, the FDD cannot utilize all allocated spectrum efficiently and spectrum efficiency becomes low. Finally, FDD is not beneficial for machine-to-machine communications because additional receiver or transmitter chains would be needed in every device for such operations. These facts evidently push towards a TDD-based solution, which is further required by the requirement for flexibility in quickly adapting to changing traffic requirements. The radio-interface designs presented in this chapter are considered only for TDD access. The discussion related to channel and traffic models applies to both FDD and TDD access.

The main difficulty in designing an ultra-low-latency radio interface is to minimize the overheads of the control channels and the reference symbols (RSs). The design of the control field and related overhead is ameliorated if we assume that we serve only a few users simultaneously per transmitted frame. This assumption is valid for a very fast radio interface in an ultra-dense SCN. Also, we should concentrate on pushing the data packet of a certain users through the PHY layer as fast as possible and this requires a maximal spectral allocation per user. The faster we can transmit a packet, the shorter is the active time of the transmitter and receiver chains, or the duty cycle, and the better is the energy efficiency of the transceiver. Flexible TDD is considered because it provides the most degrees of freedom in allocating DL and UL resources per frame for a small number of users. In the RS layout design, as we decrease the frame duration we note that the channel correlation time is relatively large when compared to the frame duration and therefore we can transmit DL and UL RSs only once per frame. This allows us to reduce the RS overhead and more importantly enables efficient pipelined receiver architectures. These topics are further discussed in Section 13.3.

In the 5G flexible TDD-based local area (5GETLA) reference design we considered 100-MHz channel bandwidth as the baseline. Even with this bandwidth it is difficult to achieve the multi-gigabit link throughput envisioned for 5G communications. The solutions for multi-gigabit throughput are carrier aggregation or even wider bandwidths. So far, we have not considered the carrier aggregation scheme. Therefore, in Section 13.4, we will extend our design considerations to mm-wave communications [6] and propose a mm-5GETLA frame structure that can provide PHY layer RTT of the order of 0.1 ms with 1-GHz channel bandwidth. This kind of PHY layer latency, a fraction of a millisecond, allows us to achieve two major goals. The first achieved goal is that we can support higher layer latencies, of the order of 1 ms, if the network design is sufficiently flat, the core network delays are reduced from the current solutions, and the content we are accessing is sufficiently close to the end user [7]. The second goal is that with this kind of PHY layer RTT combined with ultra-fast synchronization and simplified association, we can provide high-throughput services even for high mobility users in a mm-wave-based ultradense SCN.

In Section 13.5, we will wrap up the design principles described in this chapter and also discuss the most important open topics related to ultra-low-latency radio-interface design for cm-wave and mm-wave communications.

13.2 Small-cell Channel Environment Considerations and Expected Traffic

One of the key enablers for ultra-low PHY latency is the small cell, and small-cell development is driven by increased system capacity and area-throughput requirements. As we reduce the cell size the probability for line-of-sight (LOS) connection increases and the propagation times and experienced delay spreads become shorter. This allows us to reduce the cyclic prefix (CP) used to protect transmitted symbols from inter-symbol-interference and also to reduce the required guard periods (GP). The GP is used to allow reception of DL transmission in all of the scheduled user equipments (UE) before switching to UL mode and to provide additional time for early transmissions controlled by timing alignment (TA), to provide a transition time for hardware to switch from DL to UL or vice versa, and also to protect the receivers from hearing echoes of their own transmissions in an ultra-low-latency TDD-based radio access system.

In a small-cell environment, TDD access is considered to be superior because it can utilize unpaired frequency bands for both DL and UL and it allows agile adaptation to the traffic orientation inside a small cell [8]. In a small cell, there are typically only a few simultaneously active devices and therefore the traffic demand can vary from only-DL to only-UL. We emphasize that as we speak of simultaneously active devices, we are considering the frame timescale (″0.5 ms). The cell may also contain devices in short or long sleep cycles, and devices in an active state but with currently empty transmission (Tx) and reception (Rx) buffers. Flexible and dynamic TDD-based radio interfaces are the best solution to tackle this problem. On the other hand, using per-frame adaptation in the DL and UL allocation can cause severe interference problems between neighboring small cells. The interference types encountered in a TDD network are presented in Figure 13.1. The 5GETLA access nodes

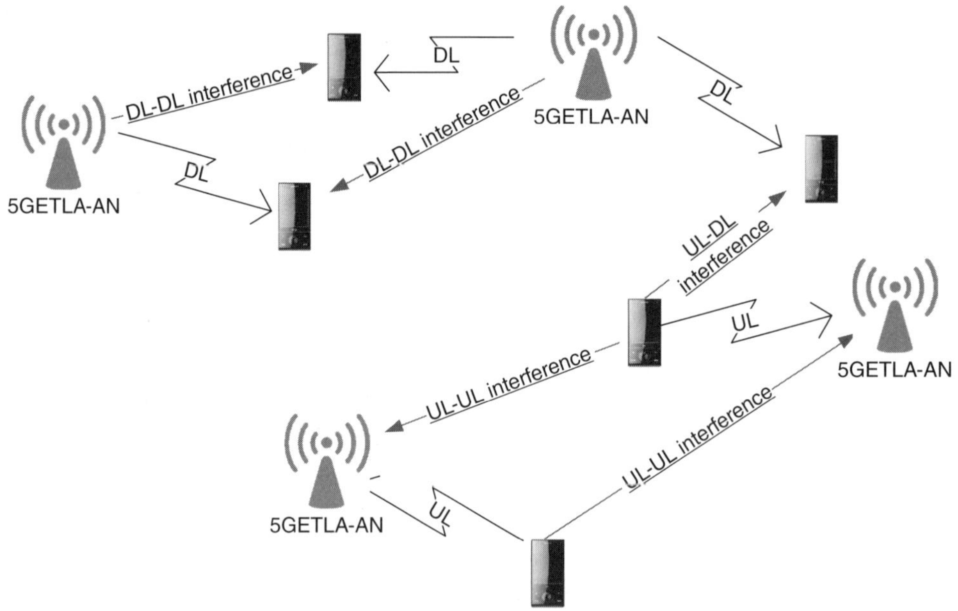

Figure 13.1 Examples of the interference types encountered in a small-cell network based on agile TDD radio access.

(ANs) are assumed to operate without centralized control but can have over-the-air communication with their closest neighbors for local cooperation and coordination. The DL–DL and UL–UL interference types are common for all networks, but the UL–DL interference type can be a severe problem in uncoordinated networks. In this case it is possible that the UL transmission will completely destroy DL reception. The simplest and most commonly used solution is to add a network controller to manage transmissions and interference among neighboring cells. Another way to cope with this kind of interference is a novel PHY layer design combined with AIC techniques. The AIC techniques typically work well in the case that the desired or interfering signal is clearly stronger, as long as the channel-estimation accuracy is sufficient for both components, as shown for example by Halperin et al. [9]. However, AIC introduces significant processing requirements, as the interfering packets need to be detected and subtracted before our own signal can be detected, and this might lead to problems with very stringent latency requirements. The situation is even worse if the transmissions are not time- and frequency-synchronized.

13.2.1 Centimeter-wave Channel Models

To design a new PHY layer numerology, we have to understand the channel environment in which we are supposed to communicate. The 5GETLA reference design [10] is designed for cm-wave communications with emphasis on channel models available for carrier frequencies below 6 GHz. Currently, the most commonly used channel models for SCN or wireless local area networks (WLAN) are the IEEE WLAN 802.11n channel models [11] and their refinement for IEEE 802.11ac channel models [12], and the WINNER channel models [13]. Current IEEE WLAN 802.11ac [14–16] capable devices have been designed based on simulations on the IEEE WLAN 802.11n and 802.11ac channel models. In Table 13.1 the main parameters for the IEEE WLAN 802.11n channel models are shown. Typically, for indoor scenarios, channel model D is assumed. Channel models E and F are for larger open indoor areas and outdoor scenarios, respectively [11]. Another commonly used set of channel models, especially in 3GPP LTE-related modeling are the WINNER channel models. The WINNER channel models and related measurements form the basis for the channel models provided 3GPP's E-UTRA paper [17]. The most interesting WINNER channel environments and their relevant parameters are provided in Table 13.2. The results contain different values for LOS and non-line-of-sight (NLOS) for certain scenarios. For the amplitude coherence bandwidth we have assumed an exponentially decaying channel-impulse response with Rayleigh fading.

From Tables 13.1 and 13.2, we can see that a RS interval of 2 MHz in the frequency domain provides a good channel estimation accuracy for most of the considered channels and that the RMS delay spreads are typically very short, of the order of 100 ns or less. If we consider small cells where the maximum supported radius is of the order of 50 m, the propagation delay is in maximum approximately 167 ns. With these values we can already state that the minimum CP duration for our new radio interface has to be, at a minimum, approximately 300 ns, assuming that coarse TA is used. In IEEE WLAN 802.11n, the short CP duration is 400 ns and it has been shown that with sufficient synchronization accuracy it works well even with IEEE WLAN 802.11n channel model F [16]. Also, using carrier frequencies between 5 GHz and 30 GHz allows us to use larger antenna arrays and narrower beams for data transmission. As we decrease the beamwidth, the experienced delay spreads become shorter and this effectively allows us to reduce the CP used for data transmission [18].

Table 13.1 Main parameters related to WLAN 802.11 TGn channel models.

Channel model	B	C	D	E	F
RMS delay spread (ns)	15	30	50	100	150
Coherence bandwidth with 50% correlation (MHz)	18.4	9.2	5.5	2.8	1.84
Maximum excess delay (ns)	80	200	390	730	1,050
Minimum reference symbol interval (MHz)	12.5	5	2.6	1.4	1.0

Table 13.2 WINNER small cell related channel model parameters.

Channel model	$A1_{LOS}$	$A1_{NLOS}$	$B3_{LOS}$	$B3_{NLOS}$	*A2*	*B4*
RMS delay spread 50% fractals (ns)	26.1	25.2	26.5	39.3	21.4	45.4
50% correlation bandwidth (MHz)	10.6	10.9	10.4	7.01	12.9	6.07
Max. delay 50% fractals (ns)	180	146	125	175	175	239
Min. reference symbol interval (MHz)	5.6	6.9	8.0	5.7	5.7	4.2
RMS delay spread 90% fractals (ns)	47.3	44.6	40.9	47.2	65.5	109
50% correlation bandwidth (MHz)	5.8	6.2	6.7	5.8	4.2	2.5
Max. delay 90% fractals (ns)	377	252	175	250	362	487
Min. reference symbol interval (MHz)	2.6	4.0	5.7	4.0	2.8	2.1

A1, indoor office/residential; B3, large indoor hall; A2, indoor to outdoor; B4, outdoor-to-indoor.

Table 13.3 Mobility related parameters for 3.5-GHz and 5.0-GHz carriers.

Carrier frequency (GHz)	3.5		5	
Velocity (km/h)	3	30	3	30
Coherence time with 50% correlation (ms)	24.9	2.5	17.4	1.7
Coherence time with 90% correlation (ms)	10.5	1.0	7.3	0.7
Max. Doppler frequency (Hz)	9.8	97.3	13.9	139.0
Min. reference symbol interval in time (ms)	51.3	5.1	35.9	3.6

In Table 13.3, we present the main mobility-related parameters for two carrier frequencies that could be available to 5G small-cell networks: the 3.5-GHz and the 5-GHz carrier. We have also considered two different device mobilities: 3 km/h, which would correspond to normal walking speed, and 30 km/h, which we consider to be the maximum supported mobility for the 5GETLA reference design. For the coherence-time evaluation in Table 13.3, we have assumed Jakes spectrum and the minimum RS interval is defined as half of the inverse of the maximum Doppler frequency based on Nyquist sampling theorem. In Tables 13.1–13.3, the minimum RS interval allows us to acquire all essential information about the channel-impulse response. From Table 13.3, we can see that in all scenarios the minimum RS interval in time is 3.6 ms and 90% time correlation is 0.7 ms. If we can design a PHY layer frame structure with frame duration shorter than these, we can transmit DL and UL RS only once per frame and obtain a snapshot of the channel state that is valid over the whole frame. This facilitates the design of a spectrally efficient, low-latency radio interface. If desired, some simple linear extrapolation based on previous and current channel estimates may be used to improve the system performance.

13.2.2 Millimeter-wave Channel Models

The idea of using mm-waves in the range of 30–300 GHz for wireless communications is not new, although for a long time people were considering it from the wide-area network or macro-layer point of view, which made it look impossible. Also, the development of hardware components to enable low-power implementation for this frequency range has gained momentum since the release of the 60-GHz spectrum by the Federal Communications Commission [19]. Now that we are talking about ultra-dense SCNs, where the link distances are of the order of tens of meters, heavy rain attenuation or air absorption are not an issue [6]. In addition, the experienced path loss does not directly increase with the frequency if we keep the physical size of the antenna fixed or use large antenna arrays of the same effective size. This means that we will go towards large antenna arrays with mm-wave communications, where large refers only to the number of antennas and not to the actual physical size. As our practical experience on antenna-element design and 2D-antenna element arrays has improved, we are prepared to support the large antenna arrays that can achieve the beamforming (BF) signal-to-noise-ratio (SNR) gains that are required to support mm-wave communications.

In Tables 13.4 and 13.5, the expected and 99th percentile values are given for the measured RMS delays in four different locations within the University of Texas at Austin campus. The detailed description of the measurement setup and specific measurement locations noted as

Table 13.4 Measured RMS delay-spread values for a 25-dBi Rx antenna at 38 GHz.

	Measurement location			
	WRW – A (ns)	*ENS – A* (ns)	*ENS – B* (ns)	*ECJ* (ns)
Expected value	13.5	15.1	5.3	16.5
99th percentile value	65.4	129	27.6	133

The expected RMS delay-spread value over all realizations is 13.7 ns and the 99th percentile is 129 ns.
Source: Rappaport et al. [20].

Table 13.5 RMS delay spread values for a 13.3-dBi Rx antenna at 38 GHz.

	Measurement location			
	WRW – A (ns)	*ENS – A* (ns)	*ENS – B* (ns)	*ECJ* (ns)
Expected value	10.1	17.3	4.8	17.5
99th percentile value	57.1	81.2	18.5	135

The expected RMS delay spread value over all realizations is 12.7 ns and the 99-percentile is 110 ns.
Source: Rappaport et al. [20].

WRW-A, ENS-A, ENS-B, and ECJ are provided in the paper by Rappaport et al. [20]. There it is also mentioned that street canyons are especially good environments for yielding high RMS delay spreads, since they have many reflective objects spaced at nearly regular intervals for long distances and very few obstructions that block reflected or scattered waves. The maximum measured NLOS RMS delay spreads were 185 ns and 166 ns for the 25- and 13.3-dBi receiver antennas, respectively. To provide a comparison, 25- and 13.3-dBi antenna gains correspond approximately to 256 and 16 element uniform linear antenna arrays, respectively. In the same paper it was indicated that as the distance between the transmitter and the receiver increases, the RMS delay spreads typically decrease due to increased link attenuation [20]. These initial measurements inform us of the channel delay spreads to be expected in outdoor environments for mm-wave communications. These results are not all-encompassing but are used as a starting point for our mm-5GETLA NLOS PHY layer numerology proposal (see Section 13.4).

When it comes to LOS RMS delay spreads, the maximum measured value mentioned by Rappaport et al. was 15.5 ns and most of the measured values were close to 1 ns [20]. Kyrö provided mean RMS delay spreads in different indoor environments for the 60-GHz carrier frequency [21, Table 3.2]. Among the 23 values presented, the maximum mean delay spread was detected from a large hall in an NLOS scenario with a value 22 ns. In most of the indoor measurements provided the mean RMS delay spread was around 20 ns. These measurements show that at least in LOS communications and in indoor communications at 60-GHz carrier frequency the expected RMS delay spreads are very small, typically below 20 ns. This special aspect is noted in our proposal for a mm-5GETLA LOS PHY layer numerology.

Table 13.6 Mobility-related parameters for 70-GHz carrier.

Carrier frequency (GHz)	70	
Velocity (km/h)	30	360
Coherence time with 50% correlation (μs)	124.4	10.4
Coherence time with 90% correlation (μs)	52.4	4.4
Max. Doppler frequency (Hz)	1,945.8	23,349.5
Min. RS interval in time (μs)	257.0	21.4

Let us next look at the UE mobility-related parametrization for the 70-GHz carrier. In Table 13.6, we have evaluated the coherence times and minimum RS intervals in time based on assumptions similar to those in Section 13.2.1. These results indicate that a UE mobility of 30 km/h is rather easily supported by the mm-5GETLA design proposed later on. If we want to support very high mobility users with a mm-wave-based SCN, then based on these results we would have to increase the number of RS transmitted inside the frame to allow sufficient channel estimation accuracy. It should be emphasized that these values are based on simple assumptions that might not hold in strongly beamformed mm-wave communications and this question needs to be carefully studied before reaching conclusions. Especially in the case of highly beamformed transmissions the Doppler spread will likely reduce to Doppler shift, which can be easily tackled by the synchronization algorithms. In general, the channel models for mm-wave communications have not been extensively studied and the amount of information we have about the delay spreads and coherence times in mm-wave channels is still relatively scarce. Therefore, the mm-5GETLA frame structure discussed in Section 13.4 is considered as a proposal for mm-wave communications; as new information on the propagation environment becomes available we may have to reconsider the proposed PHY layer numerology.

13.2.3 Comments on Expected Traffic and Traffic Modeling

In the Cisco forecast for 2018 [22], it is indicated that over two-thirds of the world's mobile data traffic will be video as soon 2018 and most of this content will be streamed through small cells, mostly due to offloading but also due to the increase in cloud-based services used at home by the end user. In particular, the development of ultra-high definition television is pushing the required bit rates for video streams. At the same time, more and more content is downloaded through IP-based services to satisfy the on-demand requirements of the end users, instead of being watched during the actual broadcasting time or from local media. Therefore, updating the traffic models used for video traffic is one critical step in providing correct modeling for the main traffic that will be generated in future SCNs.

Another important aspect typically not sufficiently noted in the PHY layer design is the significant fraction of small-packet traffic generated by the UE. The expected traffic for local area communications has been described by the IEEE 802.11 working group, with emphasis on the high dynamics of packet sizes expected in future small cells [23, 24]. These results are backed up by the measurements provided by Palit et al. [25]. Therefore, in addition to providing high throughput, the system design has to be able to cope with large quantities of small packets. Our solution for the cm-wave design is the best-effort contention-based data channel, which allows users to transmit small packets without scheduling requests, thus minimizing the delays and providing an opportunistic, ultra-low-latency transmission channel.

When it comes to modeling the traffic in future low-latency SCNs, there are two important modeling principles that are currently not well noted. First of all, in addition to the actual high-data-rate traffic, such as video streams, the devices have to generate small packet streams modeling the higher-layer acknowledgments (ACK) related to high-data-rate traffic, and the UE has to generate low-rate, sporadic small-packet streams to model the existing background processes: from social media applications, stock exchange status updates, email status indicators, and so on. TDD radio interface performance is not properly evaluated if only the high–data-rate stream is modeled because the small packets transmitted in both directions also require resources, and in the case of higher-layer ACK they are especially susceptible to the latency experienced.

The second important modeling aspect is to use finite buffer traffic when modeling the system performance. From the system point of view, the actual active data transmission phase is relatively short and the interference statistics vary significantly with respect to time in the small-cell environment. Additionally, when using a finite buffer model the initial queue time – the time between receiving a packet to transmit in the medium access control (MAC) layer and transmitting the first part of that packet – affects the total achieved bitrate. This emphasizes the system's capability to rapidly provide transmission resources to the UE having UL data to transmit. The finite buffer nature of the data transmission may also affect the optimal scheduler design [26]. Furthermore, if full buffer-like traffic is modeled, with each device always having something to transmit in DL or UL, the fixed DL–UL ratio corresponding to the average traffic distribution in the cell would typically give the best performance assuming receivers without AIC capabilities.

13.3 New Radio-interface Design for Low-latency 5G Wireless Access

When the design of 5GETLA PHY layer was initiated, the main target was to minimize latency. Minimized latency inevitably leads to reduced duty cycles and improved energy efficiency with a proper MAC layer design. In addition to ultra-low latency, we wanted to keep the spectral efficiency at least on the same level as in the current LTE [5, 27]. The 5GETLA design is basically a combination of:

- the AN-controlled, synchronized, and scheduled frame-based access scheme, as in mobile networks
- optimized PHY layer numerology for local area and SCNs, as in IEEE WLAN 802.11.

By combining the two, we can get the best of both worlds: a low-latency, spectrally, and energy-efficient radio interface optimized for small cells, which is capable of fairly serving a large number of users. The main design parameters for the 5GETLA system are provided in Table 13.7. Because we see 5GETLA as a revolutionary step from LTE and we believe that some support for LTE-X generation will be supported by future mobile devices, we use as a sampling rate $F_s = 92.16$ MHz, which is an integer multiple of the current LTE sampling rate of 30.72 MHz. The assumed channel bandwidth is 100 MHz and the bandwidth utilization factor is $\alpha = 0.9$, meaning that the effective transmission bandwidth is 90 MHz. The frames are

Table 13.7 Main parameters for the 5GETLA radio interface.

Parameter	Value
F_s: Sampling frequency (MHz)	92.16
B: Channel bandwidth (MHz)	100
α_{new}: Bandwidth utilization factor	0.9
N_{slot}: Number of slots per frame	5
N_{RB}: Number of resource blocks	5
Δf_{RB}: Frequency allocation per RB (MHz)	18
$T_{longframe}$: Long frame duration (ms)	10
$T_{masterframe}$: Master frame duration (ms)	100

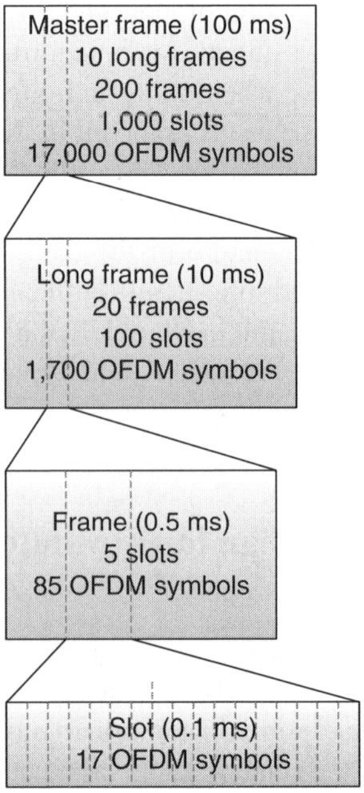

Figure 13.2 Frame hierarchy for the 5GETLA design.

divided in time into five slots, if slot wise DL–UL allocation is preferred, and in frequency into five resource blocks (RB). Note, that in our notation the RB defines only the frequency allocation and the DL–UL allocation varies per frame basis. For interoperation with existing LTE networks, we have also defined a higher-level frame structure of durations $T_{longframe} = 10$ ms and $T_{masterframe} = 100$ ms, as indicated in Figure 13.2. These higher-level frame structures also provide a basis cycle for short and long sleep times. The master frame duration also defines the minimum repetition time for the synchronization channel and for 5GETLA MIB transmission [10].

The main design principles of the 5GETLA are discussed in detail in the literature [10], as is the reduced latency numerology design based on 256-point FFT [28]. In this section, we will revise the main principles and results, and concentrate in particular on the achievable

Table 13.8 PHY layer numerologies for low-latency 5GETLA.

N_{FFT}	T_{FFT} (μs)	Δf_{SC} (kHz)	N_{Act}	$N_{sym/slot}$	T_{CP} (ns)	N_{CP}	CP overhead (%)
512	5.56	180	500	17	347/326	32/30	5.6
256	2.78	360	250	16	347	32	11.1

PHY layer RTT. In Table 13.8, the proposed PHY layer numerology for the 5GETLA reference model using 512-point FFT and the ultra-low-latency version using 256-point FFT are given. With the 512-point FFT-based design the symbol duration is 5.56 μs and the CP overhead is 5.6%; with the 256-point FFT-based design the OFDM symbol duration is 2.78 μs and the CP overhead is 11.1%. For the 512-point FFT design, each slot has one OFDM symbol, with 347 ns CP and 16 symbols with 326 ns CP, whereas the shorter design has 347 ns CP for all 16 OFDM symbols per slot. As discussed in Section 13.2, the minimal CP duration for a small-cell environment is approximately 300 ns. In our design, we have pushed the CP duration close to this limit to achieve extremely low latency in the radio interface. In the ultra-low-latency design with 256-point FFT, we have sacrificed some of the spectral efficiency through increased CP overhead to achieve PHY RTT in the UL of below 1 ms.

The basic layout of the 5GETLA frame design is shown in Figure 13.3. Our frame design has clearly defined logical channels, which are listed next for clarity. For more a detailed description of the logical channels, please see Levanen et al. [10].

Downlink Common Reference Symbol: The DLCRS is used to estimate the downlink control channel and in the case of single-stream transmission also the downlink data channel. DLCRS can also be used for synchronization [10]. The DLCRS has a frequency reuse factor of 10 with the 512-point FFT design and 5 with the 256-point FFT based design. This frequency reuse factor reduces the interference among the closest neighbors in the 5GETLA network, improves the channel estimation and synchronization quality, and enables simultaneous measurement of multiple closest-neighbor 5GETLA-AN control channels and alleviates the DL AIC implementation for the DL control channel.

Downlink Control Channel: The DLCCH carries the minimum amount of broadcast information to define scheduled UEs, used modulation and coding scheme (MCS), the number of data streams, and so on. The DLCCH also provides the ACKs for the UL transmissions of the previous frame. If a contention-based channel is active, it also contains ACKs for detected packets, and indicates detected sequence identifiers from the random access channel.

Downlink Shared Channel: The DLSCH contains the downlink data channel and possibly downlink channel state information reference symbols (DLCSIRS) or downlink demodulation reference symbols (DLDMRS). The possible allocation for RSs in the DLSCH is explicitly indicated by the demodulation reference symbol (DMRS) or channel state information reference symbol (CSIRS) frame indicator and by the number of DL streams.

Downlink Data Channel: The DLDCH carries the actual data packets for the UE. Each data packet is appended with a control header for non-broadcasted control information.

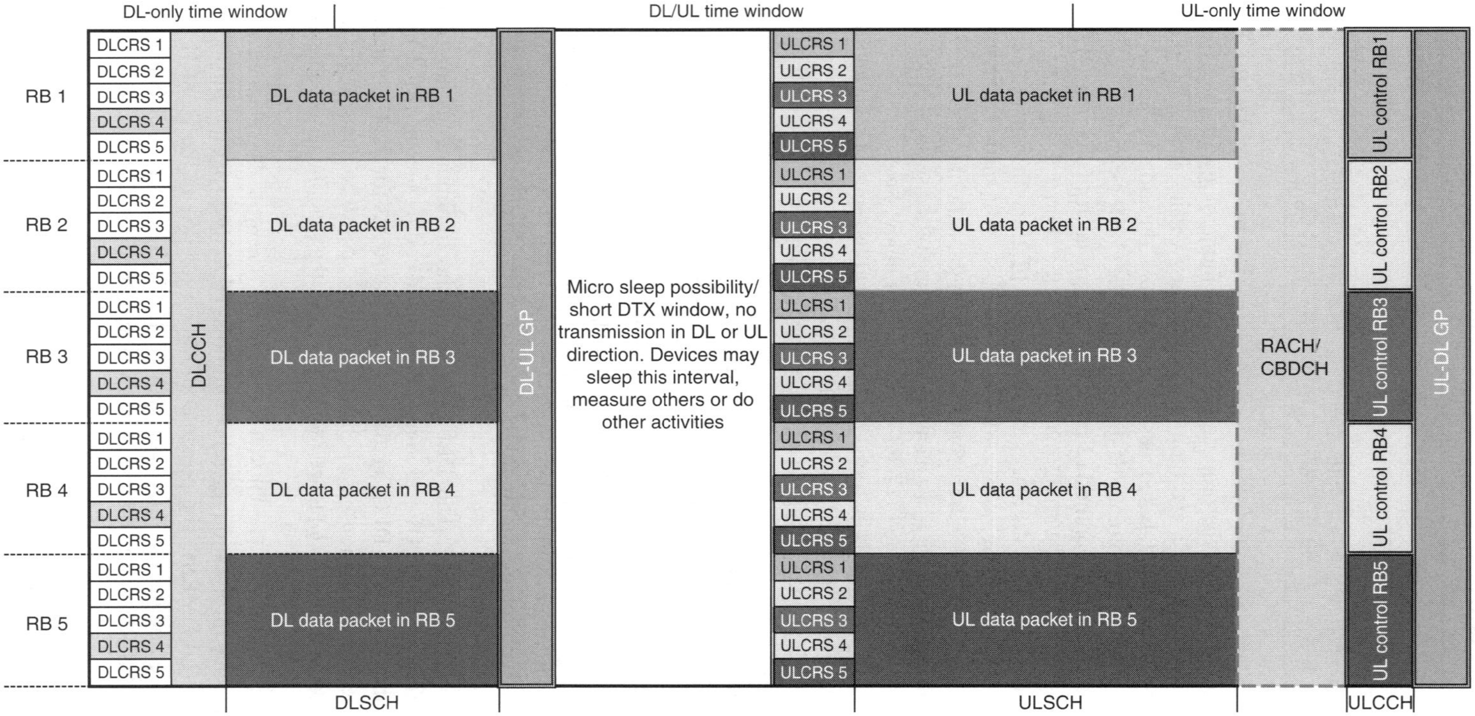

Figure 13.3 Reference frame design for 5GETLA. DMRS or CSIRS are not shown for clarity. In the case of no DMRS or CSIRS, DLSCH (ULSCH) is completely allocated for DLDCH (ULDCH).

Downlink Channel State Information Reference Symbol: DLCSIRS are used to estimate the channel between the 5GETLA access node (5GETLA-AN) and UE, and also measure signal-to-interference-and-noise (SINR) statistics related to DL–DL interference.

Downlink Demodulation Reference Symbols: DLDMRS are used to estimate the effective channel including transmitter precoder between the 5GETLA-AN and UE, and related SINR statistics.

DL-to-UL Guard Period: DL–UL GP is used to allow all UEs to end their reception before switching to transmission, allows time for the UE hardware to switch from Rx to Tx, and provides time for early transmissions by cell-edge UEs.

Uplink Common Reference Symbols: ULCRS are used to estimate the UL control channel and in the case of single-stream transmission also the UL data channel. ULCRS are also used to measure SINR statistics, possibly containing UL–UL and DL–UL interference, and track the synchronization and TA of the UE.

Uplink Shared Channel: The ULSCH is used to transmit the UL data channel and possibly uplink channel state information reference symbols (ULCSIRS) or uplink demodulation reference symbols (ULDMRS). The RS existence in ULSCH is explicitly indicated by the DMRS or CSIRS frame indicator, and the actual allocation is implicitly indicated by the number of served UEs per RB and number of streams per UE.

Uplink Data Channel: The ULDCH is used to transmit UL data packets, which can also contain measurement information requested by the 5GETLA-AN.

Uplink Channel State Information Reference Symbols: ULCSIRS are used to estimate the uplink channel between the UE and 5GETLA-AN and also collect SINR statistics related to the UL-UL interference.

Uplink Demodulation Reference Symbols: ULDMRS are used to estimate the effective channel between precoded UL transmission and the 5GETLA-AN receiver, and related SINR statistics, possibly containing either UL–UL or DL–UL interference, or both.

Random Access Channel: RACH is used by the UE to either initiate the association process or to ask for resource allocation. The random access indicators are pseudo-random sequences selected based on the DLCRS used by the 5GETLA-AN. By selecting a proper sequence the UE may indicate is it asking for association or for resource allocation. Also the symbol time index may be used to separate new users from associated one, if the RACH allocation is larger than one OFDM symbol. Similarly, based on the assigned sequences for resource allocation and used symbol time index, the 5GETLA-AN can identify UEs to a certain degree based only on the RACH sequence used.

Contention-based Channel: CBDCH is an optional channel that may or may not be active. Based on the cell congestion and number of served UEs the 5GETLA-AN can decide to activate or disable the CBDCH. The CBDCH is an low-latency, best-effort channel used by the UE to transmit small data packets to the 5GETLA-AN. As indicated in Section 13.2.3, the number of small packets and especially the latency in higher layer ACKs is a critical issue when minimizing the latencies in all layers, and therefore we have added a best-effort channel to boost the system performance in low- or medium-congestion scenarios.

Uplink Control Channel: ULCCH is used used to provide DL ACK for the DL code words and inform 5GETLA-AN about the buffer status and report SINR statistics or other measurement information requested by the 5GETLA-AN.

Uplink-to-downlink Guard Period: UL–DL GP is used to allow small timing inaccuracies in the timing advance parameter used by the UE; it can be used to provide additional receiver signal processing time for the 5GETLA-AN, and it can also protect the UE Rx from strong echoes of its own Tx. As we push the latency and transition times towards zero, the reflections of the UE transmission have to be considered and this guard period can alleviate the problem without additional signal processing in the UE.

Each frame contains DL-only, DL/UL, and UL-only sections. The DL-only section contains at a minimum the DLCRS and DLCCH, and can be extended by the 5GETLA-AN in the case of MIMO transmission, when the DLCSIRS or DLDMRS need to be transmitted, as will be described in Section 13.3.2. The DL/UL section is the flexible part of the frame, in which the DL–UL allocation is freely adjusted per frame basis. The UL-only section at the end of the frame contains at a minimum the RACH and ULCCH. The UL-only section can be adjusted in the case of SU- or MU-MIMO, when the ULCSIRS need to be transmitted, or by activating the CBDCH. The layout of the MIMO-transmission-related RS is discussed in detail in Section 13.3.2.

In addition to reducing the symbol and frame durations, we have to be ambitious in minimizing the processing-time requirements between receiving a packet and providing an ACK for it. Although our design has significantly reduced OFDM duration and processing times when compared to LTE, the numerology is relatively close to IEEE WLAN 802.11ac [15] when considering channel bandwidths of 80 MHz and 160 MHz. For example, in IEEE WLAN 802.11n the short interframe space time used between end of the received packet and transmission of an ACK is only 10 μs when operating in the 2.4-GHz band and 16 μs when operating in the 5-GHz band [14]. This is a 250–400 times faster processing-time requirement than in LTE, facilitated by the fact that in WLAN no time-domain interleaving or channel interpolation is required. In 5GETLA the 10 μs corresponds approximately to two or four OFDM symbols with 512- and 256-point FFT, respectively. In Figure 13.4, we show the strict timing relations between the DL and UL block ACKs. All DL codewords have to be decoded and ACK provided inside the same frame in the ULCCH, and all UL codewords have to decoded and ACK provided in the DLCCH of the next frame. For DL transmission the RACH/CBDCH provides some additional time for detecting the codewords. It has been proposed that with multistream MIMO transmission the 5GETLA-AN can increase the RACH/CBDCH allocation just to provide the UE additional time to decode all the streams before providing the ACK [29]. The detection time for DL transmission also depends on the only-UL fraction that is strictly reserved for UL transmissions and therefore usable for receiver signal processing. For UL transmission the UL–DL GP and DLCRS provide processing time for detection. It should

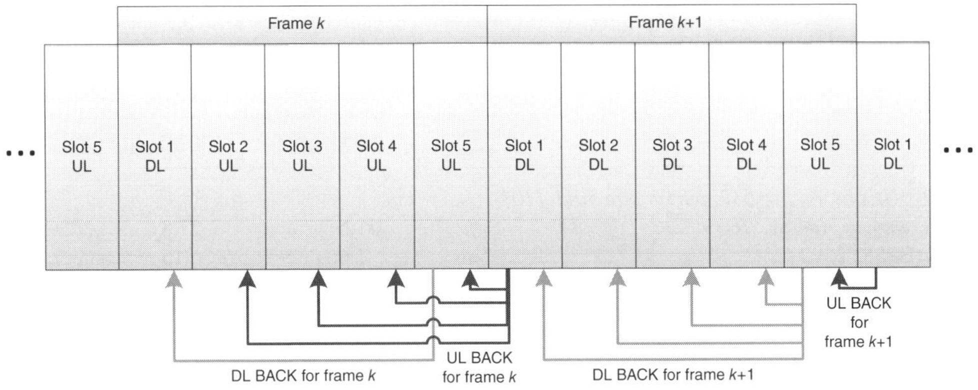

Figure 13.4 Strict timing dependency between DL and UL acknowledgments.

be noted that the current reference model has full OFDM symbols allocated for DL–UL GP and UL–DL GP, and that we can shift time from DL–UL GP to UL–DL GP to allow more processing time for 5GETLA-AN. In Section 13.3.1, the achievable PHY layer RTTs in DL and UL transmission are discussed in more detail through examples.

To provide a simplified comparison of IEEE WLAN 802.11ac, 3GPP LTE-A, and 5GETLA, we have modeled the overheads of different control channels and evaluated the available resources with different systems for single-input single-output (SISO) transmission and for 8×8 DL SU-MIMO for LTE-A and 5GETLA. The results shown assume that maximum resource allocation is given to the UE studied and that no packet failures take place, and no additional contention times are required in WLAN. These results thus give the optimal performance of each system. We evaluate the different systems with respect to the total control and RS overhead and the maximum RTT of transmitting one MAC-layer packet, starting the time count from the first DL data packet and ending with the last received DL ACK while ignoring the processing time of the last DL ACK.

What is important in the results, is that the overhead of the 5GETLA system is the smallest in all of the modeled cases and that the PHY layer latencies are clearly lower than with LTE-A. For LTE-A the RTT times assume TDD operation with frame type 2 and configuration 5 with maximal DL allocation and 10 ms DL–UL switch-point periodicity. For TDD-LTE an RTT interval, which is shown in Table 13.9 as min(RTT)/max(RTT), is given for each case. The minimum time of the interval is limited by the three subframe processing time requirement for an LTE device and the number of subframes required to transmit the given packet with the given MCS. The maximum of the time interval is obtained as a result of the worst-case scheduled subframe location, from which the UE has to wait 9 ms to be able to send the DL ACK. With the 256-point FFT we are achieving the WLAN 802.11ac RTT with 256-QAM and code rate $R = 5/6$, giving us RTT of 0.25 ms for transmitting a MAC packet of size 50 kb. These initial results indicate that we can truly provide sub-millisecond PHY layer RTT for special industry applications or haptic control with the ultra-low-latency cm-wave 5GETLA design. The DL PHY RTT is always less than or equal to the frame duration in the 5GETLA design, as we will see in the next section.

Table 13.9 Comparison of IEEE WLAN 802.11ac, 3GPP LTE-A, and 5GETLA spectral efficiency and MAC layer round-trip times.

		WLAN 802.11 ac (80 MHz)	LTE-A (100 MHz)	5GETLA (100 MHz)
SISO, 5GETLA $N_{FFT} = 512$, packet size 50 kb [10]				
Total overhead (BPSK, $R = 1/2$)	%	31.6	27.8	21.4
RTT (BPSK, $R = 1/2$)	ms	1.66	6/15	1.5
Total overhead (256-QAM, $R = 5/6$)	%	78.5	27.8	21.4
RTT (256-QAM, $R = 5/6$)	ms	0.234	5/14	0.5
SISO, 5GETLA $N_{FFT} = 256$, packet size 50 kb [10]				
Total overhead (BPSK, $R = 1/2$)	%	31.6	27.8	29.0
RTT (BPSK, $R = 1/2$)	ms	1.66	6/15	1.5
RTT (256-QAM, $R = 5/6$)	ms	0.234	5/14	0.25
8×8 DL SU-MIMO, 5GETLA $N_{FFT} = 512$, packet size 2.5 Mb [29]				
Total overhead	%	–	37.4	26.4
RTT (BPSK, $R = 1/2$)	ms	–	16/25	8.5
RTT (256-QAM, $R = 5/6$)	ms	–	5/14	1

13.3.1 Achieving Ultra-low Latency with Strict Timing Requirements

In this section, the strict timing requirements are further elaborated through several examples and the achievable PHY layer RTT in DL and UL transmission, with or without the use of CBDCH. Let us start with the basic DL transmission, as indicated in Figure 13.5. In the basic scenario of scheduling DL data for an associated user, the scheduling indicator is transmitted in the DLCCH, indicating the used RB for selected UE. The selected UE can transmit a UL packet in the same RB if UL time resources are available and if it has something to transmit. The DL block ACK (BACK) is transmitted inside the same frame in the ULCCH. Thus, independent of the FFT size, the PHY RTT from DL allocation to DL BACK is always approximately one frame in duration. Also, in the case of MIMO transmission, it might be required that the 5GETLA-AN updates its channel state information and schedules in the preceding frame channel measurement and update procedures to improve the scheduling accuracy. Remembering that we are targeting mainly low-mobility users, the channel state information is valid over several frames and highly correlated for several long frames and therefore may be used directly for scheduling. To improve reliability, a reduction in the MCS used may be used to incorporate our knowledge of the channel state information inaccuracies.

In Figure 13.6, we have indicated two different RTT metrics for basic UL access. In the UL RTT evaluations we assume that the UE requesting resources or associating to 5GETLA-AN is always prioritized in the scheduler and served in the first possible frame to reduce the connection-opening latencies. The first RTT is evaluated as the time from synchronizing to the DLCRS to the reception of the first ACK. In this case it takes one frame for (re)synchronization and transmitting the RACH indicator, one frame to obtain resource allocation and transmit the UL packet, and two OFDM symbol duration to receive the DLCCH in which the UL BACK is transmitted. Thus the time from synchronizing to the DLCCH to receiving BACK

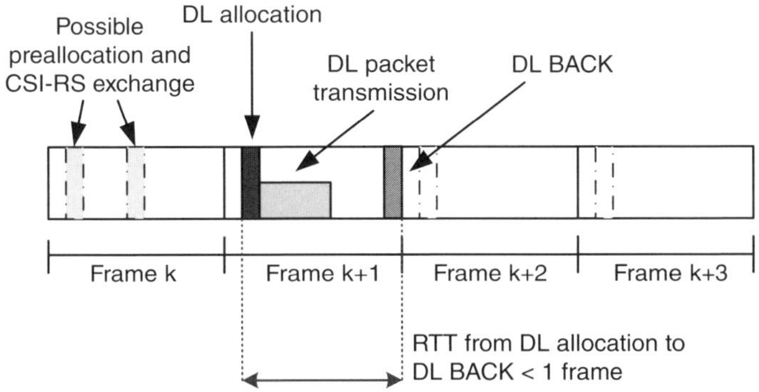

Figure 13.5 Example of the PHY RTT from DL allocation to DL BACK, where the time measure corresponds to the frame duration.

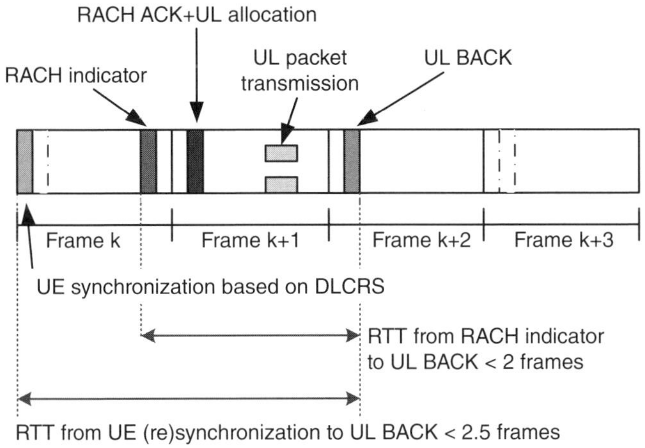

Figure 13.6 Example of the PHY RTT from (re)synchronization or from resource request to UL BACK, where the time measure corresponds to the frame duration.

for the transmitted UL packet is always less than 2.5 frame durations. The second RTT is evaluated as the time from transmitting a RACH indicator and using it as a scheduling request to the reception of the first UL BACK. As seen from Figure 13.6, this takes less than two frame durations. Using the proposed numerology for 5GETLA, the first RTT is less than 1 ms for the 256-point FFT and the 1-ms goal is also achieved with 512-point FFT for the second RTT.

Next we will look at the achievable RTT with the proposed CBDCH. When we truly want to minimize the packet delays in the 5GETLA network, we have to provide best-effort contention-based access. In theory, it is possible to set up the 5GETLA network in such a manner that most of the resources are allocated for CBDCH and in this case the access is mostly asynchronous, and best-effort like. Let us again use the same RTT scenarios as described for basic UL

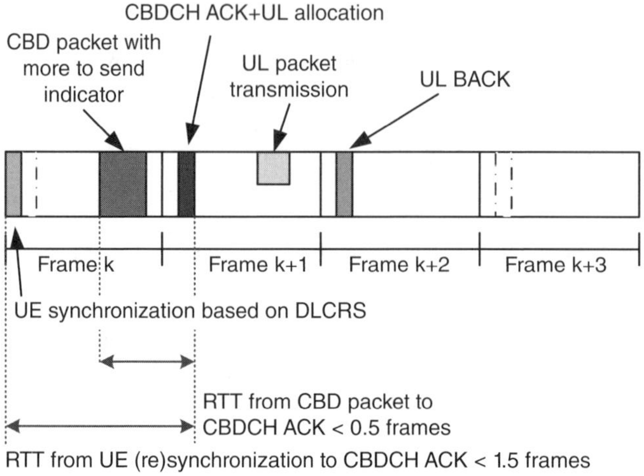

Figure 13.7 Example of the PHY RTT from CBDCH transmission to CBDCH ACK, where the time measure corresponds to the frame duration.

access. As shown in Figure 13.7, the RTT from synchronizing to the DLCRS, transmitting a packet in the CBDCH, and receiving BACK in the next frame's DLCCH takes less than 1.5 frame durations and the RTT from transmitting a packet in the CBDCH (without any resource request) to receiving BACK in the next DLCCH typically takes less than 0.5 frame durations. In the case that most of the resources are allocated for CBDCH it may take up to one frame duration to receive the BACK Although this scenario is assumed to be rarely used or to only occur in certain special use cases, the latency is only 0.25 ms or 0.5 ms for 256- or 512-point FFT-based numerology, respectively.

13.3.2 Reference-symbol Layout Design for Spectrally Efficient MIMO Communications in 5GETLA

In this section, we describe the RS layout for supporting SU- and MU-MIMO transmission in the 5GETLA. Additional discussion on this topic can be found in the literature [29]. To support SU- or MU-MIMO in a system, we have to enable channel estimation between all the Tx–Rx antenna pairs. This requires per-antenna RSs to be transmitted from the 5GETLA-AN and UE. For the beamforming approach, we rely on using the channel reciprocity as part of the design. Therefore, the DL precoder is based on the UL channel measured from the ULCSIRS, from the latest SINR statistics received from the UE, and a correction factor providing the relative difference in the estimated performance of the UE hardware when compared to the performance of the 5GETLA-AN hardware. In the UL, the precoder is defined in similar manner. Instead of using code books and reporting the used precoder to either direction, we use the DMRS to convey the information of the used precoder in both directions. This increases the degrees of freedom in the precoder design and allows improved spectral efficiency.

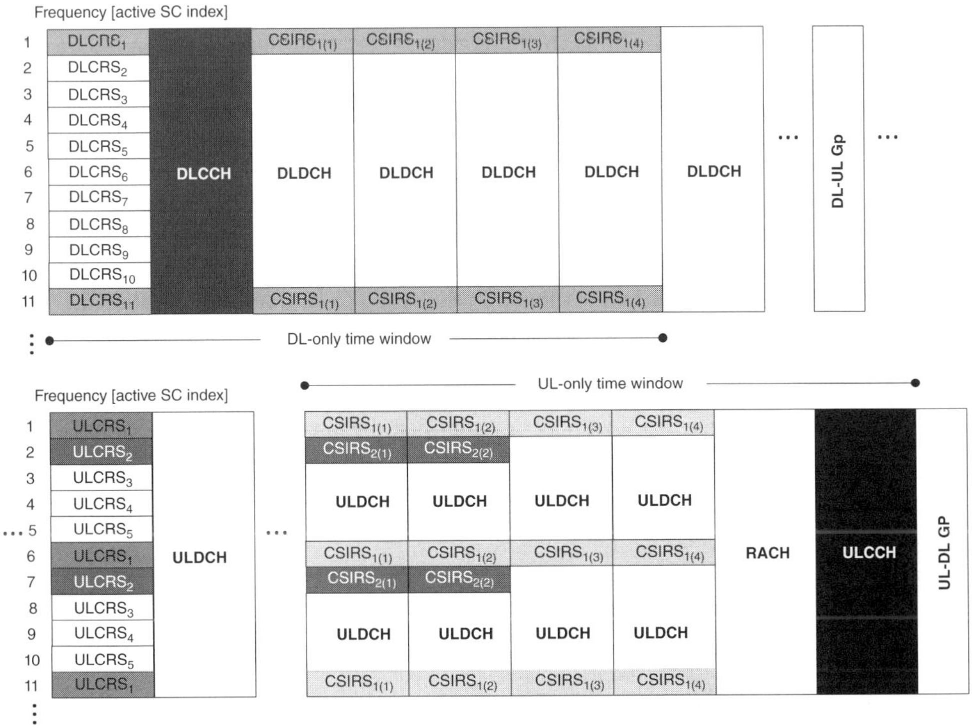

Figure 13.8 Example of the CSIRS mapping for four antennas in the 5GETLA-AN, the UE in RB1 having four antennas and the UE in RB2 having two antennas.

In Figure 13.8, we present the proposed CSIRS layout for supporting MIMO communications. The DLCSIRS uses the same fractional frequency allocation as is used by the DLCRS. This provides frequency reuse among the closest neighbors in the 5GETLA network and reduces the interference on top of DLCRS and possibly on top of DLCSIRS. The DLCSIRS are time multiplexed and have antenna-wise orthogonal codes to improve antenna separation in the receiver. Depending on the number of DLCSIRS, the DL-only window in the frame changes. It should be noted that the DL-only definition does not have to be network-wise and can be adjusted on a cell basis. On the UL side, UEs scheduled in different RBs respond in different orthogonal frequency resources. For the 512-point FFT design the ULCSIRS resolution is twice as large to allow better precoder design accuracy in the 5GETLA-AN based on reciprocity [30]. In the 256-point FFT design, the DLCRS frequency-diversity order is decreased to five and in this case the ULCRS resolution is the same as for DLCRS. The ULCSIRS are transmitted inside the UL-only window preceding the RACH/CBDCH, and this UL-only limitation can also be defined per cell. If some level of cooperation is possible, the 5GETLA-ANs can negotiate with each other the minimum durations for the DL- and UL-only time windows. In the case of MU-MIMO, the UE in the same RB schedule their ULCSIRS into the ULSCH with TDD using the information obtained from the DLCCH. The examples in Figures 13.8–13.10 are given for the 512-point FFT based design.

After the CSIRS have been exchanged and the devices have defined their precoders, they send the data frame combined with DMRS precoded with the latest precoder. In Figure 13.9, the proposed DMRS layout for SU-MIMO is presented. The main goal in the design is to achieve pipelined implementation for the receiver. This requires that the DMRS are transmitted in the beginning of the DL and UL sections. In the DL section, depending on the FFT size used and the number of precoded steams, the OFDM symbols are filled starting from the first, and by using distributed DLDMRS having the same frequency separation as DLCRS. If the number of DL streams is larger than the available orthogonal resources in the OFDM symbol (ten for 512-point FFT and five for 256-point FFT), the DLDMRS are distributed evenly among the two first OFDM symbols. In the case of UL transmission, the ULDMRS symbols are not placed at the end of the frame, but are provided immediately after the ULCRS. In addition, depending on the number of scheduled RBs per UE, the UE may distribute its ULDMRS among the scheduled RBs to reduce the time delay in 5GETLA-AN detecting all of the ULDMRS. In Figure 13.9 we have exemplified the special case of allocating all the

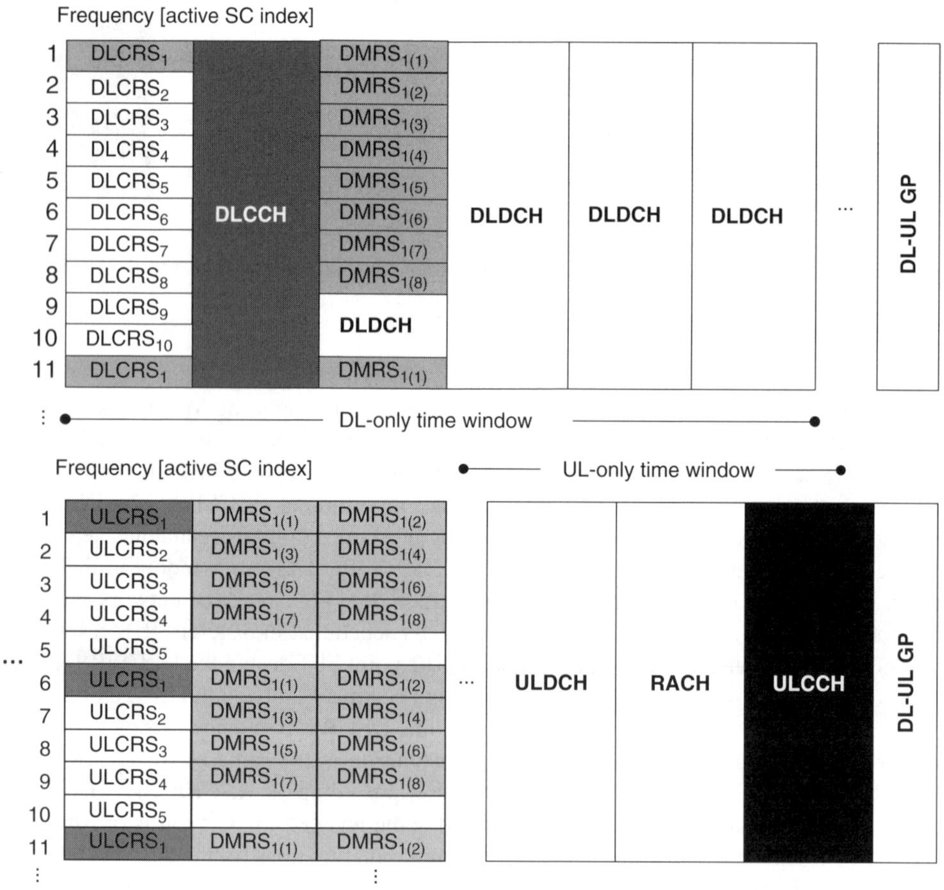

Figure 13.9 Example of the SU-MIMO DMRS mapping for eight antennas in the 5GETLA-AN and UE having eight antennas. It is assumed that only one UE is served over the whole bandwidth.

resources to a single user, which is then using the frequency resources to minimize the time duration of the ULDMRS transmission. If different users are scheduled on different RBs, each user may use only the frequency resources related to the scheduled RBs following the ULCRS allocation and may have to extend the ULDMRS transmission in the time domain.

In the case of the MU-MIMO, the ULDMRS between different UEs in the same RB are again multiplexed using TDD, as shown in Figure 13.10. In this figure, the subindices in the ULDMRS fields are defined as follows: the first index describes the RB index, the second index is the UE index in this RB, and the value in parenthesis is the stream index for this UE in this RB, assuming orthogonal coding is not used to spread the RS over OFDM symbols. As discussed earlier, the ULDMRS are in the beginning of the UL section to minimize detection delay, users scheduled in different RBs use only the allowed resources related to the ULCRS, and the different users in the same RB are separated by TDD. The possible downside of the flexible TDD-based design is that the interference statistics on top of the ULDMRS might not be valid for the whole ULDCH and this can degrade the performance. Another design option would be to transmit the updated ULDMRS in the UL-only window in frame k and start using the updated precoder for UL data in frame $k+1$. This would remove the effect of DL–UL interference from the estimated effective channel. The downside for this design is that it increases the delay of updating the precoder used for the UL data by one frame, thus increasing

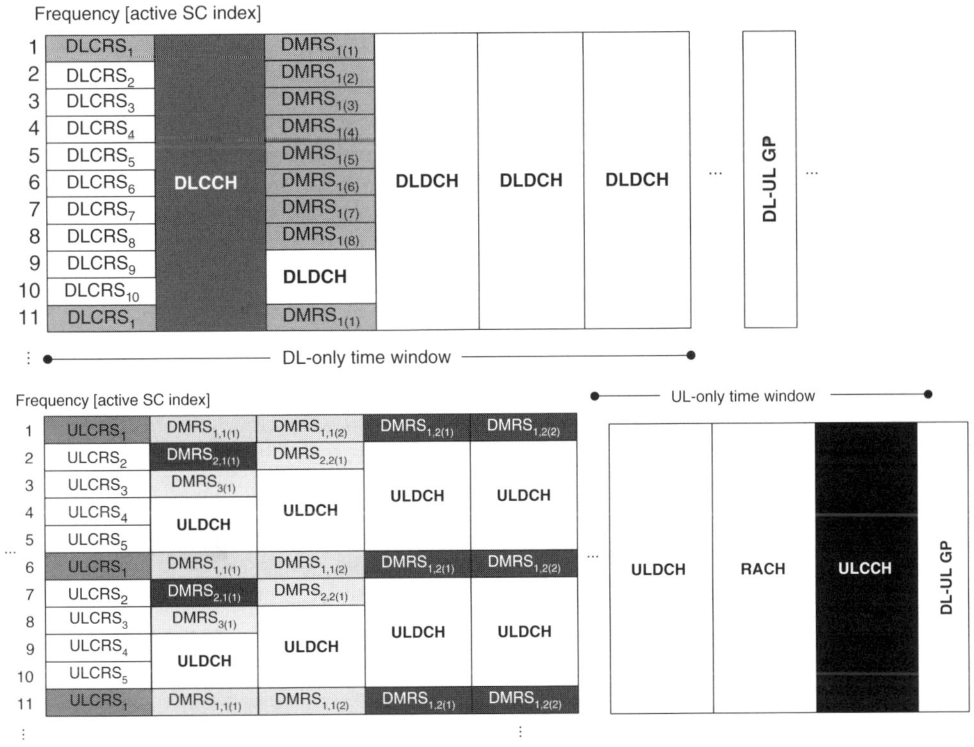

Figure 13.10 Example of the MU-MIMO DMRS mapping for eight antennas in the 5GETLA-AN and UE1 and UE2 in RB1 having two antennas, UE1 and UE2 in RB2 having one antenna, and UE1 in RB3 having one antenna.

the number of frames required for full MIMO capability in DL and UL to three. A dynamic TDD-based radio interface for beyond-4G communications has been presented, with fixed DL or UL frames allowing fixed interference environment per frame basis [31–33]. The numerology obtained is very similar to the 256-point FFT-based 5GETLA design, and it provides an interesting reference design for future system-level evaluations.

13.4 Extending the 5GETLA Reference Design to Millimeter-Wave Communications

The mm-wave 5GETLA (mm-5GETLA) design introduced in this section is discussed in more detail in the literature [34]. The design follows the same principles as the cm-wave 5GETLA, but certain modification to the frame design are proposed to better suit the special characteristics of mm-wave communications. The PHY layer parametrization proposed for mm-5GETLA is shown in Table 13.10. The most notable difference is in the assumed channel bandwidth $B = 1000$ MHz. A mm-wave radio interface using null cyclic prefix single carrier (NCP-SC) has been proposed elsewhere [35–37]. As a reference, this used a carrier frequency $F_C = 72$ GHz, channel bandwidth $B = 2$ GHz, and symbol rate $F_{sym} = 1.536$ GHz, which corresponds to 50 times the LTE sampling rate. In general, the assumptions and values used are similar to the ones presented here. We have designed the mm-5GETLA radio access using the OFDM waveform. However, there is no limitation on changing the waveform to single carrier or some novel waveform if desired.

The two designs, 512-point and 1,024-point FFT based design, are proposed for indoor/LOS and NLOS communications, respectively. As indicated in Section 13.2.2, the measured mean delay spreads in LOS links and indoors are of the order of 20 ns and therefore the 512-point design has CP of duration 31 ns. The DL–UL GP and UL–DL GP are 51 ns and 52.1 ns, respectively. Without TA we can support link distances up to 10 m in LOS. For long LOS links,

Table 13.10 The parametrization proposed for the mm-5GETLA PHY layer.

Parameter		Value	
F_s	MHz	921.6	
B	MHz	1,000	
N_{FFT}		**512**	**1,024**
T_{FFT}	ns	556.0	1,111.0
N_{CP}		29	150
T_{CP}	ns	31.47	162.76
$N_{DL-ULGP}$		47	294
$T_{DL-ULGP}$	ns	51.0	319.0
$N_{UL-DLGP}$		48	294
$T_{UL-DLGP}$	ns	52.1	319.0
$N_{sym/slot}$		85	78
T_{Frame}	µs	50	100
N_{Act}		500	1,000

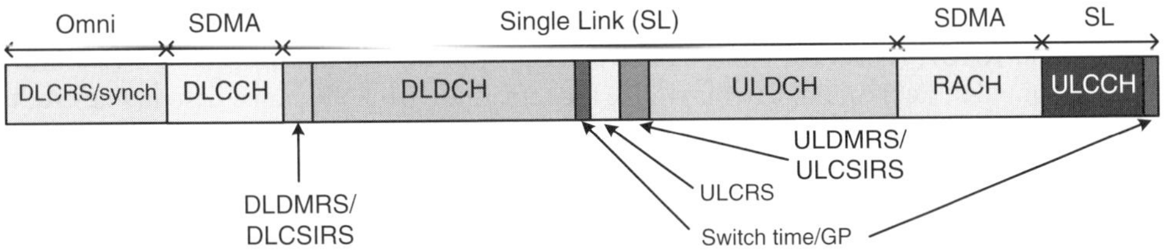

Figure 13.11 Frame-design proposal for mm-5GETLA. Source: Reprinted from [34] with permission from IEEE.

reasonably accurate TA has to be used combined with extended GP. For example, by converting one data symbol into GP time we can extend the supported LOS link distance up to 100 m. The NLOS design has a CP of length 162.76 ns, which is chosen to be larger than the 99th percentile RMS delay spread values shown in Tables 13.4 and 13.5. The GP used is also longer, to allow larger propagation delays imposed by the NLOS communications. The achievable throughput with both schemes is the same [34]. As a practical example, if we assume single-stream transmission with 16-QAM modulation and coding rate $R = 1/2$, we can get goodput of $G_{PHY} = 1.42$ Gbps. By goodput we mean the achievable throughput above the PHY layer.

The frame design of the mm-5GETLA is presented in Figure 13.11. The logical channels used in mm-5GETLA have the same basic properties and usage as presented for 5GETLA in Section 13.3. The main difference between the 5GETLA reference design and mm-5GETLA is that in mm-wave communications we have dropped the assumption of serving multiple UEs per frame. The reason for this is twofold. Firstly, if we serve only one device per frame we can maximize the BF SNR gain for this user and simultaneously maximize the experienced SINR and used MCS. Secondly, because the frame duration is only 50 μs or 100 μs, we can use TDD to schedule different users in different frames while maintaining sub-millisecond scheduling delays. Also, as we push the frame durations towards zero and assume very fast resynchronization and scheduling delays, the probability of simultaneously active devices is reduced. Another notable difference is that we have removed the CBDCH from the mm-5GETLA radio frame. The reason is that the PHY layer in mm-wave is so fast that even including the transmission of the RACH sequence into the physical-layer RTT, we are considering delays of the order of 0.1 ms, which should be sufficient for 5G systems.

There are basically three different transmission modes inside one mm-5GETLA frame. The first is the omni-directional-like transmission, labelled "omni" in the Figure 13.11. In this mode, the DLCRS is transmitted with minimal directivity. The DLCRS sequence used is clearly longer than the one used in cm-wave communications and we have also discarded the assumption of spectral reuse for DLCRS because of the lower transmission power and higher attenuation in the mm-wave path loss. The DLCRS is used by UE to detect mm-5GETLA-AN and synchronize to it. It is also used to estimate the main angle-of-arrival to be used in defining the transmitter beamformer in the UE. Some initial considerations on the length of this sequence based on simplified link budget evaluation have been presented elsewhere [34]. Because the DLCRS sequences are longer in the mm-wave design, we can also more freely distribute them among the mm-5GETLA-ANs and obtain good code diversity for all the received DLCRS sequences in the UE.

The second transmission mode is the spatial division multiple access mode, in which at least DL or UL BF is used. The DLCCH is transmitted, possibly to multiple UE. It contains

ACKs for the previous frame UL for certain UEs, scheduling indications for other UEs, and may contain RACH ACKs for other detected UEs. Still, the number of served UEs during DLCCH should be kept minimal to maximize the remaining BF SNR gains. Also in the RACH channel, the UE has defined the optimal angle-of-departure for its transmission and uses it to beam its RACH sequence transmission. Therefore, we assume that the RACH sequences can be shorter than the DLCRS because of the UE-side BF.

The third transmission mode is single-link (SL) transmission. SL transmission is used for the DLSCH, ULSCH, and ULCCH. In this mode, Tx and Rx BF is used to maximize link SINR. The flexible TDD is especially suitable for mm-wave communications due to the increased degree of freedom in the DL–UL allocation and because the active links are basically isolated from each other due to the narrow Tx and Rx beams and higher attenuation in the mm-wave channel.

13.4.1 High Mobility Support in mm-Wave Communications

As the frame durations become extremely short in mm-wave communications, new ultra-low-latency applications, such as augmented reality, self-driving cars, robotics, and haptic control are enabled. In addition it becomes possible to provide small-cell coverage for high-mobility devices. Let us assume that the maximum cell range is 50 m and the cell is approximately circular. Let us also assume a UE with speed of $v = 360$ km/h or $v = 100$ m/s. Now, let us define the average chord length inside a circle, so the distance traveled under the cell averaged over all possible straight lines passing through the circle when entering from a certain point, is equal to $l_{chord,ave} = 4r / \pi$, where r is the cell radius. From the speed of the UE and the average chord length we get the average service time under this cell to be $T_{serv,ave} = l_{chord,ave} / v = 636.6$ ms. Then, assuming that on average the synchronization and association delay is approximately 100 ms, we end up with roughly 500 ms of active time under the cell for a high-mobility user. From this value we can approximate the maximum achievable average throughput. Typically in 5G systems the target is to achieve average throughputs of 1 Gbps in the cell, so this value would provide approximately 500 Mbps throughput for the high-mobility user. This kind of throughput is not easily achievable, even from the macro layer, which is commonly assumed to provide the coverage for the high-mobility users. In addition, special care in the BF design alongside highways and train lines can improve the performance, and in these scenarios the estimates of the direction of movement directly implies to which mm-5GETLA-AN the UE will next synchronize and associate.

To release ourselves from exchanging packets between mm-5GETLA-ANs, it is better to minimize trunk-network latency and the synchronization and association latency, and use the combined value of these latencies as a time window during which the UE may not transmit inside the current cell when leaving the cell. This allows the UE to receive all packets that it has requested from the network and synchronize to the next cell before requesting or transmitting new packets. The downside in serving high-velocity users with simple channel estimators allowing pipelined detection is that the required RS time interval is four OFDM symbols, as indicated by the 90% correlation value (4.4 μs) given in Table 13.6. This leads to overhead close to 47% in the current frame design. If future measurements and analyses indicate that the coherence times could be larger (say, in the case that the Doppler spread is measured to be more like a Doppler shift) or if more complex channel-interpolation-based receivers are allowed, then high-mobility support by dense mm-wave SCNs will become reality.

In addition to a novel PHY layer, from the network side novel re-routing concepts are required to support cell changes every second while still providing constant connectivity. One possibility for DL is to multicast traffic related to a single device at the border of two cells to both corresponding 5GETLA-ANs. This can be combined with the possibility of simultaneously transmit DL data from both nodes. From the UE point of view this would require multi-beam reception capability and synchronization to two 5GETLA-ANs. In the special case of a highway or train line the neighboring 5GETLA-AN can negotiate cooperation over in-band mm-wave links. The UL traffic would be transmitted only to the strongest 5GETLA-AN. The initial ideas presented here on high-mobility user support require more elaborated studies and provide several interesting topics for a wide variety of research.

13.5 Conclusions and Open Research Topics

In this chapter, we have discussed two different designs for ultra-low-latency radio interfaces in cm-wave and mm-wave communications. The main design parameter has been to minimize latency, which leads to reduced duty cycles and directly to enhanced energy efficiency on both the UE and network side. By carefully designing the control information broadcast and the RS layout, we have also managed to achieve high spectral efficiency.

There are several reasons to emphasize the minimization of the radio-access latency. Firstly, Rx and Tx energy consumption in mobile devices is directly proportional to the active time. Doubling the data rate and halving the active on-time leads to reduced power consumption, even without improvements in the hardware energy efficiency. Secondly, if we want to support up to 1,000 times more traffic per device, we have to be careful to design physical-layer and receiver algorithms that provide similar or, if possible, reduced detection complexity. As an example, if the receiver processing-energy consumption per bit remained constant and we transceived 1,000 times more traffic, we would go from 10-h operation time to 36-s operation times. In addition, it is not only the per-device data rate that is expected to increase, but also the number of devices in the network or unit area. Physical-layer design with a reduced latency approach and pipelined receiver architecture is one key enabler for reduced complexity receivers. Thirdly, the overall experienced latency is the sum of trunk-network latency and radio-interface latency. Even though significant improvements on the trunk network side wil take place by 2020, unless we improve the PHY layer the benefits are not visible to the end user in the form of improved user experience or new services.

There are several open research topics related to ultra-low-latency radio interfaces for 5G SCN. In the list that follows we mention the most important ones as open topics for researchers to solve.

SINR Statistics with Ultra-low-latency Radio Interface using Dynamic or Flexible TDD: As we reduce the cell sizes, increase the system bandwidth, and reduce the frame durations, the question remains of what happens to the SINR statistics. Can we truly see changes in the interference statistics, or will the ever-increasing throughput demand and increased device density eat away any possible gain achieved by pushing the radio-frame duration towards zero? And what is the effect of very narrow beamforming on the SINR statistics? Furthermore, can we use varying UL–DL allocation sizes per frame for cm-wave communications or should we consider designs in which the DL–UL direction can change per frame but the data allocation size is always fixed? With only DL or UL frames in the system the interference statistics inside one frame are fixed and this enables simpler estimation and cancellation schemes. However, in the case that very narrow beamforming isolates neighboring cells, there is no harm from the SINR point of view on the varying DL–UL allocation per frame.

Synchronization and Association Delay: As discussed, especially in the mm-5GETLA design but also in cm-wave communications, the synchronization and association delays need to be minimized. This allows devices to synchronize and associate to the network when required and otherwise be in a dormant state, with minimal power consumption. For mm-wave communications, if these delays can be pushed below 100 ms, we can consider providing small-cell coverage also for high-mobility users. The trunk network delays play a significant role in this problem and need to be re-evaluated for the 5G systems. This item has not yet been commonly considered and is open for research and innovation.

Millimeter-wave Channel Modeling: The time correlation statistics for highly beamformed transmissions are of great interest when designing a mm-wave radio interface capable of also supporting high-mobility users.

Traffic Models: Will the 5G network traffic still be based TCP/IP traffic or will there be something better suited to wireless communications? Video streaming in both directions is going to apparently be the main application, so we need up-to-date models for traffic generated by different services used by the end user. In addition, one aspect of traffic typically ignored is the background small-packet traffic generated by social media applications, email status checks, and so on. When modeling the system performance, each UE will as a minimum generate three traffic streams: high-rate UL or DL data connections, higher-level ACK packets for both directions, and low-rate small-packet streams modeling the background processes running in the UE. All these models together affect the system performance, and the ultra-low-latency TDD-based radio interface needs to be carefully evaluated in a realistic environment like this. As the last point, the evaluations should be always done with finite buffer models to capture the time-varying interference statistics observed in small cells and because in the case of infinite buffer statistics the fixed UL–DL allocation provides the most stable SINR environment and is typically considered as the best choice for receivers without advanced interference cancellation capabilities.

The 5GETLA radio interface can provide sub-millisecond physical-layer latencies for cm-wave communications and an additional factor of 10 reduction in mm-wave communications. The proposed design takes into consideration the small-cell environment and carrier frequency used, and provides an optimized radio interface facilitating ultra-low latency, high spectral efficiency, and greatly improved energy efficiency when compared to existing solutions. Therefore, the 5GETLA radio interface potentially enables new types of services, such as augmented reality, haptic control, and robotics.

References

[1] Ericsson, "5G Radio Access – Capabilities and Technologies," April 2016. Available: www.ericsson.com/res/docs/whitepapers/wp-5g.pdf (accessed August 15, 2016).

[2] Nokia Solutions and Networks, "Looking ahead to 5G," 2013. Available: http://brooklyn5gsummit.com/wp-content/uploads/2014/01/NSN_5G_White_Paper.pdf?download (accessed August 9, 2016).

[3] G. Fettweis and S. Alamouti, "5G: Personal mobile internet beyond what cellular did to telephony," *Communications Magazine, IEEE*, vol. **52**, no. 2, pp. 140–145, 2014.

[4] Nokia Siemens Networks, "2020: Beyond 4G: Radio Evolution for the Gigabit Experience," 2011. Available: http://docplayer.net/13573948-2020-beyond-4g-radio-evolution-for-the-gigabit-experience-white-paper.html.

[5] 3rd Generation Partnership Project, "Technical Specification Group Radio Access Network; Evolved Universal Terrestrial Radio Access (E-UTRA) and Evolved Universal Terrestrial Radio Access Network (E-UTRAN); Overall description; Stage 2 (Release 12)," *3GPP TS 36.300 V12.2.0 (2014-06)*, pp. 1–215, 2014.

[6] T. Rappaport, S. Sun, R. Mayzus, H. Zhao, Y. Azar, K. Wang, G. Wong, J. Schulz, M. Samimi, and F. Gutierrez, "Millimeter wave mobile communications for 5G cellular: It will work!" *Access, IEEE*, vol. **1**, pp. 335–349, 2013.

[7] X. Wang, M. Chen, T. Taleb, A. Ksentini, and V. Leung, "Cache in the air: exploiting content caching and delivery techniques for 5G systems," *Communications Magazine, IEEE*, vol. **52**, no. 2, pp. 131–139, 2014.

[8] Huawei, "The second phase of LTE-Advanced – LTE-B: 30-fold capacity boosting to LTE," 2013. Available: www.huawei.com/ilink/en/download/HW_259010.

[9] D. Halperin, T. Anderson, and D. Wetherall, "Taking the sting out of carrier sense: interference cancellation for wireless LANs," in *Proceedings of the 14th ACM International Conference on Mobile Computing and Networking*, MobiCom '08, New York, pp. 339–350, 2008. Available: http://doi.acm.org/10.1145/1409944.1409983.

[10] T. Levanen, J. Pirskanen, T. Koskela, J. Talvitie, and M. Valkama, "Radio interface evolution towards 5G and enhanced local area communications," *Access, IEEE*, vol. **2**, pp. 1005–1029, 2014.

[11] V. Erceg, L. Schumacher, P. Kyritsi, et al., "TGn Channel Models," *IEEE 802.11-03/940r4*, pp. 1–45, 2004. Available: https://mentor.ieee.org/802.11/documents?is_dcn=940&is_group=000n&is_year=2003.

[12] G. Breit, H. Sampath. S. Vermani, et al. "TGac Channel Model Addendum," *IEEE 802.11-09/0308r12*, pp. 1–20, 2010. Available: https://mentor.ieee.org/802.11/documents?is_dcn=0308&is_group=00ac.

[13] P. Kyösti, J. Meinilä, L. Hentilä, X. Zhao, T. Jämsä, C. Schneider, M. Narandzic, M. Milojevic, A. Hong, J. Ylitalo, V.-M. Holappa, M. Alatossava, R. Bultitude, Y. de Jong, and T. Rautiainen, "WINNER II Channel Models," *IST-4-027756 WINNER II D1.1.2 V1.1*, pp. 1–82, 2007. Available: projects.celtic-initiative.org/winner+/WINNER2-Deliverables/D1.1.2v1.1.pdf.

[14] IEEE "Standard for Information Technology – Telecommunications and information exchange between systems local and metropolitan area networks – specific requirements Part 11: Wireless LAN medium access control (MAC) and physical layer (PHY) specifications," *IEEE Std 802.11-2012 (Revision of IEEE Std 802.11-2007)*, pp. 1–2793, 2012.

[15] IEEE "Standard for Information Technology – Telecommunications and Information Exchange between systems Local and Metropolitan Area Networks – Specific requirements. Part 11: Wireless LAN medium access control (MAC) and physical layer (PHY) specifications – Amendment 4: Enhancements for very high throughput for operation in bands below 6 GHz." *IEEE Std 802.11ac-2013 (Amendment to IEEE Std 802.11-2012, as amended by IEEE Std 802.11ae-2012, IEEE Std 802.11aa-2012, and IEEE Std 802.11ad-2012)*, pp. 1–425, 2013.

[16] E. Perahia and R. Stacey, *Next Generation Wireless LANs: 802.11n and 802.11ac*, 2nd edn. Cambridge University Press, 2013.

[17] 3rd Generation Partnership Project "Technical Specification Group Radio Access Network; Evolved Universal Terrestrial Radio Access (E-UTRA); Further advancements for E-UTRA physical layer aspects (Release 9)," *3GPP TS 36.814 V9.0.0 (2010-03)*, pp. 1–104, 2010.

[18] R. Ertel, P. Cardieri, K. Sowerby, T. Rappaport, and J. Reed, "Overview of spatial channel models for antenna array communication systems," *Personal Communications, IEEE*, vol. **5**, no. 1, pp. 10–22, 1998.

[19] T. Rappaport, J. Murdock, and F. Gutierrez, "State of the art in 60-ghz integrated circuits and systems for wireless communications," *Proceedings of the IEEE*, vol. **99**, no. 8, pp. 1390–1436, 2011.

[20] T. Rappaport, F. Gutierrez, E. Ben-Dor, J. Murdock, Y. Qiao, and J. Tamir, "Broadband millimeter-wave propagation measurements and models using adaptive-beam antennas for outdoor urban cellular communications," *Antennas and Propagation, IEEE Transactions on*, vol. **61**, no. 4, pp. 1850–1859, 2013.

[21] M. Kyrö, *Radio Wave Propagation and Antennas for Millimeter-wave Communications*. Aalto University Publication Series Doctoral Dissertations, 179/2012, 2012. Available: http://urn.fi/URN:ISBN:978-952-60-4951-9.

[22] "Cisco Visual Networking Index: Global Mobile Data Traffic Forecast Update, 2013–2018," 2013. Available: docplayer.net/1029647-Cisco-visual-networking-index-global-mobile-data-traffic-forecast-update-2013-2018.html.

[23] W Carney, K Agarde, H Suzuki et al. "Simplified traffic model based on aggregated network statistics," *IEEE 802.11-13/1144r1*, pp. 1–10, 2013. Available: https://mentor.ieee.org/802.11/documents?is_dcn=1144&is_group=0hew.

[24] W Carney, K Agarde, and H Suzuki "Network optimization for expected HEW traffic patterns," *IEEE 802.11-13/0728r1*, pp. 1–10, 2013. Available: https://mentor.ieee.org/802.11/documents?is_dcn=0728&is_group=0hew.

[25] R. Palit, K. Naik, and A. Singh, "Anatomy of WiFi access traffic of smartphones and implications for energy saving techniques," *International Journal of Energy, Information and Communications*, vol. **3**, no. 1, pp. 1–16, 2012. Available: http://www.sersc.org/journals/IJEIC/vol3_Is1/1.pdf (accessed August 10, 2016).

[26] P. Ameigeiras, Y. Wang, J. Navarro-Ortiz, P. Mogensen, and J. Lopez-Soler, "Traffic models impact on OFDMA scheduling design," *EURASIP Journal on Wireless Communications and Networking*, vol. 2012, no. 1, p. 61, 2012. Available: http://jwcn.eurasipjournals.com/content/2012/1/61.

[27] E. Dahlman, S. Parkvall, and J. Skold, *4G: LTE/LTE-Advanced for Mobile Broadband*, 1st edn. Academic Press, 2011.

[28] T. Levanen, J. Pirskanen, T. Koskela, J. Talvitie, and M. Valkama, "Low latency radio interface for 5G flexible TDD local area communications," in *Communications Workshops (ICC), 2014 IEEE International Conference on*, pp. 7–13, 2014.

[29] T. Levanen, J. Talvitie, J. Pirskanen, and M. Valkama, "New spectrally and energy efficient flexible TDD based air interface for 5G small cells," in *2014 IEEE 79th Vehicular Technology Conference (VTC2014-Spring) – The 2nd International Workshop on 5G Mobile and Wireless Communication System for 2020 and Beyond*, Seoul, Korea, pp. 7–13, 2014.

[30] J. Haartsen, "Impact of non-reciprocal channel conditions in broadband TDD systems," in *Personal, Indoor and Mobile Radio Communications, 2008. PIMRC 2008. IEEE 19th International Symposium on*, pp. 1–5, 2008.

[31] P. Mogensen, K. Pajukoski, B. Raaf, E. Tiirola, E. Lahetkangas, I. Kovacs, G. Berardinelli, L. Garcia, L. Hu, and A. Cattoni, "B4G local area: High level requirements and system design," in *Globecom Workshops (GC Wkshps), 2012 IEEE*, pp. 613–617, 2012.

[32] E. Lahetkangas, K. Pajukoski, E. Tiirola, G. Berardinelli, I. Harjula, and J. Vihriala, "On the TDD subframe structure for beyond 4G radio access network," in *Future Network and Mobile Summit (FutureNetworkSummit), 2013*, pp. 1–10, 2013.

[33] P. Mogensen, K. Pajukoski, E. Tiirola, E. Lahetkangas, J. Vihriala, S. Vesterinen, M. Laitila, G. Berardinelli, G. W. Da Costa, L. G. Garcia, F. M. Tavares, and A. F. Cattoni, "5G small cell optimized radio design," in *Globecom Workshops (GC Wkshps), 2013 IEEE*, pp. 111–116. 2013.

[34] T. Levanen, J. Pirskanen, and M. Valkama, "Radio interface design for ultra-low latency millimeter-wave communications in 5G era," in *IEEE Global Communication Conference, Exhibition & Industry Forum 2014 (IEEE Globecom 2014) – International Workshop on Ultra-Low Latency and Ultra-High Reliability in Wireless Communications*, Austin, Texas, USA, 2014.

[35] S. G. Larew, T. A. Thomas, M. Cudak, and A. Ghosh, "Air interface design and ray tracing study for 5G millimeter wave communications," in *Globecom Workshops (GC Wkshps), 2013 IEEE*, pp. 117–122, 2013.

[36] A. Ghosh, T. Thomas, M. Cudak, R. Ratasuk, P. Moorut, F. Vook, T. Rappaport, G. MacCartney, S. Sun, and S. Nie, "Millimeter wave enhanced local area systems: A high data rate approach for future wireless networks," *Selected Areas in Communications, IEEE Journal on*, vol. **32**, no. 6, pp. 1152–1163, 2014.

[37] M. Cudak, A. Ghosh, T. Kovarik, R. Ratasuk, T. Thomas, F. Vook, and P. Moorut, "Moving towards mmwave-based beyond-4G (B-4G) technology," in *Vehicular Technology Conference (VTC Spring), 2013 IEEE 77th*, pp. 1–5, 2013.

Part III

Candidate Technologies – Revolutionary

14

New Physical-layer Waveforms for 5G

Gerhard Wunder,[1] Martin Kasparick,[1] Peter Jung,[1] Thorsten Wild,[2] Frank Schaich,[2] Yejian Chen,[2] Gerhard Fettweis,[3] Ivan Gaspar,[3] Nicola Michailow,[3] Maximilian Matthé,[3] Luciano Mendes,[3] Dimitri Kténas,[4] Jean-Baptiste Doré,[4] Vincent Berg,[4] Nicolas Cassiau,[4] Slawomir Pietrzyk,[5] and Mateusz Buczkowski[5]

[1] *Fraunhofer Heinrich Hertz Institute, Berlin, Germany*
[2] *Alcatel Lucent Bell Labs, Stuttgart, Germany*
[3] *Technische Universität Dresden, Dresden, Germany*
[4] *CEA, LETI, Grenoble, France*
[5] *IS-Wireless, Piaseczno, Poland*

14.1 Why OFDM Fails

The main hypothesis of this chapter is that the underlying design principles – synchronism and orthogonality – of the PHY layer of today's LTE-A radio access network constitute a major obstacle for the envisioned service architecture. Synchronism means that the senders operate with a common clock for their processing. Orthogonality means that no crosstalk occurs in the receiver's waveform detection process. Often, both are related, such that some rough synchronization is required to establish orthogonality. LTE-A OFDM modulation keeps the subcarrier waveforms orthogonal even after the channel, provided the discrete Fourier transform (DFT) window can be properly adjusted by a suitable synchronization mechanism, which is then near-optimal processing in a single cell if a capacity-achieving scheme such as superposition coding is used per subcarrier. However, as soon as the orthogonality is destroyed (for example, due to random channel access or multi-cell operation), the distortion accumulates without bounds in OFDM. This is due to the so-called reproducing Dirichlet kernel $\sin(Nx)/\sin(x)$ of OFDM, which quickly approaches the $\sin(x)/x$ kernel for large N where N is the number of

Towards 5G: Applications, Requirements and Candidate Technologies, First Edition.
Edited by Rath Vannithamby and Shilpa Talwar.

Companion website: www.wiley.com/go/vannithamby/towards5g

subcarriers. For such a kernel, it is well known that the amplification of small errors (for example, due to sampling or frequency offsets) is not independent of N and can grow with order $\log(N)$. Hence, for certain applications, we believe it is better to abandon strict orthogonality, partially or altogether, and instead control the impairments.

Let us discuss several intriguing examples.

14.1.1 Sporadic Traffic

Devices that generate sporadic traffic (for example, machine-type communication (MTC) devices in the IoT) should not be forced to be integrated into the bulky synchronization procedure of the LTE-A PHY layer random access, which has been deliberately designed to meet orthogonal constraints. Instead, ideally, they awake occasionally, and then should transmit their messages right away and only be coarsely synchronized. By doing so, MTC traffic would be removed from standard uplink data pipes, with drastically reduced signaling overhead. Therefore, alleviating the synchronism requirement can significantly improve operational capabilities and network performance as well as user experience and the lifetimes of autonomous MTC nodes. Interestingly, sporadic access poses another significant challenge to mobile access networks due to an operation known as fast dormancy. Fast dormancy is used by smartphone manufacturers to save battery power by using the feature that a mobile can break ties to the network individually, and as soon as a data piece is delivered the smartphone changes from an active into an idle state. Consequently, when the mobile has to deliver more pieces of data it will always go through the complete synchronization procedure again. Actually, this can happen several hundred times a day, resulting in significant control-signaling growth and network-congestion threat. A rough estimate suggest that 2000 control resource elements (that is, a subcarrier) can be necessary to deliver one data resource element. We conclude that sporadic traffic must be carried by non-orthogonal waveforms for asynchronous signaling in the uplink and specifically in an uplink random access channel (RACH). We outline later in this chapter a suitable sparse signal processing concept together with new waveforms to efficiently deal with the sporadic traffic and control signaling problem. In fact, the ratio of control and data can actually be reversed by such a concept, approaching a value below 5% within a single subframe.

14.1.2 Spectral and Temporal Fragmentation

Due to fragmentation, spectrum is scarce and expensive but also underutilized: this is commonly referred to as the spectrum paradox. Therefore, carrier aggregation will be implemented to achieve much higher rates by variably aggregating non-contiguous frequency bands (Noguet et al. 2011). Carrier aggregation implies the use of separate RF front-ends accessing different channels, thereby reinforcing the attraction of isolated frequency bands such as the L-band. Actually, the search for new spectrum is very active in Europe and the United States, in order to provide mobile broadband expansion. It includes the opportunistic use of spectrum, which has been an interesting research area in wireless communications in the past decade. Moreover, techniques to detect and assess channel vacancy using cognitive radio could well make new business models possible in the future. The first real implementation will start with the exploration of TV white spaces in the United States. Combined with the preparation of the ongoing regulatory framework in Europe, opportunistic use of spectrum can address a 5G market if it overcomes, with spectrum agility, the rigorous implementation requirements of low out-of-band radiation for protection of legacy systems (Noguet et al. 2011). The LTE-A waveform imposes generous guard bands on other legacy networks to satisfy spectral-mask requirements, which either severely deteriorate spectral efficiency or even prevent band usage at all, which is again an artifact of strict orthogonality and synchronism constraints within the PHY layer. Moreover, in a scenario with uncoordinated interference from pico- or femtocells and highly overlapping coverage, it seems illusive to provide the degree of coordination to maintain synchronism and orthogonality in the network while calling for new waveforms as well. In addition to spectral fragmentation, temporal fragmentation is another key issue. This occurs, for example, due to sporadic access in the asynchronous uplink RACH. Notably, asynchronous signaling also matters in the downlink in the context of cooperative multipoint (CoMP). In conclusion, such 5G scenarios where multiple users are allocated a pool of frequencies with relaxed (or even no) synchronization in time must be addressed by new waveforms. Such waveforms must implement sharp frequency notches and tight spectral masks in order not to interfere with other legacy systems, and must be robust to asynchronous signaling and handle uncoordinated interference. Traditional OFDM schemes are not suited due to the inflexible handling of guard intervals (GIs) cyclic prefixes (CPs) or cyclic suffixes, as well as poor spectral localization. In a later section, we discuss waveforms achieving 100 times better localization. For example, 35-dB side lobes can be achieved with LTE-A OFDM compared to 55-dB side lobes with filter bank multicarrier (FBMC) (Farhang-Boroujeny 2011a). New waveforms thus make a real difference in fragmented spectrum and CoMP scenarios.

14.1.3 Real-time Constraints

Fourth-generation systems offer latencies of several tens of milliseconds between terminal and BS, which originate from resource scheduling, frame processing, retransmission procedures, and so on. However, future application scenarios such as the tactile Internet scenario require ultra-low latency matched with the human tactile sense (Fettweis 2012). In such an environment, a massive number of distributed sensors and actuators will be connected to enable real-time tactile interaction in an augmented way. Sharing the medium becomes an additional challenge and imposes short wake-up cycles on the nodes and the use of burst transmission. Instead of consuming spectrum and power resources by introducing sophisticated algorithms to reach synchronism, an asynchronous approach appears promising. In order to achieve ultra-low latency, each and every element of the communication and control chain must be optimized.

Focusing on the PHY layer, an LTE-A system supports different granularity of scheduling resources in a fixed transmission time interval (TTI) of 1 ms. TTI represents an inherent lower bound of the LTE-A system's PHY latency. Clearly, as the time budget on the PHY layer in the tactile Internet scenario is 100 μs maximum, frame duration must be reduced, and LTE-A with its OFDM symbol duration of 67 μs is not an option.

In order to discuss possible alternatives, assume 20 μs symbol duration. This means that a frame is composed of five symbols, allowing for an appropriate frame structure for random channel access. Considering, say, a 1-km cell range, the expected delay spread is around 3 μs; thus, 4-μs CP is required to ensure an intersymbol interference (ISI)-free scenario. Hence, use of conventional OFDM entails a 20% loss in spectral efficiency. A non-orthogonal waveform that allows for transmitting multiple symbols with a single CP relaxes such strict time-domain requirements. Another major drawback of short frames is the fixed bandwidth increment required to keep a given throughput. A flexible nonorthogonal multicarrier waveform also allowing for intercarrier interference (ICI) can use nonproportional subcarrier spacing to accommodate the necessary bandwidth. Alternatively, non-contiguous spectrum can be aggregated, again enabled by the low out-of-band emissions of the non-orthogonal waveform.

Short frames also have a positive impact on mobility support and operational frequencies. LTE-A has been designed to support Doppler spreads of 100 Hz caused by 50 km/h mobility. By reducing the frame duration, it is possible to either support higher mobility or operate in a higher frequency range. Finally, a short frame brings benefits to upper protocol layers: although the latency requirements of real-time applications demand a robust PHY layer to avoid retransmissions of the frame, applications may desire acknowledged signaling. A short frame will enable the implementation of less time-consuming retransmission algorithms. Summarizing, although OFDM could be tuned to address a different granularity of scheduling resources, there is no mode in the current LTE-A standard that can adapt to the latency requirements of real-time services running on top. If the symbol duration is reduced to achieve very short round-trip delays, the GIs cannot be scaled accordingly without severely compromising spectral efficiency or cell size. The required flexibility can only be achieved with new waveforms, as shown in a later section.

14.2 Unified Frame Structure

A 5G approach must be able to efficiently support different traffic types, which all have to be part of future wireless cellular systems. Our vision of a unified frame structure, depicted in Figure 14.1, aims to handle the large set of requirements of a single 5G system. A filtered multicarrier approach will enable the mix of synchronous/asynchronous and orthogonal/non-orthogonal traffic types, where the reduced side-lobe levels of the waveform seek to minimize ICI and ISI. In Figure 14.1, the different areas represent time-frequency resource elements of different traffic types and different classes of synchronicity. Radio-resource control can adjust the assigned bandwidth semi-statically based on the respective service loads. The third dimension in this unified frame structure is the use of multiple superimposed signal layers. The principle of interleave-division multiple access (IDMA), initially published by Ping et al. (2003), is a very appealing approach to generating these signal layers, and an elegant receiver and coding concept for it. High-volume broadband traffic (Type I), typically human-initiated, will operate as in LTE-A with synchronicity, whenever possible, using scheduled access. At cell edges, with coordinated multi-point (CoMP) transmission and reception, it is not always possible to establish synchronicity to all cells, so the system has to operate with relaxed synchronicity requirements (Type II traffic). For high-volume data applications in those cell areas (Type II), a multi-cell multi-user transceiver concept is required.

For sporadic small-packet services, for example in MTC, the general relaxation of time–frequency alignment reduces signaling overhead and battery consumption (Type III traffic). Multiple signal layers may superimpose, which is handled by advanced multi-user/multi-cell receivers. A contention-based access technique is attractive, saving overhead by dropping the strict synchronism requirement. For low-end sensor devices, it may be helpful to spread the transmissions over a longer time and to allow completely asynchronous transmission (Type IV) traffic. For such sensor-type traffic it can be shown that, from an energy-efficiency

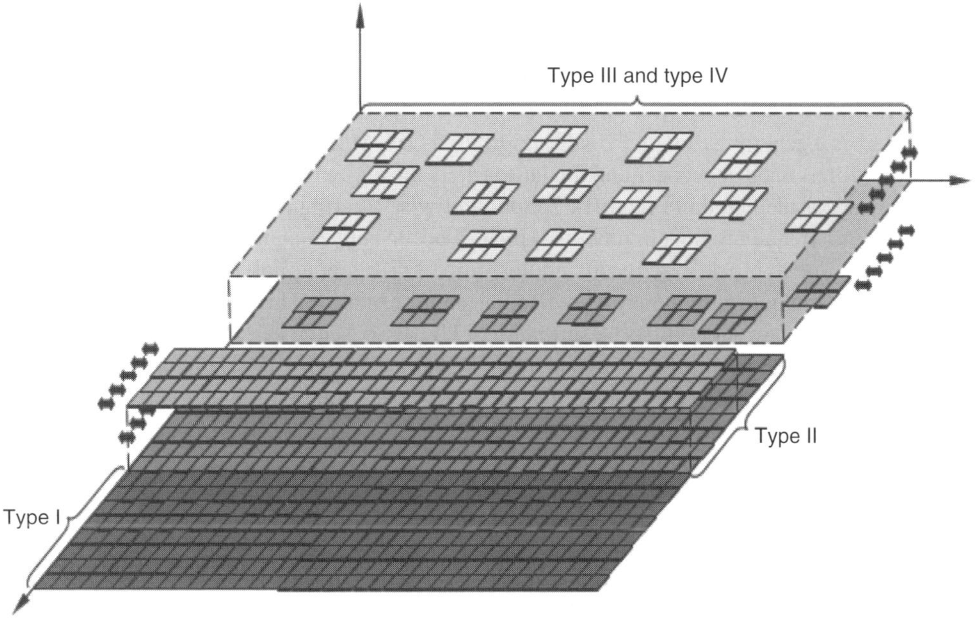

Figure 14.1 5G vision of unified frame for different types of traffic. Types I and II are high-rate data for video with possible advanced receiver processing for cell-edge or CoMP users. Type I possibly also carries real-time traffic. Types III and IV are sporadic MTC traffic, possibly containing a spreading element. The third dimension illustrates the superimposed signal layers.

perspective, it is beneficial to stretch the transmissions in time by spreading. This additional signal layer, again, can be handled by an IDMA-like approach. The unified frame concept will be main part of the standardization processes to be initiated in the future.

In that context, exemplary multiple access schemes are defined as follows:

- Dynamic, channel-adaptive resource scheduling for traffic Type I using standard resource scheduling mechanisms.
- Semi-static/persistent scheduling for traffic Type II. From a MAC layer point of view it is necessary to decide on the amount of resources allocated for this type of traffic, since schedulers will not adapt to specific parts of the frequency (may also be used for high speed terminals).
- "One-shot" transmission (low amount of data and pilots) with contention-like approaches for random access (Type III and IV) enabling payload transmission in the physical layer random access channel (PRACH).

Notably, traffic types II and III rely on open-loop synchronization. The device listens to the downlink and synchronizes itself coarsely, based on synchronization channel and/or reference symbols, similar to 4G systems. Furthermore, the devices may apply an autonomously derived timing advance, which we call autonomous timing advance (ATA); it is relevant particularly for MTC.

14.3 Waveform Candidates and Multiple-access Approaches

14.3.1 Universal Filtered Multicarrier

UF-OFDM, also known as universal filtered multicarrier (UFMC), can, roughly speaking, be interpreted as a generalization of FBMC and filtered OFDM (Schaich and Wild 2014c). While the former filters each subcarrier and the latter filters the entire band, UF-OFDM filters blocks of subcarriers, as depicted in Figure 14.2. Subband-wise filtering is motivated by the observation that time–frequency misalignments typically occur between entire blocks of subcarriers (for example due to block-wise resource allocation of different uplink users). Furthermore, as the filters are broader in frequency, they become shorter in time. This provides a good support for communication in short bursts. A summary of UFMC-based 5G air interface design advantages and the available results is provided by Wild et al. (2014). Matlab code of the basic UFMC transceiver can be downloaded from Schaich and Wild (2014b).

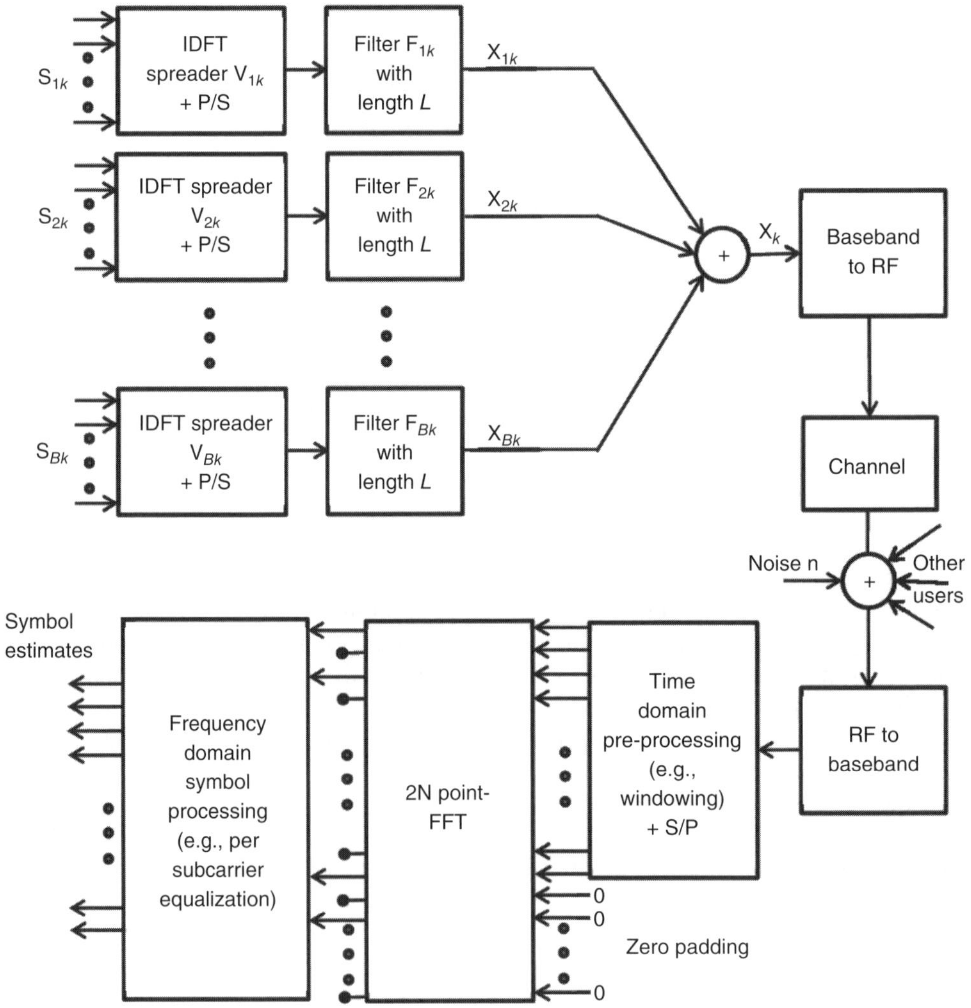

Figure 14.2 UF-OFDM transceiver.

The time-domain transmit vector $\vec{x}_k$ for a particular multi-carrier symbol of user k is the superposition of the subband-wise filtered components, with filter length L and FFT length N, and can be represented by

$$\underbrace{\vec{x}_k}_{[(N+L-1)\times 1]} = \sum_{i=1}^{B} \underbrace{\mathbf{F}_{i,k}}_{[(N+L-1)\times N]} \cdot \underbrace{\mathbf{V}_{i,k}}_{[N\times n_i]} \cdot \underbrace{\vec{s}_{i,k}}_{[n_i\times 1]} \cdot \tag{14.1}$$

For the ith subband ($1 \le i \le B$), the n_i complex QAM symbols are transformed to time-domain by the tall IDFT-matrix $\mathbf{V}_{i,k}$. Its mth row, for subcarrier index n, consists of the elements $V_{i,k}(m,n) = \exp(j2\pi(m-1)n/N)$. $\mathbf{F}_{i,k}$ is a Toeplitz matrix, composed of the filter impulse response $\mathbf{f}_{i,k}$, performing the linear convolution. $\mathbf{f}_{i,k}$ is a design parameter, adjustable to propagation conditions and time–frequency offset requirements (Wang et al. 2014). For $L = 1$, UF-OFDM converges to (non-CP-)OFDM.

There are several receiver implementations possible for UFMC. A very efficient approach is to zero-pad the $N + L - 1$ receive samples of a multi-carrier symbol to the next larger power of two (or efficient FFT implementation length) (Chen et al. 2014). Each second value in frequency represents a subcarrier output, which can be treated by a 1-tap scalar equalizer in order to account for the transmit-filter frequency response. This makes UFMC almost as low complexity as CP-OFDM.

14.3.1.1 Frequency- and Time-domain Properties

The spectrum of one subband is depicted in Figure 14.3. The OFDM drawbacks of high sidelobe levels can be removed with UFMC. This makes UFMC suitable for fragmented spectrum and increases robustness against all sources of ICI. This ICI robustness was shown for uplink coordinated multi-point (CoMP) joint reception in the presence of carrier frequency offsets of

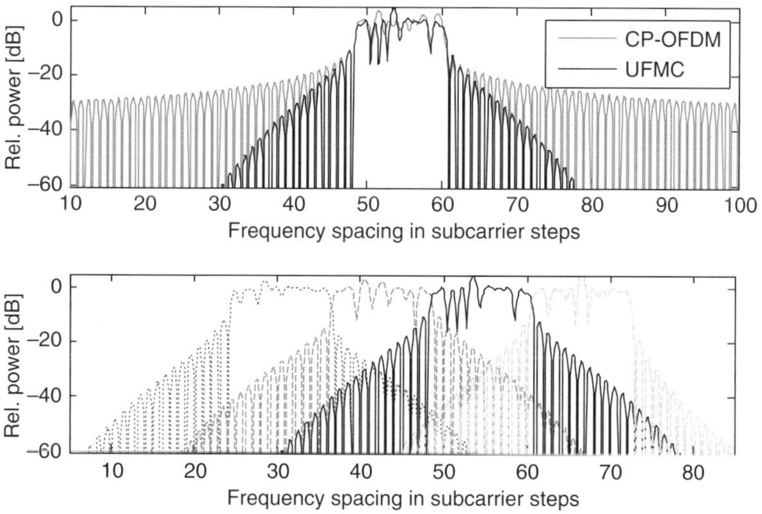

Figure 14.3 UF-OFDM spectrum for subbands with 12 subcarriers: left, one UFMC subband compared to CP-OFDM; right, multiple superimposed UFMC subbands.

multiple users by Vakilian et al. (2013). UFMC thus becomes an enabling waveform for the unified frame structure, supporting various traffic types and devices, including time–frequency misalignments (Schaich and Wild 2014a).

As long as the same frequency-shifted prototype subband filter is used, the subcarriers in UFMC are fully orthogonal in the complex domain. This means, in contrast to FBMC-SMT (Farhang-Boroujeny 2011b), which uses offset-QAM, UFMC can use QAM modulation symbols. This complex orthogonality eases the usage of MIMO and CoMP, supports complex-valued pilot sequences and allows re-use of all the knowledge gained from CP-OFDM.

In the time domain, UFMC in its default form is characterized by soft transitions between different multi-carrier symbols. Subsequent multi-carrier symbols, characterized by Eq. (14.1), are non-overlapping. This means the enveloping amplitude of the subcarriers ramps up, stays constant for the symbol body, and ramps down again, as shown in Figure 14.4. This provides a soft ISI protection. For example, the mean-square error (MSE) caused by the delay spread of a Vehicular A channel, when UFMC operates with LTE-like parameters is far below −40 dB (when the filter length $L = L_{CP} + 1$ accounts for the CP length L_{CP} in LTE). This allows for scalar symbol-wise equalization and omission of the CP. For example, with $N = 1024$, $L = 74$, UFMC operates with 10 MHz LTE-like parameters, on the same symbol rate and subcarrier

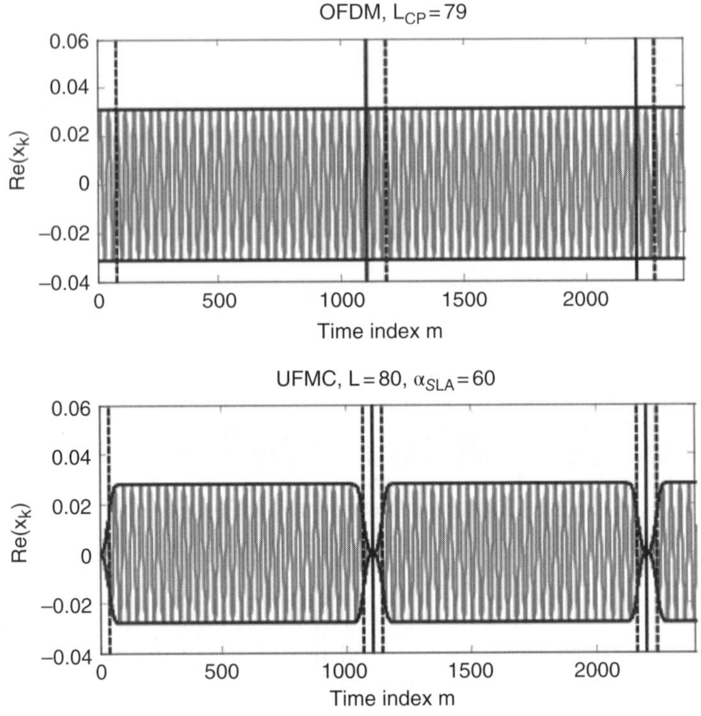

Figure 14.4 Time-domain signal of CP-OFDM and UF-OFDM, illustrated by real part of one subcarrier. CP-OFDM is characterized by the CP, UF-OFDM is characterized by soft transitions between subsequent symbols.

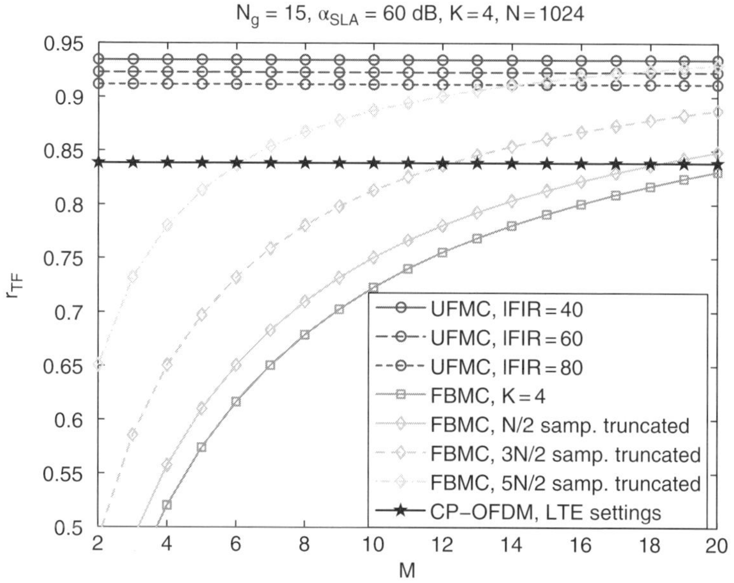

Figure 14.5 Time–frequency efficiency of FBMC, CP-OFDM, and UFMC as a function of burst length *M* (in number of multi-carrier symbols): number of usable resource elements per unit time-frequency bin.

spacing as CP-OFDM in LTE, but being able to allocate roughly 10% more subcarriers than CP-OFDM (Schaich and Wild 2014c). This is demonstrated by Figure 14.5. Due to the long filter length, FBMC has a time-overhead dependency as a function of the burst length, which is not the case for CP-OFDM and UFMC. This means that UFMC, like CP-OFDM, allows for the efficient insertion of small control elements occupying one or a few resource elements. Examples include sounding symbols and control information such as ACK/NACK feedback.

14.3.1.2 Relaxed Synchronization Support and Autonomous Timing Advance

The unified frame structure concept supports various types of traffic and devices. For sporadic small-packet transmission, our vision is to use open-loop synchronization for battery and signaling-overhead saving reasons. Available downlink-synchronization sequences and/or pilots are used to establish a coarse downlink-based synchronization. The remaining timing offsets result from the unknown propagation round-trip times. The idea of autonomous timing advance (ATA) (Schaich and Wild 2014a) is that the device uses estimates of its propagation delay (for example based on the receive power levels of the pilot signals and/or some additional broadcast information by the BS for transmit power levels or cell sizes) and coarsely pre-compensates its round-trip delay by adapting its transmission timing autonomously. This allows skipping closed-loop timing adjustments, which eases direct (for example contention-based) data transmission via random access.

Two variants of ATA are possible:

Type 1. On the one hand the single devices may estimate the round-trip time their signals are suffering and apply a respective timing advance. When doing so, performance of the respective transmission is improved.

Type 2. Another simpler variant is for all type II/III/IV devices to apply the same timing advance, for example to half of the highest round-trip time $N_{OL;max}$ appearing within the cell, or alternatively the 95th percentile of all existing round-trip times. When using this technique, inter-traffic interference at the borders is reduced (in Figure 14.6 results for the second variant are presented). Various means are available for the devices in order to get knowledge on $N_{OL;max}$. The most simple one is for the BS to broadcast the value based on measurements (for example during closed-loop synchronization of type I devices).

Figure 14.6 shows simulation results for an FDMA multi-user scenario. In the absence of CFO and guard subcarriers, UFMC supports cell radii 33–60% larger than CP-OFDM for the same MSE. When using two guard subcarriers, the cell radii supported by UFMC are more than double the ones from OFDM. With a CFO of 10%, OFDM cannot support the target MSE of −30 dB.

This enhanced waveform property of UFMC over OFDM can be used for an integrated flexible and scalable air interface where different service and device classes coexist: fully synchronous traffic in parallel to traffic with relaxed synchronization, using ATA.

14.3.1.3 Supporting Multiple Signal Layers with Interleave Division Multiple Access

Using multiple signal layers on top of the frequency and time dimensions adds further degrees of freedom for multiple access. Superposition of layers is required to achieve the capacity of the multiple access channel (MAC). A well-known mechanism to separate layers is via code; in other words, using CDMA. A more modern approach is to use different per-layer interleavers for separation, which is known as interleave division multiple access (IDMA), and was first addressed by Ping et al. (2003). IDMA fully benefits from code design and can inherently leverage the turbo principle for efficient multiuser detection (MUD). In 5G networks those strengths are very desirable. MUD can be applied to data transmissions via a random access channel to provide receiver-side collision resolution. This so-called multipacket reception (MPR) is a very desirable feature for efficient contention-based access, which can be beneficial for sporadic small-packet traffic, as for example occurring in massive machine communication (MMC) (Nizamuddin et al. 2014).

Furthermore IDMA can be used on top of advanced 5G waveforms. Chen et al. (2014) combined IDMA with either CP-OFDM or UFMC and compared to FDMA, resulting in user allocations as depicted in Figure 14.7. In low code rates, improvements of 2 dB in terms of E_b/N_0 could be shown. IDMA's naturally larger allocation sizes with lower code rates (due to superposition), compared to FDMA, lead to fewer edge effects of ICI and a larger frequency diversity. Both effects are valuable in contention-based transmission with relaxed time–frequency alignment. The combination of UFMC and IDMA turned out to be the best choice (Chen et al. 2014).

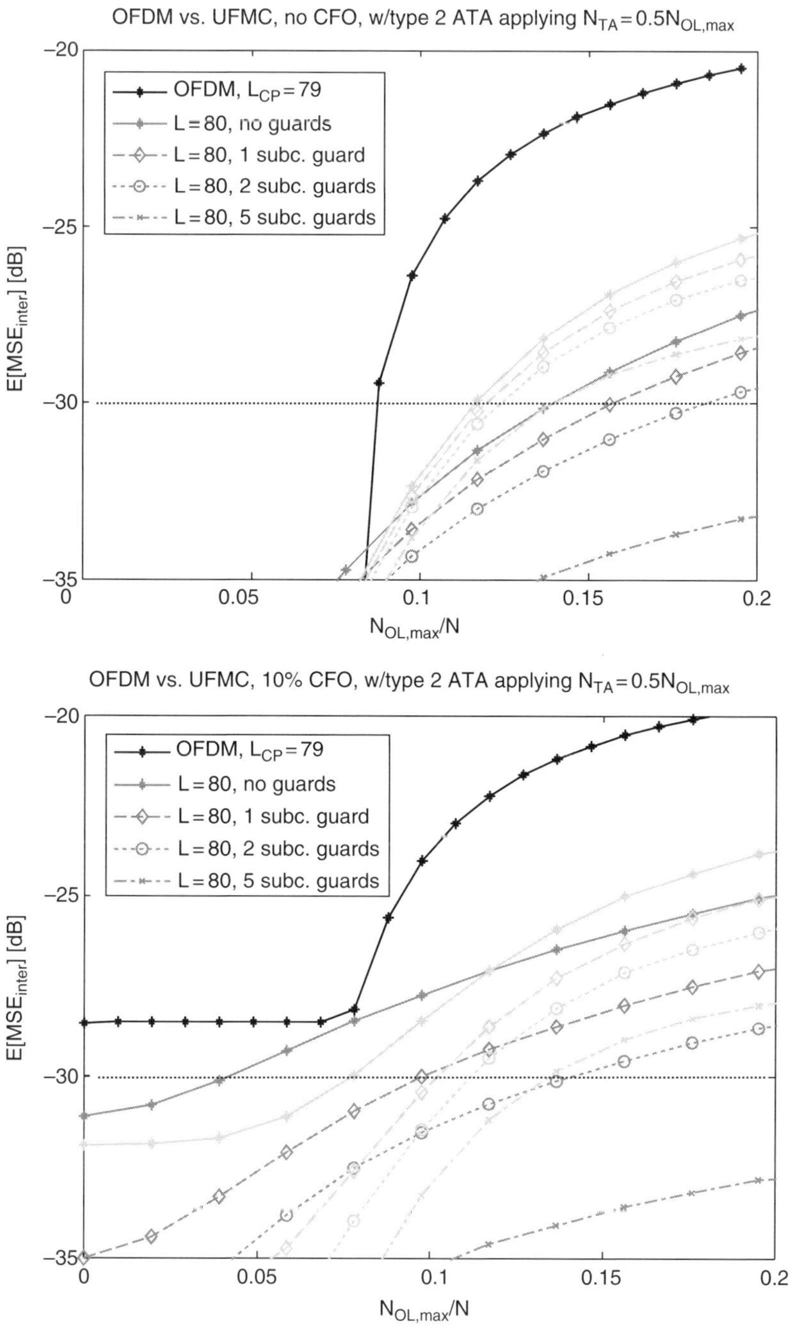

Figure 14.6 CP-OFDM vs UFMC using autonomous timing advance (ATA) type 2; average distortion from frequency division multiple access (FDMA) multiuser interference as a function of cell radius normalized to multi-carrier symbol duration: left, no carrier frequency offset (CFO); right, CFO 10% of subcarrier spacing. The dotted line represents a targeted maximum distortion of −30 dB, determining the highest supported normalized delay. The dark gray curves represent UFMC with RC-shaped windowing, and the light gray curves represent UFMC without windowing.

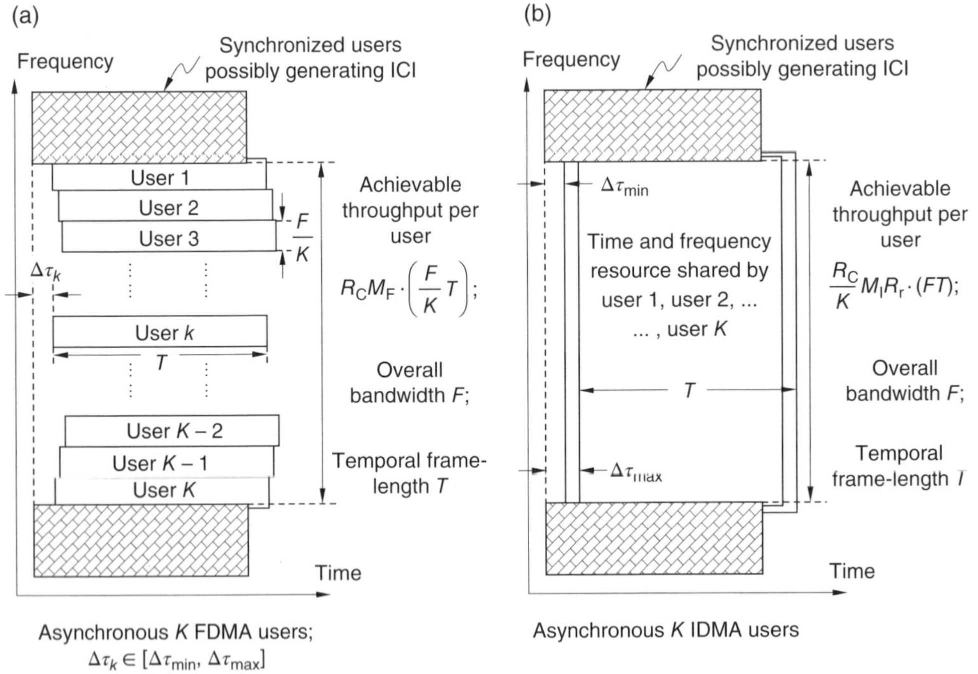

Figure 14.7 Resource allocation setting for either FDMA or IDMA with relaxed synchronicity in conjunction with multi-carrier waveform (either UFMC or CP-OFDM).

14.3.2 Generalized Frequency Division Multiplexing

Future wireless communication systems will face a multitude of new applications. Flexibility is a key feature for any waveform to be used in 5G networks, if the PHY layer must adapt itself to several different scenarios and channels. In this section we will describe how generalized frequency division multiplexing (GFDM) can add flexibility to deal with the requirements of modern communication systems.

14.3.2.1 Principles

Consider the block diagram in Figure 14.8. A data source provides the μ-QAM symbol data vector, which is decomposed into K subcarriers with M subsymbols each, according to $\vec{d} = (\vec{d}_0^{\mathrm{T}}, \ldots, \vec{d}_{M-1}^{\mathrm{T}})^{\mathrm{T}}$ with $\vec{d}_m = (d_{0,m}, \ldots, d_{K-1,m})^{\mathrm{T}}$. The total number of symbols follows as $N = KM$. Therein, the individual elements $d_{k,m}$ correspond to the data transmitted on the kth subcarrier and in the mth subsymbol of the block.

Each $d_{k,m}$ is transmitted with a circular time–frequency shift of a prototype filter $g[n]$ such that the transmit samples $x[n]$ are given by the superposition of all data symbols

$$x[n] = \sum_{k=0}^{K-1} \sum_{m=0}^{M-1} g\left[(n - mK) \bmod N\right] \cdot \exp\left(-j2\pi \frac{k}{K} n\right) \cdot d_{k,m}, \quad n = 0, \ldots, N-1. \quad (14.2)$$

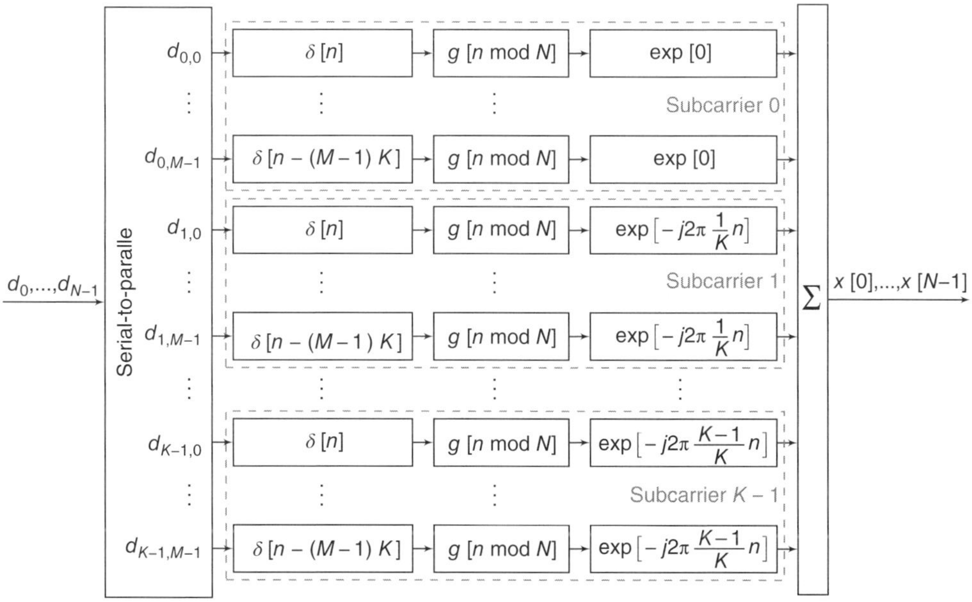

Figure 14.8 GFDM modulator block diagram.

with n denoting the sampling index. In Eq. (14.2), the modulo operation performs the circular time-shifts while the complex exponential performs the shifting operation in frequency. The linear relation in Eq. (14.2) can be simply formulated as

$$\vec{x} = \mathbf{A}\vec{d} \tag{14.3}$$

where $\mathbf{A}$ is a $KM \times KM$ transmitter matrix as described by Michailow and et. al. (2014). At this point, $\vec{x}$ contains the transmit samples that correspond to the GFDM data block $\vec{d}$. Lastly, a cyclic prefix of N_{CP} samples is added to produce $\vec{\tilde{x}}$.

Transmission through a wireless channel is modeled by $\vec{\tilde{y}} = \tilde{\mathbf{H}}\vec{\tilde{x}} + \vec{\tilde{w}}$, where $\vec{\tilde{y}}$ is the received counterpart of $\vec{\tilde{x}}$. Here, $\tilde{\mathbf{H}}$ denotes the channel matrix with band-diagonal structure based on a channel impulse response $h = (h_0, \ldots, h_{N_{\mathrm{ch}}-1})^{\mathrm{T}}$. Lastly, $\vec{\tilde{w}} \sim \mathcal{CN}(\mathbf{0}, \sigma_w^2 \mathbf{I})$ denotes additive white Gaussian noise (AWGN). At the receiver, time and frequency synchronization is performed, yielding $\vec{\tilde{y}}_{\mathrm{s}}$. Under the assumption of perfect synchronization – that is, $\vec{\tilde{y}}_{\mathrm{s}} = \vec{\tilde{y}}$– the cyclic prefix can be utilized to simplify the model of the wireless channel to

$$\vec{y} = \mathbf{H}\vec{x} + \vec{w}, \tag{14.4}$$

by replacing $\tilde{\mathbf{H}}$ with the $N$by$N$ matrix $\mathbf{H}$, which is the corresponding circular convolution matrix. This allows use of zero-forcing channel equalization. The overall transceiver equation can be written as $\vec{y} = \mathbf{HA}\vec{d} + \vec{w}$. Introducing $\vec{z} = \mathbf{H}^{-1}\mathbf{HA}\vec{d} + \mathbf{H}^{-1}\vec{w} = \mathbf{A}\vec{d} + \vec{\bar{w}}$ as the received signal after channel equalization, linear demodulation of the signal can be expressed as

$$\hat{\vec{d}} = \mathbf{B}\vec{z}, \tag{14.5}$$

where $\mathbf{B}$ is a $KM \times KM$ receiver matrix. Several standard receiver options for the GFDM demodulator are readily available in the literature:

- The matched filter (MF) receiver $\mathbf{B}_{\mathrm{MF}} = \mathbf{A}^{\mathrm{H}}$ maximizes the signal-to-noise ratio (SNR) per subcarrier, but with the effect of introducing self-interference when a non-orthogonal transmit pulse is applied.
- The zero-forcing (ZF) receiver $\mathbf{B}_{\mathrm{ZF}} = \mathbf{A}^{-1}$ completely removes any self-interference at the cost of enhancing the noise. Also, there are cases in which $\mathbf{A}$ is ill-conditioned and thus the inverse does not exist.
- The linear minimum mean square error (MMSE) receiver $\mathbf{B}_{\mathrm{MMSE}} = \left(\mathbf{R}_w^2 + \mathbf{A}^{\mathrm{H}}\mathbf{H}^{\mathrm{H}}\mathbf{H}\mathbf{A}\right)^{-1}\mathbf{A}^{\mathrm{H}}\mathbf{H}^{\mathrm{H}}$ makes a trade-off between self-interference and noise enhancement. Here, $\mathbf{R}_w^2$ denotes the covariance matrix of the noise.

Note that in case of MMSE reception, the received signal is jointly demodulated and equalized in the receiving process, hence the zero-forcing channel equalizer block is not required and $\hat{\vec{d}} = \mathbf{B}\vec{y}$. Finally, the received symbols $\hat{\vec{d}}$ are demapped and delivered to the sink.

14.3.2.2 GFDM in a Gabor Transform Setting

If $g[n]$ is identified as a discrete Gabor prototype window where the data symbols $d_{k,m}$ are the Gabor expansion coefficients, then, clearly, the GFDM signal presented in Eq. (14.2) is a critically sampled Gabor expansion (Matthe et al. 2014) and $\gamma[n]$ is the corresponding dual window to $g[n]$ with the same time–frequency structure. The receiving filters can be obtained by cyclic shifting $\gamma[n]$ in time and frequency domains, similar to Eq. (14.2). An important practical application of this theory is the computation of the prototype receiver filter. Since $\gamma[n]$ is the dual to $g[n]$, the prototype receiver filter can be obtained by

$$\gamma[n] = \left(\mathcal{Z}^{-1}\right)^{(K,M)}\left(\frac{1}{K\left(\mathcal{Z}^{(K,M)}g\right)^{*}[n,k]}\right), \tag{14.6}$$

Where

$$\left(\mathcal{Z}^{K,M}g\right)[n,k] = \sum_{l=0}^{M-1} g[n+lK]\exp-j2\pi\frac{l}{M} \tag{14.7}$$

is the discrete Zak transform (DZT) of a periodic sequence $g[n]$ and $(\mathcal{Z}^{(K,M)})^{-1}$ is the inverse DZT. Using Eq. (14.6) and Eq. (14.7), the evaluation of the receiver filters does not require a complex computation based on all possible waveforms, and the efficient DZT transform pair can provide the prototype receiver filter only based on the transmit prototype filter.

14.3.2.3 Time-reversal Space–Time Coding for GFDM Access

Time-reversal space-time coding (TR-STC) has been proposed by Al-Dhahir (2001) to allow the use of STC for single-carrier transmission over frequency-selective channels. In this approach, K data symbols are combined in data blocks $x_i[n]$, where $n = 0,1,\ldots,K-1$. Two subsequent blocks are used to build the space–time codeword as:

	Antenna 1	Antenna 2
Block i	$\vec{x}_{1,i}[n] = x_i[n]$	$\vec{x}_{2,i}[n] = -\vec{x}^{*}_{i+1}[<-n>_K]$
Block i+1	$\vec{x}_{1,i+1}[n] = x_{i+1}[n]$	$\vec{x}_{2,i+1}[n] = \vec{x}^{*}_{i}[<-n>_K]$

(14.8)

where i is an even number. Note the property of the Fourier transform $\mathcal{F}\{x^{*}[<-n>_K]\} = X^{*}[k]$, which explains why the time-reversing operation at the transmitter is necessary. With its help, at the receiver the signal can be transformed into the frequency domain and combined using standard Alamouti combining in order to gain diversity. After combining, the signal is transformed back into the time domain and demodulated.

TR-STC can be directly applied to GFDM. Consider two GFDM frames obtained from two sets of data vectors $\vec{d}_i$:

$$\vec{x}_i = \mathbf{A}\vec{d}_i. \tag{14.9}$$

Clearly, $\vec{x}_i[n]$ and $\vec{x}_{i+1}[n]$ obtained from Eq. (14.9) can be space–time encoded as stated in Eq. (14.8). Then, at the receiver, the signals can be Alamouti-combined in the frequency domain to detect the overall GFDM frames. Subsequently, $\hat{x}_i$ and $\hat{x}_{i+1}$ can be directly sent to either the linear demodulators described earlier in the section.

TR-STC-GFDM demands that two GFDM frames be transmitted. This amount of data can be very high for a single user, which will have to transmit dummy information. Of course, this would lead to a waste of resources and inefficiency. A solution for this problem is to introduce a multiple-access scheme combined with TR-STC-GFDM. Frequency division multiple access is an efficient solution, although other multiple-access schemes can be applied.

TR-STC generalized frequency division multiple access (GFDMA) can be achieved by sharing the subcarriers between different users. The K subcarriers can be equally distributed between the U users, hence the uth user will receive $K_u = K/U$ adjacent subcarriers. Assuming $K = 4$, $M = 2$, and $U = 2$, the data matrix for each user is given by

$$\mathbf{D}^{(1)} = \begin{pmatrix} d^{(1)}_{0,0} & d^{(1)}_{0,1} \\ d^{(1)}_{1,0} & d^{(1)}_{1,1} \\ 0 & 0 \\ 0 & 0 \end{pmatrix}, \quad \mathbf{D}^{(2)} = \begin{pmatrix} 0 & 0 \\ 0 & 0 \\ d^{(2)}_{0,0} & d^{(2)}_{0,1} \\ d^{(2)}_{1,0} & d^{(2)}_{1,1} \end{pmatrix}. \tag{14.10}$$

Therefore, each user u generates a GFDM signal based on two successive data matrices $\mathbf{D}^{(u)}_i$, which are used to create two data vectors $\vec{d}^{(u)}_i = \text{vec}(\mathbf{D}^{(u)}_i)$. The data vectors are modulated as in Eq. (14.3) and then encoded as in Eq. (14.8). The signals from the two transmitting antennas are sent to the users through time-variant frequency-selective channels. Notice that the adjacent subcarriers from two different users will interfere with each other in case a non-orthogonal prototype filter is used. Once the channels for different users are different, this will lead to high ICI that cannot be easily removed. Hence, the users must have one subcarrier as band guard interval between each other to avoid interference. Guard band subcarriers are not necessary if an orthogonal pulse-shape is used. Let us assume that all users are synchronized with the BS and the signals from different users reach the l receiving antennas within the CP duration. In this situation, the received signal in the frequency domain, after removing CP, is given by

$$\begin{aligned} \vec{Y}_{l,i} &= \sum_{u=1}^{U} \vec{H}_{1,l}^{(u)} \vec{X}_{i}^{(u)} - \vec{H}_{2,l}^{(u)} \vec{X}_{i+1}^{(u)*} + \vec{W}_{l,i} \\ \vec{Y}_{l,i+1} &= \sum_{u=1}^{U} \vec{H}_{1,l}^{(u)} \vec{X}_{i+1}^{(u)} + \vec{H}_{2,l}^{(u)} \vec{X}_{i}^{(u)*} + \vec{W}_{l,i+1}. \end{aligned} \tag{14.11}$$

The received signals can be combined with Alamouti combining and the users' data is retrieved using standard demodulators.

The proposed scheme can be used by users with one or two transmitting antennas. In case a user has only one transmit antenna, the pilot information that is sent for channel estimation will be available from only one antenna. The receiver will assume that the attenuation for the second transmitting antenna is very high and the combiner will deliver a signal without diversity gain.

14.3.2.4 Reducing Latency in LTE Time–Frequency Grid

Low latency is an important feature for future wireless communication systems and GFDM can be designed to reduce the impact of the PHY layer in the overall system latency. The main approach is to reduce the GFDM symbol duration and add only a single CP for M subsymbols. The latter property is particularly important, because prefixing each subsymbol would actually increase the latency and reduce the spectral efficiency. The short subsymbol duration will lead to wider subcarriers that might suffer from frequency = selective effects of the mobile multipath channel. However, because GFDM subcarriers have M times more samples in the frequency domain than LTE subcarriers, it is still possible to achieve acceptable symbol error rate performance. To achieve low latency, it is advantageous if each GFDM symbol can be demodulated independently, so the channel estimation can not rely on the system frame structure and each symbol must carry all the necessary information for single-shot synchronization (Gaspar et al. 2014) and channel estimation.

GFDM can be configured to reduce the latency and yet be aligned with the current LTE time–frequency grid. The main goal is to reduce the GFDM frame duration to an integer fraction of the 1-ms subframe duration and a set of GFDM subcarriers must fit in an integer multiple of 180-kHz bandwidth. Table 14.1 presents one possible set of GFDM parameters for this scenario.

From Table 14.1, it can be seen that the proposed parameters for GFDM have the same subframe duration as the LTE grid. Nevertheless, the GFDM symbol duration is 7.5 times smaller than the equivalent slot duration of the LTE system. Assuming that the GFDM system is designed for independent demodulation of each symbol, it is possible to reduce the contribution of the PHY layer modulation to the latency by a factor of 15, compared with the current

Table 14.1 GFDM configuration aligned with the LTE grid for low latency applications.

Parameter	Normal mode
Subframe duration	1 ms or 30.720 samples
GFDM symbol duration	66.67 μs or 2,048 samples
Subsymbol duration	4.17 μs or 128 samples
Subcarrier spacing and bandwidth	240 kHz
Sampling freq. (clock)	30.72 MHz
Subsymbol spacing K	128
Number of subsymbols per GFDM symbol M	15
Number of GFDM symbols per subframe	15
CP length	4.17 μs or 128 samples
Prototype filter	Dirichlet

LTE frame structure. Hence GFDM can contribute to latency reduction while keeping the same time–frequency grid as the 4G system. Notice that three GFDM subcarriers occupy the bandwidth of four LTE resource blocks. This means that each GFDM subcarrier is 16 times wider than the LTE subcarriers. Because each GFDM subcarrier has $M = 15$ times more samples than an LTE subcarrier, the spectrum resolution of both systems is approximately the same. GFDM employs a slightly smaller CP length than LTE and it does not require a larger CP for the first GFDM symbol. The Dirichlet pulse makes the system orthogonal and the subcarriers do not overlap with the surrounding subcarriers outside of the used resource blocks.

14.3.3 Filter Bank Multicarrier

Relaxed synchronization and multiple user access to fragmented spectrum are considered essential for future generations of wireless networks in order to reduce physical channel signaling. FDMA for filter bank multicarrier (FBMC) modulation provides promising performance without strict synchronization requirements, in contrast to conventional OFDM. In this section, the architecture of a FBMC receiver suitable for this scenario is considered, and its performance in terms of bit error rate (BER) and achievable capacity is compared against OFDM.

14.3.3.1 Principles

With FBMC, a set of parallel data symbols are transmitted through a bank of modulated filters. The choice of the prototype filter controls the localization in frequency of the generated pulse and provides better adjacent channel leakage performance in comparison to OFDM. Offset quadrature amplitude modulation (OQAM) combined with Nyquist constraints on the prototype filter is used to guarantee orthogonality between adjacent symbols and adjacent carriers while providing maximum spectral efficiency (Bellanger et al. 2010).

Frequency sampling techniques may be applied to design the prototype filter. The duration D of the prototype filter is a multiple of the size of the FFT, N, so that $D - KN$. K is an integer and often referred as the overlapping factor. Frequency spreading FBMC (FS-FBMC) has been recently proposed by Bellanger (2012) and Doré et al. (2014b) as an alternative to classical polyphase network (PPN) FBMC. This technique is inspired by the frequency-sampling technique used to design the prototype filter. With this approach, the number of non-zero samples in the frequency response is given by $P = 2, K - 1$. For $K = 4$, the frequency-domain pulse-response coefficients are equal to:

$$\begin{aligned} &H_0 = 1; H_1 = 0.971960 = H_{-1}; \\ &H_2 = \frac{\sqrt{2}}{2} = H_{-2}; H_3 = \sqrt{1 - H_1^2} = H_{-3} \end{aligned} \tag{14.12}$$

FS-FBMC spreads the data over P carriers by filtering the output of the OQAM process by the frequency-domain pulse response of Eq. (14.12). The output is then processed through an inverse FFT (IFFT) of size $K \times N$. OQAM precoding imposes that real and pure imaginary symbol values alternate on successive subcarrier frequencies and on successive transmitted symbols for any given subcarrier. This guarantees orthogonality between adjacent carriers since the coefficients of the prototype filter are real. The output of the IFFT is converted through a parallel-to-serial conversion and is accumulated with the following IFFT output data block stream delayed by $N/2$. This parallel-to-serial conversion is called overlap-and-sum. Once the transient period is over, $2K$ of the KNIFFT output samples are added together at any given time.

The dual operation of the overlap-and-sum operation of the transmitter is a sliding window in the time domain at the receiver that selects KNpoints every $N/2$ samples. An FFT is then applied every block of KN selected points. Figure 14.9 illustrates the principle of the FS-FBMC receiver. One advantage of this architecture is that frequency-domain time synchronization may be performed independently of the position of the FFT (Bellanger 2012). This is realized by combining timing synchronization with channel equalization. Another main benefit of FS-FBMC is that channel equalization may be limited to a one-tap complex-multiply operation while still sustaining significant channel impulse response delay spread, even compared to the conventional PPN-FBMC architecture (Berg et al. 2014a).

14.3.3.2 Multi-user Receiver Architecture

A multiuser FS-FBMC receiver architecture is depicted in Figure 14.10 where a total of B parallel users may be decoded at the same time. The key feature here is that the signal is demodulated in the frequency domain without *a priori* knowledge of the FFT timing alignment (that is, the location of the FFT block); this property is called asynchronous FFT. An asynchronous FFT of size KN is thus processed every block of $N/2$ samples generating KN points; that is, if $\mathbf{r}_\mathrm{m}$ is the mth received vector, a KN-point FFT is computed for samples $k = (n + m \times N / 2)$ with $n = 0, 1, \ldots, NK - 1$. These successive KN points are stored in a memory unit. The memory unit buffers the data for later processing.

In parallel, a multiuser frequency-domain synchronization detector detects the start of burst and estimates carrier frequency offset (CFO) for each of the B-users directly at the output of the FFT. Once the start of burst is detected, CFO is estimated on the preamble by computing the phase of the product between two consecutive FBMC symbols of any given user per active subcarrier. This synchronization information is used to estimate channel response independently for each user. On the channel-estimation datapath, CFO compensation is first

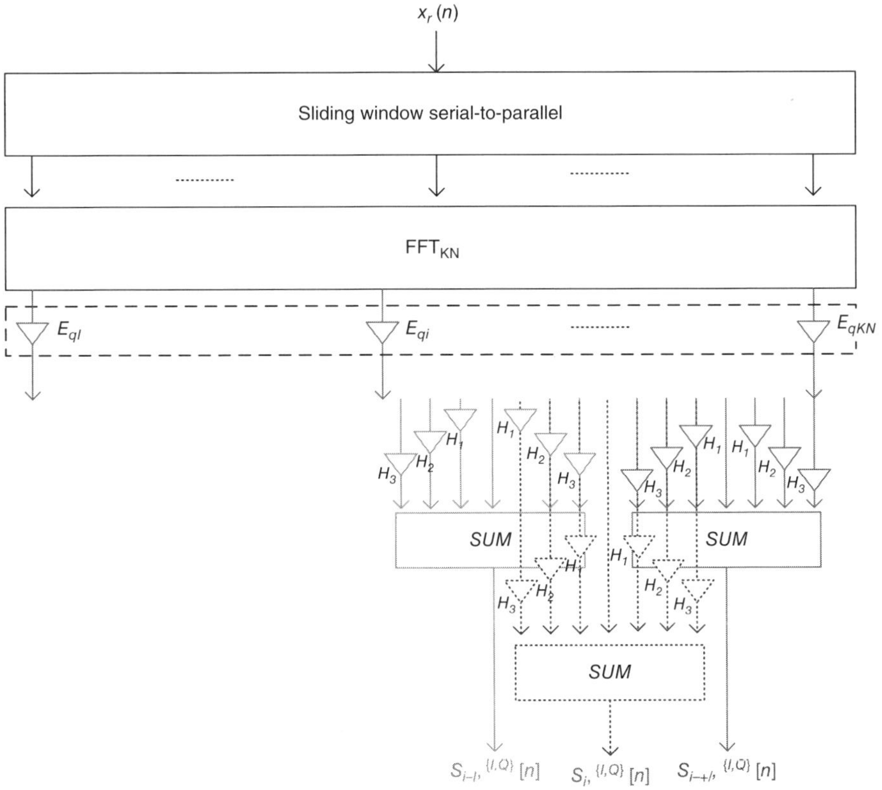

Figure 14.9 FS-FBMC receiver principles.

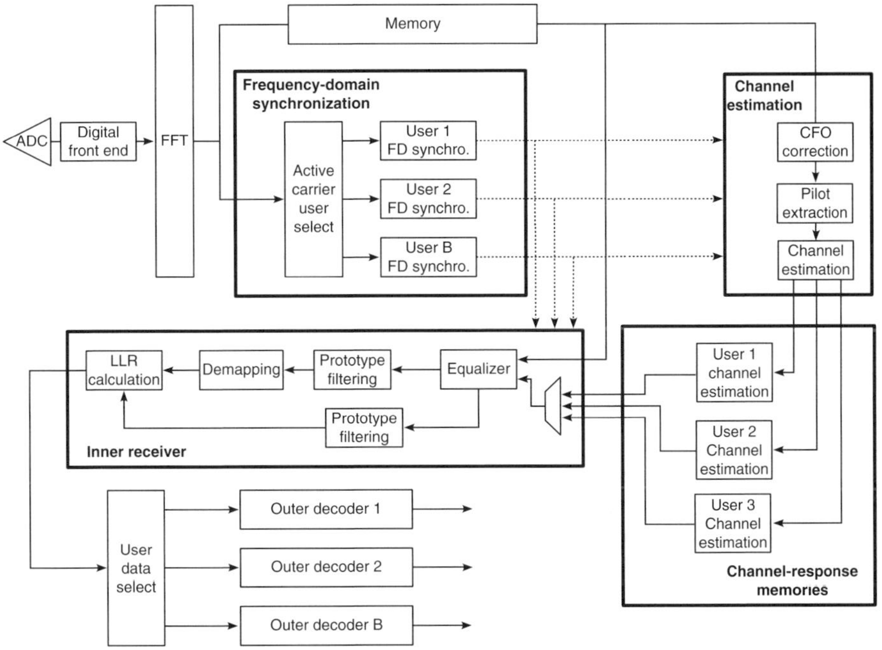

Figure 14.10 Block diagram of the proposed multiuser asynchronous receiver architecture.

performed in the frequency domain using a feed-forward approach. Then, channel coefficients are estimated on the pilot subcarriers before being interpolated for every active subcarrier. Once the channel is estimated on all the active subcarriers the response is stored for each user in a dedicated channel-response memory. The data buffered in the memory unit are then processed through a one-tap per subcarrier equalizer before filtering by the FBMC prototype filter. Demapping and log-likelihood ratio computation complete the inner receiver architecture. Per-user dedicated soft-input forward error correction (FEC) decoders finally recover the originally sent messages.

The asynchronous frequency-domain processing of the receiver combined with the high stop-band attenuation of the FBMC prototype filter provides a receiver architecture that allows for multiuser asynchronous reception. The hardware implementation of a multiuser (two users) flexible receiver has been presented by Berg et al. (2014b) based on the FS-FBMC architecture depicted in Figure 14.10. It was demonstrated that the FS-FBMC multi-user receiver could fit in a Xilinx Kintex-7 XC7K325T FPGA and it was also shown that the receiver complexity smoothly scales with the number of users, as only a limited number of building blocks need to be duplicated. Furthermore, it was demonstrated by Berg et al. (2014a) that the complexity overhead of an FS-FBMC single-user receiver implementation is kept reasonably low, at 30% overhead in logic compared to an OFDM receiver.

14.3.3.3 Robustness of the Receiver to Channel Delay Spread

The proposed FBMC receiver has been evaluated by simulation using the parameters derived from the 10-MHz LTE mode parameters. The results are summarized in Table 14.2.

A working SNR of 5 dB has been considered and the BER after Viterbi decoding for various channel delay spreads of length L time-domain samples has been measured. The channel impulse response has been defined as:

$$h_L(t) = 1 + \sum_{i=0}^{L-2} \alpha_i \delta\left(t - \frac{i}{F_s}\right) \qquad (14.13)$$

where L is the number of taps in the channel impulse response, F_s is the sampling frequency, and α_i is a set of complex coefficients following a Rayleigh distribution. Consequently, the delay spread of the channel is equal to L/F_s. The resulting BER at the output of the receiver has been calculated by averaging 10,000 channel realizations. Figure 14.11 depicts the

Table 14.2 Simulated system parameters.

Parameter	Value
Prototype filter	$K = 4$, $H = \{0.235147, \sqrt{2}/2, 0.97196, 1, 0.97196, \sqrt{2}/2, 0.235147, 0\}$
N	1024
F_s	15.36 MHz
Active carriers N_a	601 (9 MHz)
Modulation	QPSK
FEC	CC 1/2, $k = 7N_{FEC} = 1944$ bits, $R = 2/3$, SOVA algorithm, row-column interleaver

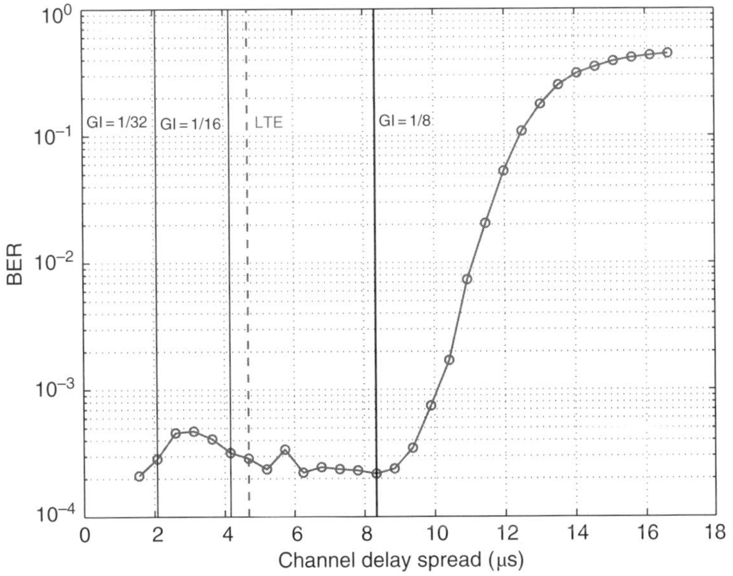

Figure 14.11 Performance of the proposed receiver as a function of channel delay spread. Equivalent guard interval (GI) durations have been added for reference.

performance of the receiver as a function of the channel delay spread. When the delay spread of the channel is greater than $0.15 \times N / F_s$(10 μs), performance is limited by the proposed equalization scheme as the one-tap equalizer becomes inefficient. On the other hand, for channel delay spread below $0.15 \times N / F_s$(10 μs), one-tap equalization is sufficient. These results compare with the performance of an OFDM receiver with a guard-interval length of at least $1/8 \times N$. In Figure 14.11, the guard interval length for an equivalent OFDM system with a FFT size of N is given. As a comparison, the normal LTE guard interval is set to 4.69 μs or approximately $0.703 \times N$.

14.3.3.4 Capacity Results and Analysis

The scenario considered is fragmented spectrum access in the context of asynchronous uplink multiuser transmissions. In particular, we consider a set of three users as illustrated in Figure 14.12, where the user of interest is located in between the other users in frequency. The UEs are considered unsynchronized. No guard carriers are considered for Scenario (a), while one guard carrier is inserted at the edges of each user spectrum bands for Scenario (b). We assume LTE-like parameters with a bandwidth of 10 MHz, the intercarrier spacing is set to 15 kHz, and the cell coverage is assumed to be 5 km.

We depict in Figure 14.13 the SINR measured at the output of the equalizer for OFDM and FBMC waveforms. No guard carrier is inserted between the edges of each user spectrum band (24 carriers per user). The SINR has been computed for each carrier location and for each timing offset from 0 to 256 samples (16.67 μs). For OFDM, thermal noise dominates the SINR when the timing offset is lower than the duration of the guard interval (72 samples). In that

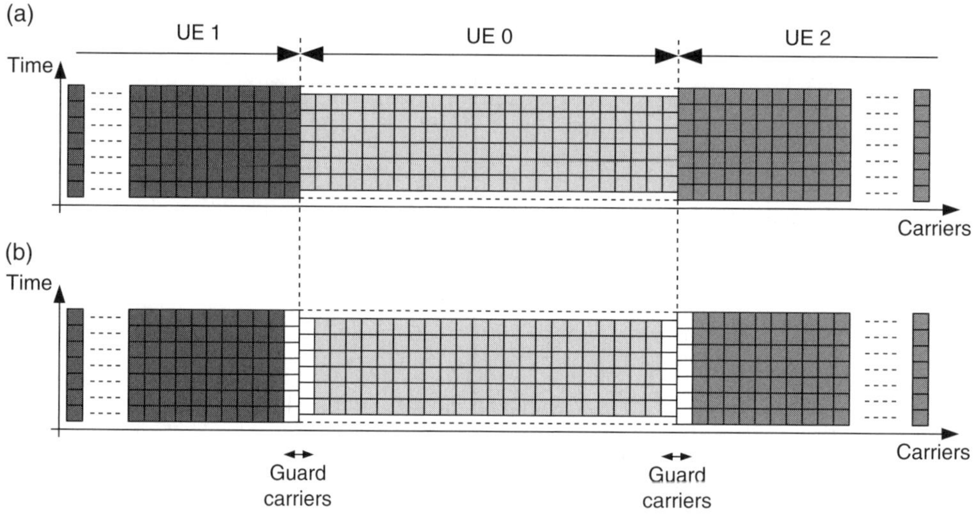

Figure 14.12 Proposed asynchronous scenario with three users: (a) without guard band; (b) with guard bands.

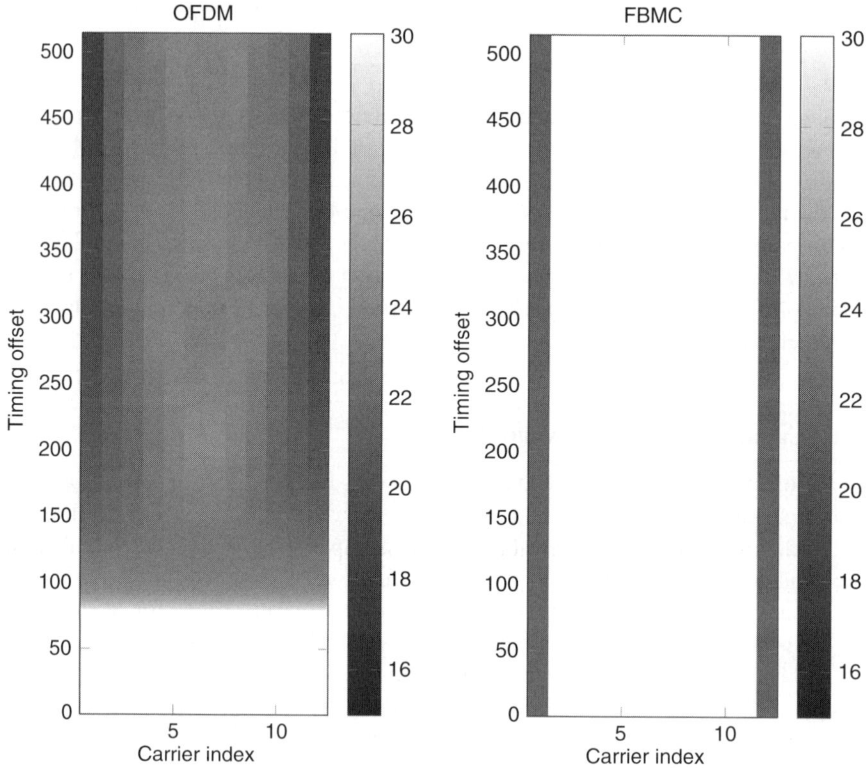

Figure 14.13 SINR in dB as a function of carrier index and timing offset for OFDM and FBMC waveforms assuming Scenario (a).

case, the received signal is not affected by interference. However, as soon as the timing offset is larger than the guard interval, SINR is dominated by the interference. This clearly illustrates that the interference level is more important on the edges of the spectrum, when the OFDM signal is not synchronized within the guard interval of its adjacent signal. This is due to the sinc frequency shape of the OFDM waveform. For FBMC, the center of the spectrum is always dominated by thermal noise, while interference at the edges of the carrier is preponderant, whatever the timing offset. This is a direct consequence of the properties of the prototype filter, which has been designed to minimize out-of-band interference. Interference affects the edges of the spectrum because the orthogonality of OQAM modulation is not preserved between asynchronous adjacent users. These results underline the benefit of FBMC compared to OFDM in this scenario.

The effect of asynchronous transmission is modeled on capacity computation by averaging the capacity over all the possible timing misalignments. In a cellular network, the time misalignment τ is the consequence of the propagation delay between UEs and the BS. If the users are equidistributed in the cell, τ may be considered uniform and bounded over a time interval $[0,\tau_{max}]$. The capacity is given by (Berg et al. 2014b):

$$C_{async}\left(\sigma_n\right)=\frac{1}{\tau_{max}}\int_0^{\tau_{max}}\frac{1}{N_a}\sum_{p=0}^{p=N_a-1}C_{M-QAM}\left(\sigma\left(\tau,p\right)\right)d\tau \tag{14.14}$$

where σ_n is the standard deviation of the thermal noise, N_a is the number of active carriers, C_{M-QAM} the channel capacity of M-ary quadrature amplitude modulation (M-QAM) constellation in the presence of AWGN, and σ the total noise variance per active carrier.

We depict in Figure 14.14 the capacity of OFDM and FBMC in the case of Scenarios (a) and (b), using Eq. (14.14) for various SNR values and modulation orders. Perfect power control is assumed. In the case of QPSK modulation (asymptotic capacity of 2), without guard carrier, the capacity is close to that for synchronous transmission. In that case the level of interference is much lower than the required SNR to allow the decoding of QPSK. For 16-QAM modulation (asymptotic capacity of 4), FBMC gives a significantly better capacity, particularly for SNR values above 10 dB. Due to the better frequency localization, only the carriers located at the border of the user spectrum are affected by interference. These results also demonstrate that performance is limited by interference, as the capacity of synchronous transmission is never reached. For 64-QAM modulation (asymptotic capacity of 6), the FBMC waveform clearly outperforms the OFDM waveform. Interference dominates the SINR for the OFDM waveform and consequently for a capacity of 5 bits/s/carrier, the SNR loss compared to FBMC is around 5 dB.

We also illustrate in Figure 14.14 the capacity comparison for Scenario (b); that is, with guard carriers. The achievable capacity is evaluated assuming a constant power on the bandwidth. In others word, the power of each carrier is boosted in the case of guard carriers. Moreover, the loss in capacity as a consequence of the use of guard carriers is taken into account in the calculation. For the OFDM case, as the power on each carrier is increased, the level of interference follows. For all the values of SNR, and all the modulation orders, FBMC outperforms OFDM. Not only does FBMC without guard carriers provide better capacity than FBMC with guard carriers but FBMC with guard carriers even gives better capacity than OFDM without guard carriers. This difference in capacity is even further emphasized when the effect on capacity of a limited power control feedback is measured (Doré et al. 2014a).

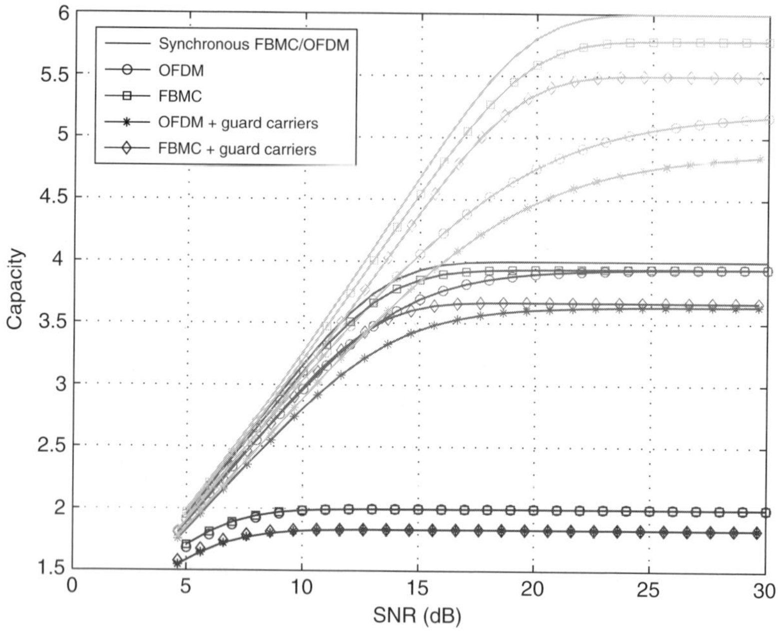

Figure 14.14 Capacity in bit/s/carrier for OFDM and FBMC waveforms for Scenarios (a) and (b). The upper, middle, and lower groups of curves represent QPSK, 16-QAM, and 64-QAM modulation, respectively.

14.4 One-shot Random Access

A conceptional approach to efficiently support sporadic traffic in 5G is to use an extended physical layer random access channel (PRACH) that achieves device acquisition and (typically small) payload transmission "in one shot" (Kasparick et al. 2014; Wunder et al. 2015b, d).

Similar to the implementation in UMTS, the goal is to transmit small user data packets using the PRACH, without maintaining a continuous connection. So far, this is not possible in LTE, where data is only carried using the physical uplink shared channel (PUSCH) so that the resulting control-signaling effort renders scalable sporadic traffic (for example, several hundred nodes in the cell) infeasible. By contrast, in our design a data section is introduced between synchronous PUSCH and standard PRACH, called D-PRACH (data PRACH) supporting asynchronous data transmission. Clearly, by doing so, sporadic traffic is removed from standard uplink data pipes, resulting in drastically reduced signaling overhead. Another issue that is closely related to the signaling overhead is the complexity and power consumption of the devices. The proposed time–frequency resource grid for the described channels is illustrated in Figure 14.15.

We assume that each D-PRACH data resource contains only a very few subcarriers, say 5–20 subcarriers. In addition, in a 5G system, we can expect that there is a massive number of MTC devices that will concurrently employ these data resources in an uncoordinated fashion. For example, in the simplest approach the D-PRACH uses the guard bands between PRACH and PUSCH so that channel estimation is achieved by PRACH. Clearly, in an extended setting

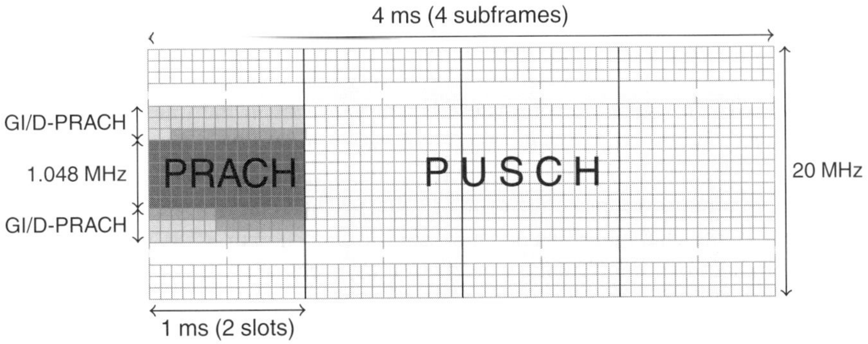

Figure 14.15 Illustration of LTE PRACH and PUSCH regions. A guard interval (GI) separates PUSCH from PRACH in LTE. Parts of this area are used for data transmissions of asynchronous users in a novel D-PRACH, whose size can be variably determined by MAC. Rows in this illustration do not represent true subcarrier quantities.

this region can be enlarged (by higher-layer parameters) but clearly at some point the question of efficient control signaling is arises again. Hence Wunder et al. (2014a; 2015a, c) also outline a sparse signal processing approach for the control signaling (for example, channel estimation) to handle such situations efficiently.

Notably, in a D-PRACH setting waveform design is necessary since the LTE OFDM waveform cannot handle the highly asynchronous access of different devices with possible negative delays or delays beyond the CP. Clearly, guards could be introduced between the individual (small) data sections but this again makes the approach very inefficient. Our results indeed show that up to six additional subcarriers can be obtained compared to the OFDM setting. Moreover, giving up guard bands for transmitting data will naturally lead to increased interference for PUSCH users, which must be also handled with waveform design.

14.4.1 Bi-orthogonal Frequency Division Multiplexing

For D-PRACH, an approach based on bi-orthogonal frequency division multiplexing (BFDM) (Kozek and Molisch 1998, Schafhuber et al. 2002) is adopted, where we replace orthogonality of the set of transmit and receive pulses with bi-orthogonality. In particular, time–frequency representations of the transmit and receive pulses are pairwise (not individually) orthogonal. Thus, there is more flexibility in designing a transmit prototype, for example, in terms of side-lobe suppression.

The underlying principle is to transmit the symbols according to a set of shifted pulses on time–frequency lattice points (kT, lF), where T is the time-shift period and F is the frequency-shift period, and $k, l \in \mathbb{Z}$. The only requirement of perfect symbol reconstruction is that the set of transmit pulses $\{g_{k,l}\}$ and the set of receive pulses $\{\gamma_{k,l}\}$ form bi-orthogonal Riesz bases (Kozek and Molisch 1998). The determining factors to meet that condition are, first, the properties of the pulses itself, and second, the time–frequency product TF being greater than one. See also Jung and Wunder (2007) for further explanations of these conditions. Here, we choose $TF = 1.25$, which is compliant with 4G.

However, instead of a matched filter, an optimally mismatched filter will be used (Kozek and Molisch 1998). The BFDM approach is well suited to sporadic traffic, since the PRACH symbols are relatively long so that transmission is very robust to (even negative) time offsets. In addition, BFDM is also more robust to frequency offsets in the transmission that, as is well-known, typically sets a limit to the symbol duration in OFDM transmission. Finally, the concatenation of BFDM and several OFDM symbols requires good tail behavior of the transmit pulse in order to keep the distortion to the payload-carrying subcarriers in PUSCH small. Conversely, the dual pulse that accounts for the distortion of PUSCH onto PRACH can be controlled by iterative interference cancellation (if necessary). This alleviates the typical problem of controlling time/spectral localization of pulse and dual pulse. The excellent and controllable tradeoff between performance degradation due to time and frequency offsets is the main advantage of BFDM with respect to conventional OFDM.

Let us now focus on transmitter and receiver design. Note that efficient implementations are available in the literature (Vangelista and Laurenti 2001).

14.4.1.1 Transmitter

In the following, let T_s denote the sampling period, which is equal to $1/f_s$, with f_s being the sampling frequency. In the following discrete model, we let all time indices be multiples of T_s and frequency indices be multiples of F. Furthermore, we use N to denote the discrete counterpart of the symbol duration and submit K symbols. Let N_{FFT} be the FFT-length in D-PRACH. Note, for numerical reasons N_{FFT} must divide into NK.

For the pulse-shaped D-PRACH signal, additional processing is needed when compared to standard OFDM. In contrast to standard processing, we process more than one symbol interval, even if we use only one symbol to carry the preamble. We refer to Schafhuber et al. (2002) for implementation details. A pulse g is used to shape the spectrum of the D-PRACH signal, for example to allow the use of PRACH guard bands with acceptable interference. Let P be the length of pulse g. We extend the output signal $s[n]$ after the IFFT stage by repeating it and taking modulo P to get the same length as the pulse g. Given K symbols, each symbol $s_k[n]$ is pointwise multiplied by the shifted pulse g and superimposed by overlap and add, such that we get the baseband pulse shaped D-PRACH transmit signal as:

$$s[n] = \sum_{k=0}^{K-1} s_k[n]\, g[n-kN]. \tag{14.15}$$

In greater detail, this can be written as

$$s[n] = \beta \sum_{k=0}^{K-1} \sum_{l \in \mathcal{F}_1 \cap \mathcal{F}_2} \tilde{X}_{k,l}\, g[n-kN]\, e^{j\frac{2\pi nl}{N_{\text{FFT}}}}, \tag{14.16}$$

where $\tilde{X}_{k,l}$ is the Fourier-transformed preamble signal (which is constructed from Zadoff–Chu root sequences (Chu 1972) in LTE-A) occupying subcarriers $l \in \mathcal{F}_1$ of length N_{ZC} and the actual data-occupying subcarriers $l \in \mathcal{F}_2$ at the kth symbol, and β is an amplitude scaling factor for normalizing the transmit power.

14.4.1.2 Receiver

The only difference to the standard PRACH receiver is the processing before the FFT. In standard PRACH processing, the CP is first removed from the received signal $r_{\mathrm{PR}}[n]$ and then the FFT is performed. In the pulse-shaped PRACH, an operation to invert the (transmitter-side) pulse shaping has to be carried out first. To be more precise, the K symbols of the received signal $r_{\mathrm{PR}}[n]$ are pointwise multiplied by the shifted bi-orthogonal pulse γ, such that we have

$$r_k^{\gamma}[n] = r_k[n]\gamma^*[n - kN]. \tag{14.17}$$

Subsequently, we perform a kind of prealiasing operation to each windowed $r_k^{\gamma}[n]$:

$$\tilde{r}_k^{\gamma}[n] = \sum_{l=0}^{P/N_{\mathrm{FFT}}-1} r_k^{\gamma}[n - lN_{\mathrm{FFT}}], \tag{14.18}$$

such that we obtain the Fourier-transformed D-PRACH signal at the kth symbol and lth subcarrier after the FFT operation:

$$\tilde{Y}_{k,l} = \sum_{n=0}^{N_{\mathrm{FFT}}-1} \tilde{r}_k^{\gamma}[n] e^{-\frac{j2\pi nl}{N_{\mathrm{FFT}}}}. \tag{14.19}$$

Although we do not employ a CP as in standard PRACH, the time–frequency product of $TF = 1.25$ allows the signal to have temporal and frequency guard regions as well. These time–frequency guard regions and the overlapping of the pulses cause the received signal to be cyclostationary (Bolcskei 2001), which gives the same benefit as the cyclostationarity made by CP. Furthermore, it has been shown that the bi-orthogonality condition of the pulses is sufficient for the cyclostationarity and makes it possible to estimate the symbol timing offset from its correlation function (Bolcskei 2001).

14.4.1.3 Pulse Design

As mentioned before, the pulses g and γ used play a key role and should therefore be carefully designed. Since we consider here the BFDM approach, we set up the transmit pulse g according to system requirements and compute from g the receive pulse γ as the canonical dual (bi-orthogonal) pulse. For this computation, here we follow the method already used, for example, by Jung and Wunder (2007 and see also the references therein). Briefly explained, bi-orthogonality in a stable sense means that g should generate a Gabor Riesz basis and γ generates the corresponding dual Gabor Riesz basis. From the Ron–Shen duality principle (Ron and Shen 1997) it follows that γ has the desired property if it generates on the so-called adjoint time–frequency lattice a Gabor (Weyl–Heisenberg) frame, which is dual to the frame generated by g. However, this can be achieved with the S^{-1}-trick explained by Daubechies (1992). Side effects such as spectral regrowth due to periodic setting when calculating the bi-orthogonal pulses are negligible.

As a rough and well-known guideline for the well-conditioning of this procedure, the ratio of the time and frequency pulse widths (variances) σ_t and σ_f should be approximately matched to the time–frequency grid ratio

$$\frac{T}{F} \approx \sqrt{\frac{\sigma_t}{\sigma_f}}, \tag{14.20}$$

and this should also be of the order of the channel's dispersion ratio (Jung and Wunder 2007). However, here we consider only the first part Eq. (14.20) of this rule since we focus on a design being close to the conventional LTE PUSCH and PRACH.

We propose constructing the pulse g based on the B-splines in the frequency domain. B-splines have been investigated in the Gabor (Weyl–Heisenberg) setting, for example, by Prete (1999). The main reason for using the B-spline pulses is that convolution of such pulses have excellent tail properties with respect to the L_1-norm, which is beneficial with respect to the overlap of PRACH on the PUSCH symbols. We also believe that the give a good trade off of the time offset for the frequency offset performance degradation, but this is part of on-going investigations and beyond the conceptional approach described here.

Because of its fast decay in time, we choose a second order B-spline (the "tent" function) in the frequency domain given by

$$B_2(f) = B_1(f) * B_1(f), \text{ where} \tag{14.21}$$

$$B_1(f) := \chi_{\left[-\frac{1}{2},\frac{1}{2}\right]}(f) \tag{14.22}$$

and * denotes convolution. It has been shown by Prete (1999) that $B_2(f)$ generates a Gabor frame for the (a, b)-grid (translating B_2 on $a\mathbb{Z}$ and its Fourier transform on $b\mathbb{Z}$) if, due to its compact support, $a < 2$ and $b \le 1/2$ and it does not generate a frame, if:

$$\{a \ge 2, b > 0\} \cup \{a > 0, 1 < b \in \mathbb{N}\}. \tag{14.23}$$

Recall that, by Ron–Shen duality (Ron and Shen 1997), it follows that the same pulse prototype $B_2(f)$ generates a Riesz basis on the adjoint $(\frac{1}{b}, \frac{1}{a})$-grid. In our setting we will effectively translate the frequency-domain pulse $B_2(f)$ by half of its support, which corresponds to $\frac{1}{b} = \frac{3}{2}$ and we will use $\frac{1}{b} \cdot \frac{1}{a} = \frac{5}{4} = 1.25$ such that $a = \frac{6}{5}$. It follows, therefore, that our operation point $(a,b) = (\frac{6}{5}, \frac{2}{3})$ is not in any of the two explicit (a, b)-regions given previously. But for $1.1 \le a \le 1.9$ a further estimate has been computed explicitly for $B_2(f)$ (Prete 1999; Table 2.3) ensuring the Gabor frame property up to $b \le 1/a$. Finally, we should mention that for $ab \le 1/2$ the dual prototypes can be expressed again as finite linear combinations of B-splines; that is, explicit formulas exist in this case (Laugesen 2009).

However, in practice g has to be of finite duration; in other words the transmit pulse in the time domain will be smoothly truncated:

$$g(t) = \left(\frac{\sin(B\pi t)}{B\pi t}\right)^2 \chi_{[c,d]}(t), \tag{14.24}$$

where B is chosen equal to F and parameters c and d align the pulse within the transmission frame. Theoretically, a (smooth) truncation in Eq. (14.24) would imply again a limitation on

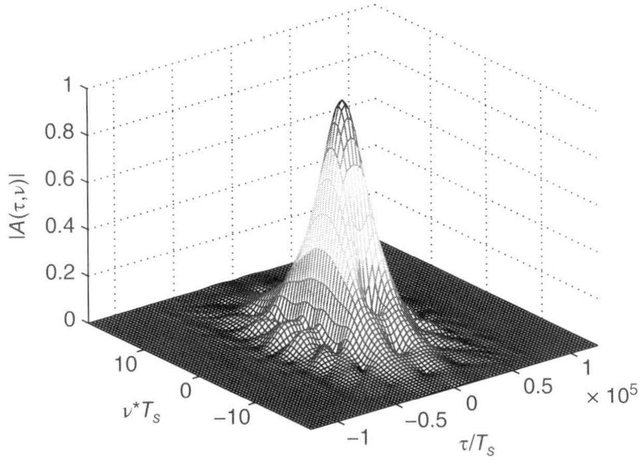

Figure 14.16 (Kasparick et al. 2014) The cross-ambiguity function $A_{g,\gamma}(\nu,\tau)$ for transmit and receive pulses on the frame length of 4 ms.

the maximal frequency spacing B (Christensen et al. 2012). Although the finite setting is used in our application, the frame condition (and therefore the Riesz-basis condition) is a desired feature since it will asymptotically ensure the stability of the computation of the dual pulse γ and its smoothness properties.

To observe the pulse's properties regarding time–frequency distortions we depict in Figure 14.16 the discrete cross-ambiguity function $A_{g,\gamma}$ between pulse g and γ, which is given as:

$$A_{g,\gamma}(\tau,\nu)=\sum_{n} g[n]\gamma^{*}[n-\tau]e^{j2\pi\nu n}. \tag{14.25}$$

It can be observed that its value at the neighboring symbol is already far below 10^{-3}. Obviously, the bi-orthogonality condition states $A_{g,\gamma}(kT,lF)=\delta_{k,0}\delta_{l,0}$ and ensures perfect symbol recovery in the absence of channel and noise. However, the sensibility with respect to time–frequency distortions is related to the shape of the slope of $A_{g,\gamma}$ around the grid points. Depending on the loading strategies for these grid points it is possible to obtain numerically performance estimates using, for example, the integration methods presented by Jung and Wunder (2007).

14.4.1.4 Numerical Results

To get an insight into the performance of the proposed approach, we investigate the influence of asynchronicities of users on the symbol error rate. In order to be compliant with 4G, we set $TF=1.25$, and construct the BFDM signal by the basic overlap–and-add method of digital signal processing, as only the concept is of importance here. However, efficient implementations are available in the literature (Vangelista and Laurenti 2001). The receiver accordingly inverts the process. The simulation parameters are chosen according to the LTE specifications for 20-MHz bandwidth. For the transmission in PRACH, we use 4-QAM modulation. We assume two IoT devices transmitting data in the D-PRACH, where one device is completely

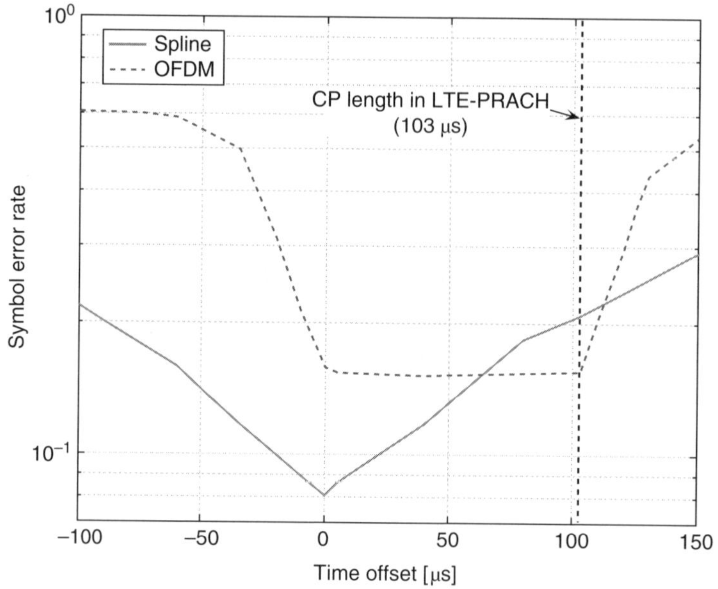

Figure 14.17 Symbol error rate in PRACH (using 4-QAM) with perfect channel knowledge, averaged over 10 out of 20 data subcarriers versus a varying time offset of a second user with a frequency offset of 62.5 Hz. The SNR is set to 25 dB. The vertical dashed line shows the CP length in LTE PRACH.

asynchronous and has a high carrier frequency offset. We assign half of the subcarriers available for D-PRACH data transmission to each. However, we only evaluate the performance of a "user of interest", which is assumed to transmit at the "inner" subcarriers close to the control PRACH. Figure 14.17 illustrates the advantages of the spline-based transmission scheme over OFDM for both frequency offsets and time delays. We plot the PRACH symbol error rates over the time offset of a second asynchronous user. In addition, a constant small frequency offset of 62.5 Hz is applied. The SNR is fixed at 25 dB. Moreover, we assume perfect channel knowledge for the user of interest. It can be observed that the additional frequency offset has a detrimental effect on both schemes, however, the performance loss of OFDM is significantly higher.

14.4.2 System-level Performance

With increasing number of devices in the network, which frequently communicate without human interaction, traffic characteristics tend to change. More devices transmit small data packets, which come with with high signaling overhead. A possible improvement is the modification of the random access procedure to enable small data packages to be sent along with the preamble. With this assumption, an improved procedure is proposed as follows (see Figure 14.18):

- The D-PRACH preamble is sent with data. The UE sends the data scrambled or with appended C-RNTI for the eNB to verify which UE has transmitted the data.
- The BS sends ACK/NACK confirmation to acknowledge to the UE if it can go to sleep or needs retransmission.

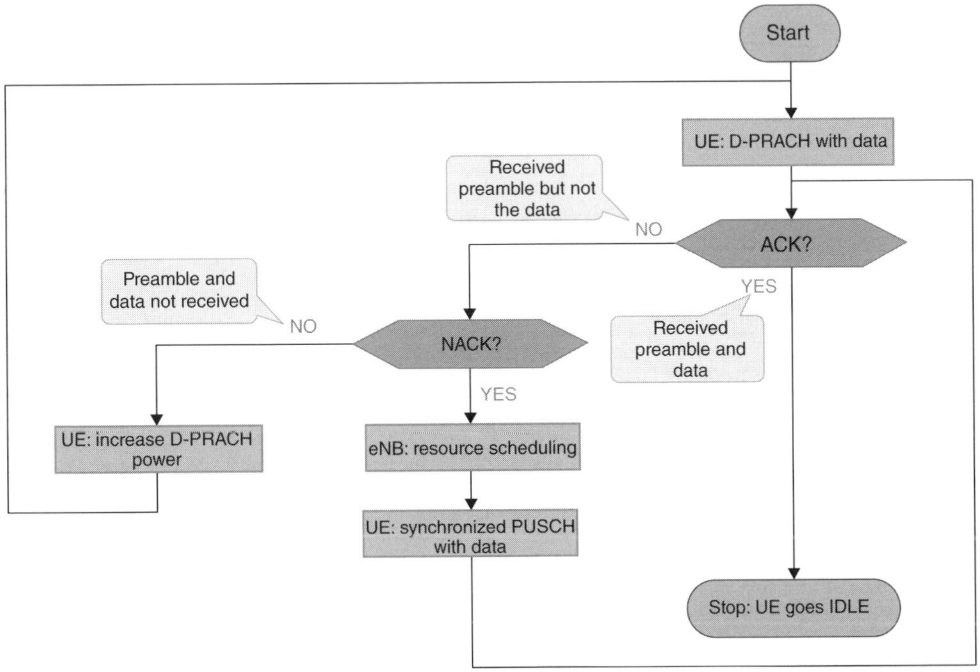

Figure 14.18 Improved random access procedure.

- If NACK is received, the receiver failed to decode the data, but received the preamble sequence. In that case, a notification is sent together with resource allocation and then retransmission of the data is sent over PUSCH.
- If ACK is received, the BS received the data and preamble. In that case nothing else needs to be sent and the MTC device will idle.
- If nothing (no ACK/NACK) is received, the D-PRACH preamble needs to be sent with higher power.

Figure 14.19 shows the signaling flow for the LTE case and the new D-PRACH case. The signaling flow is based on a best-case scenario, where no retransmissions are required. Simulation results and analytical calculations show that the new procedure leads to a significant improvement in an MTC scenario over the LTE case:

- a decrease of the signaling overhead required to convey small messages in an MTC scenario
- thus increased available system capacity
- additionally increased standby time of a UE, which now needs to wake up for a shorter time to transmit the small amount of data, thus saving battery power and decreasing latency.

System-level simulations were conducted to evaluate the performance of proposed concept as well as the ATA mechanism, which allows avoidance of closed-loop synchronization. The ATA mechanism, which was described in Section 14.3.1.2, enables transmission in open-loop

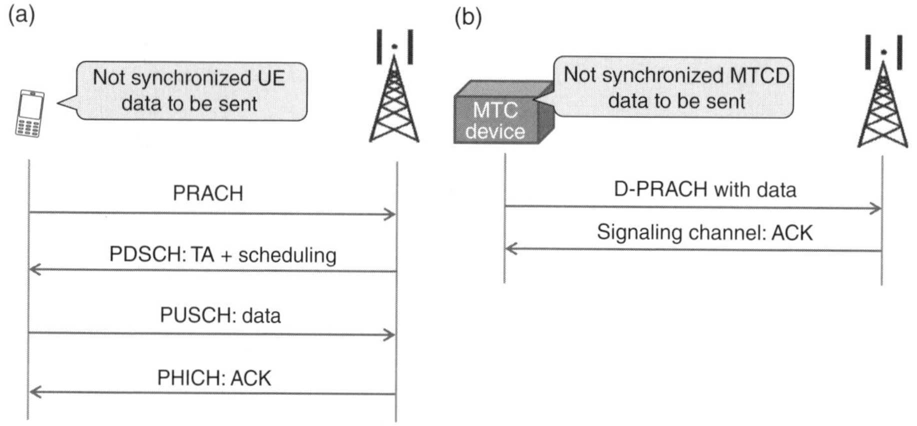

Figure 14.19 Signaling comparison of (a) LTE case and (b) the new D-PRACH case.

Table 14.3 Main simulation parameters.

Parameter	Setting
Intersite distance	500 m
System	1 omnidirectional cell
System bandwidth	5 MHz
Carrier frequency	2 GHz
Distance path loss	$L = 40(1 - 4 \cdot 10^{-3} D_{hb}) log_{10}(R)$ $-18 log_{10} D_{hb} + 21 log_{10} f + 80\ dB$
Modulation	QPSK R 1/3
Simulation time	60 min
System duplex	FDD
Number of devices	1,000–30,000
PRACH configuration index	6
N_{TA}	$0.5 N_{OL,max}$
Carrier frequency offset	0
Synchronization	Open-loop

synchronization by reducing inter-user interference. The overall simulation includes an MTC traffic model proposed by 3GPP, with a single cell supporting a large number of devices (up to 30,000) in a specific scenario of stationary energy meters. The MTC devices are assumed to be in RRC idle. Application-layer packets with a size of 24 bytes are transmitted to the device using UDP/IP. The simulation includes an implementation of the random access procedure presented in Figure 14.19(a) and of the ATA mechanism. The main simulation parameters are gathered in Table 14.3.

Figure 14.20 shows improvements in processing latency of the new procedure, which were obtained by analytical calculations based on the signal propagation time and message processing in the terminal and BS. By combining the random access procedure with data transmission, the communication between BS and UE can be reduced from four messages down to two. This approach can significantly shorten transmission times. Even the worst case in D-PRACH is still better than the best case in LTE (with respect to the number of retransmissions). The shortest possible transmission in case of LTE is 11.5 ms, whereas the new

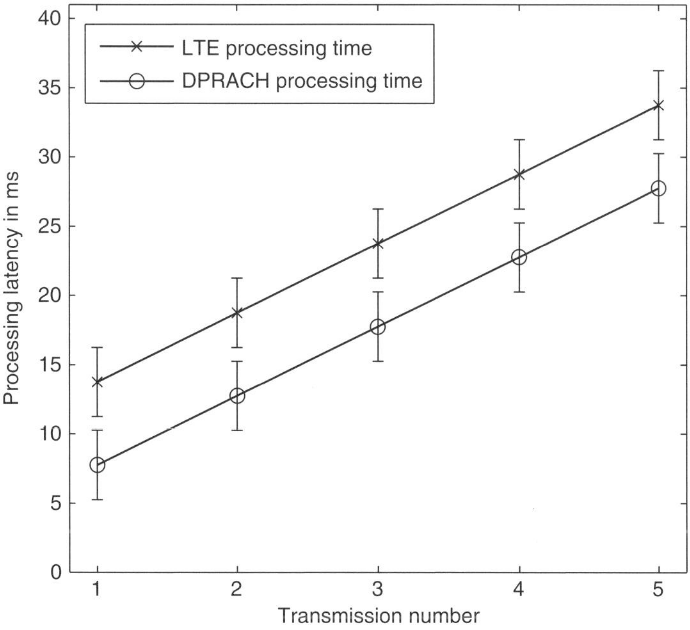

Figure 14.20 Processing latency comparison of LTE and D-PRACH.

procedure, in the best case scenario, can deliver the data in 5.5 ms, which is more than twice as fast. The variation of the time of processing is caused by the need to wait for the first available random access opportunity.

Shorter and simpler procedures also leads to decreased signaling overhead. Table 14.4 shows results of analytical calculations of the number of resources used while transmitting 24 bytes of data. The resource elements shown in Table 14.4 include all signaling and data required to transmit the data in a scenario with no retransmission. The efficiency (the ratio between useful data and the total amount of resource elements used for transmission, including signaling) can be increased from 18.5% up to 27.4%.

During the RACH procedure, where resources are not allocated by the BS, collisions may occur. Collisions cause preamble retransmissions, which lengthen the whole procedure, and thus its probability should be optimized. Reducing the collision probability by increasing the number of PRACH slots decreases the overall throughput by occupying more resource elements for signaling. The increasing number of terminals connected to a single BS affects the collision probability in the manner shown in Figure 14.21. If the collision probability exceeds a threshold assumed for the network, it can be lowered by increasing either the number of possible preambles dedicated to the MTC devices, or RACH slots. During simulations, occurrences of collisions are calculated in the same manner for both LTE case and the new procedure. Also the SINR values used for evaluating ACK/NACK response are the same. System-level simulations were also used to evaluate how interference, caused by lack of closed-loop synchronization, can be reduced using ATA. The simulation case includes the random access procedure and packet transmission over PUSCH channel. Results obtained during the simulations show that the performance of the network can be kept on a high level using the ATA mechanism. Figure 14.22 shows how the average transmission time of the

Table 14.4 Signaling overhead of LTE and D-PRACH.

	LTE procedure	New procedure
Total number of used resource elements	2,076	1,404
Ratio of data over all resources	0.185	0.274
Ratio of data over signaling	0.227	0.376

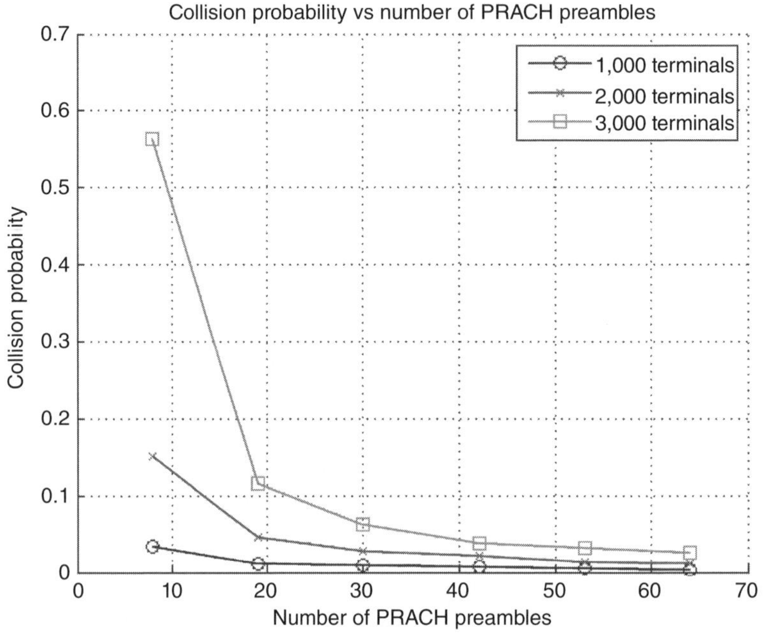

Figure 14.21 Collision probability.

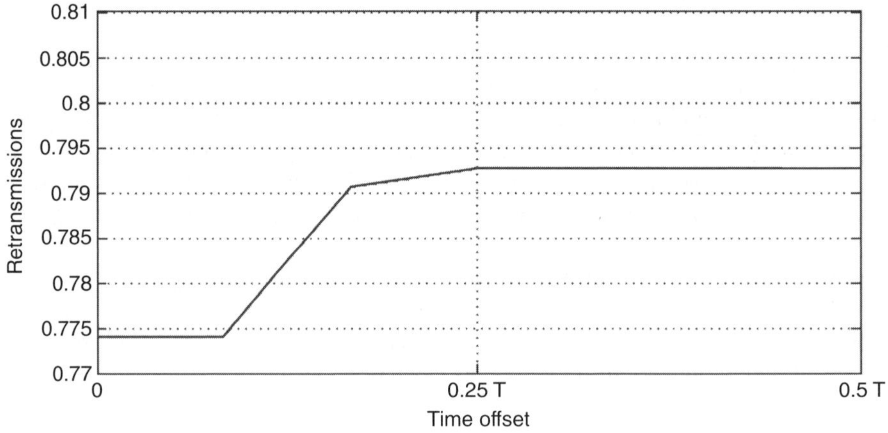

Figure 14.22 Average transmission time for different timing offsets.

single packet is affected by the interference caused by the open-loop synchronization in UFMC, while using ATA.

As one can see, even users that are located far from the eNB (where time offset = 0.5 T) does not suffer much from lack of closed-loop synchronization. The average number of retransmissions in this case rises from 0.774 to 0.793. Such a value shows how efficiently MTC-like transmission can be completed, where timing is more important than throughput.

14.5 Conclusions

LTE and its evolution LTE-A are standardized via the 3GPP. The foreseen diversification of the service and device-class mix of future telecommunications and the related expansion of the requirement space (Wunder et al. 2014b) require a revolutionary step. This step from 4G to 5G implies a backward compatibility drop. New 5G waveforms and the use of the 5G unified frame structure with a mixture of synchronous and asynchronous traffic are major building blocks for supporting those goals. Starting from the main drivers of 5G, sporadic traffic, spectral and temporal fragmentation, and real-time constraints together with the vision of supporting a single unified air interface, this article presents such building blocks. General waveform considerations are discussed in the light of the Gabor theory. Then the four 5G waveform candidates – GFDM, UFMC, FBMC and BFDM – are described in detail, summarizing the available results. These waveforms are supporting and enabling the unified frame sturcture. The unified frame structure concept is the heart of the 5G frame design, designed for supporting various heterogeneous traffic and device types in parallel. We argue that each multicarrier waveform has properties that make it advantageous in specific scenarios and will be operated in parallel with others by proper re-parameterization. Performance results for the candidate transceiver approaches are provided, including optimized and/or reasonable waveform parameters.

Those new signal formats require standardization, as they need to be known on both ends of the link. The wireless industry as a whole has to build up consensus on the technology candidates for 5G standardization.

References

Al-Dhahir N 2001 Single-carrier frequency-domain equalization for space-time block-coded transmissions over frequency-selective fading channels. *IEEE Communications Letters* **5**(7), 304–306.

Bellanger M 2012 FS-FBMC: A flexible robust scheme for efficient multicarrier broadband wireless access. *2012 IEEE Globecom Workshops*, pp. 192–196.

Bellanger M 2010 FBMC physical layer: a primer. Technical Report, PHYDAS Project.

Berg V, Doré JB and Noguet D 2014a A flexible FS-FBMC receiver for dynamic access in the TVWS. *9th International Conference on Cognitive Radio Oriented Wireless Networks and Communications (CROWNCOM)*, pp. 285–290.

Berg V, Doré JB and Noguet D 2014b A multiuser FBMC receiver implementation for asynchronous frequency division multiple access. *Digital System Design (DSD), 2014 Euromicro Conference on*, pp. 16–21.

Bolcskei H 2001 Blind estimation of symbol timing and carrier frequency offset in wireless OFDM systems. *IEEE Transactions on Communications* **49**(6), 988–999.

Chen Y, Schaich F and Wild T 2014 Multiple Access and Waveforms for 5G: IDMA and universal filtered multicarrier. *Vehicular Technology Conference (VTC Spring), 2014 IEEE 79th*, pp. 1–5.

Christensen O, Kim HO and Kim RY 2012 Gabor windows supported on [−1,1] and dual windows with small support. *Advances in Computational Mathematics* **36**(4), 525–545.

Chu D 1972 Polyphase codes with good periodic correlation properties (corresp.). *IEEE Transactions on Information Theory* **18**(4), 531–532.

Daubechies I 1992 *Ten Lectures on Wavelets*. SIAM.

Doré JB, Berg V and Kténas D 2014a Performance of FBMC multiple access for relaxed synchronization cellular networks. *10th International Workshop on Broadband Wireless Access (BWA'14) co-located with IEEE Globecom'14*,.

Doré JB, Berg V, Cassiau N and Kténas D 2014b FBMC receiver for multi-user asynchronous transmission on fragmented spectrum. *EURASIP Journal on Avances in Signal Processing, Special Issue on Advances in Flexible Multicarrier Waveform for Future Wireless Communications*, **2014**(1), 1–20.

Farhang-Boroujeny B 2011a OFDM versus filter bank multicarrier. *IEEE Signal Processing Magazine* **28**(3), 92–112.

Farhang-Boroujeny B 2011b OFDM versus filter bank multicarrier. *IEEE Signal Processing Magazine* **28**(3), 92–112.

Fettweis G 2012 A 5G Wireless Communications Vision. *Microwave Journal* **55**(12), 24–39.

Gaspar IS, Mendes LL, Michailow N and Fettweis G 2014 A synchronization technique for generalized frequency division multiplexing. *EURASIP Journal on Advances in Signal Processing* **2014**(1), 67–76.

Jung P and Wunder G 2007 The WSSUS pulse design problem in multicarrier transmission. *IEEE Trans. on Communications*, **55**(10), 1918–1928.

Kasparick M, Wunder G, Jung P and Maryopi D 2014 Bi-orthogonal waveforms for 5G random access with short message support. *Proc. European Wireless Conference '14*, pp. 1–6.

Kozek W and Molisch A 1998 Nonorthogonal pulseshapes for multicarrier communications in doubly dispersive channels. *IEEE Journal on Selected Areas in Communications* **16**(8), 1579–1589.

Laugesen RS 2009 Gabor dual spline windows. *Applied and Computational Harmonic Analysis* **27**(2), 180–194.

Matthe M, Mendes L and Fettweis G 2014 Generalized frequency division multiplexing in a Gabor transform setting. *IEEE Communications Letters*, **18**(8), 1379–1382.

Michailow M, Matthe M, Gaspar IS, Caldevilla AN, Mendes LL, Festag A, and Fettweis G. 2014 Generalized frequency division multiplexing for 5th generation cellular networks. *IEEE Transactions on Communications*, **62**(9), 3045–3061.

Nizamuddin H, dos Santos AF and Wild T 2014 A random access and multiuser detection approach for massive access of low-complexity machine communications in cellular networks. *Wireless Communications and Networking Conference (WCNC), 2014 IEEE*, pp. 1774–1778.

Noguet D, Gautier M and Berg V 2011 Advances in opportunistic radio technologies for TVWS. *EURASIP Journal on Wireless Communications and Networking* **2011**(1), 1–12.

Ping L, Liu L and Leung WK 2003 A simple approach to near-optimal multiuser detection: interleave-division multiple-access. *Wireless Communications and Networking, 2003. WCNC 2003*, vol. **1**, pp. 391–396.

Prete VD 1999 Estimates, decay properties, and computation of the dual function for Gabor frames. *Journal of Fourier Analysis and Applications*, **5**(6), 545–562.

Ron A and Shen Z 1997 Weyl–Heisenberg frames and Riesz bases in L_2(Rd). *Duke Mathematics Journal* **89**(2), 237–282.

Schafhuber D, Matz G and Hlawatsch F 2002 Pulse-shaping OFDM/BFDM systems for time-varying channels: ISI/ICI analysis, optimal pulse design, and efficient implementation. *13th IEEE International Symposium on Personal, Indoor and Mobile Radio Communications*, vol. **3**, pp. 1012–1016.

Schaich F and Wild T 2014a Relaxed synchronization support of universal filtered multi-carrier including autonomous timing advance *Wireless Communications Systems (ISWCS), 2014 11th International Symposium on*, pp. 203–208.

Schaich F and Wild T 2014b UFMC OFDM transceiver chain. URL: www.5gnow.eu/?page_id=424 (accessed August 10, 2016).

Schaich F and Wild T 2014c Waveform contenders for 5G #x2014; OFDM vs. FBMC vs. UFMC. *Communications, Control and Signal Processing (ISCCSP), 2014 6th International Symposium on*, pp. 457–460.

Vakilian V, Wild T, Schaich F, ten Brink S and Frigon JF 2013 Universal-filtered multi-carrier technique for wireless systems beyond LTE. *Globecom Workshops (GC Wkshps), 2013 IEEE*, pp. 223–228.

Vangelista L and Laurenti N 2001 Efficient implementations and alternative architectures for OFDM-OQAM systems. *IEEE Transactions on Communications* **49**(4), 664–675.

Wang X, Wild T, Schaich F and dos Santos A 2014 Universal filtered multi-carrier with leakage-based filter optimization. *European Wireless 2014; 20th European Wireless Conference; Proceedings of*, pp. 1–5.

Wild T, Schaich F and Chen Y 2014 5G air interface design based on universal filtered (UF-) OFDM. *Digital Signal Processing (DSP), 2014 19th International Conference on*, pp. 699–704.

Wunder G, Boche H, Strohmer T and Jung P 2015a Sparse signal processing concepts for efficient 5G system design. *IEEE Access* **3**, 195–208.

Wunder G, Gorgani SA and Sid-Ahmed S 2015b Waveform optimization using trapezoidal pulses for 5g random access with short message support. *16th IEEE International Workshop on Signal Processing Advances in Wireless Communications (SPAWC)*, pp. 76–80.

Wunder G, Jung P and Ramadan M 2015c Compressive random access using a common overloaded control channel *IEEE Globecom Workshops (GC Wkshps), San Diego, CA*, pp. 1–6.

Wunder G, Jung P and Wang C 2014a Compressive random access for post-LTE systems. *IEEE ICC Workshop on Massive Uncoordinated Access Protocols,* pp. 539–544.

Wunder G, Jung P, Kasparick M, Wild T, Schaich F, Chen Y, et al. 2014b 5GNOW: Non-orthogonal, asynchronous waveforms for future mobile applications. *IEEE Communications Magazine* **52**(2), 97–105.

Wunder G, Kasparick M and Jung P 2015d Interference analysis for 5G random access with short message support. *Proc. European Wireless Conference '15*, pp. 1–6.

15

Massive MIMO Communications

Frederick W. Vook, Amitava Ghosh, and Timothy A. Thomas
Nokia Networks, Illinois, USA

15.1 Introduction

Over the last two decades, multiple input multiple output (MIMO) technology has been successfully deployed on a wide scale in cellular communication systems all over the world. MIMO technology involves the use of multiple antennas at one or both ends of a communication link to boost the performance and reliability through strategies, such as beamforming, diversity transmission, spatial multiplexing, and interference suppression. The currently-deployed 4G/LTE cellular standards (LTE Rel-8/9/10) support a comprehensive suite of MIMO techniques for up to eight antenna ports in a single sector on the downlink and up to four transmit antennas at a mobile station. An antenna port is defined as a logical entity distinguished by a separate reference signal sequence. Later LTE releases (Rel-11/12) enhance these MIMO techniques by coordinating the transmissions from multiple transmission points through the use of coordinated multipoint (CoMP) technology. Even with the comprehensive suite of MIMO techniques already defined for LTE, the limitation to a maximum of eight antenna ports provides an upper limit to the performance benefits of antenna array technology.

For 5G cellular communications, massive MIMO, sometimes called full dimension MIMO (FD-MIMO), is a promising technology for enhancing system performance. Massive MIMO involves the use of antenna arrays with a large number of antenna elements, for example 32, 64, or even hundreds of antennas, for the purpose of applying high gain adaptive beamforming and multi-user spatial multiplexing on a large scale. Massive MIMO promises to provide very high data rates to a large number of users in dense deployment scenarios. Also, for 5G systems deployed in higher frequency bands such as cmWaves (6–30 GHz) and mmWaves (30–100 GHz), large-scale antenna arrays will be a system prerequisite for overcoming the poor propagation characteristics in those bands.

Towards 5G: Applications, Requirements and Candidate Technologies, First Edition.
Edited by Rath Vannithamby and Shilpa Talwar.

Companion website: www.wiley.com/go/vannithamby/towards5g

This chapter will overview the basics of massive MIMO and how it will satisfy the high data rate demands of 5G cellular systems. The current state-of-the art of MIMO technology will be reviewed, and the application of large-scale antenna arrays to 5G will be described. The chapter will survey the current trends in massive MIMO technology and system concepts with a focus on methodologies for significantly enhancing cellular system performance. Various trends and promising concepts will be identified, and various practical issues will be highlighted.

The discussion of applying Massive MIMO to 5G will be split into two parts. The first part will deal with systems deployed with carrier frequencies below 6 GHz, and the second will deal with systems deployed with carrier frequencies above 6 GHz. The 6 GHz dividing line is somewhat arbitrary, but is chosen since it is above 6 GHz where the propagation characteristics and available spectrum start to look very different from current cellular deployments operating between 700 MHz and 3.5 GHz. The abundance of spectrum available in the cmWave and mmWave bands and the degraded propagation characteristics compared with current cellular bands has necessitated a focus for massive MIMO technology that is different from what is envisioned for the evolutions of LTE that are expected to be deployed in existing cellular bands. Whether for evolved LTE systems deployed below 6 GHz or for brand new systems deployed in the cmWave or mmWave bands, massive MIMO will be ideally suited to meet the 5G data demands through the use of high-gain beamforming and high-order spatial multiplexing.

15.2 Overview of Multi-Antenna Techniques in LTE

3GPP LTE Release 12 currently supports a wide variety of MIMO transmission modes on the downlink, all of which cover the major types of antenna array techniques. In general, MIMO transmission technology can be classified in a number of ways, and reviewing these techniques will be helpful for discussing MIMO solutions for 5G. First, MIMO transmission methods can be classified into either open-loop or closed-loop methods. Closed-loop methods use a significant level of channel knowledge in order to beamform (or precode) one or more spatial streams to a user (closed-loop techniques), whereas open-loop methods do not exploit channel knowledge in the transmission process. Second, MIMO technology can be classified into either diversity techniques or spatial multiplexing techniques. Diversity techniques form a single spatial channel between the transmitter and a single user, whereas spatial multiplexing techniques form multiple spatial channels to either one user at a time (single-user or SU) or multiple users at a time (multi-user or MU). The rank of the transmission technique is a commonly used term to indicate the number of parallel spatial channels formed by the transmitter, whether those parallel channels are all intended for one user (SU-MIMO) or multiple users (MU-MIMO).

Two methodologies are used to support closed-loop transmission. The first methodology is codebook feedback where both the base station and mobile station maintain a pre-defined codebook consisting of a set of beamforming vectors (for rank 1) or beamforming matrices (for ranks greater than 1), and the mobile station feeds back an indicator of the best beamforming vector/matrix of the codebook based on measurements of the downlink channel. In LTE, the mobile station can estimate the downlink channel either by leveraging the common reference signals (CRS), which are multiplexed in time/frequency with the data symbols, or the channel state information reference signals (CSI-RS), which are reference signals dedicated to enable the mobile station to estimate the downlink channel. For demodulating the transmitted data, the CRS can be used, but to estimate the channel over which the data is being transmitted, the mobile station will require knowledge of the beamforming vector/matrix that the base is using. Alternatively, the mobile station can use the dedicated modulation reference signals (DMRS), which are beamformed along with the data streams and therefore enable the use of vendor-specific beamforming vectors/matrices, but the DMRS is only supported in certain transmission modes, as will be discussed next.

With this type of classification, Table 15.1 lists the transmission modes currently in the downlink of LTE Rel-8 through Rel-11 for both Frequency Division Duplex (FDD) systems and Time Division Duplex (TDD) systems. As can be seen, LTE supports open-loop schemes (TM1–TM3), closed-loop schemes (TM4–TM10), diversity schemes (TM2), spatial multiplexing schemes (TM3–TM5, TM8–TM10), single user schemes (TM1–TM4, TM6–TM10), and multi-user schemes (TM5, TM8–TM10). Complementing the modes listed in Table 15.1 are the CoMP methodologies, which combine MIMO techniques with various levels of coordination between base sites or transmission points. CoMP defines a transmission capability where user data for a given user equipment (UE) is available at multiple cells. From a UE point of view, CoMP defines a capability where it can demodulate data transmitted from one of multiple cells or transmitted in the same resource from multiple cells even though it is

Table 15.1 Downlink MIMO transmission modes.

Transmission mode	Description	LTE release	FDD/TDD
TM1	Single transmit antenna	8	FDD and TDD
TM2	Transmit diversity with two or four antenna ports using a space–frequency block code	8	FDD and TDD
TM3	Open-loop spatial multiplexing with rank adaptation through rank indication feedback from the UE	8	FDD and TDD
TM4	Closed-loop single-user spatial multiplexing with codebook-based feedback	8	FDD and TDD
TM5	Multi-user spatial multiplexing	8	FDD and TDD
TM6	Closed-loop Rank 1 single-user MIMO (no spatial multiplexing)	8	FDD and TDD
TM7	Beamforming with dedicated reference signals (optimized to exploit TDD reciprocity)	8	Intended for TDD
TM8	Dual stream beamforming with dedicated reference signals (also optimized to exploit TDD reciprocity); supports SU and MU-MIMO	9	Intended for TDD
TM9	Multi-stream beamforming using CSI-RS and codebook feedback and dedicated reference signals (SU-MIMO and MU-MIMO)	10	FDD and TDD
TM10	Multi-stream beamforming using multiple CSI-RS processes (SU-MIMO and MU-MIMO)	11	FDD and TDD

attached to one particular cell. Different flavors of CoMP transmission using ideal backhaul were introduced and standardized in LTE Rel-11. They include:

- coordinated scheduling/coordinated beamforming, which involves coordinating the scheduling between two adjacent sites to mitigate interference and also encompasses the strategy of beamforming with nulls steered towards potential victim UEs in adjacent cells
- dynamic point selection, in which the data targeted for that UE may be transmitted from any one of those cells or the same data may be transmitted in the same resource from multiple cells; the decision as to which cell or cells participate in data transmission to that UE is made dynamically meaning every scheduling instant (1 ms in LTE)
- joint processing or joint transmission, which involves two adjacent sectors or sites simultaneously transmitting data to the same user to improve coverage and capacity; in 3GPP, Rel-12 a CoMP scheme with non-ideal backhaul was designed based on dynamic cell muting and distributed architecture.

The choice of downlink transmission mode is often heavily influenced by the characteristics of the multipath propagation channel. Closed-loop methods leverage channel information and therefore provide the best performance when the transmit weights can accurately track variations in the spatial channel response across time and frequency. As a result, multipath angular spread, delay spread, and Doppler spread are key characteristics, since they determine how fast the channel changes across space, frequency, and time, respectively. Delayed channel information in velocity channels will degrade the performance of closed-loop methods. Moreover, the combination of angle spread and delay spread causes the spatial channel response to vary across frequency, which degrades the performance when beamforming weights are fixed across the signal bandwidth. On the other hand, in low-angle-spread channels, the channel becomes highly directional, and a fixed beamformer can often provide near-optimal beamforming gain even in velocity channels. The level of angle spread in the channel also influences the performance of spatial multiplexing methods. Single-user spatial multiplexing transmits multiple data streams to a single user, and rich multipath scattering (for example, high angle spread) can enable the user to decode the multiple streams. In contrast, multi-user spatial multiplexing transmits multiple data streams to multiple users through multiple orthogonal spatial channels, which requires precise up-to-date channel knowledge. Increasing the multipath angle spread tends to cause the spatial channel to vary more quickly across time and frequency, which tends to degrade MU-MIMO performance.

For the uplink, LTE Rel-8 incorporates MU-MIMO, where multiple UEs simultaneously transmit to the base station on the same resources. Transmit antenna selection is an optional feature for UEs. Rel-10 introduced codebook feedback-based precoding (beamforming/spatial multiplexing) on the uplink for two and four transmit antennas at the mobile.

15.3 Moving to 5G Cellular with Large-scale Antenna Arrays

Current 4G systems have been designed for high spectral efficiency due to the relatively scarce spectrum available for mobile communications below 6 GHz. Current LTE systems operate at carrier frequencies below 6 GHz, with system bandwidths up to 20 MHz, although carrier aggregation can be used to achieve an overall effective bandwidth of up to 100 MHz. The design challenge with 5G systems below 6 GHz will be to enhance the spectral efficiency of evolved LTE systems when the current LTE systems are already highly spectrally efficient. For 5G systems, the telecommunication industry is converging on a common set of 5G requirements of 10-Gbps peak rates, latency of less than 1 ms, and guaranteeing connectivity "everywhere" at 100 Mbps. The use of large-scale arrays [1–4] with the appropriate modifications to the air interface promises to provide additional gains in capacity and coverage (and thus meet the 5G requirements) beyond what is possible in LTE Rel-12, with the main focus of MIMO technology being the enhancement of the system spectral efficiency.

When considering the benefits of supporting additional antenna ports in an evolved LTE system, it is helpful to recall exactly how multiple-antenna techniques can improve system performance. Two important techniques for improving system capacity and coverage are beamforming and spatial multiplexing. Beamforming increases the link SNR through the coherent addition of the signal transmitted from the antenna array, and the result is increased coverage and capacity. Spatial multiplexing increases the system capacity by forming multiple parallel spatial channels between the base station and one or more UEs at a time. However, the benefits of additional antennas at a base station depend on whether the spatial multiplexing is single-user or multi-user.

A well known characteristic of practical single-user spatial multiplexing is that the maximum number of spatial streams that can be formed between a transmit array and a receive array is equal to the minimum of the number of transmit and receive antennas. As a result, increasing the number of antennas at the base station will not increase the maximum number of spatial streams that SU-MIMO can support. Increasing the number of base station antennas would improve the quality of those streams through the additional beamforming gain that can be provided from the additional antennas. However, as the number of base antennas increases, the peak data rate per stream provided by the air interface would eventually be reached (example, 64QAM with rate 5/6 coding, as in LTE Rel-10), and further additional base-station antennas would have only a marginal effect on system capacity.

However, in multi-user spatial multiplexing on the downlink, the number of spatial streams that can be formed is primarily limited by the number of transmit antennas at the base station. Therefore, increasing the number of antennas at the base station can indeed increase the number of spatial streams by multiplexing more and more users simultaneously through multi-user spatial multiplexing. Therefore, as more and more antennas are added at the base station, there will come a point where multi-user spatial multiplexing must be used to take full advantage of the additional antennas. Therefore, given the scarcity of spectrum below 6 GHz, the MIMO technology emphasis for 5G systems deployed below 6 GHz will continue to be the maximization of spectral efficiency through higher-order spatial multiplexing, with the focus eventually being on high-order MU-MIMO. For dense user deployments, the use of large-scale arrays and multi-user MIMO is an ideal match.

In contrast, the situation for MIMO technology for systems above 6 GHz will be quite different, at least initially. Recent research has pointed towards the use of higher frequency bands for 5G cellular communications [5–8], for example the cmWave (6–30 GHz) and the mmWave bands (30–100 GHz). A vast quantity of spectrum is available for use in those bands. For example, in the 70–85-GHz frequency range, 10 GHz of spectrum is available, and it will not be inconceivable to have system bandwidths of perhaps 1 GHz or 2 GHz. Such huge system bandwidths pose hardware implementation challenges, but also present opportunities for vast increases in per-user data rates that the 5G system will require. With large system bandwidths, high user data rates can be achieved with relatively simple methods simply due to the large bandwidth. For example, with a system bandwidth of 2 GHz, a user data rate of 10 Gbps can be achieved quite simply with two spatially multiplexed streams of 16QAM. As a result, the initial focus for antenna-array technology above 6 GHz may not be on spectral efficiency.

Operating in the cmWave and mmWave bands is particularly challenging due to the poor path-loss conditions in those bands. However, recent measurement campaigns have characterized the cmWave and mmWave propagation environment [8] and have pointed out the viability of operating a micro-cellular system in those bands and even macro systems in bands from 6–20 GHz. Consider the free-space reference-distance path-loss model [8]:

$$PL[dB](d) = 20\log_{10}\left(\frac{4\pi d_o}{\lambda}\right) + 10n\log_{10}(d/d_o) + X$$

where d_o is a reference distance, d is the transmitter–receiver separation distance in meters, n is the path-loss exponent, and X is a shadow fading term that is a zero-mean Gaussian variable with a specified standard deviation. The path-loss exponents and shadow-fading standard deviations measured for access links in New York City at 73 GHz are given in Table 15.2 [8]. It is particularly important to note that these path-loss exponents and shadow-fading terms are very similar to today's cellular systems that operate below 6 GHz. Therefore, according to the above free-space path-loss model, the difference between today's cellular systems and the cmWave or mmWave systems is the result of the difference in the wavelength that shows up in the denominator of the first term of the path-loss equation. The difference in wavelength between, for example, 2 GHz and the cmWave and mmWave bands results in an extra 20–30 dB of extra path loss in the cmWave and mmWave bands compared to today's cellular bands. Key to operating in the cmWave and mmWave bands will be the use of large-scale antenna arrays to overcome the poor path-loss situation in those bands. Fortunately, the size of the antenna array is dictated by the size of the individual antenna elements, which is dictated by the wavelength of the center frequency at which the system is operating. As the carrier frequency increases, the size of the antenna array will correspondingly decrease, so that it will be feasible from a form-factor point of view to have mmWave antenna arrays with hundreds of elements.

As a result, due to the abundance of spectrum and the poor propagation conditions, the initial main purpose for deploying MIMO technology, especially above 20 GHz, will likely be to provide additional beamforming gain to mitigate large path losses rather than for enhancing spectral efficiency. Although massive arrays will be a prerequisite for operating in the mmWave bands, recent studies have leveraged the latest path-loss and multipath fading models and have

Table 15.2 Propagation characteristics at 73 GHz for access links.

	Path-loss exponent	Shadow fading (dB)
Line-of-sight	2.0	5.2
Non-line-of-sight	3.3	7.6

Source: Ghosh et al. [8].

shown the viability of single-user spatial multiplexing as well as multi-user spatial multiplexing for ultradense small-cell deployments in the mmWave bands [31,32]. Consequently, initial deployments with beamforming-only solutions can ultimately be enhanced to support single-user and multi-user spatial multiplexing.

15.4 Antenna-array Architectures for 5G Cellular

For 5G cellular systems, it is important to review various implementation issues surrounding antenna-array technology, particularly the overall hardware architecture of the base-station antenna array. There are three classes of hardware architectures for antenna arrays being considered for 5G cellular systems: baseband-oriented architectures, radio-frequency (RF) oriented architectures, and hybrid architectures. Figure 15.1 shows a mathematical diagram of multi-stream transmit beamforming, where K streams are beamformed over Q transmit antennas. Each data stream is split and multiplied by an antenna-specific complex weight before being transmitted out of the antennas. In an antenna array implementation, the multi stream beamforming operation of Figure 15.1 can occur at baseband in the digital domain (Figure 15.2), or in the RF domain after baseband-to-RF upmixing has occurred (Figure 15.3),

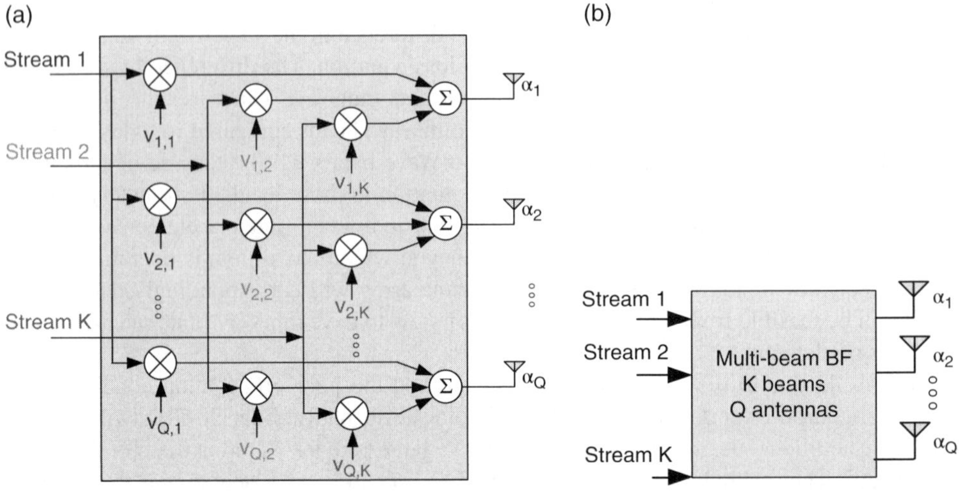

Figure 15.1 Linear multi-beam transmit beamforming: (a) detailed view; (b) block diagram view.

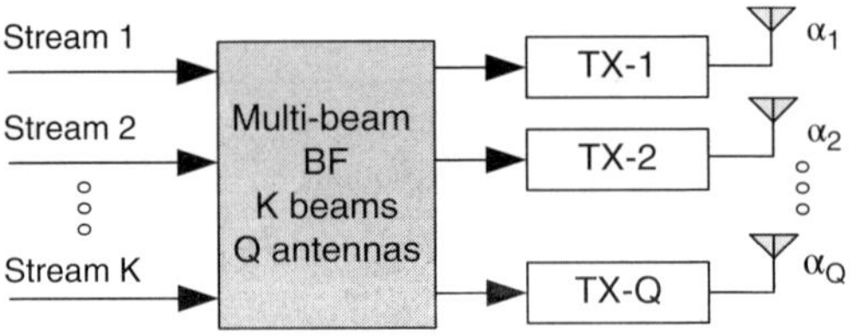

Figure 15.2 Baseband hardware architecture for transmit arrays.

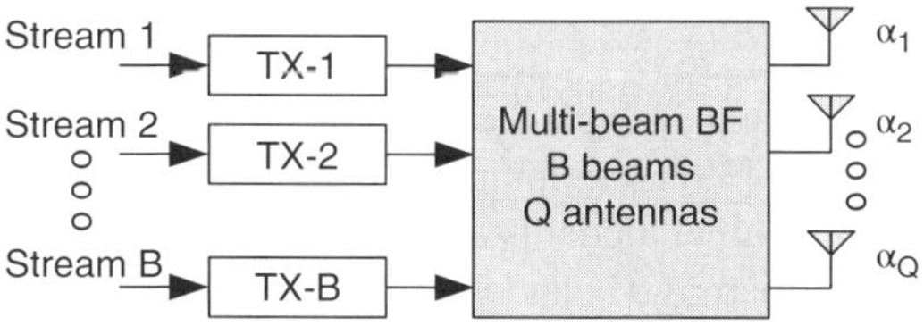

Figure 15.3 RF-oriented hardware architecture for transmit arrays.

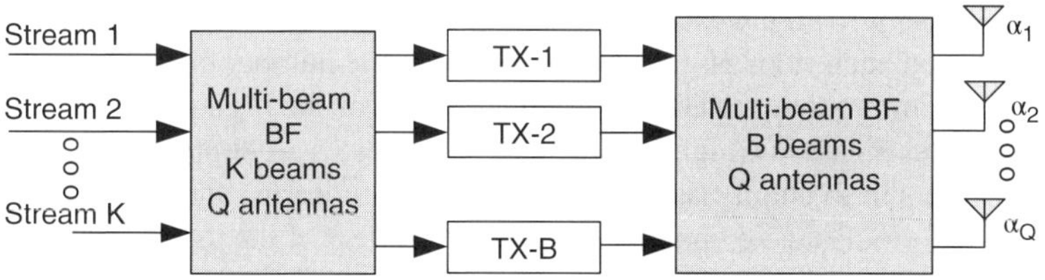

Figure 15.4 Hybrid hardware architecture for transmit arrays.

or in both domains (Figure 15.4). Figure 15.2 shows the baseband architecture where the beamforming occurs at baseband and there is a single transmit chain feeding each transmit antenna. The digital nature of the beamforming allows for high flexibility and high performance, but with potentially high cost and high power consumption. Alternatively, Figure 15.3 shows an RF architecture where each data stream feeds a single transmit chain, and the beamforming weights are applied at RF. A typical implementation assumption for the RF architecture is that the beamforming weights are implemented as phase-shifters, with limited or no ability to control the gain on each transmit branch. As a result, the transmit weight computation strategy has to account for the phase-only nature of the transmit weights and the associated performance degradation relative to the full baseband approach of Figure 15.2. Figure 15.4 shows a hybrid scheme where the transmit beamforming operation is performed both at baseband and at RF. With the hybrid architecture of Figure 15.4, performance close to that of the full baseband approach can be achieved with a significant reduction in the number of transmit chains [29, 30].

15.5 Massive MIMO for Evolved LTE Systems (Below 6 GHz)

Given that the current LTE systems have been designed for high spectral efficiency and have performance that is largely interference limited, the focus on massive MIMO for evolved LTE systems deployed below 6 GHz will be on enhancing spectral efficiency beyond what is possible in existing LTE releases. For massive MIMO systems intended for evolutions of the current Rel-12 LTE systems, we divide the discussion into three categories: transmission techniques for uplink, transmission techniques for downlink, and the concept of "massive subsectorization", which can be applied on both uplink and downlink. All of the architectures discussed in the previous section are applicable to systems operating below 6 GHz, although the initial emphasis will likely be towards the baseband architectures, where the adaptive-transmission schemes will be controlled at baseband in the digital domain.

15.5.1 3D Channel Models

In order to accurately analyze and evaluate the performance of massive MIMO, 3GPP recently developed a 3D channel model [39] that captures both the azimuth and elevation characteristics of the spatial channel. This channel model is an evolution of their existing 2D channel model. It covers frequencies of less than 6 GHz and includes user placement in 3D. This 3D channel model considers three environments: the urban microcell (200-m inter-site distance) with 80% indoor users, the urban macrocell (500-m inter-site distance) with 80% indoor users, and the urban macrocell with a single high-rise building per sector (300-m inter-site distance) and again with 80% indoor users. With the exception of users in a high-rise building, the indoor users are placed at random on each floor of each building, where the number of floors in any particular building is a uniform random variable between 4 and 8. For the high-rise building case where 50% of the indoor users are in a high-rise building, those users are dropped randomly from 1 to the number of floors in the high-rise building, which is a uniform random variable from 20 to 30. The path loss formulas were updated to include the effect of the user height.

The basic channel parameters, such as delay and azimuth angle spreads, were kept similar to the 2D channel model. The most important addition made in the 3D channel model was the careful characterization of the elevation angles of departure (from the base station) including a distance and height dependence that was taken from the literature [40]. What was observed was that as distance is increased, the elevation spread of departure and elevation angle bias both decrease (the elevation angle bias is the deviation from the line-of-sight (LOS) elevation angle). For example, in the urban-micro non-line-of-sight (NLOS) environment, the elevation angle spread decreases with distance and increases with height as shown in Figure 15.5. While

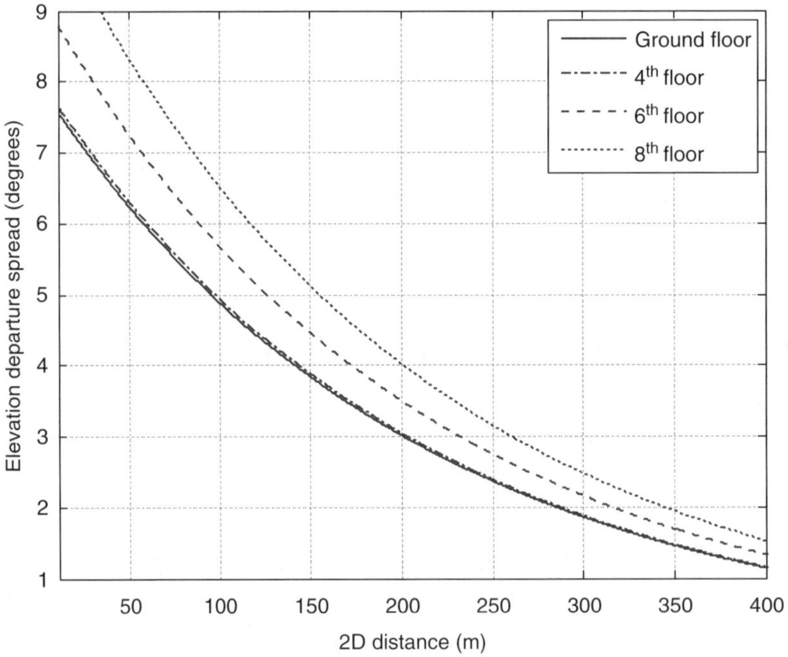

Figure 15.5 Elevation angle spread for the 3GPP urban-micro NLOS environment.

there is also a distance and height dependence of the elevation angles of arrival (at the UE), this effect was not modeled, as large 2D arrays are not envisioned by 3GPP for the UE at the frequencies covered by the channel model. Note that this observation will not be true of frequencies above 6 GHz and the channel models for those frequencies will have to incorporate those effects, as we will describe later.

In addition to the other changes, care was also taken in the 3D channel model to characterize antenna affects such as mechanical downtilts through careful translation of the antenna-element pattern and polarization characteristics.

15.5.2 Antenna-array Configurations

As for antenna-array configurations, several 2D array geometries are being considered for evolved LTE systems below 6 GHz [37]. Figure 15.6 shows an example of the 2D ULA of vertically polarized elements where there are N elements in azimuth and M elements in elevation for a total of $M \times N$ elements. Figure 15.7 shows an example of a 2D array of cross-polarized elements, where there are N columns and M rows of cross-pol elements for a total of $2 \times M \times N$ antenna elements. With $N=4$ columns, this array can be deployed in a form factor that is roughly equivalent to the form factor of the eight antenna cross-pol arrays used for LTE Rel-10. With the baseband architecture shown in Figure 15.2, each antenna shown in the diagrams would represent a single physical antenna connected to a transceiver.

15.5.3 Uplink Transmission Techniques

For the uplink, large-scale antenna arrays are generally aimed at improving coverage reliability and range through higher-gain adaptive beamforming/SU-MIMO. However, large-scale arrays also promise to improve capacity through the use of MU-MIMO and uplink

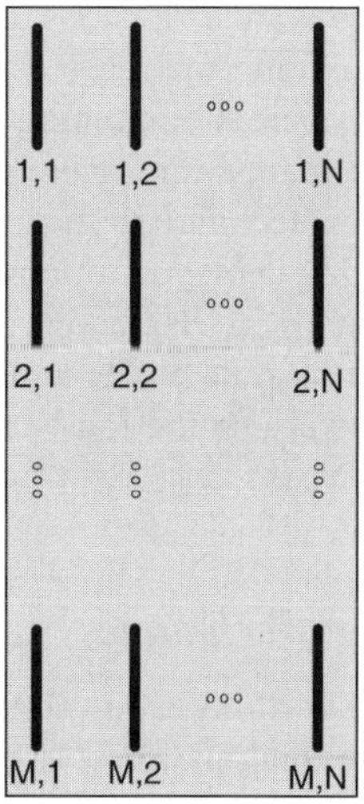

Figure 15.6 Two-dimensional-uniform linearly vertically polarized antenna array.

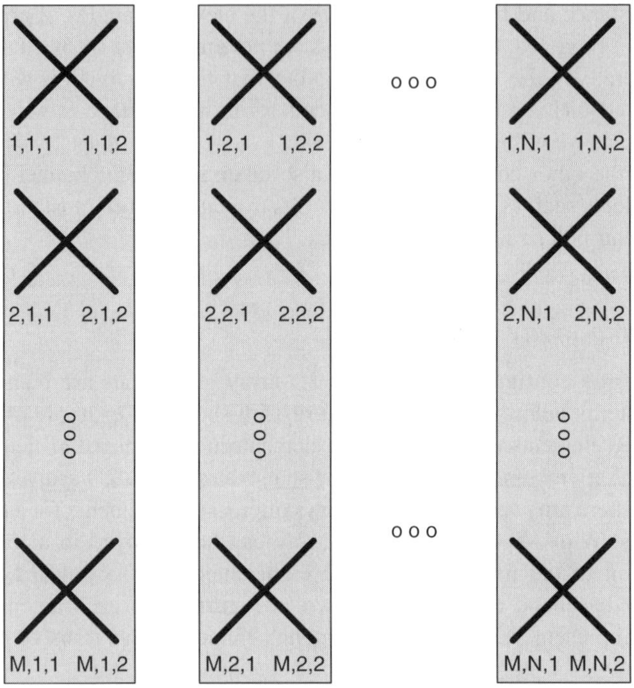

Figure 15.7 Two-dimensional-cross-polarized antenna array.

CoMP. With MU-MIMO on the uplink, large-scale arrays can serve large numbers of users simultaneously through the use of multi-user receive-array algorithms. For a large-scale array operating on the uplink, many of the existing methodologies for UL MIMO are applicable and require generally straightforward extensions to large numbers of receive antennas. One concern when increasing the number of receive antennas is the channel-estimation performance. To fully realize the coherent combining gain of a large-scale array at low SNRs, the channel-estimation performance can severely limit the gain from the additional antennas. Robust channel-estimation techniques that can operate a very low SNRs are required when massive MIMO is being deployed for coverage or range enhancement on the uplink. For evolved LTE systems below 6 GHz and/or with dense deployments, however, extended range and coverage reliability may not be a major concern, and increasing uplink capacity through MU-MIMO will likely be the most valuable operating mode with large-scale arrays on the uplink.

15.5.4 Downlink Transmission Techniques

Massive MIMO transmission techniques for the downlink will generally involve closed-loop transmission schemes where one or multiple data streams are beamformed towards one (SU-MIMO) or multiple (MU-MIMO) users at a time. When considering the benefits of large-scale arrays on the downlink, it is important to note that the performance of SU-MIMO techniques is ultimately limited by the capabilities of the mobile, most importantly in terms of the number of receive antennas and the highest modulation and coding order of the system. The highest data rate that can be delivered to a user is dictated by the number of spatial streams and the data rate per stream that a mobile can support. As the base array size increases, there will reach a point where the performance of SU-MIMO will saturate (that is, almost all users operating at the highest number of streams and the highest data rate per stream that the users can support). Further increases in the array size will then provide little benefit [20]. Therefore as the array size increases, there will reach a point where MU-MIMO transmission must be employed to increase system performance. For large-scale arrays therefore, MU-MIMO will be an essential transmission technique.

Key to the operation of downlink closed-loop SU- and MU-MIMO is the computation of the transmit-weighting coefficients that form the spatial channels between the transmit array and the users in the system. MU-MIMO requires precise channel state information at the base station in order to create multiple parallel non-interfering spatial channels between the base and the multiple users. Massive MIMO transmission techniques for the downlink can then be divided into several categories based on how the transmit array obtains the channel knowledge needed to compute those transmit weights.

15.5.4.1 Reciprocity-based Transmission Methods

The first category of transmission techniques involves leveraging the reciprocity of a static time-division-duplex channel where the multipath channel on the uplink is identical to that on the downlink due to the fact that the uplink and downlink occupy the same frequency band. Reciprocity-based techniques involve estimating the uplink channel from a user and converting that channel into an estimate of the downlink channel from which the transmit weights are calculated. Reciprocity-based techniques involve little extra signaling as the UE can simply transmit an uplink sounding waveform to enable the base station to estimate the uplink channel. However, complicating reciprocity techniques in a dense deployment is the problem of pilot contamination [11], where a large number of users simultaneously transmit uplink sounding waveforms and interfere with the accurate channel estimation of the desired users' channels. Mitigating the pilot contamination problem can be achieved by leveraging orthogonality in pilot sequences in the time, frequency, and sequence domains, and also by incorporating pilot re-use strategies where a high path loss exists between cells that use the same pilot sequences in time and frequency.

Once the uplink channel is estimated for each user, various well-known transmission schemes can be used. For SU-MIMO, a simple methodology involves computing the transmit-weight vectors to be the largest eigenvectors of the transmit covariance matrix computed from the received uplink signals [34]. For MU-MIMO, there are various zero-forcing-type algorithms for simultaneously transmitting to multiple users [35].

15.5.4.2 Codebook Feedback-based Methods

The second category of transmission technique involves feedback messages from the user that enable the base station to compute the transmit weights for the multiple users. A wide variety of methodologies has been developed over the years (see Love et al. [12] and the references therein). Here we briefly describe two types of feedback that are applicable to massive MIMO systems: codebook-based feedback is described in this subsection, and direct feedback is described in the next subsection.

A common form of feedback involves both the base station and the user devices maintaining a codebook containing a set of possible transmit-weight vectors (for rank 1 transmission) or matrices (for ranks greater than one). The user devices are responsible for estimating the downlink channel from all the base station antennas and selecting which entry (transmit-weight vector/matrix) in the codebook will perform the best for that user (where, for example, the user could simply compute which transmit-weight vector will result in the highest SNR for that user). The user device feeds back an indicator, often called a precoder matrix indicator (PMI), that informs the base station which transmit vector/matrix should be used for the user. Armed with the PMI for each user, the base station has several choices for transmission. The base can simply use the transmit-weight vector indicated by the PMI feedback as the transmit weight for that user, which is a reasonable strategy for SU-MIMO transmission in LTE TM4, for example (see the earlier discussion on LTE transmission modes). Alternatively, the base can treat the transmit-weight vector from the user as an estimate of the vector (rank 1) or matrix (rank > 1) channel between the base station and the UE, and the base can compute the transmit weights according to any of the basic transmission methods described in the earlier discussion on reciprocity-based transmission methodologies (which is commonly done for LTE TM9).

15.5.4.3 Product Codebook Feedback-based Methods

The problem of designing a codebook feedback methodology for the 2D arrays of interest to evolved LTE systems has been studied by various groups [16, 17]. A conceptually straightforward approach is to use a single-dimension codebook, where the dimension of the codebook equals the total number of antennas in the array. As the array size gets larger and larger, the number of entries in the codebooks can increase significantly, which can make the codebook search complexity unnecessarily burdensome for the UE. An alternative approach is the product codebook approach [18–20], where two separate codebooks are maintained, one for azimuth and one for elevation. The codebook feedback process is likewise decomposed into two separate feedback processes, one each for azimuth and elevation. The advantage of the codebook approach is that two smaller more manageable codebooks are leveraged for the two dimensions rather than one extremely large codebook that encompasses the entire array [20].

The product codebook approach involves several steps. First the base transmits reference signals on the downlink to enable the UE to determine the best precoder vector/matrix for the azimuth codebook and the best precoder vector/matrix for the elevation codebook. The UE feeds back the PMI for the azimuth codebook and the PMI for the elevation codebook. To recover a full dimension precoder from the two PMI values, the base can compute the Kronecker product of the two codebook entries. Based on the reconstructed full dimension precoders, the base can compute the transmit weights for the UE. The concept is shown in more detail in the literature [20], along with system-level performance results. The authors of that paper point out, however, that there is a large gap between the performance of the product codebook concept and the performance with ideal knowledge of the downlink transmit covariance matrix. Closing that gap will require alternative methodologies to the standard codebook feedback concept.

15.5.4.4 Direct Feedback Methods

The codebook feedback concept essentially involves establishing a set of vectors that span the total space of possible transmit-weight vectors that the base could use to transmit to a UE. Alternatively, codebook feedback can be thought of as establishing a set of vectors that span the range of vector channels that a UE could experience. A problem with codebook feedback is the problem of the quantization error between the true channel vector or the optimal transmit-weight vector and the entries in the codebook. Given the precise channel knowledge required by MU-MIMO transmission, this quantization error will limit the performance that can be achieved.

An alternative to codebook feedback is to use a form of feedback where each entry of the channel vector or transmit-weight vector would be individually encoded and fed back to the base station. Numerous methodologies have been proposed [12, 22, 23], and in this section, we briefly describe several key feedback concepts that might be used in 5G systems to provide improved performance over the standard codebook feedback approach. The first is analog feedback [22], where the user transmits to the base an analog representation of the downlink channel coefficients. A variation on that concept is called direct channel feedback and involves various techniques for encoding the analog channel information [23]. The channel coefficients can be encoded as a bits stream rather than as an analog waveform. However, rather than encoding the channel coefficients, an alternative is for the mobile to compute the downlink transmit covariance matrix and encode and feedback the entries of the covariance matrix to the base station [24]. Another concept is to encode the entries of the dominant eigenvectors of the transmit covariance matrix, which would involve less feedback than encoding the covariance matrix. Encoding the dominant eigenvectors can be shown to provide performance very close to the performance with ideal knowledge of the channel transmit covariance matrix.

15.5.5 Massive Subsectoring with Large-scale Arrays

Large-scale arrays can enable the use of massive subsectoring, a strategy whereby the sector that a base normally covers is further split into smaller subsectors through the use of subsector beams [28]. This concept is illustrated in Figure 15.8, which shows the creation of eight total

(a)

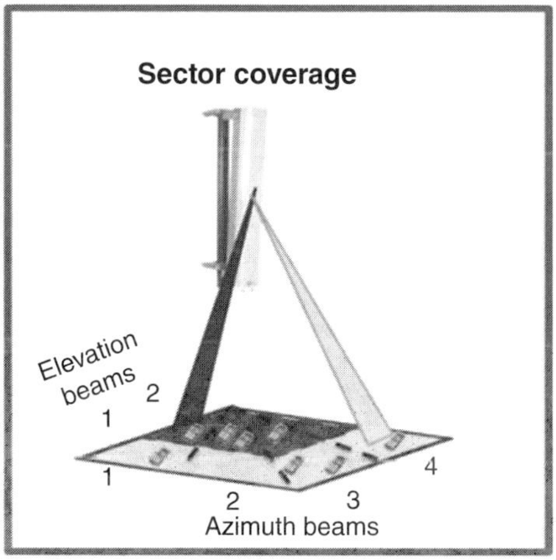

Figure 15.8 Massive subsectorization with a 4-azimuth by 2-elevation beam arrangement: (a) sector coverage and (b) beam patterns formed.

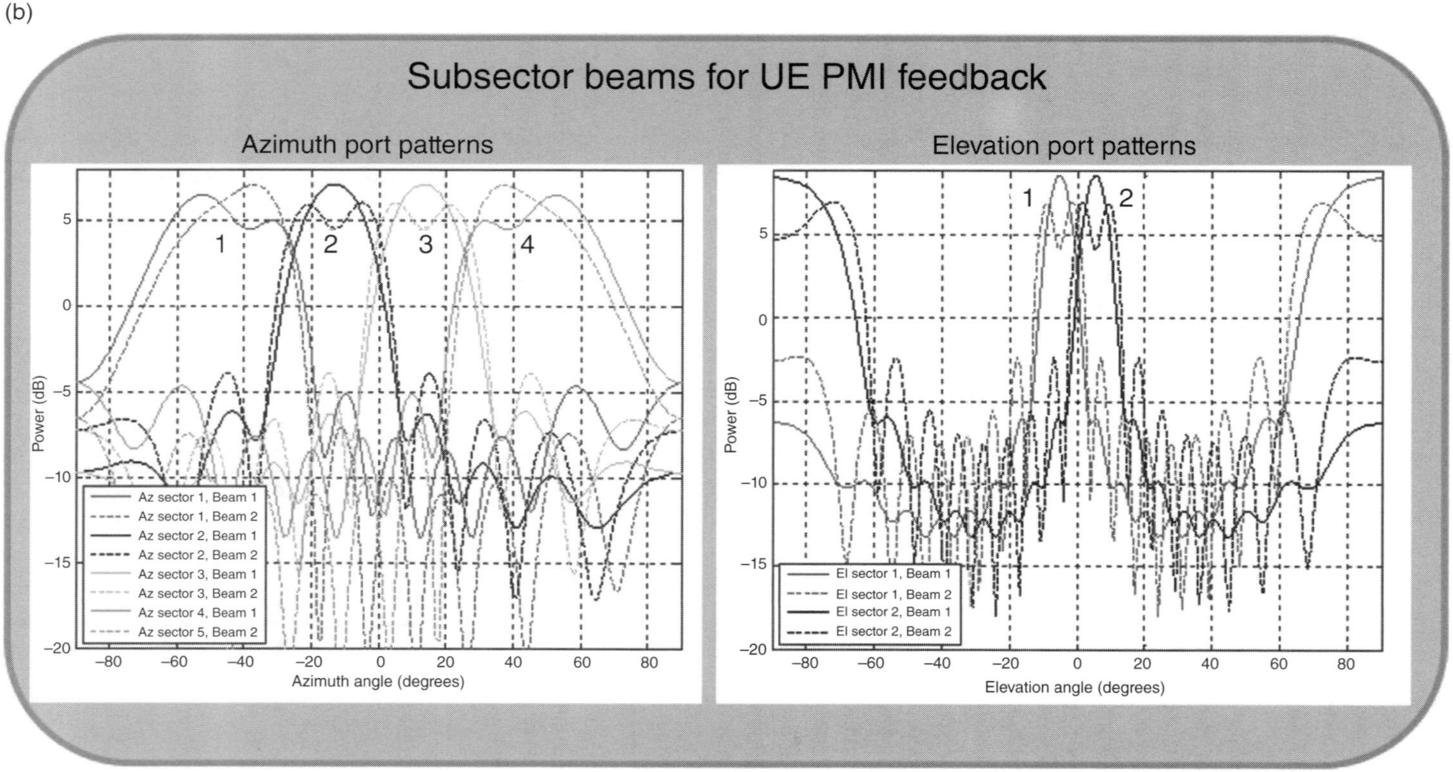

Figure 15.8 (*Continued*)

subsectors, four in azimuth by two in elevation, where a cross-polarized array with eight columns and ten rows is used that has a total of 160 antenna elements (Figure 15.7). In each subsector for each polarization, two overlapping beams are formed for each subsector in azimuth, and two overlapping beams are formed for each subsector in elevation. The result is that each subsector has a total of four ports per polarization or eight total ports with two polarizations. The transmit-weight vector for the subsector beam that forms a given port within the subsector is computed as the Kronecker product of the corresponding azimuth beam and the corresponding elevation beam within the subsector. The beams forming the ports within each subsector are envisioned to be static in nature, although it is possible to tailor the subsector beams to optimize the overall cell coverage. Figure 15.8 shows the beam patterns of the aggregation beams that form the ports in azimuth and elevation. Figure 15.9 shows the logical view of the antenna ports formed by the massive subsectoring concept. The ports are divided into a total of eight subsectors, which are arranged four in azimuth by two in elevation with two elevation subsector beams per polarization per subsector and two azimuth subsector beams per polarization per subsector. The result is eight total ports for each subsector with eight total subsectors or 64 total ports formed from the 160-element antenna panel. Within each subsector, all the usual MIMO transmission techniques can be used, such as the transmission methods currently used in LTE, which were described earlier [28]. A similar concept has been described, whereby MU-MIMO transmission is achieved through the use of a pre-beamforming matrix (similar to the subsector beams in the description previously) along with a conventional transmit precoder [25, 26].

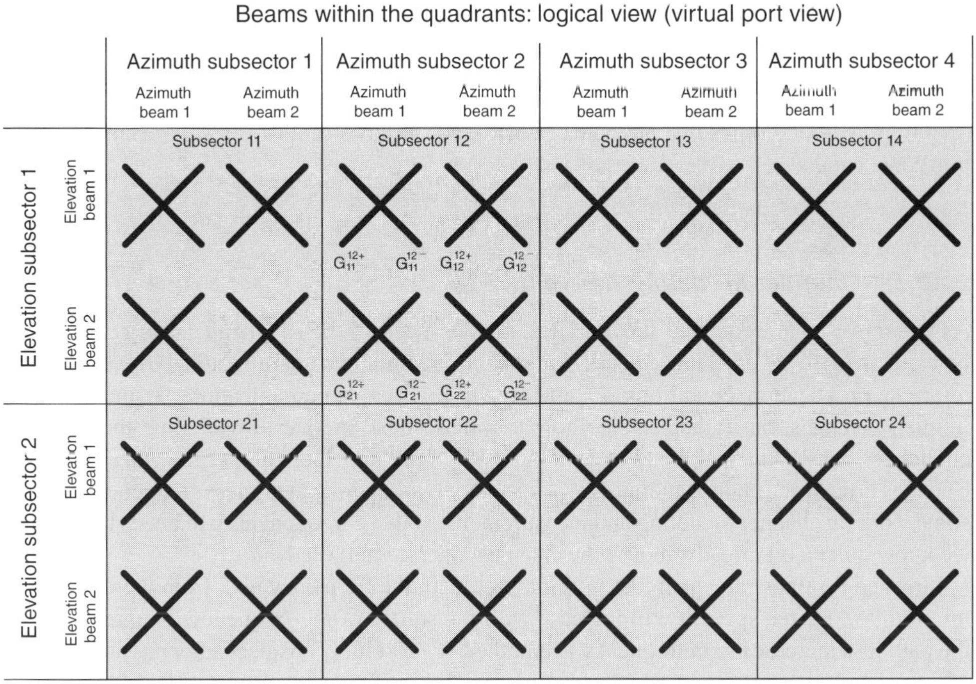

Figure 15.9 Logical antenna port view of the 4-azimuth by 2-elevation subsectorization approach. Two overlapping beams for both horizontal and vertical for a total of 64 ports. Eight total ports per subsector and eight total subsectors.

15.6 Massive MIMO for cmWave and mmWave Systems (Above 6 GHz)

The limited availability of bandwidth below 6 GHz has led to a significant interest in the use of the cmWave (6–30 GHz) or mmWave (30–100 GHz) bands for 5G cellular systems [7–10]. In the cmWave bands, system bandwidths are expected to be on the order of 100–200 MHz, whereas in the mmWave bands system bandwidths on the order of 1–2 GHz are a distinct possibility. The implication of the high system bandwidths is that user data rates much higher than those possible in today's LTE systems can be achieved with rather simple schemes mostly by virtue of the high system bandwidth.

However, a variety of challenges exist for deploying large-scale antenna arrays in the cmWave and mmWave bands. Compared to the traditional cellular systems operating at carrier frequencies around 2 GHz, systems operating in the mmWave bands will have significantly worse link budgets due to significantly lower power-amplifier output powers, higher path losses, and higher shadowing and blocking losses due to the significant decrease in diffraction and dispersion. Fortunately, with antenna sizes proportional to the carrier wavelength, more and more antennas can be packed into the same physical area as the carrier frequency increases. As a result, large-scale antenna arrays with tens or even hundreds of elements are expected to be a key enabling technology for systems operating in the mmWave bands. A large-scale antenna array can mitigate the poor path-loss conditions through the use of high-gain adaptive beamforming.

Before considering the topic of massive MIMO for systems above 6 GHz, we first consider the problem of modeling the channel above 6 GHz, as the propagation characteristics of these channels will play a large role in dictating how massive MIMO will be deployed in them. We will then discuss hardware implementation issues and describe the techniques and operating methodologies that are being considered for large-scale arrays in the cmWave and mmWave bands.

15.6.1 Channel Modeling Above 6 GHz

As you move up in frequency above 6 GHz, many aspects of the RF channel change, as would be expected. However, there are still some phenomena that remain relatively constant. The different propagation conditions do not preclude access communications at the different frequency ranges, but rather change how a system must operate to overcome the different challenges the RF channel poses at each band. A proper understanding of the channels above 6 GHz is crucial for understanding how a system would operate at at these frequencies. While some work on channel modeling and measurements at these frequencies still needs to be done, we know quite a bit already about how channels operate in these bands.

First and foremost, the path loss will increase with the frequency and, from the Friis transmission law, in free space it will increase with the square of the frequency. The main reason the path loss increases with frequency is that the Friis formula considers antennas at both ends whose size is proportional to to a wavelength. However, as you go up in frequency, more antennas can fit in the same area. What this means is that the additional path loss can be made up by increasing the number of antennas without increasing the overall dimensions of the antenna array.

Path loss can be expressed as [8]:

$$PL[dB](d) = 20\log_{10}\left(\frac{4\pi d_o}{\lambda}\right) + 10n\log_{10}\left(\frac{d}{d_o}\right) + X$$

where d is the distance between transmitter and receiver in m, d_o is a reference distance (say, 1 m), λ is the wavelength in meters, n is the path-loss exponent, and X is the shadow-fading term (a zero-mean Gaussian random variable with a given standard deviation, the shadow fading, in dB). It has been found that the path-loss exponent, n, changes little across frequency. For example, it is around 2.0 for a LOS access link at 73 GHz, and 3.4 for a NLOS access link at 73 GHz; as a comparison, at 28 GHz it was found to be 2.1 for LOS and also 3.4 for NLOS. Hence there is very little change seen in the path-loss exponent going from a carrier frequency of 28 GHz to 73 GHz.

Some early studies into the propagation in the cmWave and mmWave bands looked into the effects of both rain and oxygen absorption [42]. The oxygen loss hits a peak around 60 GHz resulting in a loss of 19 dB/km. While this loss sounds particularly bad, for the small cells that are expected for mmWaves, this loss only equates to 1.9 dB for a 100-m link. Note that this is a worst-case loss in the 6-100 GHz range since the other frequencies will see a much smaller loss from oxygen absorption. Hence oxygen absorption losses are not expected to play a significant role in these bands for small-cell access, even at 60 GHz. Rain loss is fairly constant across these bands and again for small-cell access links the loss should be relatively low even for worst case levels of rains(no more than 5.0–6.0 dB for a 100-m link distance).

One of the important effects that dramatically changes between 6 and 100 GHz is diffraction. While diffraction certainly plays a major role for channel models below 6 GHz, it will have a lesser role from around 6 to 10 GHz and then significantly decrease after that, hardly playing a role in the channel for frequencies of 20 GHz and higher.

While diffraction will not play a role at these frequencies, specular reflection still will and is relatively constant across all of these frequencies. In NLOS links it will be these reflective paths that will provide coverage at frequencies above 20 GHz.

Finally, material and foliage penetration losses will play a significant role at these frequencies and it is observed that as the carrier-frequency increases, these losses also will increase. For example, while some in-building coverage is seen at 28 GHz [43] (this also depends on the building material), it is expected that by the time you reach 73 GHz there will be no in-building coverage from outdoor access points (APs). Again the system will have to adapt to the characteristics of the different frequency ranges and at the higher end of this frequency range indoor users will have to be covered by indoor APs.

15.6.2 Hardware Implementation Issues Above 6 GHz

Key to the consideration of the transmission strategies for massive MIMO in systems above 6 GHz is a variety of hardware implementation issues. The need for tens or hundreds of antenna elements will likely lead to the use of the RF or hybrid array architectures described earlier. Cost and power-consumption constraints will make it difficult to have a transceiver behind every antenna element. A topic of ongoing research is the design and development of RF integrated circuit (RFIC) technology, which will provide the core radio components of the system. Key to enabling RF or hybrid antenna array architectures is the provision of transmit weights in the RF domain, as shown in Figures 15.3 and 15.4. The phased array is envisioned to provide a wide field of view of somewhere between the 90° and 120° azimuth range; elevation coverage range is expected to be smaller in many applications but the design should not preclude a range comparable to azimuth. The size and technology of the array needed will vary with the frequency band, actual deployment scenario or application; in other words, it will depend on the desired system gain. The phased array architecture must therefore support scalable solutions in which multiple phased array monolithic microwave integrated circuits (MMIC) and antennas can be combined to form larger or smaller arrays as needed. Phased arrays can be designed around various RF architectures, particularly in terms of the phase and amplitude steering and combining of antenna-element signals. For example, phase steering can occur at RF, intermediate frequency (IF), and baseband. Likewise, signal combining can occur at RF, IF, and baseband. Initial implementations will be constrained to provide phase-only weights in the RF domain, which has implications for the overall transmission and reception strategy that will be used. In the case of phased array solutions at bands above 70 GHz, the array of antenna elements can be produced on different material substrates, such as low-temperature co-fired ceramics or liquid crystal polymers, with the RFIC die flip-chip bonded to the substrate. Special attention must be paid to minimize the coupling losses between the RFIC and the antenna elements.

15.6.3 Acquiring Channel State Information

An important problem to solve with any downlink massive MIMO scheme in the bands above 6 GHz is the problem of acquiring the channel knowledge necessary to facilitate the high-gain adaptive beamforming or high-order spatial multiplexing that is made possible through the use of large-scale antenna arrays. The details of any antenna-array transmission scheme depend heavily on how the channel knowledge is acquired. In the LTE bands, as discussed earlier in this chapter, channel knowledge is generally acquired on a per-antenna-element basis, where reference signals are transmitted out of each antenna port/antenna element to enable the channel to be estimated for that antenna port. As a result, the transmit beamforming weights for SU- or MU-MIMO can be computed based on the knowledge of the downlink channel matrix. In contrast, in the mmWave bands, the poor path-loss conditions are expected to make it very difficult to obtain channel knowledge at the individual antenna element level.

One solution to the channel state acquisition problem is the so-called "grid-of-beams" approach, similar to what has been adopted in IEEE 802.11ad [9], in which a set of quasi-non-overlapping narrow beams are used to cover the service area. In this approach, the base station will cycle through the set of narrow beams and transmit a reference signal on each of the beams, and the mobile will feed back to the base an indicator of the beam that was determined to be the best. The base can then transmit to the mobile with the best beam. An extension of the grid-of-beams concept is the basis function approach [33], where an orthonormal set of transmit beams act as an orthonormal basis for the downlink channel, and the mobile feeds back an estimate of the complex channel gain for each orthonormal transmit beam. With knowledge of the transmit beams and the corresponding fed-back channel gains, the base can reconstruct a wideband channel matrix for each transmit antenna, thereby enabling the use of the traditional MIMO transmission schemes that leverage per-antenna channel knowledge.

15.6.4 Transmission Strategies Above 6 GHz

When considering the strategy for operating a large-scale array in the above-6 GHz bands, it is helpful to consider the performance benefits that can be obtained through the use of large-scale arrays. In bands below 6 GHz, system bandwidth is scarce as LTE systems are deployed with bandwidths up to 20 MHz per carrier with aggregation of up to five carriers. Currently a study item in 3GPP Rel-13 supports aggregation of up to 32 carriers. Due to the scarcity of bandwidth and the typically interference-limited nature of the cellular deployments, the emphasis in the design for systems below 6 GHz is generally towards maximizing spectral efficiency. Techniques such as spatial multiplexing, both SU- and MU-MIMO, will be essential for maximizing the spectral efficiency in the bands below 6 GHz, as was discussed earlier in this chapter. In contrast, in the mmWave bands, system bandwidths on the order 1 or 2 GHz are a distinct possibility, in which case, high data rates relative to LTE systems can be achieved rather simply due to the large system bandwidth. In the cmWave bands, supporting bandwidths from 100 MHz–500 MHz, 5G peak rates of 10 Gbps can be achieved using 4 × 4 MIMO and 256QAM modulation using 400-MHz bandwidth. With the challenging propagation conditions, the emphasis in the massive MIMO strategies above 6 GHz may not be spectral efficiency, but instead may simply be to provide high-gain beamforming to overcome the poor path loss.

Recent performance studies have shown how systems in the mmWave bands can provide extremely high data rates compared to current LTE systems through the use of relatively simple SU-MIMO transmission schemes [31]. Follow-up performance studies have shown the benefits of spatial multiplexing, particularly MU-MIMO, in mmWave dense deployments despite the initial expectation that such systems will be power limited [32]. In the following discussion, we first discuss single-user transmission methodologies for systems above 6 GHz; multi-user transmission schemes will be discussed in the following section.

15.6.5 SU-MIMO Transmission

For single-user transmission in the cmWave and mmWave bands, several simple schemes can be used. The grid-of-beams approach can be used simply to select the best beam out of a set of possible beams for transmission to the user. Extending the grid-of-beams approach to incorporate higher numbers of spatial streams is difficult without knowledge of the full channel matrix. However, a simple approach is to leverage orthogonal polarization in the transmit-array structure and exploit the high level of polarization discrimination present in the mmWave channels [44]. The grid-of-beams can perform two-stream spatial multiplexing with a cross-polarized array, where a grid of beams is created for each set of co-polarized elements, and the user feeds back an indicator of the best beam for each polarization. The base then transmits two streams, one with the best beam for one polarization, the other with the best beam for the other polarization. The advantage of these grid-of-beam approaches is the relative simplicity and the ability to perform the entire beamforming operation at RF with phase-only transmit weights (phase shifters behind each transmit element).

An alternative to the grid-of-beamse approach is the basis function approach described earlier wherein the base can acquire an estimate of the complex channel matrix between each transmit and each receive antenna. With knowledge of the complex channel matrix, traditional SU-MIMO approaches can be used, although a phase-only constraint in the transmit weights may be required depending on the hardware implementation of the RF transmit weights. Two-stream single-user spatial multiplexing can play a significant role in providing extremely high data rates in the mmWave bands relative to current LTE data rates [31, 32].

15.6.6 MU-MIMO Transmission

As mentioned in Section 15.6.1, the challenging propagation conditions in the bands above 6 GHz, especially in the mmWave bands, may necessitate the use of simple high-gain beamforming to overcome the high path loss. MU-MIMO involves splitting the transmit power amongst multiple users, and therefore may not be the initial focus of massive MIMO systems in the cmWave or mmWave bands. However, in a case study of the performance of MU-MIMO in an ultradense small-cell deployment in the mmWave bands, it was shown that there are cases where MU-MIMO can provide significant capacity gains over SU-MIMO [32]. The gain of MU-MIMO over SU-MIMO depends heavily on the user density – the presence of multiple users in the cell that can be paired together for simultaneous transmission – the transmission scheme, and the strategy for pairing users together. A simple MU-MIMO transmission scheme consisting of reusing the SU-MIMO transmission weights for each user can provide significant gains at mmWave bands, but only if more than one stream can be transmitted to each of the paired users [32]. The more complex zero-forcing style of transmission can provide even higher gains over SU-MIMO. MU-MIMO promises to provide significant gains over SU-MIMO even in the cmWave or mmWave bands, and the paper by Vook et al. provides an insight into the performance trends that occur when different MU-MIMO transmission strategies are used [32]. For example, using phase-only transmit weights rather than transmit weights controlled in both amplitude and phase tends to have only a small impact on SU-MIMO performance, but tends to cause large degradations in MU-MIMO performance. Another interesting aspect discussed the same paper is the often significant performance benefit of transmitting more than one data stream to each of the co-scheduled users during the MU-MIMO transmission.

15.7 Conclusion

Massive MIMO technology promises to greatly enhance system performance for below-6 GHz LTE and 5G systems and stands as a prerequisite for systems in the cmWave (6–30 GHz) and mmWave (30–100 GHz) bands. This chapter reviewed the current state of the art in MIMO technology and surveyed the trends for deploying large-scale antenna arrays in 5G cellular. The scarcity of spectrum and interference-limited nature of existing LTE deployments motivates the use of high-order spatial multiplexing with a focus on MU-MIMO in order to satisfy the demands of high-user-density deployments. In contrast, the abundance of available spectrum and the poor propagation characteristics of the cmWave and mmWave bands will lead to initial massive MIMO systems being aimed primarily at providing high beamforming gains for the simple purpose of overcoming the poor link budgets in those bands. However, as has been shown in recent studies, spatial multiplexing, both single-user and multi-user, will eventually play a role in systems deployed specifically in the cmWave bands so as to satisfy the high peak and edge rate requirements of 5G systems.

References

[1] T.L. Marzetta, "Noncooperative cellular wireless with unlimited numbers of base station antennas," *IEEE Trans. Wireless Commun.*, vol. **9**, no. 11, pp. 3590–3600, 2010.

[2] E.G. Larsson, F. Tufvesson, O. Edfors, and T.L. Marzetta, "Massive MIMO for next generation wireless systems," *IEEE Commun. Mag.*, vol. **52**, no. 2, pp. 186–195, February 2014.

[3] Y.-H. Nam, B.L. Ng, K. Sayana, Y. Li, J. Zhang, Y. Kim, and J. Lee, "Full-dimension MIMO (FD-MIMO) for next generation cellular technology," *IEEE Commun. Mag.*, vol. **51**, no. 6, 2013.

[4] F. Rusek, D. Persson, B.K. Lau, E.G. Larsson, T.L. Marzetta, O. Edfors, and F. Tufvesson, "Scaling up MIMO: Opportunities and challenges with very large arrays," *IEEE Signal Process. Mag.*, vol. **30**, no. 1, pp. 40–60, 2013.

[5] T.S. Rappaport, S. Sun, R. Mayzus, H. Zhao, Y. Azar, K. Wang, et al., "Millimeter wave mobile communications for 5G cellular: It will work!", *IEEE Access*, vol. **1**, pp. 335–349, 2013.

[6] S. Rangan, T.S. Rappaport, and E. Erkip, "Millimeter-wave cellular wireless networks: potentials and challenges," *Proc. of the IEEE*, vol. **102**, no. 3, pp. 366–385, 2014.

[7] M. Cudak, A. Ghosh, T. Kovarik, R. Ratasuk, T.A. Thomas, F.W. Vook, and P. Moorut, "Moving towards mmWave-based beyond-4G (B-4G) technology," *Proc. IEEE 77th Vehicular Technology Conference (VTC Spring)*, 2013, pp. 1–5.

[8] A. Ghosh, T.A. Thomas, M.C. Cudak, R. Ratasuk, P. Moorut, F.W. Vook, et al., "Millimeter-wave enhanced local area systems: a high-data-rate approach for future wireless networks," *Selected Areas in Communications, IEEE Journal on*, vol.**32**, no. 6, pp. 1152–1163, June 2014.

[9] IEEE P802.11ad/D5, "IEEE draft standard for local and metropolitan area networks – specific requirements – Part 11: wireless LAN medium access control (MAC) and physical layer (PHY) specifications – Amendment 3: enhancements for very high throughput in the 60 GHz band".

[10] G.R. MacCartney and T.S. Rappaport, "73 GHz millimeter wave propagation measurements for outdoor urban mobile and backhaul communications in New York City," in *Proc. IEEE ICC, June* 2014, pp. 4862–4867.

[11] J. Jose, A. Ashikhmin, T. Marzetta, and S. Viswanath, "Pilot contamination problem in multicell TDD systems," *UCSD Information Theory and Applications Workshop,* San Diego, Feb **8–13**, 2009.

[12] D.J. Love, R. W. Heath, V.K.N. Lau, D. Gesbert, B.D. Rao, and M. Andrews, "An overview of limited feedback in wireless communications," *IEEE IEEE J. Select. Areas in Commun.*, vol. **26**, no. 8, pp. 1341–1365, 2008.

[13] O. Yilmaz, S. Hamalainen, and J. Hamalainen, "System level analysis of vertical sectorization for 3GPP LTE," in *Proc. 6th International Symposium on Wireless Communication Systems (ISWCS)*, 2009, pp. 453–457.

[14] T. Thomas and F. Vook, "Transparent user-specific 3D MIMO in FDD using beamspace methods," in *Proc. IEEE Global Communications Conference (GLOBECOM)*, 2012, pp. 4618–4623.

[15] F.W. Vook, T.A. Thomas, and E. Visotsky, "Elevation beamforming with beamspace methods for LTE," *Proc. IEEE 24th International Symposium on Personal, Indoor and Mobile Radio Communications*, pp. 554–558.

[16] J. Li, X. Su, J. Zeng, Y. Zhao, S. Yu, L. Xiao, and X. Xu, "Codebook design for uniform rectangular arrays of massive antennas," in *Proc.77th Vehicular Technology Conference (VTC) Spring.* IEEE, 2013, pp. 1–5.

[17] X. Su, J. Zeng, J. Li, L. Rong, L. Liu, X. Xu, and J. Wang, "Limited feedback precoding for massive MIMO," *International Journal of Antennas and Propagation*, vol. 2013, Article ID 416352.

[18] Y. Xie, S. Jin, Y. Zhu, X. Gao, and Y. Huang, "A limited feedback scheme for 3D multiuser MIMO based on Kronecker product codebook," in *Proc. IEEE 24th International Symposium on Personal, Indoor and Mobile Radio Communications, IEEE 2013*, pp. 1130–1135.

[19] Y.Yuan, Y. Wang, W. Zhang, and F. Peng, "Separate horizontal & vertical codebook based 3D MIMO beamforming scheme in LTE-A networks," in *Proc.78th Vehicular Technology Conference (VTC) Fall.* IEEE, 2013.

[20] F.W. Vook, E. Visotsky, T.A. Thomas, and B. Mondal "Product codebook feedback for massive MIMO with cross-polarized 2D antenna arrays," *Proc. IEEE 25th International Symposium on Personal, Indoor and Mobile Radio Communications,* IEEE, 2014, pp. 502–506.

[21] D. Ying, F.W. Vook, T.A. Thomas, D.J. Love, and A. Ghosh, "Kronecker product correlation model and limited feedback codebook design in a 3D channel model," *IEEE International Conference on Communication (ICC)*, 2014, pp. 5865–5870.

[22] T.L. Marzetta and B.M. Hochwald, "Fast transfer of channel state information in wireless systems," *IEEE Trans. Signal Process.*, vol. **54**, no. 4, pp. 1268–1278, April 2006.

[23] T.A. Thomas, K.L. Baum, and P. Sartori, "Obtaining channel knowledge for closed-loop multi-stream broadband MIMO-OFDM communications using direct channel feedback," *Proc. IEEE Globecom-2005*, St. Louis, MO, November, 2005, pp. 3907–3911.

[24] Motorola, "Covariance matrix quantization for CSI feedback," R1-094685, 3GPP TSG RAN1#59, Jeju, South Korea, November 2009.

[25] A. Adhikary, J. Nam, J. Ahn, and G. Caire, "Joint spatial division and multiplexing–the large-scale array regime," *IEEE Trans. Inf. Theory*, vol. **59**, no. 10, pp. 6441–6463, Oct 2013.

[26] A. Adhikary, E. Al Safadi, M. Samimi, R. Wang, G. Caire, T. Rappaport, and A. Molisch, "Joint spatial division and multiplexing for mm-Wave channels," *IEEE Journal on Selected Areas in Communications*, vol. **32**, no. 6, pp. 1239–1255, June 2014.

[27] F.W. Vook, T.A. Thomas, and A. Ghosh, "MIMO and beamforming solutions for 5G technology," *Proc. IEEE International Microwave Symposium (IMS)*, IEEE, 2014.

[28] D. Ying, F.W. Vook, T.A. Thomas, and D.J. Love, "Sub-sector-based codebook feedback for massive MIMO with 2D antenna arrays," *Proc. IEEE Global Communications Conference (GLOBECOM)*, 2014, pp. 3202–3207.

[29] A. Liu and V. Lau, "Phase only RF precoding for massive MIMO systems with limited RF chains," *IEEE Trans. Signal Process.*, vol. **62**, no. 17, pp. 4505–4515, Sept 2014.

[30] D. Ying, F.W. Vook, T.A. Thomas, and D.J. Love, "Hybrid structure in massive MIMO: Achieving large sum rate with fewer RF chains," *IEEE International Conference on Communication (ICC)*, 2015, pp. 2344–2349.

[31] T.A. Thomas and F.W. Vook, "System level modeling and performance of an outdoor mmWave local area access system," *Proc. IEEE 25th International Symposium on Personal, Indoor and Mobile Radio Communications*, IEEE, 2014, pp. 108–112.

[32] F.W. Vook, T.A. Thomas, and E. Visotsky, "Massive MIMO for mmWave Systems," *2014 Asilomar Conference on Signals, Systems and Computers*, Nov. 2014, pp. 820–824.

[33] T.A. Thomas and F.W. Vook, "Method for obtaining full channel state information for RF beamforming," *Proc. IEEE Global Communications Conference (GLOBECOM)*, 2014, pp. 3496–3500.

[34] F.W. Vook, T. A Thomas, and X. Zhuang, "Transmit diversity and transmit adaptive arrays for broadband mobile OFDM systems," *Proc. IEEE Wireless Communications and Networking (WCNC)*, IEEE, 2003, pp. 44–49.

[35] Q.H. Spencer, A.L. Swindlehurst, and M. Haardt, "Zero-forcing methods for downlink spatial multiplexing in multiuser MIMO channels," *IEEE Trans. Signal Process.*, vol. **52**, no. 2, pp. 461–471, 2004.

[36] 3GPP TR 25.814, Physical layer aspects for evolved UTRA, v.2.0.0, June 2006.

[37] 3GPP TSG RAN: (E-UTRA): 3GPP TR 36.873 version 2.0.0 Study on 3D Channel Model for LTE (Release 12).

[38] 3GPP TSG RAN: (E-UTRA): 3GPP TR 36.819 version 11.1.0 Coordinated multi-point operation for LTE physical layer aspects (Release 11).

[39] 3GPP TR 36.873, Study on 3D channel model for LTE, v1.3.0, February 2014.

[40] B. Mondal, T.A. Thomas, H. Nguyen, E. Visotsky, and F.W. Vook, "Ray tracing results for elevation angle spread of departure and its impact on system performance," *in Proc. IEEE PIMRC 2014,* September, 2014, pp. 166–171.

[41] G.R. MacCartney Jr., M.K. Samimi, and T.S. Rappaport, "Omnidirectional path loss models in New York City at 28 GHz and 73 GHz," *in Proc. IEEE PIMRC 2014*, September 2014, pp. 227–231.

[42] M. Marcus and B. Pattan, "Millimeter wave propagation: spectrum management implications," *IEEE Microwave Magazine*, vol. **6**, pp. 106–113, June 2005.

[43] W. Roh, J. Seol, J. Park, B. Lee, J. Lee, Y. Kim, J. Cho, and K. Cheun, "Millimeter-wave beamforming as an enabling technology for 5G cellular communications: theoretical feasibility and prototype results," *IEEE Communications Magazine*, vol. **52**, pp. 106–113, February, 2014.

[44] M. Kyrö, S. Ranvier, V. Kolmonen, K. Haneda, P. Vainikainen, "Long range wideband channel measurements at 81–86 GHz frequency range," *in Proc. EuCAP 2010*, April 2010.

16

Full-duplex Radios

Dinesh Bharadia and Sachin Katti
Stanford University, Stanford, CA, USA

A long-held assumption in wireless is that radios have to operate in half-duplex mode; that is, either transmit or receive but not both simultaneously on the same channel. Recent work has attempted to invalidate this assumption. Researchers at Stanford [1, 2], Rice [3, 4], and several other groups in industry and academia [5, 6] have proposed designs to build in-band full-duplex radios. Full duplex, if possible, has tremendous implications for network design, not least of which is the fact that cellular networks could cut their spectrum needs by half. For example, LTE uses equal-width separate uplink and downlink channels to enable radios to achieve full duplex. With an in-band full-duplex system, we could use a single channel to get the same performance. Consequently, the problem has attracted significant attention, both from industry and academia and has spurred significant follow-up work.

To achieve full duplex, a radio has to completely cancel the significant self-interference that results from its own transmission on the received signal. For example, in WiFi, since signals are transmitted at 20 dBm (100 mW) average power, and the noise floor is around –90 dBm, the transmit self-interference has to be canceled by 20 dBm –(–90 dBm) = 110 dB to reduce it to the same level as the noise floor and render it negligible. If self-interference is not completely canceled, any residual self-interference acts as noise to the received signal and reduces SNR and consequently throughput. For example, if the received signal's SNR without full duplex is 25 dB but is reduced to 5 dB due to 20 dB residual self-interference, then the throughput with full duplex is that achieved using two 5-dB SNR links. This is significantly worse than using the original half duplex link with 25-dB SNR and it is better to turn off full duplex in this case. To sum up, the amount of self-interference cancelation dictates overall throughput and is a figure of merit for any full-duplex design.

Prior designs have made significant progress on the self-interference cancelation problem [1, 2, 7]. However, the best performing prior designs can at best provide 85 dB of cancelation,

Towards 5G: Applications, Requirements and Candidate Technologies, First Edition.
Edited by Rath Vannithamby and Shilpa Talwar.

Companion website: www.wiley.com/go/vannithamby/towards5g

which still leaves about 25 dB of residual self-interference and therefore reduces the SNR of each direction of the full duplex link by 25 dB. A calculation similar to the one in the previous paragraph shows that to see throughput benefits with these full-duplex designs, the half-duplex SNR of the link has to be extremely high (45 dB or higher). In terms of range, the two nodes would have to be closer than 5 m to see such high SNRs. Outside this range, it is better to turn off full duplex and use the traditional half-duplex mode. To be fair, however, these designs were intended for low-power, narrow-band, fixed rate protocols such as Zigbee, where 85 dB of self-interference cancelation is sufficient for full duplex. WiFi is far more demanding both in terms of bandwidth and cancelation.

Prior designs also need to have at least two antennas [1, 7] in place of the one used by half-duplex systems: one each for transmit and receive and possibly more [2]. However, with two or more antennas, the argument for full duplex becomes weaker since the same doubling of capacity could be obtained by using the two antennas as MIMO antennas to spatially multiplex two independent packets in half-duplex mode instead of using them for full duplex.

In this chapter, we present the design and implementation of a full-duplex WiFi radio that uses a single antenna[1] and delivers close to the theoretical doubling of throughput under all link SNR and distance ranges. Our key technical contributions are novel self-interference cancelation circuits and algorithms that provide the required 110 dB of self-interference cancelation for standard WiFi signals and thus eliminate all self-interference to the noise floor. Our design is wideband: it works with the highest bandwidths (80 MHz) and data rates used by the latest 802.11ac PHY in the 2.4-GHz spectrum. We also experimentally demonstrate a complete full-duplex communication link that uses the full WiFi PHY (OFDM, constellations up to 256QAM, and all the channel coding rates) and achieves close to the theoretically expected doubling of throughput. To the best of our knowledge, this is the first working implementation of a complete WiFi PHY single-antenna full-duplex link.

The reader might be wondering why full duplex is hard to realize. After all, as the sender knows the signal being transmitted, subtracting it should be relatively simple to implement. One of the key insights in this work is that in fact the radio does not know what it is transmitting. What it does know is the clean digital representation of the signal in baseband. However, once the signal is converted to analog and up-converted to the right carrier frequency and transmitted, the transmitted signal looks quite different from its baseband incarnation. The numerous analog components in the radio TX chain distort the signal in both linear and non-linear ways (analog circuits will create cubic and higher-order components of the signal for example), add their own noise (for example, power amplifiers add transmitter noise), are slightly inaccurate (your oscillator might be tuned slightly off 2.45 GHz), or delay it by different amounts at different frequencies and so on. In effect the transmitted signal is a complicated non-linear function of the ideal transmitted signal along with unknown noise. Unsurprisingly, naively subtracting a "known" baseband version of the transmit signal without accounting for all these analog distortions does not work. As we will show in Section 16.4, prior designs fail to account for these distortions and hence are limited to at best 85 dB of cancelation.

[1] Picasso [8] uses a single antenna, but it only allows the radio to simultaneously transmit and receive on *different* adjacent channels. Hence it fails to address the much harder problem of simultaneous TX/RX on the same channel. Our system does address this challenge, and offers novel and higher-performance analog and digital cancelation techniques than Picasso.

In this chapter, we describe two key techniques that enable us to tackle full duplex. First, we design dynamic algorithms to estimate the distortions introduced by analog circuits and accurately model the actual self-interference being experienced by the received signal. Second, we design a novel programmable analog cancelation circuit using off-the-shelf components that allow us to implement the above algorithm in "analog" and dynamically cancel the self-interference. Such analog cancelation prevents receiver saturation from strong self-interference and allows us to use commodity radios. However, the analog cancelation stage does not completely cancel the self-interference. We complement it with a novel digital cancelation algorithm and implementation that cancels any remaining self-interference. Our digital cancelation algorithm differs from all prior work because it not only models the linear distortions, but also non-linear effects and other special effects such as oscillator noise. Thus, overall we use a hybrid analog–digital design that successfully models all linear and non-linear distortions as well as transmitter noise.

We implement our design via a combination of circuit designs and software implementations. Our analog cancelation is implemented on a PCB that we designed and populated using off-the-shelf components. We integrate our board with an off-the-shelf antenna and software radio transceiver [9, 10] based on test equipment from Rohde–Schwarz (RS) as well as on commodity WARP radios. We also implement our digital cancelation algorithms as well as a fully WiFi-compliant PHY layer based on OFDM, supporting constellations up to the standard required 256QAM and all the channel coding rates. We deployed and evaluated our system in an indoor and noisy office environment in the 2.4-GHz ISM band, operating the WiFi PHY over the 80-MHz bandwidth on RS radios, and over the 20-MHz bandwidth using WARP radios.

Our experiments demonstrate that our design delivers on the promise of full duplex. Under typical indoor deployment scenarios, our system delivers a median throughput gain with WiFi radios of 87% – close to the theoretically expected doubling. Looking into the cancelation itself, we show that our design consistently delivers the required 110 dB of cancelation in a dense indoor office environment for both the RS 80 MHz radios as well as the commodity 20 MHz WARP radios. The system is robust to environmental changes and reflections, and can handle all the different constellations used in WiFi. We compare the best known prior full-duplex approaches [1, 3] and show experimentally that they can at best deliver 85 dB of cancelation and, therefore, can only reduce the SNR of the received signal by up to 25 dB.

16.1 The Problem

Full duplex, in theory, should be simple to accomplish. After all, we know the signal we are transmitting and we are only designing circuits and algorithms to subtract it from the received signal. The intuition follows from the conventional abstraction that the analog radio (also known as the RF front-end) is a black box that takes the digital baseband signal, converts it to analog, up-converts it to the carrier frequency, scales it to the right power, and sends it. In other words, the assumption has been that the radio preserves the original baseband signal except for power scaling and frequency shifting. In practice this abstraction turns out to be incorrect. Radios in fact significantly distort the signal being transmitted, relative to the digital baseband representation.

To demonstrate the distortions, we use the following experiment throughout this section. We take a software radio transceiver [9, 10] and send the following signal: two tones at 2.449 GHz

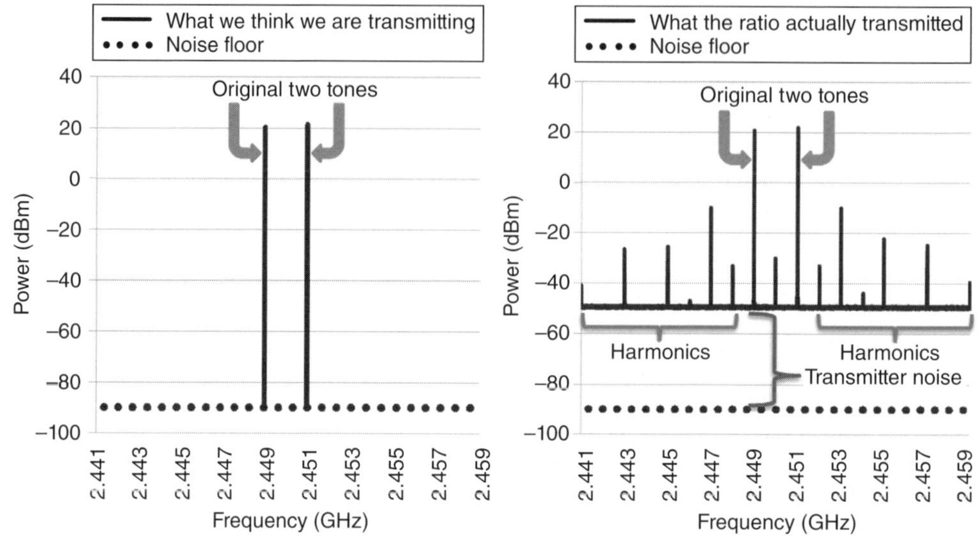

Figure 16.1 Actual and ideal transmitted signals in our experimental setup: left, what we think we are transmitting in digital; right, what the radio actually transmitted. The actual transmitted signal differs significantly from the two tones generated in digital baseband. Note that transmitter noise and harmonics are also generated to the two main transmitter tones.

and 2.451 GHz. In other words, we are sending an extremely simple signal, two sine waves with frequencies 1 MHz away from the carrier frequency of 2.45 GHz. We do this by creating a digital baseband signal with samples of the sine waves at –1 MHz and 1 MHz, which the radio then up-converts to 2.45 GHz and amplifies to 20 dBm average transmit power (the power used by WiFi radios). We then compare the signal output of the antenna to what we would ideally expect if the radio did not introduce any distortions. This experiment serves as some sort of lower bound on the quality of radios. If radios cannot transmit even this simplest of signals without distortion then more complex signals such as WiFi are likely to be significantly distorted. Figure 16.1 plots the ideal and actual transmitted signals' spectra that resulted from our experimental set-up (we ensured that this was a clean environment with no other interference present in the environment at the time of the experiment).

Ideally, we expect to see only two tones at 2.451 GHz and 2.449 GHz, as shown on the left-hand side of Figure 16.1. However, in the transmitted signal, the spectrum of which is plotted on the right-hand side of Figure 16.1, we can easily see that there are several other distortions present in addition to the two main tones that were transmitted. The main components in self-interference can be classified into three major categories:

Linear Components. This corresponds to the two main tones themselves, which are attenuated and could consist of reflections from the environment. These are linear components because the received distortion can be written as a linear combination of different delayed copies of the original two tones.

Non-Linear Components. These components are created because radio circuits can take in an input signal x and create outputs that contain non-linear cubic and higher order terms such as x^3, x^5. These higher-order signal terms have significant content at frequencies close to the transmitted frequencies, which directly correspond to all the other harmonics we see on the right-hand side of Figure 16.1. Harmonics, as the name suggests, are signal distortions that occur at equally spaced frequency intervals from the transmitted frequencies. As the right-hand side of Figure 16.1 shows, we see spikes at frequencies 2.447 GHz and 2.453 GHz that are spaced 2 MHz apart from the two transmitted tones 2.451 GHz and 2.449 GHz, on either side.

Transmitter Noise. The general increase we see in the base signal level that we can clearly see on the sides of the two main tones is noise from the radio transmitter. A radio will, of course, always have noise, which works out to a noise power level of –90 dBm [9]. However, as we can see, the power at the side-bands is significantly higher, at a level of –50 dBm, or 40 dB higher than the receiver noise floor. This extra noise is being generated from high-power components in the radio transmitter, such as power amplifiers. In the radio literature, this is referred to as broadband noise [11]. Radios also have phase noise generated by local oscillators, which is typically of level of –40 dBm, or 50 dB above the noise floor (not seen in the Figure 16.1 because hidden under the main signal component).

16.1.1 Requirements for Full Duplex Designs

This analysis suggests that any in-band full-duplex system has to be able to cancel all the above distortions in addition to the main signal component itself, since all of these are within the frequency band we are transmitting and receiving on and act as strong self-interference to the received signal. In this section, we discuss how strong each of these components are for typical transceivers, and what the requirements are for full duplex. We will state all self-interference power levels relative to the receiver noise floor. The reason is that to implement full duplex, we need to cancel any self-interference sufficiently so that its power is reduced to the same level as the receiver noise floor. There is no point in canceling beyond that since we will not see any benefits – the received signal's SNR will then be dictated anyway by the receiver noise floor, which cannot be canceled or reduced, just as it is today in half-duplex radios.

We use similar experiments for OFDM wideband signals to quantify the power levels of the different distortions, shown in the left side of Figure 16.2. In a typical WiFi radio using 80 MHz bandwidth, the receiver has a noise floor of –90 dBm (1 picowatt). First, since the main signal component is being transmitted at 20 dBm (100 mW), self-interference from the main linear component is $20-(-90)=110$ dB above the receiver noise floor. Second, we observed experimentally that the non-linear harmonics are at –10 dBm or 80 dB above the receiver noise floor. Finally, the transmitter noise is at –40 dBm or 50 dB above the receiver noise floor. Note that these numbers are consistent with other RF measurement studies reported in the literature for standard WiFi radios [12].

There are four takeaways from this analysis:

Linear Self-interference Cancelation. Any full-duplex system needs to provide 110 dB of linear self-interference cancelation to reduce self-interference to the receiver noise floor. This will ensure that the strongest component (the main signal), which is 110 dB above the noise floor, will be eliminated.

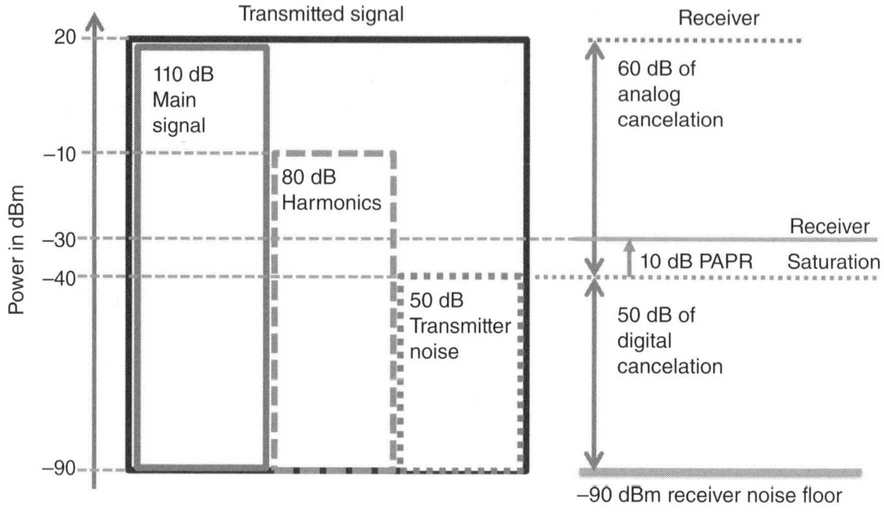

Figure 16.2 Power levels of distortions: left, transmitted signal with sub-components; right, how this impacts the requirements of analog and digital cancelation.

Non-linear Self-interference Cancelation. A full-duplex system has to reduce non-linear harmonic components that are 80 dB above the noise floor, so any full-duplex technique has to provide at least 80 dB of non-linear self-interference cancelation.

Analog Cancelation Component. Transmitter noise is by definition noise and is random. In other words, we cannot infer it by any algorithm. Hence the only way to cancel transmitter noise is to get a copy of it where it is generated – in the analog domain – and cancel it there. This implies that any full-duplex system has to have an analog cancelation component that provides at least 50 dB of analog noise cancelation so that transmitter noise is reduced to below the receiver noise floor.

Amount of Analog Cancelation. A final constraint is that RX chains in radios get saturated if the input signal is beyond a particular level that is determined by their ADC resolution. Assuming a 12-bit ADC resolution, as typically found in commodity WiFi radios, we have a theoretical 72 dB of dynamic range, which implies that the strongest signal level that can be input to the radio relative to the receiver noise floor is –90 dBm +72 = –18 dBm. However, in practice it is necessary to leave 2 bits worth of margin, so a 12-bit ADC should be used as if it were a 10-bit ADC to reduce quantization noise. So the maximum input signal level can be $20-(-90)=110$ dBm $+60$ dBm $= -30$ dBm. Since in WiFi, the transmitted self-interference can be as high as 20 dBm, a full-duplex system needs to have an analog cancelation stage that provides 60 dB of self-interference reduction (we keep a further 10 dB margin for OFDM peak-to-average power ratio where instantaneously an OFDM signal's power level can rise 10 dB above the average power).

To sum up, any full duplex design needs to provide 110 dB of linear cancelation, 80 dB of non-linear cancelation, and 60 dB of analog cancelation.

16.1.2 Do Prior Full-duplex Techniques Satisfy these Requirements?

There are two state-of-the-art designs: ones that use an extra transmit chain to generate a cancelation signal in analog [4] and ones that tap the transmitted signal in analog for cancelation [1, 2]; both use a combination of analog and digital cancelation. Note that all these designs use at least two antennas for transmit and receive instead of the normal single antenna, and the antenna geometry ones use more than two.

Designs that use an extra transmitter chain report an overall total cancelation of 80 dB (we have been able to reproduce their results experimentally). Of this, around 50 dB is obtained in the analog domain by antenna separation and isolation between the TX and RX antennas of around 40 cm; the designs also assume some form of metal shielding between the TX and RX antennas to achieve 50 dB isolation. Note that this 50 dB reduction applies to the entire signal, including linear and non-linear components as well as transmitter noise since it is pure analog signal attenuation. Next, these designs also use an extra transmit chain to inject an antidote signal [4, 13] that is supposed to cancel the self-interference in analog. However, the antidote signal only models linear self-interference components and does not model non-linear components. Further, it is incapable of modeling noise because by definition noise is random and cannot be modeled. Overall, this extra cancelation stage provides another 30 dB of linear self-interference cancelation in the best case. Thus these designs provide 80 dB of linear cancelation, 50 dB of non-linear cancelation and 50 dB of analog noise cancelation, falling short of the requirements by 30 dB for the non-linear components. Hence, if full duplex is enabled over links whose half-duplex SNR is 30 dB or lower, then no signal will be decoded. Moreover, to see any throughput improvements with full duplex, the half-duplex link SNR would have to be greater than 50 dB.

The second design [1] gets a copy of the transmitted analog signal and uses a component called the balun (a transformer) in the analog domain to create a perfectly inverted copy of the signal. The inverted signal is then connected to a circuit that adjusts the delay and attenuation of the inverted signal to match the self-interference that is being received on the RX antenna from the TX antenna. We show experimentally in Section 16.4 that this achieves only 25 dB of analog cancelation, consistent with the prior work's results. The cancelation is limited because this technique is very sensitive and requires precise programmable delays, with resolution as precise as 10 picoseconds, to exactly match the delay experienced by the self-interference from the TX to the RX antenna. Such programmable delays are extremely hard to build in practice; at best we could find programmable delays with a resolution of 100 – 1,000 picoseconds and these were, in fact, the ones used by the prior design [1]. Hence the cancelation circuit is never able to recreate perfectly the inverted self interference signal, and therefore cancelation is limited to 25 dB in analog. However, this design also uses two separate antennas separated by 20 cm for TX and RX and achieves another 30 dB in analog cancelation via antenna isolation. Hence a total of 55 dB of self-interference reduction is obtained in analog and this cancelation applies to all the signal components (linear, non-linear, and noise). The digital cancelation stage of this design also only models the linear main signal component; it does not model the non-linear harmonics that we discussed previously. Thus, we found that we obtain another 30 dB of linear cancelation from digital in this design.

Overall, the second design provides 85 dB of linear self-interference cancelation, 55 dB of non-linear cancelation, and 55 dB of analog noise cancelation. Thus this design falls short of the requirements by 25 dB (especially for the non-linear component). Hence if full duplex is enabled over links whose half duplex SNR is 25 dB or lower, then no signal will be decoded. Moreoever, to see any throughput improvements with full duplex, the half-duplex link SNR would have to be greater than 45 dB.

16.2 Our Design

In this section, we describe the design of our self-interference cancelation technique. Our design is:

- a single-antenna system: the same antenna is used to simultaneously transmit and receive)
- wideband: it can handle the widest WiFi bandwidth of 80 MHz as well as all the LTE bandwidths
- truly full duplex: it cancels all self-interference to the receiver noise floor
- a hybrid: it has both analog and digital cancelation stages.

Note that our hybrid cancelation architecture is not novel, similar architectures have been proposed in prior work [1, 14, 15]. The novelty of our design lies in the cancelation circuits and algorithms, as well as their performance. To the best of our knowledge, this is the first technique that achieves 110 dB of cancelation and eliminates self-interference to the noise floor.

16.2.1 Analog Cancelation

We introduce a novel analog cancelation circuit and tuning algorithm that robustly provides at least 60 dB of self-interference cancelation. Figure 16.3 shows the high-level design of the circuit and where it is placed in the radio architecture. A single antenna is connected to a circulator (at port 2), which is a three-port device that provides limited isolation between port 1 and port 3 while letting signals pass through consecutive ports. The TX signal is fed through port 1, which routes it to the antenna connected to port 2, while the received signal from the antenna is passed from port 2 through to port 3. The circulator cannot completely isolate port 1 and port 3, so inevitably the TX signal leaks from port 1 to port 3 and causes interference to the received signal. From our experiments, we find that the circulator only provides 15 dB of isolation; in other words, the self-interference that is leaking to the RX circuit be reduced only by 15 dB. To get to the noise floor, we still have to provide 95 dB of cancelation, and at least 45 dB of that has to come in analog to ensure the transmitter noise is sufficiently canceled and we do not saturate the receiver. We accomplish this using our novel analog cancelation circuit, which we describe next. Note that when we report analog cancelation performance numbers, for simplicity of description we include the 15 dB of reduction we get from the circulator.

Figure 16.3 shows the design of our analog cancelation circuit. We tap the TX chain to obtain a small copy of the transmitted signal just before it goes to the circulator. This copy, therefore, includes the transmitter noise introduced by the TX chain. The copy of the signal is then passed through a circuit that consists of parallel fixed lines of varying delays (essentially wires of different lengths) and tunable attenuators. The lines are then collected back and added up, and this combined signal is then subtracted from the signal on the receive path. In effect, the circuit is providing us copies of the transmitted signal delayed by different fixed amounts and programmatically attenuated by different variable amounts. The key challenge is to pick

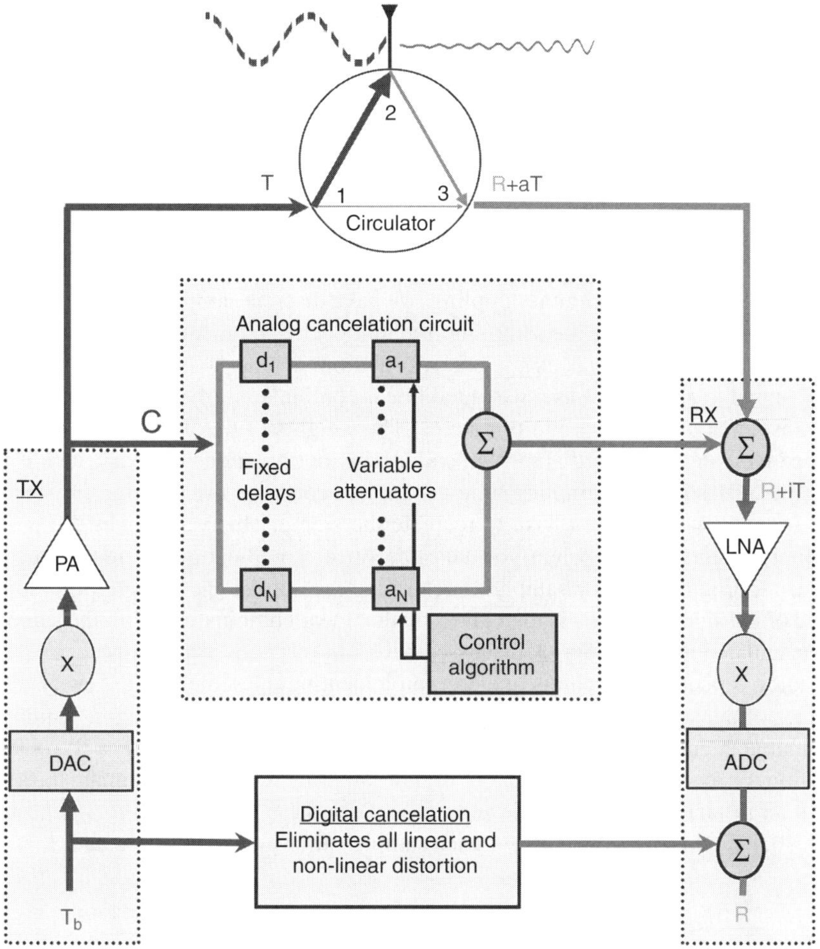

Figure 16.3 Full duplex radio block diagram. T_b is the intended baseband signal we think we are transmitting, but in fact the transmit signal is T. The intended receive signal is R, but we see strong components of the transmitted signal (T) on the RX side. Some of these signals are undesirably leaked through the circulator. The analog cancelation circuit is trying to recreate a signal that matches the leaked interference signal for cancelation. The digital cancelation stage eliminates any residual self-interference.

the fixed delays, as well as to dynamically program the tunable attenuators appropriately so that the we maximize self-interference cancelation. Note that unlike prior work our design uses components that are all available off the shelf and is, therefore, easy to manufacture; unlike prior designs, we do not need sophisticated high-resolution programmable delays that are hard to build [1].

The design of our cancelation circuit is based on a novel insight: we can view cancelation as a sampling and interpolation problem. The actual self-interference signal has a particular delay and amplitude that depends on the delay *d* and attenuation *a* through the circulator. Our insight (the reason for which will become clear shortly) is that we should pick the fixed delays in our cancelation circuit such that they straddle the delay of the self-interference signal through the circulator. So if we have N fixed delay lines, $N/2$ of those lines should be placed at equidistant intervals, all of which have delays that are less than the delay of the self-interference d, and we should do the same for the other half of the delays but greater than d. In practice it is hard to know the precise value of d since it is a function of how the circuit is put together, but we can always find the range over which it varies and place our fixed delays outside of that range on either side.

At this stage we have leading and lagging copies of the transmitted self-interference signal. How might we use them to approximate the actual self-interference at some intermediate instant? If we take a step back, this is essentially an interpolation problem, similar to Nyquist digital sampling. In Nyquist digital sampling, we have discrete samples of the signal at a time period equal to the inverse of the sampling frequency. The Nyquist theorem [16] tells us that sampling (at the Nyquist rate) does not lose information; in other words, we can always reconstruct the signal at any instant as a weighted linear combination of samples taken before and after the instant at which we want to recreate. The weights of the linear combination can be determined by using a standard algorithm called sinc interpolation. The basic idea is that you overlay sinc pulses at each sampling time instant and calculate the value of the sinc pulse at the time instant t at which you wish to recreate the signal. This value gives the weight you should apply to this sample when you take the linear combination for reconstruction. We repeat this algorithm for every sample to determine the corresponding weight to apply to it. The value of the signal at time t is then given by the linear combination of all the samples with weights calculated by the sinc trick discussed here.

Our analog cancelation circuit is in effect implementing the same trick; at every instant we have copies of the signal at different equally spaced delays, just like in digital sampling. The programmable attenuators essentially function as the weights we need to apply in the linear combination for reconstruction. Similar to digital sampling, we need to estimate the self-interference at an instant d that lies somewhere in between these fixed delays $d_1, \ldots, d_N$, as shown in Figure 16.4. To do so, the weights for each sample – the value of the attenuator that we need

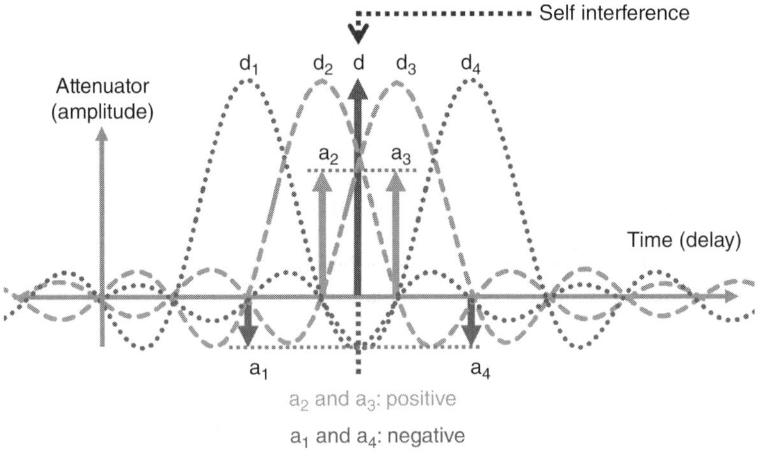

Figure 16.4 Recreating the self-interference signal. The signal is located at instant d, positioned between the fixed delay lines $\boldsymbol{d_i}$. The value of the attenuator $\boldsymbol{a_i}$ for delay $\boldsymbol{d_i}$ is given by the value taken by the sinc centered at $\boldsymbol{d_i}$ at instant d.

to set on each line i – is equal to the value of the sinc pulse centered at the fixed delay d_i at instant d. If we adjust the attenuators for each delay line to those values, then we will be able to perfectly reconstruct the self-interference and cancel it from the receive path. Figure 16.4 shows this algorithm visually in action.

In practice, however, there is an important difference with digital sampling. In digital, we can take linear combinations of a very large number of samples since memory is essentially free. To do that in analog we would need a correspondingly large number of delay lines. In practice, this is not possible for a variety of reasons, ranging from space limitations to power consumption to electromagnetic radiations. Our key insight is that in interpolation, the samples that matter most are the ones that are closest to the instant t at which the signal is being reconstructed. Intuitively, the value of a signal much further from/earlier in time than t should not affect the value of the signal at t. This is reflected in the fact that the weights in the linear combination for these further-out samples are nearly zero. This allows our analog circuit to therefore use a small number of delay lines and still approximate the self-interference fairly well. We show in Section 16.4 that 16 delay lines are sufficient to approximate the self-interference signal leaking through the circulator. Moreover, we will show in Section 16.4 that our analog cancelation delivers at least 60 dB cancelation, comfortably exceeding the requirements we developed in Section 16.1.1.

16.2.2 Digital Cancelation

The goal of digital cancelation is to clean out any remaining residual self-interference. Assuming that analog cancelation provides 60 dB, digital cancelation has to cancel the linear main-signal component by another 50 dB and non-linear components by another 20 dB. We address each of these components separately.

16.2.2.1 Canceling Linear Components

The first part of digital cancelation eliminates the residual linear components of the self-interference. This consists of the main transmitted signal that is leaking through the circulator after analog cancelation, as well as any delayed reflections of this signal from the environment. The reflections are also delayed and attenuated by different unknown amounts.

The basic idea is that this part of the self-interference can be modeled as a linear and *non-causal* function of the transmitted signal, as we know it in digital (recalling that we know the baseband IQ samples of the transmitted packet). The non-causal bit is important. Since we know the samples of the entire packet that was transmitted, we can use samples from the future to estimate the self-interference at the current instant. In other words, the received sample $y[n]$ at any instant can be modeled as a linear combination of up to k samples of the known transmitted signal $x[n]$ before and after the instant n. The parameter k is empirically chosen and is a function of the amount of memory in the channel. So we can write the equation as:

$$y[n] = x[n-k]h[k] + x[n-k+1]h[k-1] + \cdots + x[n+k-1]h[-k+1] + w[n]$$

where $h[k],\ldots,h[-k+1]$ represents the attenuations applied by the channel to the transmitted function, and $w[n]$ is the receiver noise floor.

How can we estimate the coefficients $h[n]$? We leverage the fact that most wireless transmissions have known packet preambles; for example, WiFi uses a preamble of two known OFDM symbols at the start of the packet. Let the samples representing the preamble be $x_{pr}[n]$. Let the receive samples corresponding to the preamble be $y[0],\ldots,y[n]$. Then the previous channel equations can be written specifically for the preamble as:

$$y = Ah + w$$

where A is the Toeplitz matrix of $x_{pr}[n]$.

$$A = \begin{pmatrix} x_{pr}(-k) & \ldots & x_{pr}(0) & \ldots & x_{pr}(k-1) \\ \ldots & \ldots & \ldots & \ldots & \ldots \\ x_{pr}(n-k) & \ldots & x_{pr}(n) & \ldots & x_{pr}(n+k-1) \end{pmatrix}.$$

Our goal is to find a maximum likelihood estimate of the vector h:

$$minimize\|y - Ah\|_2^2$$

Note that the matrix A is known in advance since we know the values of the preamble samples. Hence it can be pre-computed. Additionally, we know from prior work [17] that the coefficients for this problem can be computed by multiplying by the ith received sample of the preamble, as the samples arrive serially as follows:

$$h = \sum\left(y_i a_i^{\dagger}\right)$$

where $a_i^{\dagger}$, is the ith column of pseudo inverse of matrix A. Thus our estimation algorithm computes the linear distortions that the transmitted main signal has gone through for every packet, and is capable of dynamically adapting to the environment.

16.2.2.2 Canceling Non-linear Components

The second task for digital cancelation is to eliminate the residual non-linear components, the power of which is around 20 dB after being reduced by 60 dB due to analog cancelation. However, it is quite hard to guess the exact non-linear function that a radio might be applying to the baseband transmitted signal. Instead, we use a general model to approximate the non-linear function using Taylor series expansion, as this is a standard way to model non-linear functions [18]. So the signal that is being transmitted can be written as:

$$y(t) = \sum_m a_m x_p(t)^m$$

where $x_p(t)$ is the ideal passband analog signal for the digital representation of $x(n)$ that we know.

The general model here contains many terms, but the only ones that matter for full duplex are those that have non-zero frequency content in the band of interest. A little bit of analysis for passband signals (taking the Fourier transform) of the previous equation reveals that the only terms with non-zero energy in the frequency band of interest are the odd-order terms: those containing $x_p(t)$, $x_p(t)^3$, $x_p(t)^5$, and so on. We can safely ignore the even-order terms. The first term that is the linear component – the terms for $x_p(t)$ – is of course the one corresponding to the main signal and is estimated and canceled using the algorithm discussed in the previous section. In this section, we focus only on the higher-order odd-power terms. We can therefore reduce the model and write it in the digital baseband domain as:

$$y(n) = \sum_{m \in odd\ terms, n=-k,\ldots,k} x(n)\left(\left|x(n)\right|\right)^{m-1} * h_m(n)$$

where $h_m[n]$ is the weight for the term that raises the signal to order m and is the variable that needs to be estimated for cancelation, and k is the number of samples in the past and future that significantly influence the value of the signal at instant n.

To estimate these coefficients, we can use the same WiFi preamble. The WiFi preamble is two OFDM symbols long of length $8\mu s$, and assuming a sampling rate of 160 MHz, it consists of a total of 1280 digital samples at the Nyquist sampling rate. However, if we look at the previous equation, the number of variables $h_m(n)$ that we need to compute is a function of $2k$ (that is, how far in the past and future the influences on the current self-interference signal are) and the highest value of m that exhibits strength greater than the receiver noise floor. A naïve model assuming that:

- just the 1, 3, 5, 7, 9, 11th order terms matter
- up to 128 samples from both the future and the past influence the self-interference signal at any instant.[2]

would require us to estimate $128 * 2 * 6 = 1536$ variables using 1,280 equations. Clearly, this under-determined system would increase the noise floor significantly.

In practice we found empirically that many of these variables do not matter; that is, their values are typically zero. The reason is that higher-order terms have correspondingly lower power since they are created by the mixing of multiple lower-order terms and each mixing reduces power. So the 7th-order term has lower power than the 5th-order term, which has lower power than the 3rd-order term. Figure 16.5 shows a plot of the strength of the main signal and higher-order non-linear terms relative to the receiver noise floor. As we can see, higher-order terms have weaker strength relative to the main signal, and consequently their multipath components also decay quickly below the receiver noise floor. In other words, far fewer than 128 samples from the past and future impact the value of the self-interference harmonic component at this instant. We find empirically that for indoor WiFi systems, across all the non-linear higher orders, a total of only 224 such variables is needed, which we can easily accomplish using the WiFi preamble (an over-determined system of linear equations). Hence our digital cancelation algorithm calculates all these coefficients using the WiFi preamble and applies them to recreate the harmonics and cancel them. The method for

[2]The number of samples required is a function of the amount of multipath; the higher the multipath, the higher the number of samples in the past and future that matter. However, 128 is the number suggested by the WiFi standard and is equal to the length of the WiFi OFDM cyclic prefix.

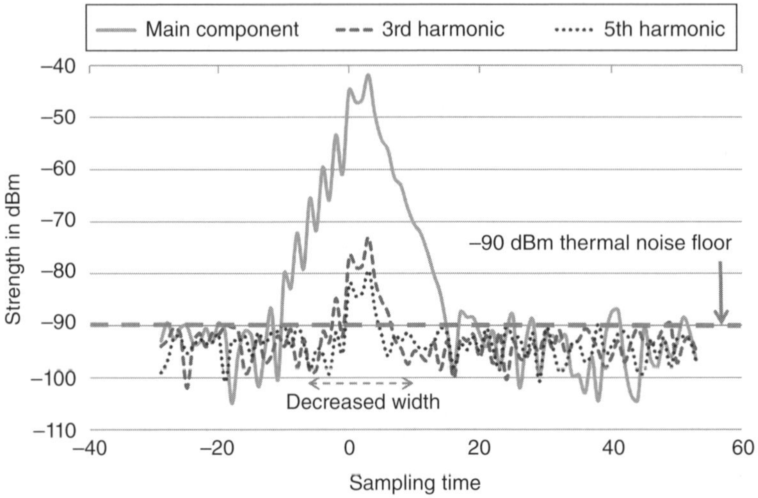

Figure 16.5 Signal strength of various harmonics that make up the transmitted signal. Note that higher order harmonics are much weaker relative to main component and, therefore, any reflections of these harmonics have to be quite closely spaced in time for them to be stronger than the receiver noise floor.

estimating the coefficients is the same as the one used in the linear digital cancelation step, but the matrix A is formed using the higher order odd powers of the preamble samples.

16.2.2.3 Complexity

The complexity of digital cancelation is the same as solving 1,280 (say W, the width of preamble in general) linear equations with 224 unknowns. The matrix that forms the linear equations is known in advance (this is the known preamble trick, as discussed earlier). Hence the pseudo-inverse of this matrix can be pre-computed and stored. Thus the complexity of digital cancelation reduces to $O(W)$ multiplications. The design is therefore relatively simple to implement and can be efficiently realized in hardware. The incremental gains from increasing n beyond 3–4 are small and expensive in terms of implementation.

16.2.3 Dynamic Adaptation of Analog Cancelation

To provide a robust full-duplex link, we need to ensure that sufficient cancelation is maintained to reduce self-interference to the noise floor, even as things such as environment, transmit power, and temperature change. These changes would clearly reduce the cancelation achieved by any static configuration, since they change the distortions that are imposed by the self-interference. Digital cancelation can cope, since it essentially estimates these distortions on a per-packet basis, but analog cancelation might be degraded and hence performance might be worsened. In this section, we describe how we can quickly tune the analog circuit to provide the required amount of cancelation (60 dB at least).

The goal of tuning is to pick the attenuation values $a_1, \ldots, a_N$ such that self-interference is minimized. More formally,

$$\min_{(a_1,\ldots,a_N)} \left(y(t) - \sum_{i=1}^{N} a_i c(t - d_i) \right)^2$$

where $c(t)$ is the reference signal that is tapped from the transmit path, $y(t)$ is the self-interference, and $d_1, \ldots, d_N$ are delays associated with the taps, as shown in Figure 16.3.

A simple and obvious technique to solve this problem in practice is an iterative gradient descent algorithm, which other prior works in full duplex have also used to tune their analog cancelation [1]. However, we found that this algorithm is extremely slow (it requires nearly 40 ms) because of the larger number of variables (16) that need to be estimated in our design, unlike prior work. That is an unacceptable overhead, since we found empirically that we need to re-tune the analog cancelation once every 100 ms on average. So taking 40 ms to tune implies a 40% overhead.

Our key contribution here is an approach that solves the tuning problem in the frequency domain. The idea is that the self-interference $y(t)$ can be modeled in the frequency domain as a function of the tapped signal $c(t)$ as

$$\boldsymbol{Y}(f) = \boldsymbol{H}(f)\boldsymbol{C}(f)$$

where $\boldsymbol{H}(f)$ is the frequency-domain representation of the distortion introduced by the circulator, antenna, and the environment, and $C(f)$ is the frequency-domain representation of the tapped signal. Recall that the tapped signal is essentially a scaled replica of the transmitted signal input to the circulator. The above equation can therefore be written in terms of the tapped signal. This frequency response $\boldsymbol{H}(f)$ is easier to measure; it is essentially an FFT of the self-interference channel, which can be measured using the WiFi preamble. In fact, standard OFDM is doing exactly this: it is estimating the frequency domain channel using the preamble and pilot symbols.

The goal of the optimization problem then is to pick the attenuator values such that the overall frequency-domain response of the cancelation circuit approximates $\boldsymbol{H}(f)$ as closely as possible. So the optimization problem can be restated as

$$a_1,\ldots,a_N \min \left(\boldsymbol{H}(f) - \sum_{i=1}^{N} H_i^{a_i}(f) \right)^2 \qquad (16.1)$$

where $H_i^{a_i}(f)$ is the frequency response for delay line i for an attenuation setting of a_i.

How might we solve this problem? The problem is twofold. First, we have to find the frequency response of each delay line of the cancelation circuit for every attenuation value: $H_i^{a_i}(f)$. Second, once we have the frequency response of the self-interference channel $\boldsymbol{H}(f)$, we need to search on the space of possible attenuation values for every delay line (attenuator), to come up with best possible solution to the optimization problem. Each delay line can take 128 different attenuation values, and there are 16 delay lines, so in total we have $128^{16} = 2^{112}$ values, a computationally expensive search.

16.2.3.1 Modeling the Frequency Response of Delay Lines $H_i^{a_i}(f)$

$H_i^{a_i}(f)$ Measuring the frequency response of an individual delay line is impossible: the entire circuit is well connected, thus isolating an individual delay line is impossible. Our key observation, is that if we can measure the frequency response of a delay line at one attenuation value, then the datasheet of the attenuator provides measurements called S parameters (specifically frequency-response measurements between different ports of a device) that can be used to extrapolate the frequency response of the delay line for all attenuation values. The S-parameter data provides the relative change in frequency response with changing attenuator value. To calculate the frequency response at this initial point, we use the following trick. We set the attenuators for all the lines to their highest attenuation setting, except the one being measured. The idea is to essentially emulate a board where none of the delay lines, except the one being measured, let any signal through. The highest attenuation value approximates that setting but does not fully accomplish it, so we apply a second least-squares fit to find a more accurate response (collecting more data for different attenuations for this delay line, keeping the all others at the highest attenuation setting). Then the frequency response of this delay line for all 128 attenuator values can be calculated. We repeat this process for all the delay lines in the circuit. Note that all of this has to be done once and can be stored, since this frequency response of the delay line and attenuation is independent of the environment or other such changing parameters.

16.2.3.2 Optimization Algorithm

To actually find the attenuation settings in real time to optimize the cancelation, we use the following algorithm.

1. Measure the frequency response of the self-interference $\boldsymbol{H}(f)$ using the WiFi preamble. This is relatively simple since we have two OFDM symbols and as part of the baseband decoding we can perform an FFT to measure the frequency response.
2. Solve the frequency-domain integer linear optimization problem posed in Eq. (16.1) by relaxing it to a linear program and then use random rounding to find a solution for attenuator settings, which achieves required cancelation of 60 dB.

The intuition behind the algorithm is that it reduces the search space of attenuator values to a polynomial set rather than an exponential search space. This is due to the fact that we are looking for a point that provides the required cancelation, instead of the optimal point (achieving the optimal point is a NP-hard problem). Note all the aforementioned calculations are offline and are implemented using the frequency-response model. Essentially the model is used for looking up the frequency response of the circuit, for any combination of attenuator values. This offline algorithm implementation is therefore extremely fast; a non-optimized C++ implementation takes less than 1μs to converge.

In practice, we find the offline solution might yield a point that provides an analog cancelation of only 45–50 dB, due to manufacturing variations of attenuator (the S-parameter data provided is accurate to 2%, so every attenuator has its own response, different from the provided standard data. To further improve the cancelation, we use an additional gradient-descent step. Typically, gradient descent takes several hundred iterations, but here, since we are starting the descent from a much more accurate starting point, the gradient descent converges to the required point in 10–12 iterations. So in the worst case, we show experimentally that analog cancelation tuning can take around 900–1000 μs. Assuming we have to do such tuning once every 100 ms, which is what we needed in our testbed, that represents less than a 1% overhead for tuning.

16.3 Implementation

Figures 16.6 shows the prototype of a single-antenna full-duplex radio. To implement it we designed our own analog circuit boards for cancelation and integrated them with existing software radios. We also implemented the digital cancelation algorithms in the software radio. Next, we discuss the different parts.

Analog Cancelation Board. The analog cancelation board is a 10×10 cm PCB board designed and built using Rogers 4350 material. The fixed delay lines are implemented using micro-strip trace lines of different fixed lengths. The attenuators are programmable step Peregrine PE43703 [19] attenuators that can be programmed in steps of 0.25 dB from 0 to 31.5 dB for a total of 128 different values.

Radio Transceiver and Baseband. Our goal was to design and implement a full-duplex system that was capable of supporting the latest WiFi protocol 802.11ac with least 80 MHz of bandwidth in the 2.4 GHz range and 20 dBm average TX power. Unfortunately, none of the widely used software radios, such as USRPs or WARPs (Figure 16.6b) support such high performance; at best they are capable of supporting 20 MHz bandwidths. For that reason, we also prototyped our design using radio test equipment from Rohde and Schwarz. For our transmitter, we used an SMBV 100A vector signal generator [10] to send our desired WiFi signals. Since the SMBV is not capable of generating 20 dBm power, we use an external power amplifier [20]. For the receiver, we use the RS spectrum analyzer [9].

A practical concern is how to kick-start re-tuning of analog cancelation. Specifically, if analog cancelation drops below a threshold, then the receiver might get saturated and the feedback needed to tune is distorted. To tackle this, we implemented an automatic gain control via a digital tunable attenuator in front of the low-noise amplifier. The idea is that if the baseband detects that the receiver is getting saturated, then it programs the attenuator to a large value, which brings the whole signal down to within the dynamic range. After cancelation is tuned, this attenuation is turned off. The RS spectrum analyzer is capable of receiving 100 MHz signals at 2.45 GHz, down-converting and digitizing it to baseband, and then giving us access to the raw IQ samples, which we can then freely process using our own baseband algorithms. The noise floor of this receiver is −90 dBm at 100 MHz bandwidth. It has a 16-bit ADC capable of sampling a 100-MHz signal, but to ensure that we are only using resources found in commodity WiFi cards, we configure the ADC only to use 12 bits of resolution.

The IQ samples are transported via ethernet to a host PC, on which we implement our cancelation and baseband software. We implemented a full WiFi-OFDM PHY that could be configured to operate over all the standard WiFi bandwidths (20 MHz, 40 MHz, and 80 MHz). We support all the WiFi constellations from BPSK to 64QAM for 40 MHz, and 256QAM for 80 MHz. We also support all the channel codes with coding rates (1/2, 2/3, 3/4, and 5/6

(a)

(b)

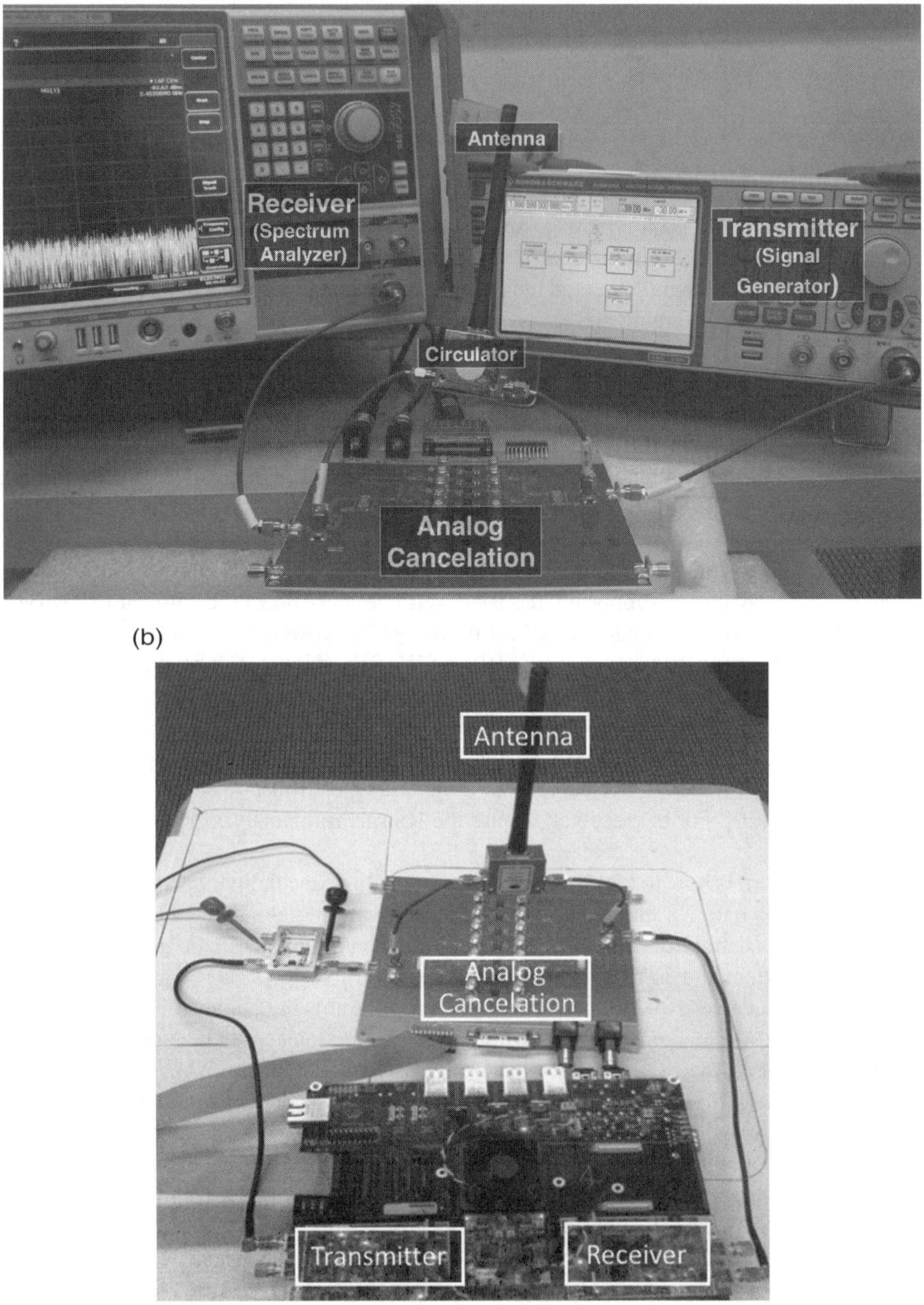

Figure 16.6 Experimental set-up of single-antenna full-duplex radio: (a) using specialized equipment; (b) using off-the-shelf WARP software-defined radio.

convolutional coding). Finally, we also implement our digital cancelation algorithm in software on the same host PC.

However, to show that our design is general and does not benefit from using expensive test equipment, we also develop an implementation using standard WARP radios as shown in Figure 16.6(b). Due to their radio limitations, these results will be for 20 MHz signals, which is the widest that the WARP supports.

16.4 Evaluation

In this section we show experimentally that our design delivers a complete full-duplex WiFi PHY link. We prove the claim in two stages. First, we show that our design provides the 110 dB of self-interference cancelation required to reduce interference to the noise floor. We also show experimentally that the received signal is received with almost no distortion in full-duplex mode (the SNR of the received signal is reduced by less than 1 dB on average), and that these results are consistent across a wide variety of bandwidths, constellations, transmit powers, and so on. Second, we take the next step and design a working full-duplex communication WiFi link. We show experimentally that it delivers close to the theoretical doubling of throughput expected from full duplex.

We start with an experimental evaluation of the cancelation system. We define two metrics we use throughout this section:

Increase in Noise Floor. This is the residual interference present after the cancelation of self interference, which manifests itself as an increase in the noise floor for the received signal. This number is calculated relative to the receiver noise floor of the radio of –90 dBm. For example, if after cancelation we see a signal energy of –88 dBm, it would imply that we increased the noise floor by 2 dB.

SNR Loss. This is the decrease in SNR experienced by the received signal when the radio is in full-duplex mode due to any residual self-interference left after cancelation. To compute this we first measure the SNR of the received signal when the radio is in half-duplex mode and there is no self-interference, and then with full-duplex mode. The difference between these two measured SNRs is the SNR loss.

We compare our design against two state-of-the-art full-duplex systems presented in prior work.

Balun Cancelation. This design [1] uses a balun transformer to invert a copy of the transmitted signal, adjusts its delay and attenuation using programmable attenuators and delay lines, and cancels it. The design also uses two antennas separated by 20 cm, one each for TX and RX, which automatically provides 30 dB of self-interference reduction. We implement this design and optimize it.

Extra TX Chain (Rice Design). This design uses an extra transmit chain in addition to the main transmit chain. The extra chain generates a cancelation signal that is combined with the signal on the receive chain to cancel self-interference. This design also uses two antennas and to make a fair comparison we use a 20-cm separation as in the balun-based design. However, we also provide results with 40-cm separation, since that was the value used in the prior work. We implement this design using an extra signal generator as an extra transmit chain for cancelation.

Note that our design uses a single antenna and therefore does not have the benefit of the 30 dB of self-interference reduction that prior schemes enjoy from using two physically separate antennas.

16.4.1 Can We Cancel all of the Self-interference?

The first claim we made in this chapter is that our design is capable of canceling all of the self-interference for the latest operational WiFi protocols. To investigate this assumption, we experimentally test if we can fully cancel a 80 MHz WiFi 802.11ac signal up to a maximum transmit power of 20 dBm – all of which are the standard parameters used by WiFi access points(APs) – as well as the smaller bandwidths of 40 MHz and 20 MHz. We conduct the experiment by placing our full-duplex radio in different locations in our building. We also increase the transmit power from 4 dBm to 20 dBm (a typical transmit power range). For each TX power and location (in total 100) we conduct 20 runs and compute the average cancelation across those runs and locations. The goal is to show that we can cancel to the noise floor for a variety of transmit powers up to and including the maximum average TX power of 20 dBm. Figure 16.7 plots the average cancelation as a function of TX power. It also plots the corresponding observed increase in noise floor on the other axis.

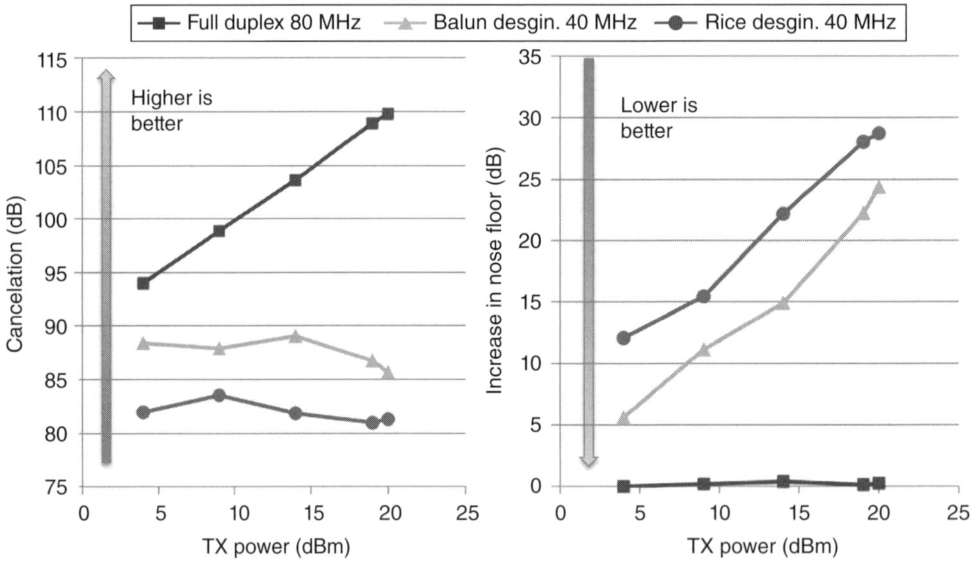

Figure 16.7 Cancelation and increase in noise floor vs TX power for different cancelation techniques with transmission of WiFi 802.11 signal. Our full-duplex system can cancel to the noise floor standard WiFi signals of **20** dBm at highest WiFi bandwidth of **80** MHz, while prior techniques still leave **25** dB of self-interference residue, even for the narrower bandwidth of **40** MHz.

Figure 16.7 shows that our design essentially cancels the entire self-interference almost to the noise floor. In the standard case of 20-dBm transmit power, the noise floor is increased by at most 1 dB over the receiver noise floor. The amount of cancelation increases with increasing TX power, reaching the required 110 dB for the 20 dBm TX power. The takeaway is that as the TX power increases, self-interference increases at the same rate and we need a correspondingly larger amount of cancelation, which our design provides.

Note that these are average cancelation numbers; in practice our WiFi transmissions exhibit transient peak-to-average power ratio as high as 10 dB, so the peak transmit power we see is around 30 dBm. We do not report the specific numbers for these due to lack of space, but our cancelation system scales up and also cancels these temporary peaks in the self-interference signal to the noise floor.

The prior balun and extra-TX designs, however, fare far worse. Further, since these designs perform very poorly at 80 MHz, we only report their results for the smaller 40 MHz WiFi bandwidth and 20 dBm TX power. As we can see, these designs can at best provide 85 dB and 80 dB of cancelation, respectively. In other words they increase the noise floor by 25 dB and 30 dB, respectively. The reasons for this are the ones we discussed in Section 16.1.2: the inability to adequately cancel transmitter noise in analog and the inability to model non-linear distortions produced by radios. To check if these designs could be made to work with larger antenna separations, we repeated the experiment with an antenna separation of 40 cm instead of 20 cm. We found that even with an impractical rough half-meter separation in antennas, the noise floor increase is at least 20 dB.

16.4.1.1 Does Our Design Work with Commodity Radios?

We repeat the previous experiment, but instead of the Rohde–Schwarz test equipment, we use off-the-shelf WARP radios in the setup. The goal is to show that our design can work with cheap commodity radios and does not depend on the precision of the test equipment. Since the widest bandwidth that the WARP can support is 20 MHz, we only report results for that bandwidth. Figure 16.8 shows the spectrum plot of canceled signals after different stages of cancelation. For comparison, we also show the spectrum plot of cancelation using the Rohde–Schwarz equipment.

As can be seen, our cancelation completely eliminates self-interference even with commodity WARP radios. The WARP has a noise floor of –85 dBm, worse than the –90 dBm of the RS equipment. Hence if we used 20-dBm transmit power, then a slightly smaller 105 dB of self-interference cancelation is required to eliminate it to the noise floor. However, for consistency, for the WARP experiments we increase the transmit power to 25 dBm to show that our design can still achieve 110 dB of cancelation and eliminate self-interference to the noise floor.

16.4.1.2 SNR Loss of the Received Signal in Full-duplex Mode

The previous section provided evidence for the amount of cancelation and increase in noise floor. However, the experiments had only one radio transmitting. A natural question is how well the system works when we are in true full-duplex mode; in other words, when the radio is transmitting and simultaneously receiving a signal. In this section, we evaluate the SNR loss for the received signal when operating in full-duplex mode.

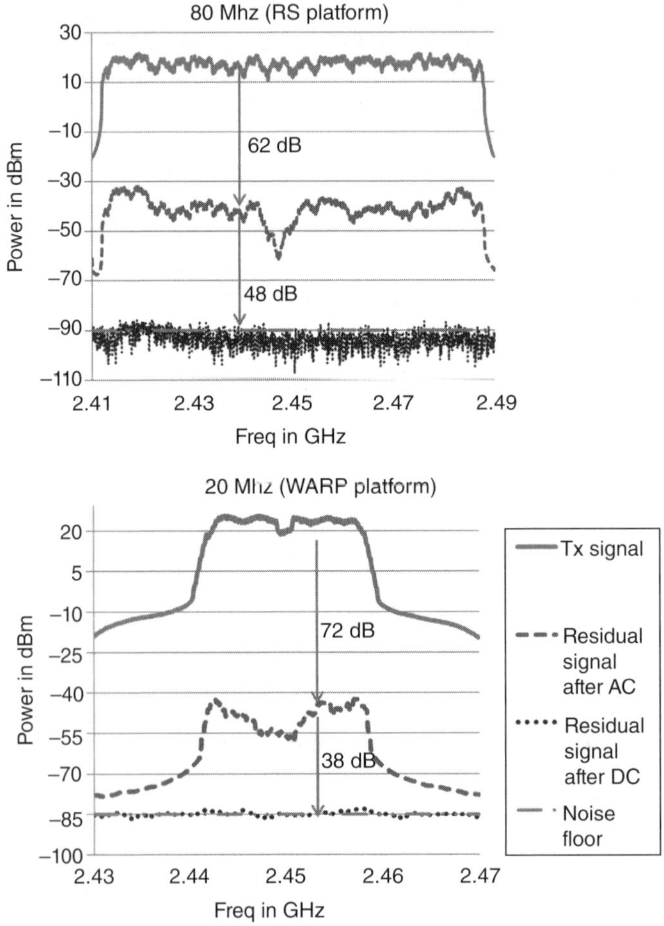

Figure 16.8 Spectrum response for our cancelation with the Rohde–Schwarz (RS) radios and the WARP radios. The figure shows the amount of cancelation achieved by different stages of our design. It also shows that our design provides the same 110 dB of cancelation even with WARP radios.

The experiment is conducted as follows. We setup two nodes capable of full-duplex operation in our building. The two nodes first send 20 WiFi packets – with the following PHY parameters: 80 MHz bandwidth, 20 dBm TX power, 64QAM constellation – to each other one after the other; in other words, they take turns and operate in half-duplex mode. They then send 20 WiFi packets to each other simultaneously; that is, they operate in full duplex mode. For each run we measure the average SNR of the received packets across the 20 packets in half-duplex mode, and then with full-duplex mode. We then compute the SNR loss, which is defined as the absolute difference between the average half-duplex SNR and full-duplex SNR measured previously. We repeat the experiment at several different locations of the two nodes in our testbed. We plot the SNR loss as a function of the half-duplex SNR in Figure 16.9.

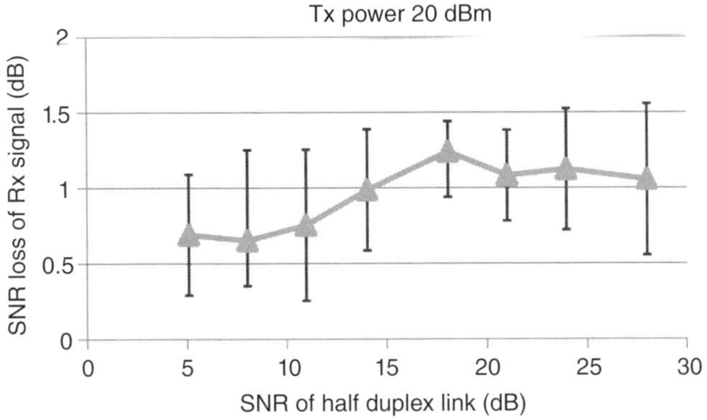

Figure 16.9 SNR loss versus half duplex SNR at fixed TX power. TX power = 20 dBm, constellation = 64QAM, bandwidth = 80 MHz, transmission of WiFi 802.11 signal. Our full-duplex system ensures that the received signal suffers negligible SNR loss regardless of the SNR it was received at.

As Figure 16.9 shows, the SNR loss is uncorrelated with the half-duplex SNR value and is almost identical to the increase in noise-floor value we saw in the previous experiment. The takeaway is that self-interference cancelation is not impacted by the received signal's strength, whether it is weak or strong. Further, the SNR loss is typically around or less than 1 dB, which implies that even in full-duplex mode the received signal should retain almost the same throughput as in clean half-duplex mode.

16.4.2 Digging Deeper

16.4.2.1 Impact of Constellation and Bandwidth

We conduct two experiments. First we use the same setup as the SNR loss experiments and fix the bandwidth to 80 MHz, but vary the constellation for the transmitted signal for the full-duplex node from QPSK to the densest constellation in WiFi 256QAM. Once again we calculate the SNR loss of the received signal across different measurements and locations from the half-duplex node. In the second experiment we fix the constellation to 64QAM but vary the bandwidth from 20 to 40 to 80 MHz and once again calculate the SNR loss of the received signal. We repeat this experiment for different locations of the two nodes. Figure 16.10 plots the cumulative distribution functions (CDFs) of the SNR losses for different choices of constellations and bandwidth.

As the figures show, our design performs consistently well for all constellation choices and bandwidths. Our cancelation technique makes no assumptions about what constellation and other parameters the PHY is using: for us, all of them are self-interference signals and hence the design is unaffected by constellation choice. Our design also works equally well for all the bandwidths used by 802.11ac in the 2.4 GHz band. The reason is that our analog cancelation, as we will show in the next section, has sufficient flexibility to provide an almost flat wideband cancelation, while prior designs are extremely narrowband and cancelation tapers off quickly with wider and wider bandwidths.

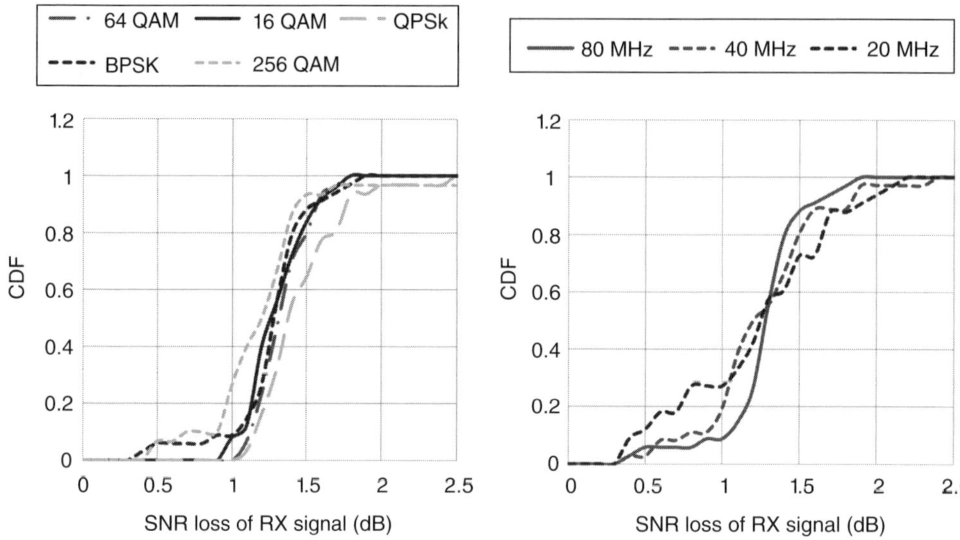

Figure 16.10 Shows CDF of SNR loss with changing bandwidths and constellations. Left, SNR loss for different constellations with TX power = 20 dBm and bandwidth = 80 MHz; right, SNR loss for different bandwidths (20 MHz, 40 MHz, and 80 MHz) for TX power = 20 dBm and constellation = 64QAM. We can support all WiFi modulation schemes and bandwidths with low SNR loss.

16.4.2.2 Deconstructing Analog Cancelation

In this section we dig deeper into the analog cancelation component of our design. The key parameter in our analog cancelation circuit board is the number of fixed delay lines, as discussed in Section 16.2.1. We conduct an experiment to examine the impact of the number of such lines. However, since these are circuit boards, we do not have the flexibility to vary the number of lines in increments of one. The granularity of our board design allows us to only test two configurations: one with 8 lines and one with 16 lines. We conduct the same self-interference cancelation experiment as described in Section 16.4. We measure the signal after analog cancelation (without digital cancelation) and plot the frequency response of the canceled signal for the two cases (see Figure 16.11). The plot should be read as the power of the self-interference signal after analog cancelation as a function of the frequency.

As Figure 16.11 shows, with 8 lines we can achieve 45 dB of cancelation over 80 MHz, while we can achieve 63 dB of cancelation with 80 MHz. The reason for the difference is the higher capability of 16 lines in canceling signal reflections in addition to the main self-interference component that is leaking through the circulator. When the full-duplex node is transmitting, the response from the circulator and antenna in the RX chain has two primary leakage components from the TX signal: one due to the direct leakage from the TX port of the circulator to the ("isolated") RX port of the circulator, and one due to reflections from impedance mismatch between the circulator and the antenna. Because these two components travel different paths in the circulator from TX port to RX port, they undergo different delays, as deduced from time-domain measurements. These delays are fixed and are a function of the particular circulator and antenna we choose to use. In our implementation we find the delay of the direct leakage component is 400 picoseconds, while the reflected component is centered

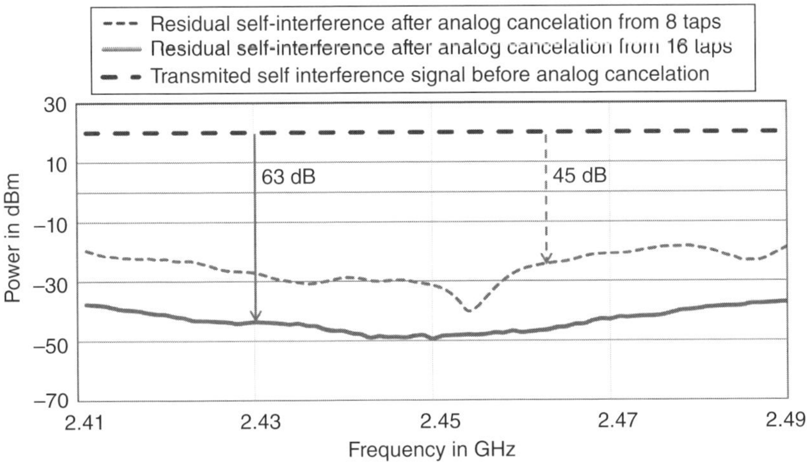

Figure 16.11 Frequency-domain representation of self-interference before and after analog cancelation using 8 taps and 16 taps. Note that with 16 taps we can provide at least 63 dB of analog cancelation over the entire 80 MHz of bandwidth.

around 1.4 nanoseconds. With 16 lines we have the capability to center the first 8 lines to have delays around 400 picoseconds, and the other 8 lines around 1.4 nanoseconds. We can then use the interpolation trick discussed in Section 16.2.1 to cancel both the direct and reflected self-interference components precisely. As expected, with 8 lines our flexibility is reduced in terms of placing our delay lines around the actual delays experienced by the self-interference and consequently cancelation is reduced.

16.4.2.3 Deconstructing Digital Cancelation

After 62 dB of analog cancelation, digital cancelation needs to clean up 48 dB and 16 dB of linear and non-linear self-interference components, respectively. In this section, we deconstruct the amount of linear and non-linear cancelation achieved by our design. To conduct this experiment, we tune our analog cancelation circuit to provide 62 dB of cancelation. We then progressively add more components to our digital cancelation design. We first implement only our "linear" digital cancelation, which cancels only the linear main self-interference components and multipath reflections from the environment. We then add the capability to model non-linear components, which we name "non-linear cancelation". We calculate the cancelation achieved by these two variants of digital cancelation techniques. For comparison with prior work, we also implement only the digital cancelation technique described in the balun-based design [1]. We plot the increase in noise floor for all the techniques as a function of transmit power (see Figure 16.12).

As we can see, our full digital cancelation technique cancels everything to the receiver noise floor. Moreover, notice that, on its own, our linear digital cancelation stage leaves 16 dB of self-interference residue above the receiver noise floor. Being able to model the non-linear harmonics allows us to reduce self-interference by a further 16 dB and cleans out the non-linear distortions almost to the receiver noise floor. In comparison, the earlier digital cancelation

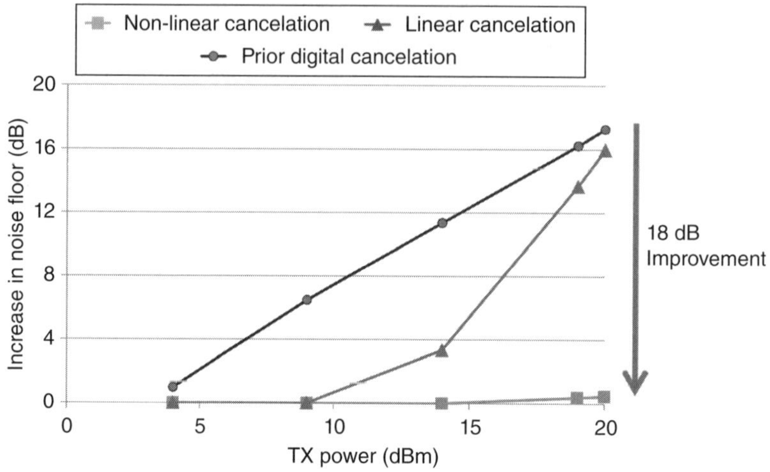

Figure 16.12 Performance of digital cancelation showing impact of different components of the algorithm vs TX power with fixed constellation = 64QAM, bandwidth = 80 MHz. Our algorithm cancels the main component, reflections, and harmonics, thus ensuring that self-interference is completely eliminated, and the increase in noise floor is less than 1 dB. Earlier techniques can not cancel harmonics, and therefore increase the noise by 18 dB.

technique falls far short, leaving nearly 18 dB of self-interference residue over the noise floor, since it cannot model non-linear distortions. Note that we have given this earlier work the benefit of an analog cancelation of 62 dB from our circuit; as we saw in Section 16.4.1, if we used their implementation of analog cancelation the numbers are worse.

16.4.2.4 Dynamic Adaptation

As environmental conditions change, the level of cancelation drops since the values of the attenuators used will be off with respect to the new conditions. In this section, we evaluate how long it takes to re-tune analog cancelation, as well as how often it needs to be re-tuned in our indoor environment. Note that digital cancelation is tuned on a per-packet basis, hence it is not a concern. Analog cancelation has to be tuned via a special tuning period during which no data is transmitted. Quantifying that overhead is therefore important.

We conduct this experiment in our busy indoor environment with other WiFi radios and students moving around. Note that an indoor environment is the worst-case scenario for full duplex, because of the presence of a large number of reflectors near the transmitter. Outdoor LTE scenarios are less likely to have such strong near-field reflectors, hence we believe our design extends relatively easily to outdoor LTE scenarios. We place the full-duplex node and conduct analog cancelation tuning as described in Section 16.2.3. Specifically, we use the WiFi preamble to determine the initial settings of the attenuators to be used to match the frequency response of the circulator and antenna. Next we run a gradient-descent algorithm to further improve the cancelation from that initial point. Each iteration of the gradient descent consumes 92 μs, since we have 16 different directions to compute the gradient on (corresponding to the 16 different attenuators). We compute the time it takes for the analog cancelation to

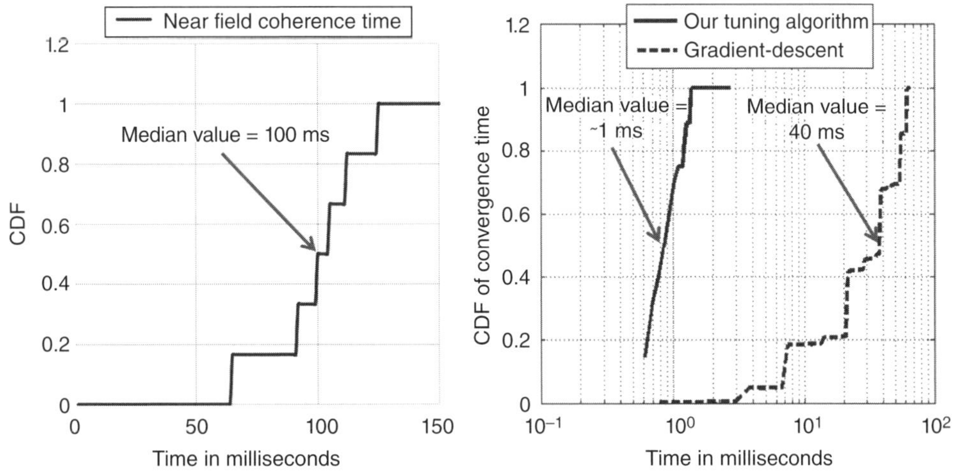

Figure 16.13 Retuning: left, CDF of near field coherence time, which implies that we have to retune analog cancelation on an average every 100 ms; right, time taken for our tuning algorithm and a gradient-descent algorithm to converge to the required cancelation, after the initiation of tuning. We observe an exponential improvement compared to the gradient descent algorithm, which takes an order of magnitude longer.

converge. We repeat this experiment several times for different node placements and environmental conditions and plot the average convergence time. We also conduct an experiment where we do not use the initial frequency-based tuning and only use gradient descent from a random starting point for the attenuator values. We show the cancelation achieved as function of tuning time on right-hand side of Figure 16.13.

As we can see from the right-hand side of Figure 16.13, our analog tuning converges in around 920 μs, compared to more than 40 μs for a pure gradient-descent approach. The reason is that the frequency-based initial point estimation is very close to optimal, and from so only a few gradient descent iterations are needed to find the optimal point. Our cancelation algorithm therefore tunes an order of magnitude faster than a simple gradient-descent approach.

But an important question is how often do we have to tune? Analog cancelation has to retuned when there is a change in the near-field reflections, since it cancels only the strong components (components 50 dB above noise floor; farther out reflections are weaker than this 50-dB threshold). Hence the question is, how often do the near-field reflections change? As expected, this depends on the environment. For the indoor office deployments we used in our experiments we found that we needed to retune once every 100 ms on average. Outdoor scenarios would be easier since changes in near field occur less frequently; we leave mobile hand-held scenarios to future work. We show the indoor results in Figure 16.13, where the left-hand plot shows the amount of cancelation observed as a function of time after we have found the optimal operating point from a large collection of different experimental runs in our testbed. We define the "near field coherence time" of analog cancelation as the time up to which the receiver remains unsaturated from when it was tuned, which we also use as the trigger to rerun the tuning algorithm. As we can see, the near field coherence time for the cancelation is roughly 100 milliseconds. In other words, we have to retune the analog cancelation once every 100 milliseconds, which leads to an overhead of less than 1%.

16.4.3 Does Full Duplex Double Throughput?

This section demonstrates experimentally that our design delivers close to the theoretically expected doubling of throughput for a full-duplex WiFi link. Note that this is a PHY-layer experiment; a full MAC design for full-duplex WiFi is beyond the scope of this chapter.

We conduct these experiments as follows. We place the two full-duplex nodes at different locations and send trains of 1,000 packets in full-duplex mode, and then similar trains for each direction of the half-duplex mode. Each train uses a particular bitrate (from WiFi) and we cycle through all the bitrates for each location. We pick the bitrate with the best overall throughput for full duplex, two antenna full duplex and half duplex, respectively. We repeat this experiment for different locations. We found that the SNRs of the links varied uniformly between 0 – 45 dB across locations, as we would find in a typical indoor deployment. We plot the CDF of the throughput for the half-duplex and full-duplex links in Figure 16.14. Note that all of these throughput numbers account for the overhead introduced by the periodic analog cancelation tuning. As we can see, our full-duplex system achieves a median throughput gain of 1.87 over the standard half-duplex mode. As we know from the experimental analysis in Section 16.4.1.2 that there is a small SNR loss due to a small amount of self-interference residue. This SNR loss is the reason that instead of the theoretical doubling, we see a slightly reduced gain of 1.87×.

How do earlier designs perform? We found that in 60% of the scenarios, the throughput with prior full-duplex techniques was zero. This is because these designs leave at least 25 dB of self-interference residue that acts as noise and if the link SNR is below 30 dB no signal is

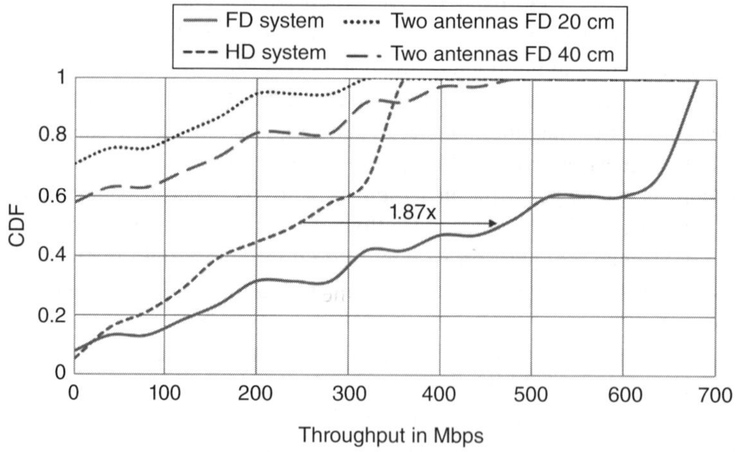

Figure 16.14 CDF of throughput for full-duplex link using TX power = 20 dBm, bandwidth = 80 MHz. We see a median gain of 87% using full duplex as compared to half duplex. An earlier full-duplex technique with two antennas separated by 40 cm shows gains only in 10% of cases compared to half duplex.

decoded (WiFi requires a minimum of 4–5 dB to decode even the lowest rate packet). As the half-duplex link SNR increases, performance improves but is still not sufficient to beat the system throughput achieved by half duplex. The reason is that even if the link half-duplex SNR is 35 dB, it implies that we only have two 10 dB links for full duplex. The throughput achieved with a single 35-dB half-duplex link is still higher than two 10-dB links. Consequently, the only region where we could find improvement for full duplex over half duplex with prior techniques was when the link SNR was greater than 40 dB.

16.5 Discussion and Conclusion

We believe this chapter shows that full duplex is not only possible, but also feasible and practical. Moreover, it can be deployed with no overhead in terms of antennas used and yet achieve the theoretical doubling of throughput. Next, we discuss the current design's limitations, potential avenues of future work and then conclude.

Size of Circuit. The current analog cancelation circuit is large; our prototype board measures 10×10 cm. Such a design is fine for APs and base-stations, which is our initial focus, but this design is not implementable on phones and other portable devices where size is at a premium. To realize full duplex on such devices, we need to design a radio frequency integrated circuit (RFIC) that is sufficiently small (at best 20–30 mm^2 for current phones). The key consumers of space on our circuit are delay lines, which we currently realize via traces on the board. For an RFIC we expect to use different techniques to realize the same delays, such as LC ladders and acoustic technologies such as SAW and BAW [11]. These techniques operate by slowing the speed of light, and thus true time delays are obtained in very short form factors that can be integrated on chips. However, this discussion is speculative and is part of our future work.

LTE. Our current prototype targets WiFi frequencies in the 2.4-GHz band. However, our prototype can also be used for the 2.3-GHz and 2.5-GHz LTE bands found in Asia and Europe. However, the general design of our system is frequency independent; the dependence in our prototype comes from the fact that several analog components in our cancelation board work only in specific frequency ranges (our tunable attenuators operate only between 2 and 2.6 GHz). However, the same design can be used for different frequencies, with corresponding components that work in those frequency ranges. Moreover, unlike WiFi, LTE uses smaller channels, the widest of which is 20 MHz and this makes the cancelation problem somewhat simpler. Hence we believe our current design can be adapted to work with LTE and this remains work for the future.

References

[1] M. Jain, J. I. Choi, T. Kim, D. Bharadia, S. Seth, K. Srinivasan, et al. 2011 Practical, real-time, full duplex wireless. In *MobiCom '11*, pp. 301–312.

[2] J.I. Choi, M. Jain, K. Srinivasan, P. Levis, and S. Katti 2010 Achieving single channel, full duplex wireless communication. In *Proceedings of the Sixteenth Annual International Conference on Mobile Computing and Networking, MobiCom '10*, pp. 1–12.

[3] E. Everett, M. Duarte, C. Dick, and A. Sabharwal 2011 Empowering full-duplex wireless communication by exploiting directional diversity. In *Signals, Systems and Computers (ASILOMAR), 2011 Conference Record of the Forty Fifth Asilomar Conference on*, pp. 2002–2006.

[4] M. Duarte and A. Sabharwal 2010. Full-duplex wireless communications using off-the-shelf radios: Feasibility and first results. In *Forty-Fourth Asilomar Conference on Signals, Systems, and Components.*
[5] B. Radunovic, D. Gunawardena, P. Key, A. Proutiere, N. Singh, V. Balan, and G. Dejean 2010 Rethinking indoor wireless mesh design: Low power, low frequency, full-duplex. In *Wireless Mesh Networks (WIMESH 2010), 2010 Fifth IEEE Workshop on*, pp. 1–6.
[6] D.W. Bliss, P.A. Parker, and A.R. Margetts 2007 Simultaneous transmission and reception for improved wireless network performance. In *Proceedings of the 2007 IEEE Workshop on Statistical Signal Processing.*
[7] M. Duarte, C. Dick, and A. Sabharwal 2011 Experiment-driven characterization of full-duplex wireless systems. *CoRR*, abs/1107.1276.
[8] S.S. Hong, J. Mehlman, and S. Katti 2012 Picasso: flexible RF and spectrum slicing. In *Proceedings of the ACM SIGCOMM 2012 Conference on Applications, Technologies, Architectures, and Protocols for Computer Communication, SIGCOMM '12*, pp. 37–48.
[9] Rohde & Schwarz 2012 FSW Signal and spectrum analyzer user manual.
[10] Rohde & Schwarz 2012 SMBV 100A signal generator user manual.
[11] T. Lee 2004 *The Design of CMOS Radio-Frequency Integrated Circuits*. Cambridge University Press.
[12] J. Bardwell 2005 WiFi radio characteristics and the cost of WLAN implementation. URL: http://www.connect802.com/download/techpubs/2005/commercial_radios_E0523-15.pdf.
[13] S. Gollakota, H. Hassanieh, B. Ransford, D. Katabi, and K. Fu 2011 They can hear your heartbeats: non-invasive security for implantable medical devices. In *SIGCOMM Comput. Commun. Rev.*
[14] US Patent 5444864. URL: http://www.google.com/patents/US5444864.
[15] US Patent 6539204. URL: http://www.google.com/patents/US6539204.
[16] A.V. Oppenheim, R.W. Schafer, and J.R. Buck 1999 *Discrete-time Signal Processing*, 2nd edn. Prentice-Hall.
[17] S. Boyd and L. Vandenberghe 2004 *Convex Optimization*. Cambridge University Press.
[18] L. Ding 2004 *Digital Predistortion of Power Amplifiers for Wireless Applications*. PhD Thesis, Georgia Institute of Technology.
[19] Peregrine Semiconductor 2009 PE47303 datasheet. URL: www.psemi.com/pdf/datasheets/pe43703ds.pdf (accessed August 10, 2016).
[20] Mini-Circuits 2012 Monolithic power amplifier PGA105+. URL: www.minicircuits.com/pdfs/PGA-105+.pdf.

17

Point to Multi-point, In-band mmWave Backhaul for 5G Networks

Rakesh Taori and Arun Sridharan
Samsung Research America, Mountain View, CA, USA

17.1 Introduction

The next generation of cellular systems (5G) is expected to meet the rapidly growing demands for data through densification of networks using high-capacity small cells [1]. As networks become dense (through small cell deployments):

- the cost of bringing fiber to every small cell becomes prohibitively expensive [2]
- inter-cell interference increases and becomes a limiting factor in achieving higher cell capacities [3]
- handover is more frequent, causing the associated handover signaling overhead to increase [4].

Using wireless technologies to lower the cost of wiring (digging/trenching to lay wire/fiber in the ground) a cell, for instance by deploying point-to-point microwave links, is already a well-accepted approach [5]. However, even to meet the 4G cellular data rate requirements, the number of point-to-point microwave links is already very high [6, 7] and 5G networks are likely to be even more dense. The first identifiable issue therefore is a cost-effective method for connecting 5G base stations (BSs) to the network (BS–to-network communication).

Inter-cell interference management necessarily involves inter-BS communication. Extensive studies in 3GPP working groups has revealed that latency is a particularly sensitive issue for the achievable performance of cell-edge throughput and/or interference mitigation schemes

Towards 5G: Applications, Requirements and Candidate Technologies, First Edition.
Edited by Rath Vannithamby and Shilpa Talwar.

Companion website: www.wiley.com/go/vannithamby/towards5g

such as coordinated multi-point (CoMP) [8]. The second issue in densification is therefore low-latency inter-BS communication (BS-to-BS communication).

The third issue in network densification is reducing the number of handovers. To reduce the handovers in a dense small-cell network, ongoing studies consider user-centric virtual cells such as cloud cells [9] requiring very low-latency communication between BSs (BS-to-BS communication).

In summary, providing low latency and cost-effective backhauling mechanisms for connecting 5G BSs to each other and to the network forms the key challenge for the dense deployment of 5G networks.

Here, we consider a point-to-multipoint (PMP) architecture as the basis for a scalable approach to fundamentally address the cost and latency issues described previously. In particular, a PMP approach will lower:

- per-link hardware costs: compared to a dedicated point-to-point link approach, savings in the annual total cost of ownership (TCO) of up to 50% [7]
- inter-BS communication latency: BSs with PMP capability can communicate with multiple BSs directly, creating a mesh topology where in *not all* the links are simultaneously active, but *any* link can be established on demand (cost effective compared to n^2 dedicated links to interconnect n BSs).

To further lower the costs, we consider an in-band approach. The term "in-band" here means that the access link (BS to mobile station (MS) link) as well as the backhaul link (BS-to-BS links or BS to access gateway links) are multiplexed on the same frequency band. The advantages of an in-band approach are:

- cost to the operator of buying separate frequency licenses for backhauling is eliminated (facilitates spectrum reuse)
- enables a single radio unit to serve the backhaul link as well as the access link (facilitates hardware reuse).

In the current wireless backhaul market [6,7], out-of-band solutions are prevalent. Even with the higher efficiencies offered by the 4G/LTE (advanced) system, out-of-band solutions are chosen due to the extreme capacity crunch experienced by the operators in the expensively licensed frequencies used for access [10, 11]. At mmWave frequencies, however, much wider channel bandwidths can be deployed, and an in-band backhaul approach becomes particularly attractive. The combination of availability of large bandwidths, advanced RF-beamforming capabilities using high-gain advanced antennas as well as MIMO digital beamforming makes the mmWave cellular system viable for an in-band backhaul solution. Accordingly, in this paper, we will consider the use of mmWave frequency bands for cellular access [12] and backhaul.

This paper is organized as follows. In Section 17.2 we provide a persuasive analysis to establish that an in-band backhaul solution is feasible. We discuss the deployment assumptions and the system design constraints applicable in a practical system in Sections 17.3 and 17.4, respectively. Finally, in Section 17.5 we present scheduling mechanisms for realizing inter-BS communication and the multiplexing of backhaul links and access links based on the aforementioned assumptions and deployment scenarios.

17.2 Feasibility of In-band Backhaul

The term in-band, as mentioned in Section 17.1, means that the access link (BS–MS link) and the backhaul link (BS–BS links or BS–network links) are multiplexed on the same frequency band. In contrast to the commonly deployed out-of-band wireless backhaul solutions (where dedicated frequency resources can be used for backhaul and access separately), in an in-band wireless backhaul system, the resources used for backhaul are taken from the access-link resources regardless of the multiplexing method used. An important task therefore is to analyze whether enough resources are available for backhauling without compromising the access needs. We will begin by setting up the requirements before discussing the feasibility. In the analysis provided next, we assume a typical cellular deployment set up where a single BS deployment site consists of multiple (typically three) sectors. If α is the fraction of resources used for backhauling, then, an in-band solution is considered feasible if:

$$R_b\alpha \geq R_a(1-\alpha),\ldots \tag{17.1}$$

where R_a denotes the sum data rate of the access links in all sectors and R_b denotes the sum data rate of all the backhaul links supporting the cell. This condition can be further expressed in terms of the backhaul and the access-link capacities as:

$$C_b\alpha\beta_b \geq C_a(1-\alpha)\beta_a,\ldots \tag{17.2}$$

where C_a, and C_b respectively denote the maximum uplink (UL) and downlink (DL) access capacity of a BS, while β_b is the efficiency on the backhaul link, and β_a is the efficiency on the access link.

We investigate the achievable link capacities through Monte Carlo simulations for the scenario illustrated in Figure 17.1 where we consider a mmWave beamformed system at 28 Ghz with uniform linear antenna arrays (multiple subarrays) at both the BS and MS sides. The simulation configuration, as illustrated in Figure 17.1, assumes an inter-site distance of 200 m between BSs.

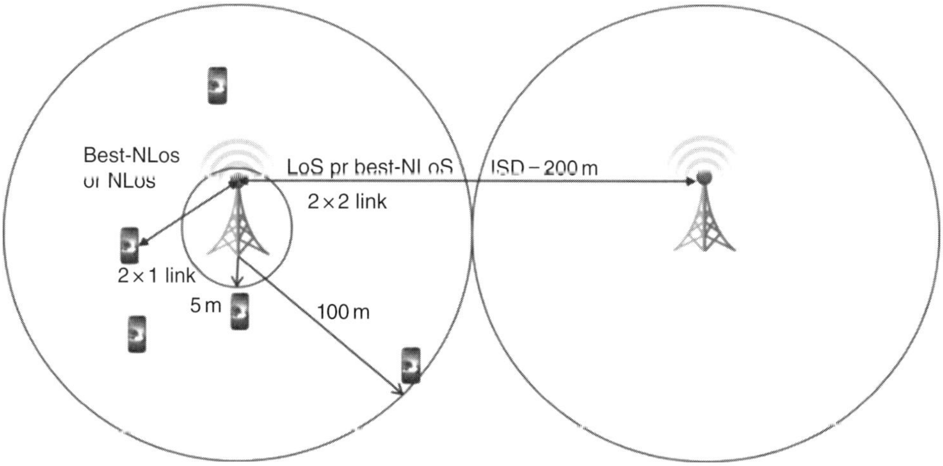

Figure 17.1 Monte Carlo simulation set up for evaluating wireless backhaul feasibility.

MSs are dropped randomly at a distance (5, 100 m) from the BS within a (−24°, 24°) angular region. We assume 2×2 subarrays for the backhaul link (BS–BS link) and 2×1 subarrays for the access link (BS–MS link). The BS and the MS transmit powers are assumed to be 43 dBm and 23 dBm, respectively. The antenna gains are based on current practical antenna capabilities (23 dBi and 6 dBi at the BS and the MS, respectively). For the BS beams, we assume a uniform angle space between (−30°, 30°) and a sector size of 60°, while a uniform angle space between (−45°, 45°) is assumed for the MS beams. We have considered two scenarios:

1. line of sight for the backhaul link and best non-line-of-sight (best NLoS) for the access link
2. best NLoS for the backhaul link and just NLoS for the access link.

The term "best non-line-of-sight" refers to the best NLoS path amongst all the NLoS paths. For the results reported here, we have only considered scenario (2), which is much closer to a real deployment in that mobility constraints will prohibit the best NLoS path from being used for access links all the time, while the backhaul link can be assumed to be best NLoS even in practical deployments (owing to the static nature of the BSs and largely static nature of the reflectors). The parameters used in the simulation are shown in Table 17.1.

Based on the deployment scenario and the simulation parameters here, we obtained the CDFs of the instantaneous cell capacities, shown in Figure 17.2.

The average capacities in Table 17.2 show that the backhaul link capacity is about 2.1 times higher than the access-link capacity. Based on these capacity numbers, we can analyze the feasibility. But first let us look at the other factors affecting the backhaul and the access links.

Although the resources used for backhaul link are taken from access link, we argue that the backhaul link (BS–BS link) is likely to be much more efficient than the access link (BS–MS link) even when the same BS hardware is reused. The main reasons are:

- the backhaul link can be assumed to have high-gain antennas on both the Tx and Rx side while the access link is likely to has fewer antennas on the MS side
- MS transmit power (uplink) for the access link is much smaller; BSs at one or both ends of the backhaul links can use higher transmit power than the access link

Table 17.1 Parameters used for Monte Carlo simulations.

System parameters	Values
Channel model	4G SCM UMi
Inter-site distance	200 m
BS Tx power	43 dBm
MS Tx power	23 dBm
BS antenna gain	23 dBi
MS antenna gain	6 dBi
Number of sub-arrays for backhaul link (BS–BS link)	2×2
Number of sub-arrays for access link (BS–MS link)	2×1
Number of antennas per subarray at the BS (MS)	8 (2)
Number of RF beams supported at the BS (MS)	9 (5)

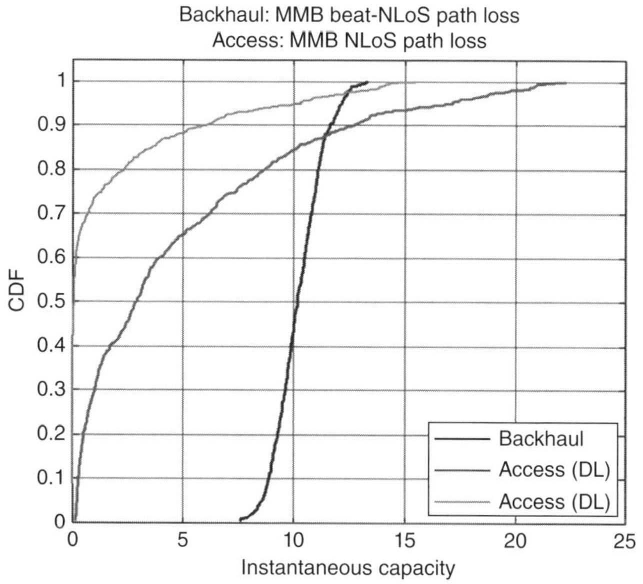

Figure 17.2 Monte Carlo simulation set up for evaluating wireless backhaul feasibility and observations.

Table 17.2 Average link capacities.

	bits/s/Hz	Gbits/s
Backhaul	10.25	5.91
Access (DL)	4.8	2.76
Access (UL)	1.59	0.91

- since backhaul links can be assumed to have less channel fluctuations compared to the access link (which needs to deal with mobility), the backhaul link is likely to incur lesser overheads than the access link (think of control signaling and reference signals)
- BSs are at fixed and at higher locations compared to the MS, which also impacts the performance positively.

Since the backhaul link is specifically designed for the wireless backhaul system, we have assumed the PHY overheads in the backhaul link to be about 5%; largely static links do not require frequent reference signal transmissions or large scheduling information maps (see scheduling-scheme description in the Section 17.3). The overhead is assumed to be around 30% for the access link (close to the access link overheads observed in the cellular systems). Using these values of efficiency, Equation (17.2) can be evaluated for different values of α – recall that α is the fraction of resources/bandwidth taken from the access link to serve the backhaul link – and different downlink-to-uplink (DL/UL) ratios. The link capacity results for the best NLoS path on the backhaul link, and the NLoS path for the access link are captured in Table 17.3, where "DL/UL" indicates the DL/UL ratio for resource allocation in a TDD system. DL/UL ratios of 1 (symmetric link) up to 3 (DL heavy) are considered on the access link. The first row (Link capacity C) in Table 17.3 shows the link capacity for the backhaul

Table 17.3 Feasible access link (AL) and backhaul link (BL) rates.*

	Backhaul link	Access link (NLoS)		
	(best NLoS) Mbps	DL/UL = 1 Mbps	DL/UL = 2 Mbps	DL/UL = 3 Mbps
Link capacity C	3,301		DL = 1817, UL = 343	
Data rate R	3,136	784	968	1,061
Data ($\alpha = 1/5$)	627	627	775	849
Data ($\alpha = 1/5$)	784	588	726	796
Data ($\alpha = 1/5$)	1,045	522	646	707

*Assuming non-line-of-sight (NLoS) conditions on the access link and best NLoS conditions on the BL for various DL/UL ratios. A best NLoS condition can be assumed on the backhaul link as there is ample time to choose the best NLoS setting (which does not change very often) compared to the access link.

link and access link (DL and UL are shown separately). The second row shows the data rate R that is obtainable for the backhaul link (5% overhead assumed) and the access link for various DL/UL ratios (assuming ~30% overhead). The subsequent rows show the obtainable data rates for various fractions of α.

It can be observed that using only 25% of the access link resources for the backhaul link is sufficient to support user data rates as high as 0.8 Gb/s. Next, we look at the deployment assumptions used in this paper.

17.3 Deployment Assumptions

We begin with the 4G cellular deployment as a basis. For simplicity, but without compromising the essence, it is assumed in this paper that the 4G BSs are deployed on a hexagonal grid (as shown in Figure 17.3), and that every existing 4G BS is connected to the network using a wired backhaul link (say, fiber). It is further assumed that when the much higher capacity 5G deployment commences, the operator is likely to re-use the existing BS sites and utilize the existing wiring first, to avoid additional cost of digging/trenching. This will result in the 5G BSs providing a much higher cell capacities but the radio coverage is very likely to shrink (as shown for illustration purposes in Figure 17.3).

To fill these coverage gaps, the operator will need additional 5G BS sites. For backhauling these additional sites, the operator has three choices:

1. bring wire/fiber to these additional sites, which incurs additional costs for digging/trenching
2. connect the new 5G BSs to the 5G BSs at the old sites using wireless backhaul ("no additional fiber" scenario)
3. A combination of (1) and (2).

These choices result in different fiber densities, the most dense resulting from option (1) and the least dense for option (2), and with the cost being directly proportional to the density. Assuming the availability of fiber-backhaul to the original 4G cell sites, and no additional fiber backhaul to the additional 5G sites, depending on how the cells are stacked, the resulting fiber density can either 1/3 (one in three 5G cells is fibered) as shown in bottom half of Figure 17.4, or the resulting fiber density can be 1/4 (one in four 5G cells is fibered) as shown

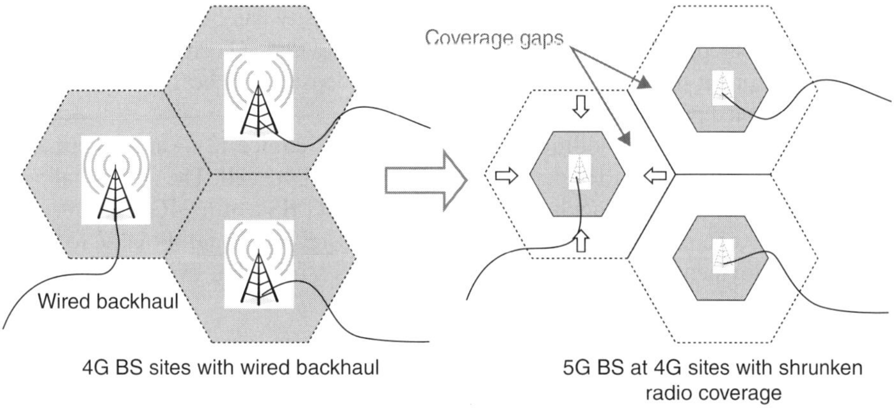

Figure 17.3 5G BSs will be deployed at the already-backhauled 4G sites.

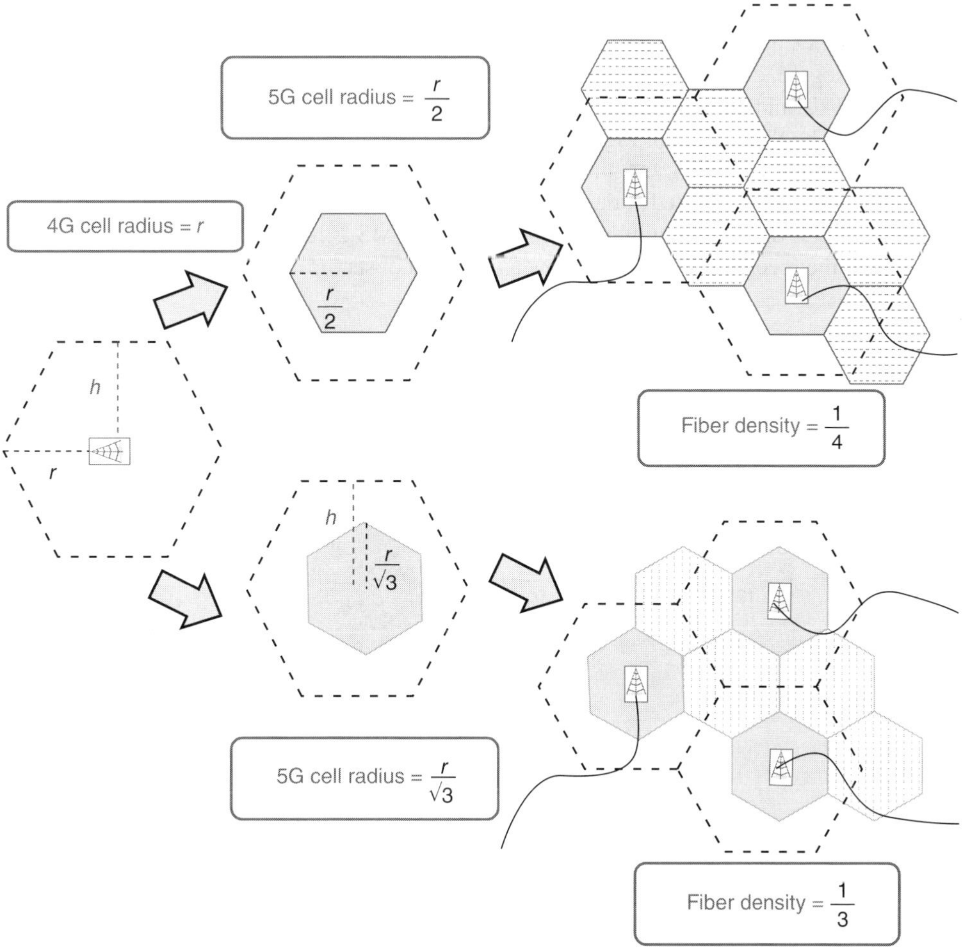

Figure 17.4 Example deployments with different fiber injection densities.

in upper half of Figure 17.4. It should be clear that the density can be raised to an arbitrarily higher ratio (as desired) by bringing fiber to some of the unwired 5G cells. In the extreme case, where fiber is brought to all the 5G BSs, the resulting density will be 1 and this is equal to choice (1) mentioned previously.

Here, we focus on the "no additional fiber" scenario – choice (2) – and where the coverage of each 5G cell shrinks to about 57.7% of the original 4G cell. The additional 5G cells are served solely by wireless backhaul provided by the 5G BSs at the 4G sites, resulting in a situation where one out of every three 5G cells has wired backhaul. Finally, to complete the cellular deployment scenario description, we will assume that every BS cell site has three sectors.

17.4 In-band Backhaul Design Considerations

To avoid interference, one can be tempted to think of using spatial division multiplexing (SDM) to serve access link and backhaul link simultaneously. However, the interference from the transmission on the backhaul link to the transmission on the access link, and vice versa, is considerable. Depending on the hardware configurations at the transmitter and receivers, significant degradation in SIR (signal-to-interference ratio) can be observed at the receiving MS on the AL and the receiving BS on the other end of the backhaul link as shown in Figure 17.5. The SIR curves of Figure 17.5 are obtained at the MS for a BS, equipped with an 8×8 phased array antenna, transmitting simultaneously on the AL and the backhaul link. It is assumed that the BS is located at the bore sight (0° azimuth steering), while the SIR at the MS is observed for a different azimuth steering. When the MS is located along the same azimuth direction as the receiving BS, SIRs as low as 0 dB are observed.

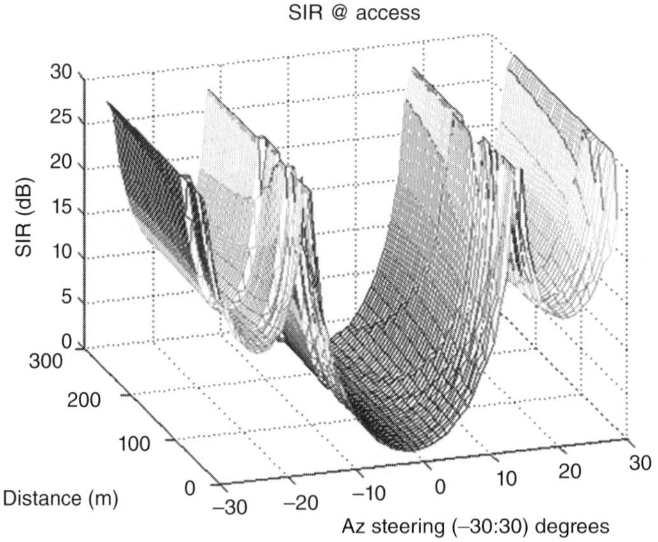

Figure 17.5 Signal–to-interference ratio as a function of azimuth beam when backhaul link is scheduled simultaneously with the access link.

Another challenge is the self-interference at a single BS site from an antenna panel in one sector to the other antenna panel serving the other sector. When one panel is set to receive, while the other panel is set to transmit, sophisticated signal processing (interference cancellation) is required for cancelling the self-interference.

In this paper, we will work with a minimal (baseline) hardware configuration that does not require SDMA capabilities or self-interference cancellation capabilities. Further, we assume that at a given BS site, all the antenna panels are either set to transmit or set to receive. To further reduce hardware requirements, it is assumed that one BS sector has one digital chain and can perform one transmission/reception in one sector at any one time. PMP functionality on the backhaul link is only achieved by adaptively/dynamically steering the beams to different neighboring BSs at different times. Lastly, we assume that the backhaul link transmissions are transparent from an MS perspective (that is, the MS operations remain unchanged and existing MSs can be used in the network deploying in-band wireless backhaul).

17.5 TDM-based Scheduling Scheme for In-band Backhauling

In a TDM-based scheme, the access and the backhaul links are time-multiplexed by reserving a portion of the access frame for backhaul transmission/reception. The idea is to facilitate that a backhauled or wired 5G BS (W-BS) can transmit backhaul data to another 5G BS that is unwired (U-BS). The U-BS is served by the W-BS such that a fraction of the DL and the UL access portions are reserved for in-band backhaul transmissions over the backhaul link. Figure 17.6 is an illustrative example for the TDD case – where the DL and the UL portions are time multiplexed and a fraction of the downlink αT_d and uplink αT_u access is reserved for in band backhaul transmissions over the backhaul link.

On the one hand, increasing α allows more data to be backhauled to/from the U-BS cell, which in turn will result in enabling higher access capacities for the U-BS. On the other hand, high values of γ directly reduce the access portion of the frame thereby reducing the capacity of the wired as well as unwired BS cells. This tradeoff makes it particularly non-trivial to gauge the overall gain. The numerical capacity analysis results in Table 17.3 provide an initial insight/engineering feel on how the various DL/UL ratios and the fraction of bandwidth allocated for backhauling (α), impacts the access/backhaul link throughput and the feasibility of the overall approach.

In discussing the TDM-based scheduling schemes for backhauling, let us first understand what needs to be done for the backhauled or the wired BSs (W-BSs) or the unwired BSs (U-BSs) to communicate with the peer BSs (which themselves can be a W-BS or a U-BS). These are:

- A U-BS needs to transmit and receive backhaul traffic from one or more W-BSs.
- All BSs need to exchange control traffic with the neighboring BSs for interference coordination or handover coordination.

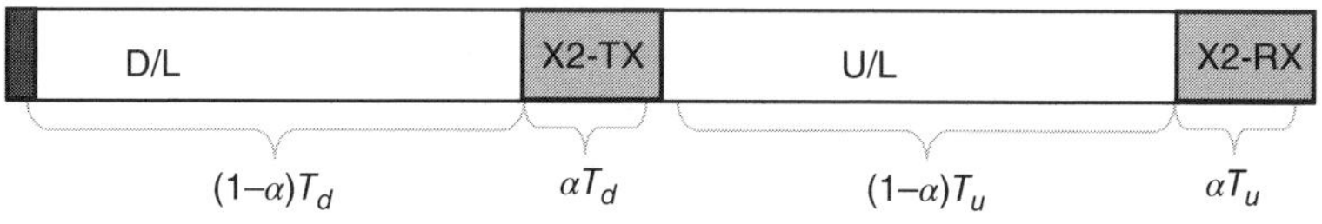

Figure 17.6 Multiplexing Wireless X2 and access links using TDM.

Figure 17.7 shows BSs with three sectors, as in typical cellular deployments. One of the 5G BSs (labelled W-BS) is connected to the core network with a wired backhaul and serves as the bandwidth injection point. The other two BSs are U-BSs and are labelled 5G-BS 1 and 2. The U-BSs are connected to the network via the wireless backhaul provided by the W-BS. Each sector in the W-BS serves one sector of two other 5G U-BSs.

Any backhaul transmission between neighboring BSs may be composed solely of control information, control information and data, or just data, as shown in Table 17.4.

The TDM-based scheme maps these X2 transmissions to a portion of the DL and the UL subframe. We illustrate the design of our scheduling scheme for a deployment of 5G cells with a fiber density of 1/3, and the same design principles may be applied to obtain scheduling schemes for other fiber densities as well. Table 17.5 lists the main parameters of the scheduling scheme.

The overall scheme is that each W-BS communicates with each of its six neighboring BSs. If each BS cell consists of three sectors, three backhaul links (one in each sector) can be served

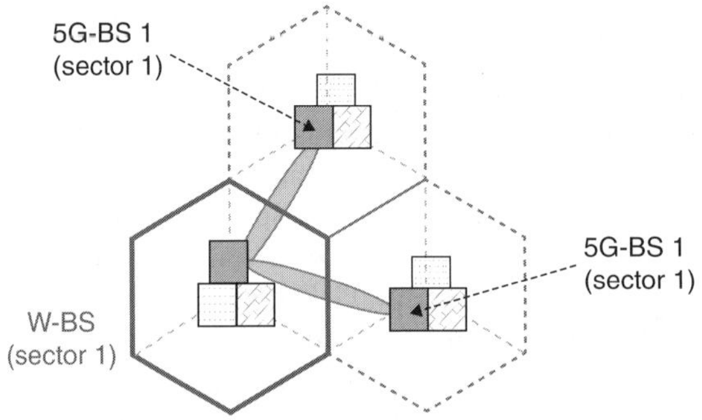

Figure 17.7 Sector 1 of a wired BS (W-BS) communicates with Sector 1 of 5G-BS 1 of an unwired BS (U-BS) in one time slot and then in the next time slot, the beam is steered to communicate with 5G-BS 2 (another U-BS).

Table 17.4 Amount of control/data traffic and who transmits what.

	Control	Data	Control plus data
Transmission duration	Very small	Significant	Significant
Transmitting BSs	All neighboring BSs	Unwired and wired BSs	Unwired and wired BSs

Table 17.5 Parameters of the scheduling scheme.

Parameter	Details
αt_{frame}	Duration of the frame used for backhaul transmissions
$\alpha' t_{frame}$	Duration of each subframe used for backhaul transmissions
t_{x_u}, t_{x_d}	Duration that an unwired BS transmits and receives backhaul data over backhaul links, respectively
$c_{\alpha_u}, c_{\alpha_d}$	Maximum UL and DL access capacity of a U-BS
r_f	Transmission rate of the backhaul link
c_{f_u}, c_{f_d}	Backhaul throughput leaving and entering the U-BS

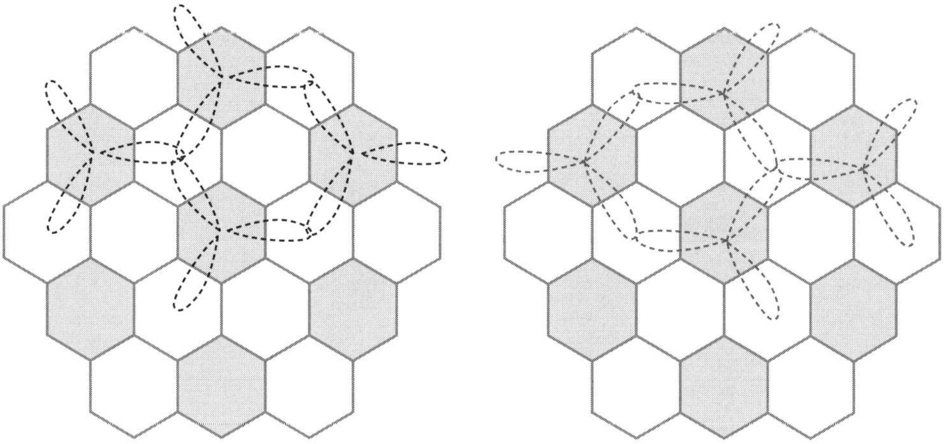

Figure 17.8 Simultaneous transmission over backhaul links to three neighboring BSs. The scheduler then schedules the next three remaining neighbors (each BS has six neighbors). This constitutes the transmit-to-3 portion of the Tx-3, Rx-3 scheduling scheme.

such that when sector 1 of W-BS transmits to sector 1 of 5G-BS 1, at that time sectors 2 and 3 of W-BS transmit over backhaul links to their associated 5G-BS sectors of the neighboring BSs (as shown in Figure 17.7). The beam direction is then steered to target the other three neighboring cells, so that sector 1 of W-BS now transmits to sector 1 of 5G-BS 2, with the other two sectors of W-BS following the same procedure. This enables transmission to all six neighboring BSs. Similarly, sectors 1, 2, and 3 of 5G-BS 1 receive over wireless backhaul links from their associated W-BSs during the Rx portion of the frame. This "transmit to 3, receive from 3" (Tx-3, Rx-3) scheduling scheme is used as a basic building block to describe the inter-BS scheduling schemes. Figure 17.8 illustrates the overall concept of a hypothetical deployment scenario consisting of hexagonal stacking of cells.

Finally, Figure 17.9 summarizes the entire in-band access and backhaul scheduling system, with which we can outline a BS-to-BS scheduling algorithm capable of scheduling the backhaul link while maintaining the usual BS-to-MS link (access) scheduling. In the description that follows, we assume:

- a time division duplexed (TDD) frame structure consisting of subframes
- each subframe further comprising a DL and a UL portion
- a few subframes (say, five) constitute a frame
- parts of the control signaling are transmitted per frame, while some other control signaling may be transmitted per subframe.

Figure 17.9 illustrates the mapping of the Tx-3, Rx-3 transmission scheme to a portion of a subframe. The sequence of transmissions can be summarized as follows.

1. The 5G W-BS transmits to all of its six neighbors, by transmitting to three neighbors at a given time using the procedure described earlier (illustrated in the figure). Since all the six 1-tier neighbors of the W-BS are 5G U-BSs, control as well as backhaul data is transmitted

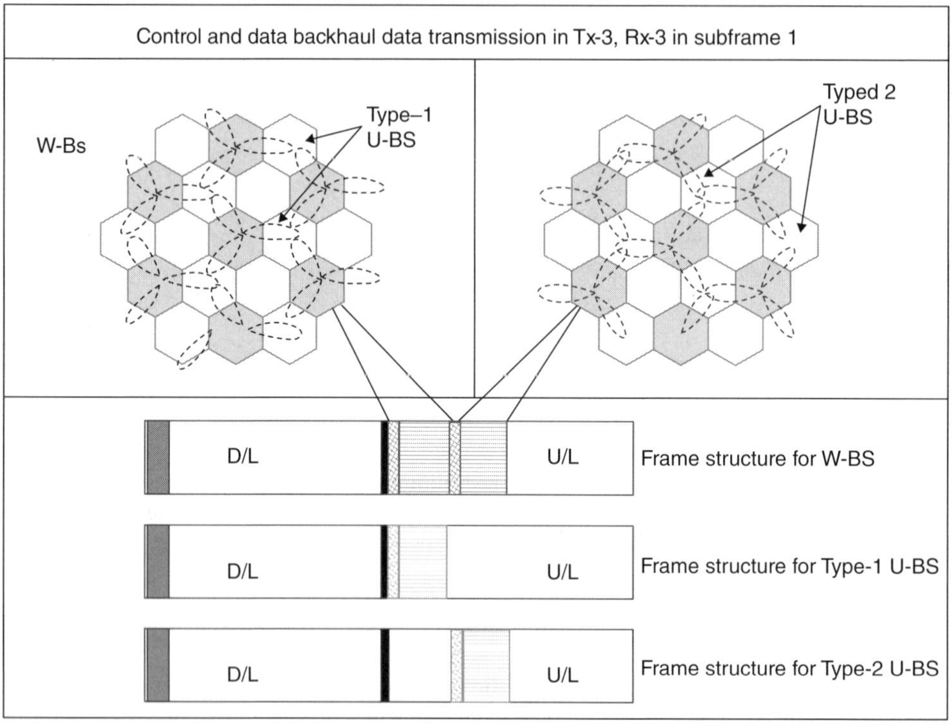

Figure 17.9 Tx-3, Rx-3 BS-to-BS scheduling scheme mapped to the frame structure for BSs.

to the U-BSs. Note that the transmissions do not require any additional Tx–Rx switching other than the DL to UL switch of the TDD systems. Step 1 can, for instance, be scheduled in a portion of subframe 1 of a frame. The remaining portion can be used for data transmission on the access link.

2. U-BS type 1 (BSs to the right of a W-BS) transmits to all six neighbors (not illustrated in the figure). It first transmits only control messages to the three U-BS type 2 neighbors (BSs to the left of W-BS). It then transmits control as well as data to its W-BS neighbors. Step 2 can be accomplished, for instance in a portion of subframe 2.
3. U-BS type 2 transmits to all six neighbors (not illustrated in the figure). Step 3 can be accomplished, for instance in a portion of subframe 3.

If scheduled as described here, by the end of the three subframes, all BSs finish exchanging one round of control messages with their neighbors. Each of the unwired BSs have transmitted and received backhaul data over only a fraction of the subframe. In practice, the backhaul data directed into the unwired BS is much higher, often twice the backhaul traffic leaving the unwired BS to the core network. To increase the backhaul throughput, a fraction of the uplink portion of subframes 4 and 5 can be used exclusively for delivering backhaul traffic to the unwired cells.

The scheme outlined here can serve as a baseline scheduling mechanism, in that very minimal hardware capabilities are assumed. Performance on the backhaul can be further improved if the transmission and receive equipment at the BS supports advanced functions such as SDMA, self-interference cancellation, and so on. Improved performance will in turn improve the attractiveness of the in-band solution.

17.6 Concluding Remarks

Cost-effective and low-latency solutions for wireless backhaul (including BS to network and BS to BS communication) will be essential for supporting the envisaged densification in high capacity 5G networks. In this paper, we have presented an initial analysis and solution framework for supporting an in-band, PMP, non-line-of-sight, mmWave backhaul.

We have shown using simulations that an in-band solution is feasible at mmWave frequencies for tolerable losses in access capacities and assuming modest hardware capabilities; it does not require SDMA, full duplexing, or multiple RF chains per sector. BS-to-BS scheduling is not a trivial matter; unlike BS-to-MS scheduling, there is no established hierarchy among BSs. In this paper, we outlined a BS-to-BS scheduling scheme for the backhaul link that can be multiplexed with the usual BS-to-MS scheduling for access, and can be considered for an in-band backhaul and access deployment scenario.

Future work should look into detailed system-level simulations for analyzing the performance of the proposed system. Other avenues of interest are investigations into SDMA and full-duplexing capabilities for further spectral-efficiency enhancements.

Acknowledgments

The authors are grateful to their colleagues at Samsung Research America for their valuable feedback and discussions. The authors would like to especially thank Md. Saifur Rahman for his help with the Monte Carlo simulations and Sridhar Rajagopalan for his help with SIR analysis and simulations.

References

[1] A. Bleicher, "A surge in small cell sites," *IEEE Spectrum*, Dec 2012. URL: spectrum.ieee.org/telecom/wireless/a-surge-in-small-cell-sites (accessed August 1, 2016).
[2] Senza Fili Consulting, "Crucial economics for mobile data backhaul: An analysis of the total cost of ownership of point-to-point, point-to-multipoint, and fiber options," 2012.
[3] A. Damnjanovic, et. al., "A survey on 3GPP heterogeneous networks," *IEEE Wireless Comm. Magazine*, vol. 18, no. 3, pp. 10–21, 2011.
[4] J.G. Andrews, "Seven ways that hetnets are a cellular paradigm shift," *IEEE Comm. Magazine*, vol. 51, no. 3, pp. 136–144, 2013.
[5] Infonetics Research, "Macrocell mobile backhaul equipment and services market share and forecast report," 2012.
[6] Cambridge Broadband Networks, "Spectrum-and-tech-issues-for-microwave-backhaul in Europe," White Paper, 2010.
[7] Wireless 20/20, "Rethinking small cell backhaul: a business case analysis of cost-effective small cell backhaul network solutions," White paper, 2012.
[8] 3GPP TSG RAN WG1 Meeting #66, R1-112340, "Consideration of X2 backhaul for CoMP."
[9] R. Taori, et. al., "Cloud cell: paving the way for edgeless networks in 5G," *Proc of IEEE Globecom*, 2013.
[10] Maravedis Market Research and Analysis, "A practical look at LTE backhaul capacity requirements," 2011.
[11] P.J. Pietraski, "The bandwidth crunch: Can wireless technology meet the skyrocketing demand for mobile data," *IEEE LISAT (Systems, Applications and Tech. Conf.)*, 2011.
[12] Z. Pi and F. Khan, "An introduction to mmWave mobile broadband systems," *IEEE Comm. Magazine*, vol. 49, no. 6, pp. 101–107, 2011.

18

Application of NFV and SDN to 5G Infrastructure

Ashok Sunder Rajan and Kannan Babu Ramia
Intel Corporation, Santa Clara, CA, USA

18.1 Chapter Overview

Telecom network infrastructure under deployment is in its fourth generation of evolution, commonly known as Long-Term Evolution (LTE). This is the infrastructure that serves smart phones globally. Even as the 4G infrastructure is being deployed, the demand on network infrastructure is exploding. 5G projections indicate that growing demand will impact the network infrastructure significantly, while network function virtualization (NFV) and software defined networks (SDN) represent ways to simplify deployment and management of that infrastructure.

NFV promises significant cost savings, flexibility, and ease in deploying network infrastructure, drawing on the highly efficient data center infrastructure that powers many Internet application services, such as Google and Amazon. SDN eases the configuration, scaling, and management of the deployed networking infrastructure.

This chapter will present a basic overview of NFV and SDN followed by a design study of the cellular network infrastructure; specifically a virtualized enhanced packet core. This will allow the reader a very clear understanding of how NFV and SDN technologies can work to address the demands placed on the cellular network infrastructure by 5G.

18.2 Background

Unlike enterprise data centers, which have fairly straightforward switching, traditional telecom networks have different workloads with many more complex interdependencies of state and scale that can affect performance. Given this complexity, cellular wireless telecom network

Towards 5G: Applications, Requirements and Candidate Technologies, First Edition.
Edited by Rath Vannithamby and Shilpa Talwar.

Companion website: www.wiley.com/go/vannithamby/towards5g

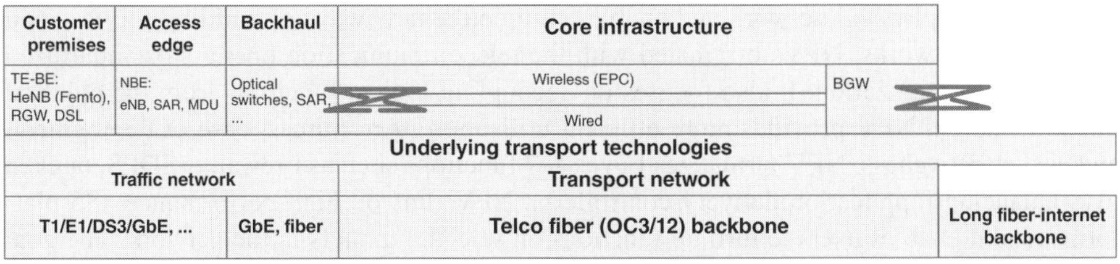

Figure 18.1 Network infrastructure topology.

infrastructure has traditionally used purpose-built hardware, each with performance optimized for a specific function in the network and with well-defined interfaces to other functions/elements, specified by the 3GPP and led by network providers and equipment manufacturers. Workloads and performance characterization of cellular network infrastructure therefore largely falls in the proprietary expertise domains of large equipment vendors, such as Ericsson, Alcatel Lucent, and Huawei. Network key performance indicators are generally specific to each operator, and the equipment is optimized by vendors to those specifications while remaining compliant at the interfaces for equipment interoperability.

As telecom services demand has continued to exponentially increase over the past few years ~100% year-on-year [1, 2], network infrastructure has started to become capacity challenged, driving communication service providers (CSPs) to transform their networks and leverage the benefits of the virtualization in data centers. While this has opened acceptance of the cellular network infrastructure to evolutions in virtualization and SDNs in the data centers, a gap remains in identifying workloads that can efficiently characterize performance requirements across different CSPs.

CSP-owned infrastructure comprises the access edge, backhaul that connects the access edge to the infrastructure core, the infrastructure core that is built over the telecom providers' backbone and the north-facing Internet peering routers. The access edge and the backhaul are generally called the traffic network, and fan out the core infrastructure to the farthest possible corners of the provider's deployment geography. The backbone ties together the provider's core infrastructure and connects the traffic network into the Internet. The core infrastructure is connected by the backbone to the Internet over border gateway routers (BGW) at peering points. The backbone with the BGW forms the transport network.

Figure 18.1 details the network infrastructure topology elements and the underlying transport technologies.

18.3 NFV and SDN

Operators have responded by getting better at managing complexity. SDNs and NFV offer tools and mechanisms to make networks simpler and more manageable. SDNs originated in the datacenter and make the network programmable by separating the control plane (telling the network what goes where) from the data plane (sending packets to specific destinations). It relies on switches that can be programmed through an SDN controller using an industry-standard control protocol, such as OpenFlow [3]. It also offers decoupling of services from the underlying physical network and enables completely new ways of building services and architecting networks. NFVs originated with the telecommunication operators standardized by the ETSI organization. It also focuses on decoupling logical functions from the physical implementation. NFV provides more efficient utilization of resources, ease of deployment, and cost effectiveness. NFV virtualizes Layer 4–7 functions such as firewall or IDPS, or even load balancing (application delivery controllers). NFV runs on high-performance x86 platforms, and it enables users to turn up functions on selected tunnels in the network. The goal is to allow people to create a service profile for a VM, or flow, and leverage x86 hardware to build an abstraction on top of the network and then build virtual services on that specific logical environment.

Combining SDN and NFV technologies leverages both the datacenter and telecommunications/IT concepts. One of the benefits that can be obtained is a reduction in hardware total cost of ownership, since there is less overall hardware footprint and fewer types of hardware. It also offers flexibility and agility, so that new services can be deployed quickly. Virtualization hides some of the complex details to service developers and provides a simplified front end. Finally, there is increased robustness and efficiency since failover and dynamic scaling of capacity (such as cloud bursting) are possible.

NFV addresses separating the software from the underlying hardware in the infrastructure. SDN's thrust is to separate control and data planes in order to better manage networks. It would not be an overstatement to say that NFV and SDN are now one of the most critical initiatives in telecom networking [4, 5].

The cellular wireless infrastructure is a very large, widely distributed network bringing together many disciplines of telecom engineering. We will look at the core aspects of network engineering to get a clear understanding of application areas of NFV and SDN technologies.

18.4 Network Planning and Engineering

Traffic networks aggregate traffic from the far corners of the deployed infrastructure into core equipment on the transport network. The transport network funnels this aggregate traffic into the Internet through hubs or exchanges. Capacity dimensioning of traffic, transport network and the core infrastructure equipment is based on data volume or the tonnage (gigabytes) that needs to be handled in a given time (Gbps). Telecom network providers take the historical tonnage data for a given market to plan the required infrastructure capacity of the network at the time of deployment. Network deployments – installation of the physical communication links and equipment – are traditionally capital-intensive heavy engineering efforts. The network planning phase generally envisages future capacity requirements, in a similar way to the planning of city transportation systems. Excess capacity will create network operational cost inefficiencies while underprovisioning will cause issues with service availability and compromise service response times, leading to a poor end-user experience. Network planning therefore relies on traffic engineering principles to time and frequency multiplex usage. Usage is defined in units of traffic called Erlangs [6], which describes the total traffic volume in 1 hour, often referred to as traffic busy hour (BH) by network providers.

The Erlang value increases as we move from the access edge toward the network core infrastructure; that is, as aggregation increases. However, the peak capacity demand (Gbps) at the access edge is independent of the traffic aggregation. The average capacity demand would mathematically compute as the capacity demand in each of the 3,600-s time windows of BH traffic.

The BH capacity-demand variance var(C) = [(C(t)^2 - Cm]/Cm^2; where C(t) Gbps is the capacity required at time instant "t" and Cm is the mean capacity demand in the BH [7]. var(C) will be highest at the access edge and progressively fall toward the network core. Generally, therefore, traffic networks (access edge and backhaul) will be characterized by high values of var(C) while transport networks will have lower demand-variance values. Cm, the mean demand, cannot therefore be used to dimension network capacity, especially in traffic networks. The high var(C) will result in poor quality of service (QoS) at the instants when C(t) far exceeds Cm.

Telecom network providers use a value called effective bandwidth (EBW) to dimension network capacity. The installed capacity of traffic networks is overprovisioned by a factor of around 30% above the EBW. The overprovisioning factor is lower in transport networks, where var(C) is lower. In order to compensate for the overprovisioning of networks, typically the subscriber will be overprovisioned by a similar factor of 30% above the spectrum bandwidth limit at the traffic network. An interesting observation of usage of statistical tools in network engineering is var(C), which can also be thought of as the inverse of the statistical gain of the system. Traffic networks with high var(C) exhibit lower statistical gains, while transport network networks with lower var(C) demonstrate higher statistical gains, due to traffic aggregation. Thus higher statistical gain reduces the overprovisioning factor and hence the system cost.

For completeness, EBW = C = m + [k(H) * (–2*ln(e))1/2]^1/H * a (1/2H) * b(–1(1-H)/H) * m^(1/2H);

Where:

k(H) = H ^ H(1 – H) ^ (1 – H)

with b = buffer size, a = variance coefficient, e = target cell loss rate, m = offered traffic, and H = the Hurst parameter. The variance of demand across the network infrastructure is shown in Figure 18.2.

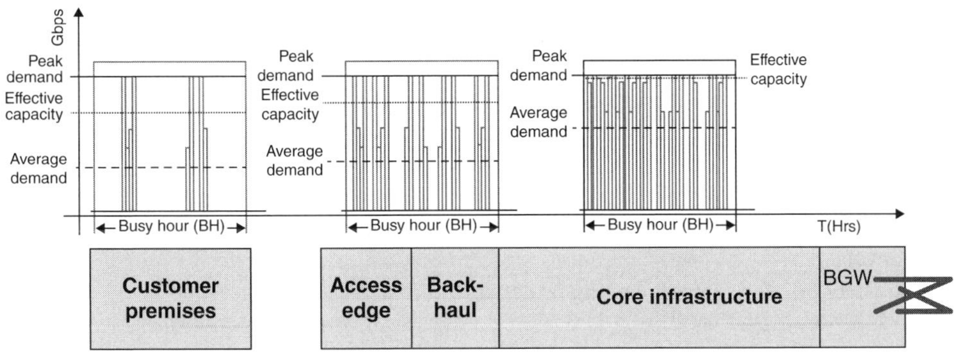

Figure 18.2 Demand variance across network infrastructure. BGW, border gateway router.

18.4.1 Cellular Network Design and Traffic Engineering

Cellular network design needs to address the market being catered for by the radio access network (RAN) infrastructure, the installed backhaul topology, the capacity available, and the backbone infrastructure with the peering points to the Internet. Typically the network-design process starts with the market design, followed by the call model, which is a statistical estimation of the expected signaling load. The traffic model is the last step and addresses the types of application services the CSP will provide to the market it caters for.

18.4.1.1 Market Design

Market design is driven by the physical geography over which the wireless provider is licensed to radiate at given frequency bands. The frequency band for which an operator is licensed could vary by region or state. Getting these bands close together or even having the same band across all the regions brings significant cost saving to operators, with the reduced complexity of the RAN. This is the reason why operators covet contiguous bands of spectrum and why they commit enormous investments in regulatory spectrum-auction processes.

Having determined the bands available in the deployment area or market, data is collected on the potential number of mobile subscribers along with the user distribution. Base-station tower (eNB) location planning follows, where the density of eNBs increases with user density. The eNBs are then organized in tracking areas (TA), which ease locating the connection end points as users move between eNBs or TAs.

The total number of bearers in the market is defined as:

$$\begin{aligned}&\text{total number of default bearers} + \text{total number of dedicated bearers}\\&\quad = \text{number of bearers / channel} \times \text{number of channels / band} \times\\&\quad\;\; \text{number of licensed bands in the market}\end{aligned}$$

The number of default bearers to be assigned to each user in the market is roughly:

the total number of bearers in the market ÷ the total number of subscribers in the market.

18.4.1.2 Call Model

The call model is an engineering estimation of the expected number of non-access stratum (NAS) signaling events per subscriber in a BH for a specific market design. Signaling events can be generated:

- by the mobile device or user equipment (UE), as defined by the 3GPP, for example attach or handovers
- by the network, for example paging or network-initiated connection requests.

The call model provides an engineering estimation of the total number of signaling events in the market and helps in the dimensioning of the signaling equipment, especially the mobility management entity. Each event consists of several short message exchanges. The key design criterion is the aggregate rate of those events in the market each second, and not the throughput, which is typically of the order of a few megabits per second, even in large markets.

18.4.1.3 Traffic Model

The traffic model is the final step in wireless network engineering and estimates usage or Erlangs for each application type envisaged in the market. The traffic model calculates the BH data demand per subscriber (DDs) in kilobits per second (kbps):

$$\mathrm{DDs} = \Sigma(1\,\mathrm{to}\,\mathrm{n})\begin{bmatrix}\text{Estimated duration of app}(\mathrm{n})(\mathrm{min}) \times \text{typical data rate}(\mathrm{kbps}) \times \\ \text{estimated usage “Erlangs”}\end{bmatrix} / 3{,}600\,\mathrm{s},$$

where app(n) is nth application of all possible application traffic types in the market and Σ(1 to n) is the sum total of the individual application type data demand per subscriber.

The total data demand for the market is:

$$\mathrm{DDs} \times \text{number of subscribers in the market}$$

The traffic model that follows provides a data demand calculation for the number of subscribers in the market design. The key point to note is that the traffic model does not factor in the specificities of the RAN itself. It essentially provides the capacity (Gbps) estimates for the application traffic.

The traffic model of the load applied to the user plane representative of a market design is shown in Table 18.1 [8].

Table 18.1 User-plane uplink and downlink traffic parameters.

Direction	Traffic		Packet	
	Type	Mix (%)	Size (bytes)	Rate (pkts/s)
Downlink	VoLTE	5.1	72	2.1 M
	Web	52.2	1,200	1.3 M
	Video	29.5	1,440	602 K
	Apps	6.7	675	290 K
	Email/other	6.7	1,440	136 K
Uplink	VoLTE	32.3	72	2.1 M
	Web	43.1	690	289 K
	Video	12.1	240	232 K
	Apps	6.5	400	76 K
	Email/other	6.2	1,000	29 K

Source: Rajan et al. [8].

18.5 Cellular Wireless Network Infrastructure

The cellular network infrastructure comprises of the eNB in the access edge, backhaul, and the evolved packet core (EPC). The EPC generally contains the service edge gateway, aggregation routers, mobility management entity (MME), serving gateway (SGW), packet gateway (PGW), policy and charging rules function (PCRF), as well as the border gateway routers (BGW) that peer with the Internet backbone. Deployments may vary, and sometimes some of the elements – especially the home subscriber server (HSS), PCRF, and BGW – may be located within the IT infrastructure of the telecom network provider.

The access edge, the eNBs grouped into TAs, the cells covered by the eNBs, and the connection bearers bundled into licensed bands together form the RAN. The EPC is a wireless industry terminology that marks a technology evolution step for the cellular wireless core infrastructure. The RAN connects into the EPC, terminating the signaling or control plane at the MME and the user plane carrying the application traffic at the SGW. The application traffic is negotiated through the PGW onto the Internet. The CSP may also offer additional services, such as IMS or VoD. Application traffic belonging to these operator-hosted services is terminated at the SGi-LAN, so called because it represents services not bound to the EPC. The SGi-LAN may exist within the operator's IT infrastructure. The SGi-LAN also presents an area for soft landing of new services before they can be moved further south into the EPC core. The EPC is the voice and data "call processing" nerve center of critical operator infrastructure assets.

18.5.1 Reference Points, Interfaces, and Protocol Stacks

A detailed schematic of the reference points between different elements of the cellular wireless network infrastructure elements and protocol stacks is presented in Figure 18.3. The three key reference points across the entire wireless infrastructure are the Uu between the eNBs and the UEs, the S1 between the eNBs and the EPC, and the SGi between the EPC and application services.

The evolved packet core (EPC) above came about as a process of network evolution, initially proposed by a 3GPP technical report that examined the key issues with the wireless network architecture in 2008–2009 [9]. The service requirements of the evolved packet system (EPS) envisaged by the report are defined in TS22.278 [10]. The EPC comprises the cellular network infrastructure elements within the EPS and is aligned with the evolved universal terrestrial radio access (E-UTRA) or simply RAN requirements defined in the 3GPP technical report TR25.913 [11].

18.5.2 Description of the EPC Main Element Interactions

The MME is the end point of the S1-MME interface from the eNBs and processes all of the UE signaling events. The NAS resides between the UE and the MME in the control plane and is transparent to the eNB; NAS messages are not interpreted by the eNB and transparently carried through the LTE-Uu and S1-MME interfaces. The NAS is responsible for the call-processing and session-management functions of creation, deletion, modification, and management of default and dedicated EPS bearers. The NAS messages

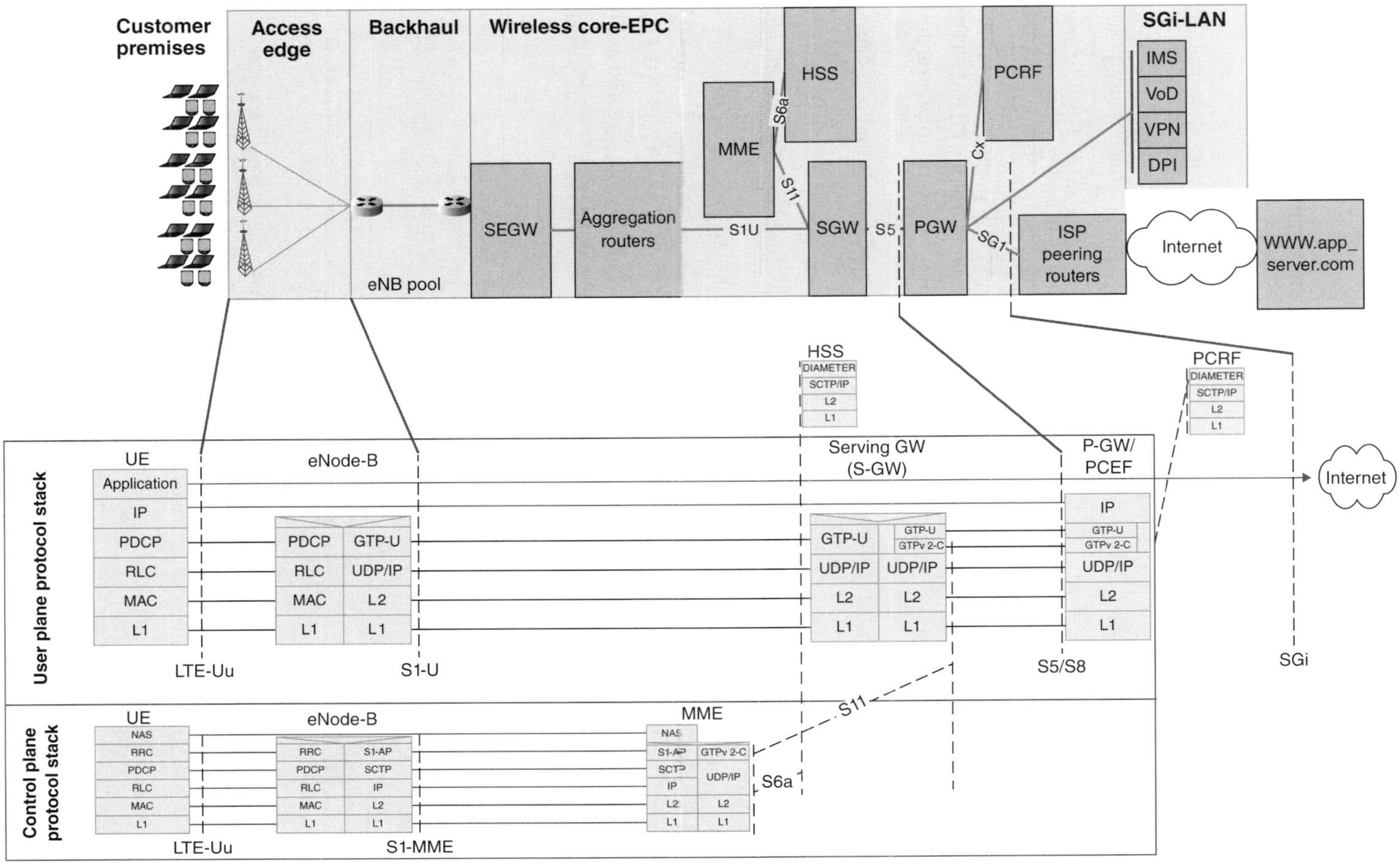

Figure 18.3 Cellular network infrastructure reference point and protocol stacks. Refer to text for definitions of acronyms.

from the different UEs connected to an eNB are assigned transmission sequence numbers and streamed as S1-AP messages over a SCTP transport that terminates at the MME to which the eNB is connected.

The Home Subscription Server is the central database and maintains a list of subscribed services for UEs registered with the network. The MME interrogates the HSS over DIAMETER for authorization of the UE bearer establishment requests.

On successful establishment of a bearer the application traffic from the UE is transported over IP and encapsulated inside GTP-U tunnels by the eNB. The GTP-U tunnels from each of the UEs connected to the eNB are terminated at the SGW over the user-plane (S1-U) interface. Each application flow is assigned a GTP-U tunnel with a tunnel endpoint identifier (TEID) at the SGW. The number of TEIDs will equal the number of application flows from the UEs attached to the EPC. Generally, each UE supports an average of two application flows. The SGW queries the PGW using messages encapsulated in GTP-C over the S5 interface. The PGW retrieves rules to be applied to the UE application flows from the policy control and rules function (PCRF) over the Gx interface using DIAMETER. The UE application flows are then de-capsulated over the SGi interface to forwarding routers to the Internet or the SGi-LAN appliances.

The downlink (DL) return traffic follows a reverse path across the elements and is encapsulated back into the GTP-U tunnels identified by the TEIDs at the SGW to be sent over to the correct eNB to which the UE is connected.

Figure 18.4 is a simplified schematic of the control-plane and user-plane interactions between the different EPC elements and the interfaces.

The NAS protocol and event response interactions are specified in the TS24.301 [12] along with the detailed listing of the different timers to be implemented on the UE and the network side. The S1-AP protocol is specified by the TS36.413 [13] and TS23.203 [14] specifies the policy and charging control architecture. These NAS events can cause the MME to trigger subsequent events in the HSS, SGW, PGW, and PCRF as required.

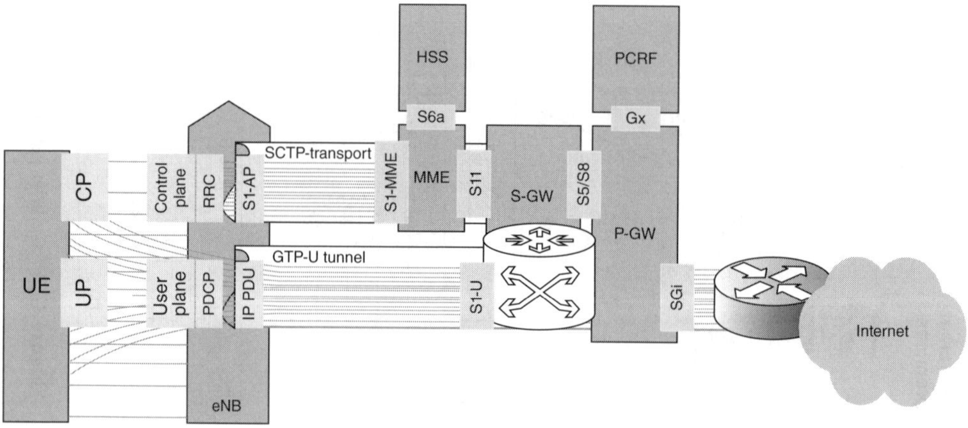

Figure 18.4 EPC control plane–user plane interactions. Refer to text for definitions of acronyms. Source: Rajan et al. [8].

Table 18.2 Transactions per NAS event per EPC element.

Event type	MME	HSS	S-GW	P-GW	PCRF
Attaches	10	2	3	2	1
Addl. default bearer setups	4	0	3	2	1
Dedicated bearer setups	2	0	2	2	1
idle-to-connected	3	0	1	0	0
connected-to-idle	3	0	1	0	0
X2-based handovers	2	0	1	0	0
S1-based handovers	8	0	3	0	0
Tracking area updates	2	0	0	0	0
Total	**34**	**2**	**14**	**6**	**3**

Source: Rajan et al. [8].

Consider the bearer establishment event in Table 18.2. For a UE to establish a bearer, the MME must authorize the UE with the HSS using the DIAMETER protocol. Out of the 10 transactions for one bearer attach event at the MME, there are two at the HSS, three at the SGW, two at the PGW, and one at the PCRF, which are propagated throughout these EPC nodes. Processing of a NAS event at the MME (in this case establishment of one bearer) is blocked as a result of this chaining and can complete only after all these dependent transactions successfully complete.

Table 18.2 counts the transactions (request–response pair) at each of the EPC nodes for different NAS event types specified in the 3GPP specification [12]. Comparing the number of transactions between the MME and the SGW for the different interactions helps us to quantify the relative distribution of the control-plane load across these two EPC nodes. We observe that 41% (14 out of the 34 transactions) of the control-plane load incident on the S1-MME interface is propagated to the SGW. However, only 18% (6 out of 34) of the transactions incident on the S1-MME are propagated to the PGW, confirming that the SGW (and not the PGW) is the critical path of the control plane, which has a direct impact on the user-plane packet processing performance. Note that this result is in contrast to prevailing wisdom that suggests that the MME or the PGW are bottlenecks [15, 16].

Other NAS events (e.g., idle-to-connected and connected-to-idle state transitions) are less "transaction" heavy on the SGW than a bearer setup or a S1-based handover, imposing a single transaction. From 3GPP specifications, "Attach" and "S1-based handover" events are the most expensive S1-MME loads on the SGW [12]. Ignoring the impact of these single-transaction NAS events, we find that 33% of the S1-MME load is propagated onto the SGW [8].

18.6 Network Workloads and Capacity Factors

It is clear from the description of the EPC main-element interactions in the previous section that the separation of the NAS control plane and user plane is unique to the cellular network. The control plane of the cellular network is completely different from control or management planes in data centers or IT infrastructure that deal primarily with route updates, flow table configuration, and QoS enforcement.

The control-plane NAS signaling procedures [17], such as call establishment, UE state transitions, and UE hand-over across towers, need to be successfully processed to establish and then maintain a call session. These messages are associated with specific activity of the UE or the application running on it and have hard response times imposed on them [12].

Each user-plane application service flow is associated to a physical bearer that has been successfully negotiated by the control plane. The call sessions or user plane flows themselves could be general data traffic, HTTP, RTP or specific types like voice over IP, voice over LTE, BiTXML MQTT, and other P2P M2M protocols. While the ratio of signaling procedures per call session would generally be higher for VoLTE, MQTT, and the like, other factors such as UE mobility, power management, and scenario-based situations such as a soccer stadium, can drive signaling up for the general data traffic as well. UE application service flows can be transported over multiple bearers with each bearer encapsulated inside a GTP-U tunnel terminating at the SGW.

Equipment implementing EPC functions are generally specified in terms of the number of control plane messages per second that can be processed or the throughput on the user plane that can be sustained. EPC system capacities in terms of ability to handle different cellular network loads are generally proprietary information.

The control plane and user plane represent the two dimensions along which cellular network load is applied. EPC capacity needs to be able to scale independently in these two dimensions.

18.6.1 EPC Workload Stress Vectors

The number of SCTP terminations on the MME and the event arrival rate in terms of NAS procedures per second are the two stress vectors on the control plane. The number of SCTP terminations of which the system is capable determines the maximum number of eNBs that can connect to it. The NAS event arrival rate capacity of the system defines the aggregate UE and application signaling activity that can be sustained over the connected eNBs.

Similarly, on the user plane, the number of GTP-U tunnels the system can handle determines the number of UEs or bearers that can attach to the system. The aggregate packet-arrival rate in terms of packets per second that can be handled by the EPC defines the sustainable system throughput for a specified application traffic mix over the established number of bearers. The cellular network-load dimensions, the stress vectors along these dimensions, their system impact and relationships are shown in Table 18.3.

Workloads are generally designed to characterize system performance under different stress conditions. Given that EPC system capacities are generally proprietary information, the key objectives of the workload design were to:

- understand EPC system performance on Intel architecture under different network load conditions
- parameterize system resources required for different network loads
- impact on virtualization on system resources for parameterized workloads.

For true realization of these design objectives, the size of the selected system configuration needs to reflect real-world operational network scales. Then, to fully characterize the

Table 18.3 Cellular network workload dimensions and stress vectors.

Network load dimension	Stress vector	System impact	Relationships
Control plane	SCTP terminations	Total number of eNBs supported	↑ SCTP ≅ ↑UE ≅ ↑ GTP-U
	NAS event rate	Aggregate UE and applications signaling	↑ NAS rate (Nps) ≅ Machine-to-machine (IoT), VoLTE applications
User plane	GTP-U terminations	Total number of supported bearers	↑ GTP-U ≅ ↑ UE or ↑ bearers/UE
	Packet rate (Pkt_Rate)	Traffic throughput	Gbps = Pkt_Rate (Mpps) × Av. Pkt Size (bytes)

configured system, each vector and meaningful combinations thereof has to be adequately stressed by the designed workload. The stress limits would need to be at least representative, if not greater than maximum network loads in the real world.

18.7 Conclusion

In this chapter we detailed the network system design, engineering, and dimensioning mechanisms required to handle the loads imposed on the infrastructure by wireless subscribers served by the CSP. Clear understanding of these network planning and design principles is crucial to realization of NFV and SDN. We also explored the interaction between the control and the user planes, which is unique to the cellular network. Dealing with the bottlenecks in the SGW in bearer establishment and sustaining user-plane traffic are problems that need to be addressed by 5G infrastructure at a network architecture level. There will need to be significant changes to the control and user-plane aggregation points and network interfaces for the network infrastructure to handle the load imposed by 5G.

In the meantime, the application of NFV and SDN to telecom networking is gaining increased industry attention [4, 5]. Many vendors have been taking existing LTE core modules, such as MME, SGW, and PGW, and porting them to a virtual machine. This kind of nodal-based approach has several drawbacks and some of which are highlighted in the literature [18, 19]. Kempf et al. [15] investigate how SDN can be applied to the EPC by deploying it in the cloud. By having control- and data-plane separation, IP flows can be routed from the eNB to the PGW via middle boxes using an OpenFlow controller [3]. While these studies provide insight into an SDN-based EPC in the cloud, it does not discuss the performance implications of moving real network workloads, especially the sustainability of control-plane event rates by the OpenFlow controller. More generally, NFV and SDN are areas of significant study outside the area of telecoms, with many applications being proposed for data centers and other contexts [20, 21].

In conclusion, the discussion and references in this chapter show that replacing EPC elements with virtualized equivalents will bring severe performance bottlenecks. Such an approach will be unable to handle network loads of 5G networks and it will not allow NFV and SDN technologies to deliver on the promise of system scalability.

Network transformation to 5G loads with built-in scalability of NFV and SDN technologies will need to be addressed at the fundamental levels of network system design, planning, and re-engineering.

References

[1] Cisco Visual Networking Index: Forecast and Methodology, 2012–2017 http://www.cisco.com/c/en/us/solutions/collateral/service-provider/visual-networking-index-vni/complete-white-paper-c11-481360.html.

[2] ABI Mobile Backhaul Market Data https://www.abiresearch.com/market-research/product/1004832-mobile-network-backhaul/.

[3] N. McKeown, T. Anderson, H. Balakrishnan, G. Parulkar, L. Peterson, J. Rexford, S. Shenker, and J. Turner (2012) Openflow: Enabling innovation in campus networks. In: *SIGCOMM Comput. Commun. Rev.*, **38**(2):69–74.

[4] AT&T's cloud future takes shape http://www.lightreading.com/atandts-cloud-future-takes-shape/d/d-id/707903.

[5] ETSI ISG on network functions virtualization (NFV) http://portal.etsi.org/portal/server.pt/community/NFV/367 (accessed August 10, 2016).

[6] Erlang definition www.erlang.com/whatis.html (accessed August 10, 2016).

[7] Y. Wang, C. Williamson, and J. Doerksen (1999) CAC performance with self-similar traffic: simulation study and performance results. http://ieeexplore.ieee.org/xpl/articleDetails.jsp?tp=&arnumber=805045 (accessed August 10, 2016).

[8] A.S. Rajan, S. Gobriel, C. Maciocco, et al. Understanding the bottlenecks in virtualizing cellular core network functions. In: *IEEE International Workshop on Local and Metropolitan Area Networks (LANMAN)*, Apr. 22–24, 2015, Beijing, China.

[9] TR 23.882 3GPP system architecture evolution (SAE): Report on technical options and conclusions https://portal.3gpp.org/desktopmodules/Specifications/SpecificationDetails.aspx?specificationId=962.

[10] TS 22.278 Service requirements for the Evolved Packet System (EPS) http://www.3gpp.org/ftp/Specs/archive/22_series/22.278/22278-d00.zip (accessed August 10, 2016).

[11] TR 25.913 Requirements for evolved UTRA (E-UTRA) and evolved UTRAN (E-UTRAN) https://portal.3gpp.org/desktopmodules/Specifications/SpecificationDetails.aspx?specificationId=1342.

[12] TS 24.301 Non-access-stratum (NAS) protocol for evolved packet system (EPS); Stage 3 https://portal.3gpp.org/desktopmodules/Specifications/SpecificationDetails.aspx?specificationId=1072.

[13] TS 36.413 Evolved universal terrestrial radio access network (E-UTRAN); S1 Application protocol (S1AP) https://portal.3gpp.org/desktopmodules/Specifications/SpecificationDetails.aspx?specificationId=2446.

[14] TS 23.203 Policy and charging control architecture http://www.3gpp.org/ftp/Specs/archive/23_series/23.203/23203-d01.zip (accessed August 10, 2016).

[15] J. Kempf, B. Johansson, S. Pettersson, H. Luning, and T. Nilsson (2012) Moving the mobile evolved packet core to the cloud. In: *IEEE Wireless and Mobile Computing, Networking and Communications (WiMob)*.

[16] M.R. Sama, S.B.H. Said, K. Guillouard, and L. Suciu (2014) Enabling network programmability in LTE/EPC architecture using OpenFlow. In: *Symposium on Modeling and Optimization in Mobile, Ad Hoc, and Wireless Networks (WiOpt)*.

[17] TS 23.401 General packet radio service (GPRS) enhancements for evolved universal terrestrial radio access network (E-UTRAN) access https://portal.3gpp.org/desktopmodules/Specifications/SpecificationDetails.aspx?specificationId=849.

[18] S. Matsushima and R. Wakikawa (2013) Stateless user-plane architecture for virtualized EPC. https://tools.ietf.org/html/draft-matsushima-stateless-uplane-vepc-00.

[19] G.-H. Tu, Y. Li, C. Peng, C.-Y. Li, H. Wang, and S. Lu (2014) Control-plane protocol interactions in cellular networks. In: *ACM Sigcomm*,.

[20] J. Hwang and K.K. Ramakrishnan and T. Wood (2014) Netvm: High performance and flexible networking using virtualization on commodity platforms. In *USENIX NSDI*.

[21] J. Martins, M. Ahmed, C. Raiciu, et al. (2014) Clickos and the art of network function virtualization. In: *USENIX NSDI*.

Index

Note: page numbers in italics refer to figures; page numbers in bold refer to tables.

Towards 5G: Applications, Requirements and Candidate Technologies, First Edition.
Edited by Rath Vannithamby and Shilpa Talwar.

Companion website: www.wiley.com/go/vannithamby/towards5g